THE WATER ENVIRONMENT
Algal Toxins and Health

ENVIRONMENTAL SCIENCE RESEARCH

THE WATER ENVIRONMENT
Algal Toxins and Health

Edited by
Wayne W. Carmichael

Wright State University
Dayton, Ohio

PLENUM PRESS • NEW YORK AND LONDON

ISBN-13: 978-1-4613-3269-5   e-ISBN-13: 978-1-4613-3267-1
DOI: 10.1007/978-1-4613-3267-1

Proceedings of an International Conference on Toxic Algae,
held June 29—July 2, 1980, at Wright State University, Dayton, Ohio

These proceedings have been approved for publication by the Health Effects
Research Laboratory, Cincinnati, Ohio, U.S. Environmental Protection Agency.
Approval does not signify that the contents necessarily reflect the views
and policies of the U.S. Environmental ProtectionAgency, nor does mention of
trade names or commercial products constitute endorsement or recommendation for use.

CONFERENCE SPONSORS

United States Environmental Protection Agency
Health Effects Research Laboratory
Cincinnati, Ohio
Cooperative Agreement R 807 309-01

Wright State University
Dayton, Ohio

CONFERENCE HOST

Wright State University
Dayton, Ohio

Editorial Committee:

WAYNE W. CARMICHAEL
Department of Biological Sciences, Wright State University,
Dayton, Ohio

PAUL R. GORHAM
Department of Botany, University of Alberta,
Edmonton, Alberta, Canada

RICHARD E. MOORE
Department of Chemistry, University of Hawaii,
Honolulu, Hawaii

JOHN J. SASNER, JR
Department of Zoology, University of New Hampshire,
Durham, New Hampshire

Conference Organizer:

WAYNE W. CARMICHAEL
Department of Biological Sciences, Wright State University,
Dayton, Ohio

Project Officer:

EDWIN C. LIPPY
U.S. Environmental Protection Agency, HERL,
Cincinnati, Ohio

PREFACE

The conference on The Water Environment: Algal Toxins and Health was held at Wright State University in Dayton, Ohio, on June 29, 30, July 1, 2, 1980. Its principal objectives were to bring together, for the first time, researchers, public officials and interested parties in order to present and discuss what is known about algal toxins. The conference concentrated almost exclusively on toxins and toxic blooms of blue-green algae (Cyanobacteria). Since the most common Cyanobacteria bloom forming species are also the ones most likely to produce toxins, they are a problem in the maintenance of safe animal and human water supplies. While poisonings by Cyanobacteria involve mainly domestic and wild animals, they may also be responsible for cases of human gastroenteritis and contact poisoning. Even though human poisonings by Cyanobacteria have historically not been a widespread problem, continued deterioration of our recreational and municipal water supplies suggests that blooms of non-toxic and toxic Cyanobacteria blooms will increase. In addition to studies on their role as disease agents, there is basic research being done on their pharmacological properties to determine their mode of action and usefulness as tools in the study of basic neuromuscular mechanisms.

These papers were centrally typed for reproduction as camera-ready copy. Each paper was reviewed and edited by at least two persons of the editorial committee. This volume is organized into five major sections: 1. A review section on algal toxins including dino-flagellate and *Prymnesium* toxins; 2. A section covering occurrence and ecology of toxic Cyanobacteria blooms; 3. Culturing methods plus studies on new toxic species including a paper on the toxin of the green alga *Pandorina morum*; 4. A section on public health problems of Cyanobacteria; 5. A section emphasizing isolation, physiology, pharmacology and detection methods for Cyanobacteria toxins.

The conference organizer would like to thank the members of the editorial committee for reviewing and editing the papers. I am also grateful to those students and faculty of Wright State University who assisted with the conference, especially Ann Blasingame who was

assistant to the conference coordinator. Laurel Carmichael typed and
proofread the entire set of manuscripts, and it was her care and persistence that permitted the Proceedings to meet its timetable. Without the financial support of U. S. E. P. A., H. E. R. L., the conference could not have been held. On behalf of the conference editors
and attendees we express our thanks for their support.

Wayne W. Carmichael
Wright State University

CONTENTS

III. CULTURING METHODS AND REPORTS OF NEW TOXIC SPECIES

IV. PUBLIC HEALTH AND TOXIC CYANOBACTERIA

V. ISOLATION, PHYSIOLOGY, TOXICOLOGY AND DETECTION OF CYANOBACTERIA TOXINS

FRESHWATER BLUE-GREEN ALGAE (CYANOBACTERIA)

TOXINS -- A REVIEW

Wayne W. Carmichael

Department of Biological Sciences
Wright State University
Dayton, Ohio 45435

ABSTRACT

Exotoxins are known to be produced by strains of at least three species of freshwater Cyanobacteria. These are *Anabaena flos-aquae*, *Microcystis aeruginosa* and *Aphanizomenon flos-aquae*. Of about 12 toxins produced by these species only one has been identified, synthesized and its toxinology determined. When water-blooms of these toxic species are present in a reservoir, lake, pond or slough, the cells and toxins can become concentrated enough to cause illness or death in almost any mammal, bird or fish which ingests enough of the toxic cells, or extracellular toxin. Major losses to animals include mainly cattle, sheep, hogs, birds (domestic and wild) and fishes while minor losses are reported for dogs, horses, small wild mammals, amphibians and invertebrates. Acute oral toxicity to humans has not been documented but there is increasing evidence that the toxins cause gastroenteritis and contact irritations in users of certain recreational and municipal water supplies. Lipopolysaccharide endotoxin is also produced by certain Cyanobacteria including *Schizothrix calcicola* and *Anabaena flos-aquae* and has been implicated in certain water-borne outbreaks of gastroenteritis among humans.

Major areas of research for toxic Cyanobacteria include: 1. structure/function analysis of the toxins, 2. biological and or physical factors which create a toxic waterbloom, 3. role of the toxins in animal and human health, 4. rapid detection methods for the presence of the toxins in water supplies.

FRESHWATER CYANOBACTERIA TOXINS

A recent literature review found over 200 world-wide references on toxic cyanophytes (see Carmichael and Bent, these proceedings). These have been published since the late 1800's but most have occurred since 1940. The early publications and reports primarily cover field cases involving the animals affected and the Cyanobacteria species suspected. In the late 1940's and early 1950's a few key workers began to investigate cases of field poisonings by a combination of bringing bloom material into the laboratory for toxinology testing and by isolating toxic strains from various field collections. These studies provided evidence on animals affected, signs of poisoning and the Cyanobacteria species responsible. In the years since these early studies, evidence has generally been slow to accumulate with regards to the actual toxins produced, their chemistry and toxicological properties. This can be attributed to the intermittent occurrence of toxic blooms which limits the availability of toxic strains, some problems with large scale culturing, and an absence of direct evidence for wide-spread human toxicities. Within the last 15 to 20 years many of the laboratory problems have been overcome and much of the new evidence has been the result of collaborative efforts involving researchers from different areas of expertise. Table 1 summarizes the main chronological events in Cyanobacteria research.

Toxic species of freshwater Cyanobacteria have been confirmed in only three out of about nine suspected toxic genera. These are *Microcystis aeruginosa* Kütz. emend. Elenkin, *Anabaena flos-aquae* (Lyngb.) de Bréb. and *Aphanizomenon flos-aquae* (L.) Ralfs. While poisonings of *An. flos-aquae* are more dramatic with respect to numbers of animals affected, toxic strains of *M. aeruginosa* appear to be more widely distributed geographically. Only one confirmed locality exists for toxic *Aph. flos-aquae* and part of the toxin it produces is similar to saxitoxin (Figure 1) which is produced by the marine red tide organism *Gonyaulax catenella* (Alam et al., 1978).

Known toxins produced by toxic strains include alkaloids, polypeptides and pteridines. Most of the toxins have been designated anatoxins (Carmichael and Gorham, 1978) because they are produced by strains of *Anabaena flos-aquae*. Only one, anatoxin-a, has been chemically identified and its toxinological/pharmacological properties investigated. It is an alkaloid, 2-acetyl-9-azabicyclo (4.2.1) non-2-ene, having a molecular weight of 165 (Devlin et al., 1977) (Figure 1). Antx-a is a potent agonist at the nicotinic receptor causing death by respiratory arrest (Carmichael et al., 1975, 1979). It has potential as a useful pharmacological tool in studying the molecular pharmacology of acetylcholine receptors (Spivak et al., 1980).

Table 1. Dates and Time Periods in Toxic Cyanobacteria Research

Date	Event	Reference
1878	Poisonous Australian Lake.	Francis, 1878.
1949 to 1960	Toxic Waterbloom Reports, Minnesota.	Olson, 1949, 1951, 1952, 1955, 1960.
1958 to 1964	Isolation and Culture of Toxic Cyanobacteria Clones.	Hughes et al., 1958; Gorham, 1964; Gorham et al., 1964.
1964 and 1968	International Conferences: Algae and Man, 1964; Algae, Man and the Environment, 1968.	Gorham, 1964; Schwimmer and Schwimmer, 1964; Schwimmer and Schwimmer, 1968.
1971 to 1980	Papers by a Number of Authors on:	
	1. *A. flos-aquae* Neurotoxin Production.	Huber, 1972; Carmichael et al., 1975, 1979; Biggs and Dryden, 1977; Spivak et al., 1980.
	2. *A. flos-aquae* Production of Several Neurotoxins.	Carmichael et al., 1978.
	3. World-wide Occurrence of Toxic *M. aeruginosa*.	Proceedings, this Conference.
	4. *Aph. flos-aquae* Neurotoxin Production.	Gentile, 1971; Alam et al., 1973, 1978.
	5. Gastroenteritis Outbreaks in Humans due to Cyanobacteria.	Lippy and Erb, 1976.
	6. Characterization of LPS Endotoxin from Cyanobacteria.	Keleti et al., 1979.
1980	International Conference on Toxic Cyanobacteria.	Proceedings, this Conference.

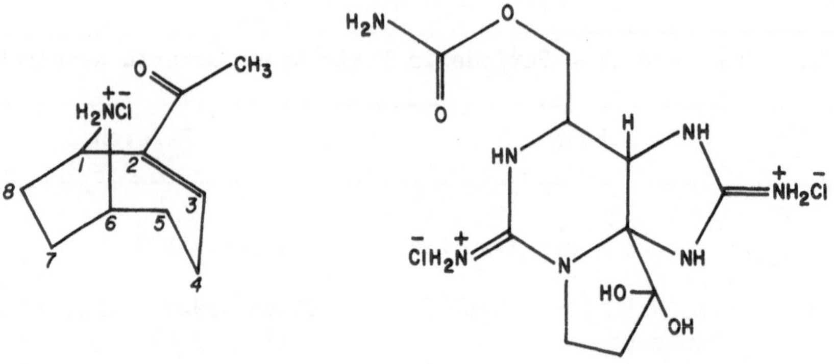

anatoxin-a hydrochloride   saxitoxin dihydrochloride

Fig. 1.   Structure of anatoxin-a hydrochloride (salt-free M.W. 165)
          and saxitoxin dihydrochloride.   Saxitoxin has been sug-
          gested to be a component of aphantoxin isolated from
          Aphanizomenon flos-aquae.

Known toxins of *Microcystis aeruginosa* are peptides.   These
peptides are small molecular weight compounds ranging in size from
500 to 1700 daltons (Gorham and Carmichael, 1979, 1980).   Until
critical structure analysis is done they are being termed microcystin, anatoxin-c and *Microcystis*-type-c.   Table 2 summarizes what
is known about these Cyanobacteria toxins.

Another source of potential environmental toxicants from Cyanobacteria is the recent confirmation of a true *Limulus* amoebocyte
lysate (LAL) reaction by species of *Schizothrix* and *Anabaena* (Keleti
et al., 1979).   This indicates the presence of lipopolysaccharide
endotoxin, which in human drinking water supplies may be the cause
of some gastroenteritis outbreaks.   In addition some other reports
indicate Cyanobacteria are capable of producing toxins causing contact dermatitis or other immunogenic reactions in users of recreational waters.

This review will now discuss the environmental effects of
Cyanobacteria toxins, methods used to detect toxic species, plus
areas that need further investigation.

Table 2. Summary of Toxins Produced by Strains of *M. aeruginosa*, *Aph. flos-aquae* and *A. flos-aquae*

Source	Toxin	Structure	Reference
Microcystis aeruginosa			
NRC-1 (Ontario)	Microcystin	Unknown-peptide	Bishop et al., 1959 Murthy and Capindale, 1970 Rabin and Darbre, 1975
A-143-a-5[a] (Alberta)	*Microcystis*-type-c	Unknown-peptide	Carmichael and Gorham, 1980 Gorham and Carmichael, 1979, 1980
Bloom (South Africa)	Microcystin-like	Unknown-two toxic factors	Toerien et al., 1976
Bloom (Australia)	Microcystin-like	Unknown-peptide	Elleman et al., 1978
Aphanizomenon flos-aquae			
Clone (Kezar Lake, New Hampshire)	Aphantoxin	Unknown-possibly contains saxitoxin = $C_{10}H_{19}N_7O_4$	Alam et al., 1973, 1978
Bloom (Klamath Lake, Oregon)	Ichthyotoxin	Unknown	Phinney and Peek, 1961 Gorham, 1964
Anabaena flos-aquae			
NRC-44-1 (Saskatchewan)	Anatoxin-a	$C_{10}H_{15}NO$	Devlin et al., 1977
NRC-525-17-b-1-e (Saskatchewan)	Anatoxin-a (s)[b]	Unknown	Carmichael and Gorham, 1978
A-52-2-n-e-7-1-2-j (Alberta)	Anatoxin-b	Unknown	Carmichael and Gorham, 1978
S-UTH-1 (Saskatchewan)	Anatoxin-b (s)	Unknown	Carmichael and Gorham, 1978
A-113-9-q-2 (Alberta)	Anatoxin-c	Unknown-peptide	Carmichael and Gorham, 1978
S-23-g, S-25-a	Anatoxin-d	Unknown	Carmichael and Gorham, 1978

[a]A-143-a-5 indicates clone "5" from clone "a" of colony number 143. "A" indicates the bloom was from Alberta. These clonings were done to establish bacteria-free (axenic) conditions.

[b](s) = salivation factor present

ENVIRONMENTAL EFFECTS OF CYANOBACTERIA TOXINS

Domestic and Wild Animals

Although poisonings by toxic Cyanobacteria blooms within a given geographical area are intermittent, the effect of a single toxic bloom on the biota can be severe. The most dramatic effects are on wild and domestic animals who water and feed in a toxic waterbloom. The numbers of dead and dying animals can be great as detailed by Rose (1953) and Schwimmer and Schwimmer (1968). As the Cyanobacteria grow within the water body, they float toward the surface, aided by gas vacuoles located inside the cell. Once near, or on the surface, gentle wind and wave action accumulates the bloom near the shore. The resulting heavy bloom and surface scum easily permits an acutely toxic amount of cells and extracellular toxin to be ingested by watering or feeding animals. It is this condition in which entire flocks of birds or herds of domestic animals die within a few hours after exposure to the toxic water-bloom.

The alkaloid toxins are the most rapidly acting toxins. They function as neurotoxins, paralyzing first peripheral skeletal muscles then respiratory muscles, causing death by respiratory arrest within a few minutes to a few hours. Different animal species response to the toxins depends on dosage, amount of the food material already in the gut and actual differences in species sensitivity to the toxins (Gorham and Carmichael, 1980; Carmichael and Biggs, 1978). The presence of more than one toxic species and toxin type in the bloom also gives different signs of poisoning and survival times. The polypeptide toxins do not cause such a rapid death (hours vs. minutes) as the alkaloid neurotoxins. While their specific mode of action is unknown, they do cause characteristic gross signs of poisoning in the liver including swelling, mottling and dark coloration. Microscopic examination of the liver tissue reveals a marked pooling of red blood cells in the hepatic units. A summary of the known signs of poisoning in domestic animals is given in Table 3, while estimated amounts to cause death, by the oral route, in selected animals are given in Table 4.

Humans

Acute poisonings in humans leading to death such as occur with red tide shellfish poisoning have never been reported for blooms of Cyanobacteria. This is probably not due to a lack of susceptibility in humans but rather the lack of ingestion by humans of a toxic bloom. There is no known vector, such as shellfish are for red tide poisoning, by which the Cyanobacteria toxins get into the human food

Table 3. Signs of Poisoning in Domestic Animals from Freshwater Cyanobacteria[a]

Toxin	Time[b]	Signs[c]	Internal Lesions
A. Neuromuscular (Alkaloid) Absorbed through Mucous Membranes	4-10 minutes	Peracute: Muscular tremors, stupor, staggering, prostration, hyperaesthesia to touch, convulsions, opisthotonus (head over back), and death.	No specific lesions.
	15-30 minutes	Acute: Very sudden onset, weakness, staggering, prostration, muscle twitching, animal goes down, labored breathing, frothy fluids from mouth and nose, wheezing, choking, convulsions, opisthotonus, subnormal temperature, coma, death. Blood is normal color. Death by respiratory failure.	Congestion of brain (cerebrum), spinal cord, and meniges. Bloody fluids in lungs and thoracic cavity. Lungs edenatous and bronchi filled with frothy slime.
B. Tissue Degeneration (Polypeptide)	30 minutes to 24 hours	Gastrointestinal: Weakness, nausea, retching, vomiting, swallowing motions, salivation, eye-watering, severe thirst, sudden purgation, blood covered hard feces, diarrhea, piloerection, lethargic and distended abdomen.	Severe hyperemia of digestive tract with hemorrhage of palate, patches of hemorrhage in intestine, sloughing of stomach lining, enteritis, blood-tinged abdominal fluid.
		Hepatic (liver): Jaundice, shock, cold extremities, anemia, pallor, abdominal pain.	Gross: Liver swollen, mottled and dark (RBC's[d] pooled in hepatic units), friable to flabby. Microscopic: Periacinous coagulation-type necrosis, loss of hepatic architecture, ruptures of hepatic cord cells, cut surface pale and fatty, cirrhosis in survivors.
		Cardiovascular: Rapid, weak pulse; congestive-type myocardial failure.	Heart dilated, blood-filled, flaccid; petechial hemorrhage of endocardium and myocardium; blood-tinged pericardial fluids.
C. Photo-dynamic Pigment[e]	Several days	Photosensitivity: Blisters and peeling on white or light spots on skin (ears, nose and tail of sheep, plus white skin of cattle, horses and goats)	

[a] Summarized from Konst et al., 1965; Gorham, 1964; Schwimmer and Schwimmer, 1964, 1968; Gorham and Carmichael, 1979, 1980; Carmichael and Gorham, 1980. Table prepared with the assistance of L. D. Schwartz, Dept. Vet. Sci., The Penn. State Univ.

[b] Survival time dose related. The greater the dose the faster onset of signs.

[c] Compilation of toxicity signs observed in all animals. Signs vary by animal species according to their individual physiology.

[d] Red blood cells.

[e] Symptomatic of toxic and non-toxic algae as well as certain drugs and poisonous weeds.

Table 4. Estimated Oral Minimum Lethal Dosage of *Anabaena flos-aquae*
          NRC-44-1[a] which would Cause Death in Selected Animals

	Minimum Lethal Dosage (mg lyophilized cells per kg body weight)	Milliliters of a 20 mg dry weight/ml[b] water-bloom to cause death
Goldfish (30 kg)	120	0.2
Mallard duck (1 kg)	350	18
Calf (60 kg)	400	1200
Pheasant (1 kg)	850	43
Laboratory rat (300 g)	1500	23
Laboratory mouse (25 g)	1800	2

[a] Produces anatoxin-a -- oral minimum lethal dosages for other anatoxins have not been determined but LD_{50} (intraperitoneal mouse)
values for all anatoxins range between 10 and 60 mg lyophilized
cells/kg body weight.

[b] Surface scum concentrations have been recorded as high as 70 mg
dry weight cells/ml from different waterblooms. Twenty mg/ml
would be considered a heavy surface scum (see Carmichael and
Gorham, these Proceedings).

chain. There are reports indicating that these toxins can cause
gastroenteritis when ingested from municipal drinking water supplies
(Lippy and Erb, 1976; Keleti et al., 1979). Reports of contact
dermatitis in humans, associated with use of recreational water, are
also known (Billings, these Conference Proceedings). The contact
dermatitis may be due to direct contact with filaments or the exotoxins secreted into the water. There may also be an immunogenic
response (i.e. dermatitis) to the cells or toxins. In areas of the
United States or other countries where water treatment facilities
do not remove toxic Cyanobacteria, the toxins can be ingested in
enough of an amount to cause severe gastroenteritis problems (Aziz,
1974, and personal communication).

    At this time the alkaloid neurotoxins of the Cyanobacteria are
not the major problem to humans. Instead it appears to be the
polypeptide toxins and lipopolysaccharide (LPS) endotoxin which,
when ingested, cause outbreaks of gastroenteritis. LPS endotoxin

is well known from Gram-negative bacteria and is present in at least
some Cyanobacteria (Keleti et al., 1979). Isolated endotoxin does
not always cause acute toxicity in mice by the intraperitoneal route
(Keleti et al., 1979), whereas all of polypeptide and alkaloid
toxins do. The effect of ingested LPS and its occurrence in drinking water is a controversial subject. While endotoxic reaction has
been observed in a hospital environment in which the affected people
were immunosuppressed (Sykora et al., 1980), there is at present
little evidence to suggest that a normal population, which already
has a gut flora possessing endotoxin, would be affected by ingested
LPS in a water supply. The alkaloid and peptide toxins are however
known to cause acute and chronic oral toxicity in animals. The
presence of both Cyanobacteria exotoxin and endotoxin should be
monitored for in open municipal and recreational water supplies.

Another effect toxic Cyanobacteria may have on humans concerns
the use of certain algae including Cyanobacteria as a source of high
protein food. If efforts are not made to qualitatively control the
strain grown for food it could adversely affect the health of both
animals and humans who might be fed the cells (Carmichael and Gorham, 1980; Lincoln and Carmichael, these Conference Proceedings).

Zooplankton

The toxins of Cyanobacteria can be considered secondary chemicals just like other secondary chemicals produced by plants and
animals. Their production and accumulation during bloom formation
might be expected to have an effect on invertebrate populations,
especially zooplankton. While it is premature to say that toxic
Cyanobacteria are selected for in a bloom by selective zooplankton
feeding behavior, or that direct inhibition of zooplankton by toxic
phytoplankton occurs, the interaction between zooplankton and
phytoplankton is important (Porter, 1977, 1978). That toxic Cyanobacteria can inhibit or kill zooplankton has been demonstrated by
Ransom et al. (1978) and Snell (1980). Because of this effect it
may be possible to use zooplankton as an assay organism to detect
the occurrence of at least some toxic blooms.

EXISTING AND FUTURE STUDIES ON CYANOBACTERIA TOXINS

Since many blooms of Cyanobacteria are non-toxic the immediate
health problem becomes how to recognize a toxic bloom. While the
most obvious form of recognition is dead or dying animals near the
source of contamination, or reports of gastrointestinal distress
and rashes among humans coming into contact with the water, there
is compelling reason for having a method or methods to predict the
occurrence of a toxic bloom or to at least detect its presence

before it reaches a toxic threshold concentration. Although some basic information exists on the formation of Cyanobacteria blooms (Mitt. Int. Ver. Limnol., 1971) nothing is known about the factors inducing a toxic bloom. The mouse bioassay is now the main method for detecting or verifying the presence of toxic Cyanobacteria in a bloom sample or laboratory culture. The method we have adapted over the years involves:

1. Coarse filtering, to remove debris, of a bloom sample.

2. Intraperitoneal injection (i.p.) of 0.5 ml of the sample into replicate male white laboratory mice (i.e. ICR-Swiss strain).

3. Death of the mice within 2 to 10 minutes with signs as listed in Table 3 (A) indicates presence of a neurotoxic Cyanobacteria.

4. Death of the mice within 1 to 3 hours with signs as listed in Table 3 (B) indicates the presence of peptide toxin(s).

5. Death of the mice in 24 to 36 hours indicates the Cyanobacteria are non-toxic but that there is an effect from bacteria (i.e. SDF).

The use of 0.5 ml of sample will be an overdose in most cases where the bloom is highly toxic but it is used to detect lower levels of toxin from accumulating blooms or from blooms having a mixture of toxic and non-toxic strains (Carmichael and Gorham, these Proceedings).

Since toxic blooms of Cyanobacteria are unpredictable, laboratory studies on the toxic species and their toxins must be carried out. Based on these studies over the past 20 years we now recognize that: 1. there are toxic and non-toxic strains of the three most common bloom forming species, 2. different strains produce a number of physiologically different toxins, 3. mixtures of toxins from different strains or from the same strain can be present in a toxic bloom, 4. different animal species differ in their susceptibility to the various Cyanobacteria toxins, 5. in order for acute poisoning to occur waterblooms or scums require a relative proportion of one or more toxic strains that equals or exceeds a poisonous threshold value (Gorham and Carmichael, 1980).

Future studies on toxic Cyanobacteria should continue these studies and expand them into the following:

1. Chemical nature and structure of toxins produced. Only antx-a has been described and synthesized. The other alkaloid and peptide toxins need their structures determined in order to understand strain differences in toxicity, and routes of biosynthesis for the toxins.

2. Effects of individual toxins and mixtures of toxins on animal and human populations. This involves laboratory studies of the toxins especially the peptide toxins. Epidemiological

studies should be done in cases of human gastroenteritis or
contact poisons. This will help establish under what conditions
Cyanobacteria toxins are the causative agent and to what degree
they may be toxic.

3. Other methods of detection for the toxins. The mouse bioassay
    has disadvantages. Newer more rapid assays using either iso-
    lated cell preparations or analytical methods such as high
    pressure liquid chromatography and fluorometry (Astrachan and
    Archer; Ikawa et al.; and Shoptaugh et al., these Proceedings)
    are needed.

4. Production of toxic vs. non-toxic strains. This includes the
    physical and/or biological factors which select for a toxic
    strain to become dominant in a waterbloom. The role of extra-
    chromosomal DNA (plasmid) would seem a good place to begin
    investigations of this problem (Hauman, these Conference Pro-
    ceedings).

5. Pharmacology of the toxins. Anatoxin-a from *Anabaena flos-aquae*
    NRC-44-1 has been shown to be a potent agonist at nicotinic
    acetylcholine receptors. It also has a unique structure for
    neuromuscular agonists allowing it to be used as a pharmacolog-
    ical tool in the study of acetylcholine receptor recognition
    sites (Spivak et al., 1980). The other toxins of Cyanobacteria
    will also prove interesting in their pharmacology and perhaps
    useful tools as well. A critical study of their pharmacologi-
    cal properties would also provide information on their mode of
    action in cases of animal poisonings.

REFERENCES

Alam, M., M. Ikawa, J. J. Sasner, Jr., and P. J. Sawyer. 1973.
    Purification of *Aphanizomenon flos-aquae* toxin and its
    chemical and physiological properties. Toxicon *11*:65-72.
Alam, M., Y. Shimizu, M. Ikawa, and J. J. Sasner. 1978. Reinves-
    tigation of the toxins from the blue-green alga, *Aphani-
    zomenon flos-aquae*, by a high performance chromatographic
    method. J. Environ. Sci. Health *A13*(7):493-499.
Aziz, K. M. S. 1974. Diarrhea toxin obtained from a waterbloom-
    producing species, *Microcystis aeruginosa* Kütz. Science
    183:1206-1207.
Biggs, D. F., and W. F. Dryden. 1977. Action of anatoxin-I at
    the neuromuscular junction. Proc. West. Pharmacol. Soc.
    20:461-466.
Campbell, H. F., O. E. Edwards, and R. Kolt. 1977. Synthesis of
    noranatoxin-A and anatoxin-A. Can. J. Chem. *55*:1372-1379.
Carmichael, W. W., D. F. Biggs, and P. R. Gorham. 1975. Toxi-
    cology and pharmacological action of *Anabaena flos-aquae*
    toxin. Science *187*:542-544.

Carmichael, W. W., D. F. Biggs, and M. A. Peterson. 1979. Pharma-
     cology of anatoxin-a, produced by the freshwater cyanophyte
     Anabaena flos-aquae NRC-44-1. Toxicon *17*:229-236.
Carmichael, W. W., and P. R. Gorham. 1978. Anatoxins from clones
     of *Anabaena flos-aquae* isolated from lakes of western Canada.
     Mitt. Int. Ver. Limnol. *21*:285-295.
Carmichael, W. W., and P. R. Gorham. 1980. Freshwater cyanophyte
     toxins: types and their effects on the use of micro-algae
     biomass. (In press) *in* G. Shelef and C. J. Soeder, eds.,
     The Production and Use of Micro-Algae Biomass. Ellsevier/
     North Holland, New York.
Devlin, J. P., O. E. Edwards, P. R. Gorham, N. R. Hunter, R. K.
     Pike, and B. Stavric. 1977. Anatoxin-a, a toxic alkaloid
     from *Anabaena flos-aquae* NRC-44h. Can. J. Chem. *55*:1367-1371.
Francis, G. 1878. Poisonous Australian lake. Nature (London) *18*:
     11-12.
Gentile, J. H. 1971. Blue-green and green algal toxins. Pages
     27-66 *in* S. Kadis, A. Ciegler, and S. J. Ajl, eds., Microbial
     Toxins Vol. VII. Algal and Fungal Toxins. Academic Press,
     London. 401 pp.
Gorham, P. R. 1964. Toxic algae. Pages 307-336 *in* D. F. Jackson,
     ed., Algae and Man. Plenum Press, New York.
Gorham, P. R. and W. W. Carmichael. 1979. Phycotoxins from blue-
     green algae. Pure Appl. Chem. *52*(1):165-174.
Gorham, P. R., and W. W. Carmichael. 1980. Toxic substances from
     freshwater algae. Prog. Water Technol. *12*:189-198.
Gorham, P. R., J. McLachlan, U. T. Hammer, and W. K. Kim. 1964.
     Isolation and culture of toxic strains of *Anabaena flos-aquae*
     (Lyngb.) de Bréb. Int. Assoc. Theor. Appl. Limnol. *15*:796-
     804.
Hughes, E. O., P. R. Gorham, and A. Zehnder. 1958. Toxicity of a
     unialgal culture of *Microcystis aeruginosa*. Can. J. Micro-
     biol. *4*:225-236.
Keleti, G., J. L. Sykora, E. C. Lippy, and M. A. Shapiro. 1979.
     Composition and biological properties of lipopolysaccharides
     isolated from *Schizothrix calcicola* (Ag.) Gomont (Cyanobac-
     teria). Appl. Environ. Microbiol. *38*(3):471-477.
Lippy, E. C., and J. Erb. 1976. Gastrointestinal illness at
     Sewickley, Pa. J. Am. Water Works Assoc. *68*:606-610.
Mitteilungen International Vereinigung Limnologie. No. 19. Factors
     that Regulate the Wax and Wane of Algal Populations. 1971.
     318 pp.
Olson, T. A. 1949. History of toxic plankton and associated
     phenomena. Algae-laden water causes death of domestic
     animals; nature of poison. Sewage Works Eng. *20*(2):71.
Olson, T. A. 1951. Toxic plankton. Pages 86-96 *in* Proceedings
     of Inservice Training Course in Water Works Problems,
     February 15-16. University of Michigan, School of Public
     Health, Ann Arbor, Michigan.

Olson, T. A.   1952.   Toxic plankton.   Water Sewage Works *99*:75.

Olson, T. A.   1955.   Studies of algae poisoning.   With special
    reference to the relationship of this phenomenon to losses
    of wildfowl and other birds.   Flicker (Minneapolis) *27*:105.

Olson, T. A.   1960.   Water poisoning – a study of poisonous algae
    blooms in Minnesota.   Am. J. Public Health *50*:883.

Phinney, H. K., and C. A. Peek.   1960.   Klamath Lake, an Instance
    of Natural Enrichment.   U. S. Public Health Service,
    Seminar on Algae and Metropolitan Wastes.   Cincinnati, Ohio.

Porter, K. G.   1977.   The plant-animal interface in freshwater
    ecosystems.   Am. Scient. *65*:159–170.

Porter, K. G., and J. D. Orcutt.   1978.   Nutritional adequacy,
    manageability and toxicity as factors that determine the
    food quality of green and blue-green algae for *Daphnia*.
    (In press) *in* Evolution and Ecology of Zooplankton Communi-
    ties.   Univ. Press of New England.

Ransom, R. E., T. A. Nerad, and P. G. Meier.   1978.   Acute toxicity
    of some blue-green algae to the protozoan *Paramecium
    caudatum*.   J. Phycol. *14*:114–116.

Rose, E. F.   1953.   Toxic algae in Iowa lakes.   Proc. Iowa Acad.
    Sci. *60*:738.

Schwimmer, M., and D. Schwimmer.   1968.   Medical aspects of phy-
    chology.   Pages 279–358 *in* D. F. Jackson, ed., Algae, Man
    and the Environment.   Syracuse Univ. Press, New York.   369 pp.

Snell, T. W.   1980.   Blue-green algae and selection in rotifer popu-
    lations.   Oecologia (Berl.) *46*:343–346.

Sykora, J. L., R. Roche, F. L. Kriess, M. A. Barath, D. Volk, J. T.
    Coyne, R. R. Jackson, and G. Keleti.   1980.   Water quality
    in open finished water reservoirs – Allegheny County, Penn-
    sylvania.   Final Report.   EPA Grant No. R805368.   Published
    by Graduate School of Public Health.   Univ. of Pittsburgh,
    Pennsylvania.

TOXINS FROM MARINE BLUE-GREEN ALGAE

Richard E. Moore

Department of Chemistry
University of Hawaii
Honolulu, Hawaii

ABSTRACT

The marine blue-green alga *Lyngbya majuscula* is the causative agent of a severe contact dermatitis known as "swimmers' itch". Two highly inflammatory agents, lyngbyatoxin A and debromoaplysia-toxin, have been isolated from this alga. Both compounds are potent tumor promoters and have the same activity as the well-known tumor promoter 12-0-tetradecanoylphorbol-13-acetate. Debromoaplysiatoxin has also been isolated from a mixture of two other marine Oscillatoriaceae tentatively identified as *Schizothrix calcicola* and *Oscillatoria nigroviridis* along with several related compounds, the oscillatoxins. The toxins from the Oscillatoriaceae are potentially dangerous to man and there is already evidence that these toxins may be entering the food chain.

TEXT

In general marine blue-green algae have not presented any serious public health problems and therefore have received little attention. Prior to 1977 when Mynderse et al. described the isolation of debromoaplysiatoxin from a deep-water variety of *Lyngbya majuscula*, there were only a few reports in the literature describing toxicity of marine blue-green algae (Hashimoto et al., 1972; Baslow, 1977). Very little progress, however, had been made on the isolation and characterization of any of the active principles. Debromoaplysiatoxin, a phenolic substance which had been isolated previously from the digestive system of a gastropod mollusk *Stylocheilus longicauda* (Kato and Scheuer, 1975, 1976)

15

along with a bromine-containing analog, aplysiatoxin, represented
the first toxin to be identified in a marine blue-green alga.

   Lyngbya majuscula had been shown to cause "swimmers' itch", a
severe contact dermatitis which occurs occasionally during the
summer months on the windward side of Oahu, Hawaii (Banner, 1959;
Grauer, 1959; Grauer and Arnold, 1961) and at the Gushikawa beach
on Okinawa (Hashimoto et al., 1976; Hashimoto, 1979). Efforts by
Moikeha et al. (1971) to identify the dermatitis-producing factor
in Hawaiian *L. majuscula* were unsuccessful. We were well aware
from the work of Kato and Scheuer (1975) that debromoaplysiatoxin
(Fig. 1) was a very highly inflammatory agent which produced
redness, blisters, and pus on contact with human skin and that
these effects were comparable to those reported for contact with
L. majuscula and its extracts (Banner et al., 1960; Moikeha and
Chu, 1971). Proof that debromoaplysiatoxin was responsible for
swimmers' itch came when an outbreak occurred in Laie Bay, Oahu in
September, 1976. Mynderse et al. (1977) collected the shallow-water
L. majuscula from Laie Bay and showed that debromoaplysiatoxin was
indeed present. In a similar study Hashimoto et al. (1976) had
isolated a phenolic substance from dermatitis-producing *L. majuscula*
from Gushikawa beach, Okinawa and had demonstrated that it produced
blisters on human skin. This vesicatory phenol was also debromoaplysiatoxin (S. Ito, Tohoku University, 1977, private communication). Solomon and Stoughton (1978) confirmed that debromoaplysiatoxin caused an erythematous pustular folliculitis in humans and
that it could be responsible for swimmers' itch.

   The dermatitis-producing agent in a shallow-water variety of
L. majuscula found at Kahala beach, Oahu, however, was not debromoaplysiatoxin, but rather an indole alkaloid, lyngbyatoxin A (Fig. 1)
(Cardellina et al., 1979), which was structurally and pharmacologi-
cally very closely related to teleocidin B (Fig. 1), a metabolite
of several strains of *Streptomyces* (Sakabe et al., 1966). The
Kahala variety of *L. majuscula* had never been implicated in
swimmers' itch cases, mainly because it grows on the leeward side
of Oahu and swimmers rarely come into contact with broken filaments
of the alga in the water since the normal tradewinds blow them out
to sea rather than into shore.

   Lyngbyatoxin A produced the same effects as debromoaplysiatoxin. Hydrogenation of lyngbyatoxin A did not alter the inflammatory action. Tetrahydrolyngbyatoxin A was as active as the
natural toxin. Teleocidin B had been reported to cause intense
irritations on rabbit skin (Takashima et al., 1962) and severe,
eruptive vesications on human skin (Nakata et al., 1966). Similarly
dihydroteleocidin B had the same activity as teleocidin B. Hydro-
lytic cleavage of the nine-membered lactam ring in lyngbyatoxin A
or teleocidin B, however, destroyed the activity.

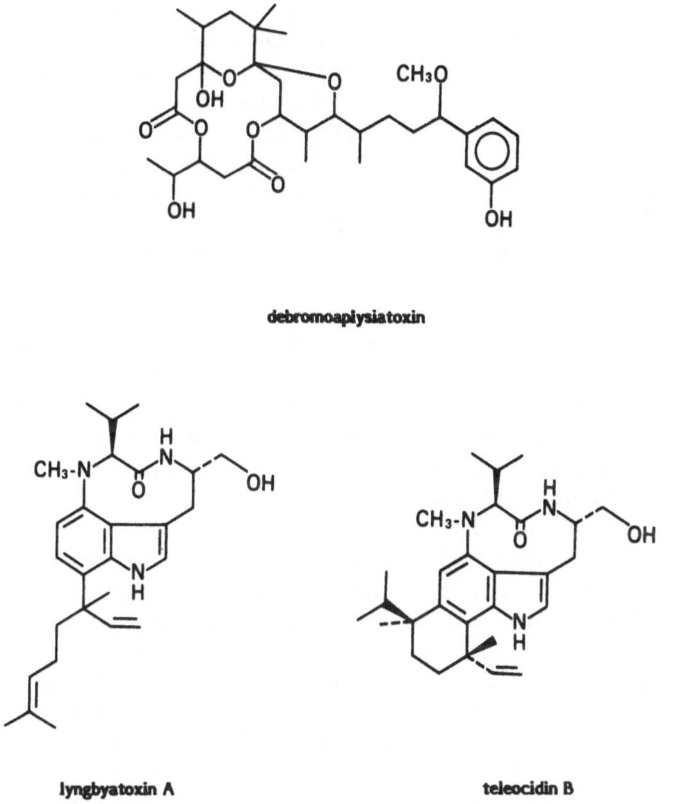

Fig. 1.   Structure of debromoaplysiatoxin, lyngbyatoxin A and
          teleocidin B, all produced by marine blue-green algae.

It is very interesting that the digestive tract of the sea hare
Stylocheilus longicauda is unaffected by the aplysiatoxins. This
mollusk accumulates these inflammatory agents in the midgut when it
ingests *L. majuscula*, its favorite food. This cyanophyte is also
very important for the biological development of the sea hare. The
veligers of *S. longicauda* settle preferentially on the surface of
L. majuscula and utilize an unknown metabolite in the blue-green
alga for metamorphosis (Switzer-Dunlap and Hadfield, 1977).

During a collection of dermatitis-producing *L. majuscula* at
Gushikawa beach in Okinawa, Hashimoto et al. (1976) noted that
rabbitfish (*Siganus fuscescens*) were feeding on sea grass entangled
with *L. majuscula*. No apparent harm to the digestive tract of the
fish resulted from this feeding. Hashimoto et al. recalled an
earlier study (1969) of human intoxication in the Ryukyus Islands
from rabbitfish, which had symptoms that were clearly different from
those of ciguatera, the most common disease from eating poisonous
fish. The Hashimoto group suggested that a relationship might exist
between toxic *L. majuscula* and rabbitfish poisoning, but no further
study of this was made.

Not all marine animals survive from eating toxic *Lyngbya*. A
very striking example occurred in 1975, 1976, and 1977 at one of
the world's largest shrimp-culture research projects in Puerto
Penasco, Mexico. Blooms of a marine species of blue-green alga,
identified first as *Spirulina subsalsa* (Lightner, 1978), then as
an *Oscillatoria* sp. (Lightner et al., 1978), and finally as a
Lyngbya sp. (D. V. Lightner, University of Arizona, 1979, private
communication), caused mass mortalities of raceway-reared blue
shrimp *Penaeus stylirostris*. The disease was characterized by
marked necrosis of the lining epithelium of the midgut, dorsal
cecum, and hindgut gland and subsequent hemocytic enteritis in the
shrimp. Lightner suspected that the necrosis was being caused by
an inflammatory agent in the ingested *Lyngbya*. A sample of the
frozen toxic alga was obtained from D. V. Lightner in 1979, but
D. Carter and M. Kashiwagi in the author's laboratory were unable
to detect any mouse toxicity in the lipophilic extract, and were
also unable to observe any activity in the lipophilic extract
against P-388 lymphocytic mouse leukemia which debromoaplysiatoxin
and lyngbyatoxin A showed at sublethal doses (Mynderse et al., 1977;
Cardellina et al., 1979). The appearance of the necrosis in the
shrimp was slow and it is possible that the causative organism was
a very small contaminant in the algae that normally grow in the
rearing tanks and that the amount of toxin was far too small to
detect in our bioassays but yet large enough to induce the disease
in the shrimp on repeated ingestion.

Fujiki et al. (1979) have shown that teleocidin B is a potent
tumor promoter. Lyngbyatoxin A and debromoaplysiatoxin also appear
to be potent tumor promoters (Fujiki et al., 1980). When topically

applied to mouse skin both compounds from *L. majuscula* induce epi-
dermal ornithine decarboxylase (ODC) activity, an activity which is
increased in fast-growing neoplasms such as L1210 leukemic cells.
The ODC activity is inhibited if 13-cisretinoic acid is applied to
the skin prior to debromoaplysiatoxin or lyngbyatoxin A. The
inflammatory agents also induce cell adhesion of human promyelocytic
leukemia cells (HL-60) and inhibit terminal differentiation of
Friend erythroleukemia cells induced by dimethyl sulfoxide. All of
these in vitro effects are identical with those produced by 12-0-
tetradecanoylphorbol-13-acetate (TPA), a well known tumor promoter
that was first isolated from the seed oil of *Croton tiglium*.
Dihydroteleocidin B and tetrahydrolyngbyatoxin A have the same
activities as teleocidin B and lyngbyatoxin A., but the acid
hydrolyzed compounds are inactive.

Tumors can be produced on mouse skin with repetitive topical
application of dihydroteleocidin B following the application of a
single subcarcinogenic dose of an initiator such as 7,12-dimethyl-
benz[a]-anthracene (Fujiki et al., 1980). Similar in vivo experiments with debromoaplysiatoxin, lyngbyatoxin A, and tetrahydrolyngbyatoxin are in progress in the laboratory of H. Fujiki and
T. Sugimura at the National Cancer Center Research Institute in
Tokyo, Japan.

The discovery of the tumor-promoting properties of debromoaplysiatoxin and lyngbyatoxin A suggests that *L. majuscula* is
potentially a dangerous seaweed and therefore an important public
health concern. There is, however, no direct evidence that *L.
majuscula* is involved in the development of human cancer. Nevertheless, we have observed that *L. majuscula* sometimes grows
epiphytically on other seaweeds. We first noted this association
on a specimen of the green alga *Tydemannia expeditionis* from
Enewetak. The lipophilic extract of the alga was cytotoxic and
showed good activity against P-388 leukemia in mice. A microscopic
examination of the alga showed that filaments of *L. majuscula* were
attached to the thalli and therefore we concluded that the toxicity
and antileukemia activity were probably due to debromoaplysiatoxin.
Toxicity and anticancer activity were also observed in the lipophilic
extracts of certain varieties of the brown alga *Ectocarpus
breviarticulatus* from Fanning Island and Hawaii (Kashiwagi et al.,
1980). Although we have not examined any of these specimens yet
for the presence of toxic epiphytic cyanophytes, microscopic blue-
green algae are known to grow on the surface of *E. breviarticulatus*
and to frequently account for the color variations of the seaweed
(Magruder and Hunt, 1979). More importantly, however, we have
found that *L. majuscula* can grow epiphytically on certain edible
seaweeds such as *Acanthophora spicifera*, a red alga that is consumed
in Indonesia and the Phillipine Islands.

Spirulina, a genus of cyanophyte that is being used for human consumption and as a fodder, belongs to the same family as *Lyngbya*. Leonard (1966) has reported that *Spirulina platensis* is collected from lakes in North Africa, dried, and pressed into flat cakes named "Dihe" which are used as a food with little or no toxicity. Farrar (1966) has suggested that a food staple of the Aztecs of Mexico was a dried cake called "Tecuitlatl" made from harvests of salt lake blooms of cyanophytes. Great caution, however, should be exercised in using members of the Oscillatoriaceae for food until more is known about the toxic strains and the requirements for toxin production.

Debromoaplysiatoxin, along with several closely related compounds, has also been isolated from other cyanophytes belonging to the Oscillatoriaceae (Mynderse and Moore, 1978; Moore, 1981). Debromoaplysiatoxin and oscillatoxin A (31-nordebromoaplysiatoxin), for example, are major toxic constituents of a mixture of cyanophytes from Enewetak identified as *Schizothrix calcicola* and *Oscillatoria nigroviridis*.

Water-soluble toxins are also present in marine cyanophytes belonging to the Oscillatoriaceae but none have been isolated and characterized. Cooper (1964) reported that the Gilbertese on Marakei atoll associated the appearance of toxic fish on the reef with a sudden bloom of an alga which they called "tan-tan". The alga was identified as *Schizothrix calcicola* and it was shown to contain both lipid and water soluble toxins which had very little or no anticholinesterase activity (Banner, 1967). *Lyngbya majuscula* also possess water-soluble toxins (Kashiwagi, 1980). Interestingly Watson (1973) found that the midgut glands of *Stylocheilus longicauda* and three other sea hares from Hawaii also contained two distinct types of toxins, lipid soluble toxins, the aplysiatoxins, and water-soluble toxins which have not been identified. The lipid and water soluble toxins were found to have clearly different physiological properties (Watson and Rayner, 1973).

Toxic aqueous extracts of marine Oscillatoriaceae frequently display good activity against P-388 lymphocytic mouse leukemia at doses showing little or no cytotoxicity (Kashiwagi et al., 1980). The lipophilic extracts of this family, on the other hand, are generally active against P-388 leukemia only at or near the cytotoxic level. Sometimes good antineoplastic activity has been observed on aqueous extracts of Oscillatoriaceae which show no toxicity at all to mice. Very little is known about the active compounds at this writing. Some of the substances appear to be of high molecular weight while others are estimated to be of low molecular weight on the basis of their behavior on gel filtration.

ACKNOWLEDGMENT

Research in the author's laboratory on toxic marine cyanophytes has been supported by a grant CA12623 from the National Cancer Institute, Department of Health, Education, and Welfare.

REFERENCES

Banner, A. H. 1959. A dermatitis-producing alga in Hawaii. Hawaii Med. J. *19*:35-36.

Banner, A. H. 1967. Marine toxins from the Pacific. I. Advances in the investigation of fish toxins. Pages 157-165 *in* F. E. Russell and P. R. Saunders, eds., Animal Toxins. Pergamon Press, New York.

Banner, A. H., P. J. Scheuer, S. Sasaki, P. Helfrich, and C. B. Alender. 1960. Observations on ciguatera-type toxin in fish. Ann. N. Y. Acad. Sci. *90*:770-787.

Baslow, M. H. 1977. Marine Pharmacology. R. E. Krieger Publishing Co., Huntingdon, New York. 315 pp.

Cardellina II, J. H., F. J. Marner, and R. E. Moore. 1979. Seaweed dermatitis: structure of lyngbyatoxin A. Science *204*:193-195.

Cooper, M. J. 1964. Ciguatera and other marine poisoning in the Gilbert Islands. Pac. Sci. *18*:411-440.

Farrar, W. V. 1966. Tecuitlatl: a glimpse of Aztec food technology. Nature (London) *211*:341-342.

Fujiki, H., M. Mori, M. Nakayasu, M. Terada, and T. Sugimura. 1979. A possible naturally occurring tumor promoter, teleocidin B from *Streptomyces*. Biochem. Biophys. Res. Commun. *90*:976-983.

Fujiki, H., M. Mori, M. Nakayasu, M. Terada, T. Sugimura, and R. E. Moore. 1980. Possible tumor promoters in *Streptomyces* and blue-green algae. Submitted for publication.

Grauer, F. H. 1959. Dermatitis escharotica caused by a marine alga. Hawaii Med. J. *19*:32-36.

Grauer, F. H., and H. L. Arnold, Jr. 1961. Seaweed dermatitis: first report of a dermatitis-producing marine alga. Arch. Dermatol. *84*:720-732.

Hashimoto, Y. 1979. Marine toxins and other bioactive marine metabolites. Japan Scientific Societies Press, Tokyo.

Hashimoto, Y., N. Fusetani, and K. Nozawa. 1972. Screening of the toxic algae on coral reefs. Pages 569-572 *in* Proc. 7th Int. Seaweed Symp., Sapporo, 1971.

Hashimoto, Y., H. Kamiya, K. Yamazato, and K. Nozawa. 1976. Occurrence of a toxic blue-green alga inducing skin dermatitis in Okinawa. Pages 333-338 *in* A. Ohsaka, K. Hayashi, and Y. Sawai, eds., Animal, Plant, and Microbial Toxins. Vol. 1. Plenum Press, New York.

Hashimoto, Y., S. Konosu, T. Yasumoto, and H. Kamiya. 1969.
    Ciguatera in the Ryukyu and Amami islands. Nippon Suisan
    Gakkaishi *35*:316-326.
Kashiwagi, M., J. S. Mynderse, R. E. Moore, and T. R. Norton. 1980.
    Antineoplastic evaluation of some Pacific basin marine algae.
    J. Pharm. Sci. *69*:735-738.
Kato, Y., and P. J. Scheuer. 1975. The aplysiatoxins. Pure Appl.
    Chem. *41*:1-14.
Kato, Y., and P. J. Scheuer. 1976. The aplysiatoxins: reactions
    with acid and oxidants. Pure Appl. Chem. *48*:29-33.
Leonard, J. 1966. The 1964-1965 Belgian transsaharan expedition.
    Nature (London) *209*:126-128.
Lightner, D. V. 1978. Possible toxic effects of the marine blue-
    green alga, *Spirulina subsalsa*, on the blue shrimp, *Penaeus
    stylirostris*. J. Invertebrate Pathol. *32*:139-150.
Lightner, D. V., D. A. Danald, R. M. Redman, C. Brand, B. R. Salser,
    and J. Reprieta. 1978. Suspected blue-green algal poisoning
    in the blue shrimp *(Penaeus stylirostris)*. Univ. Arizona,
    Environ. Res. Lab. Reprint 47.
Magruder, W. H., and J. W. Hunt. 1979. Seaweeds of Hawaii.
    Oriental Publishing Co., Honolulu, Hawaii. 116 pp.
Moikeha, S. N., and G. W. Chu. 1971. Dermatitis-producing alga
    Lyngbya majuscula Gomont in Hawaii. II. Biological properties
    of the toxic factor. J. Phycol. *7*:8-13.
Moikeha, S. N., G. W. Chu, and L. R. Berger. 1971. Dermatitis-
    producing alga *Lyngbya majuscula* Gomont in Hawaii. I. Isola-
    tion and chemical characterization of the toxic factor. J.
    Phycol. *7*:4-8.
Moore, R. E. 1981. Constituents of blue-green algae. (In press)
    in P. J. Scheuer, ed., Marine Natural Products: Chemical
    and Biological Perspectives. Vol. IV. Academic Press, New
    York.
Mynderse, J. S., and R. E. Moore. 1978. Toxins from blue-green
    algae: structures of oscillatoxin A and three related
    bromine-containing toxins. J. Org. Chem. *43*:2301-2303.
Mynderse, J. S., R. E. Moore, M. Kashiwagi, and T. R. Norton. 1977.
    Antileukemia activity in the Oscillatoriaceae: isolation of
    debromoaplysiatoxin from *Lyngbya*. Science *196*:538-540.
Nakata, H., H. Harada, and Y. Hirata. 1966. The structure of
    teleocidin B. Tetrahedron Letters 1966(23):2515-2522.
Sakabe, N., H. Harada, Y. Hirata, Y. Tomiie, and I. Nitta. 1966.
    X-ray structure and determination of dihydroteleocidin B
    monobromoacetate. Tetrahedron Letters 1966(23):2523-2525.
Solomon, A. E., and R. B. Stoughton. 1978. Dermatitis from
    purified sea algae toxin (debromoaplysiatoxin). Arch.
    Dermatol. *114*:1333-1335.
Switzer-Dunlap, M., and M. Hadfield. 1977. Observations on
    development, larval growth, and metamorphosis of four species
    of Aplysidae (Gastropoda:Opisthobranchia) in laboratory
    culture. J. Exp. Mar. Biol. Ecol. *29*:245-261.

Takashima, M., H. Sakai, R. Mori, and K. Arima. 1962. A new toxic substance, teleocidin, produced by *Streptomyces*. Part IV. Degradative studies of hydroteleocidin B and teleocidic anhydride. Agr. Biol. Chem. *26*:669-678.

Watson, M. 1973. Midgut gland toxins of Hawaiian sea hares. I. Isolation and preliminary toxicological observations. Toxicon *11*:259-267.

Watson, M., and M. D. Rayner. 1973. Midgut gland toxins of Hawaiian sea hares. II. A preliminary pharmacological study. Toxicon *11*:269-276.

POISONS PRODUCED BY DINOFLAGELLATES -

A REVIEW

Edward J. Schantz

Department of Food Microbiology and Toxicology
Food Research Institute
University of Wisconsin-Madison
Madison, Wisconsin 53706

ABSTRACT

Many of the so-called red tides are caused by excessive blooms
of dinoflagellates. Out of about 1200 species of dinoflagellates
only a few (8 or 10) are known to produce poisonous substances that
cause shellfish and fish to become poisonous or cause fish to die.
When persons eat shellfish that have consumed the poisonous dinoflagellates *Gonyaulax catenella* or *Gonyaulax tamarensis*, a disease
known as shellfish poisoning results which is often fatal. Although
shellfish poisonings have been reported as long as medical records
have been kept, the relationship between a poisonous dinoflagellate
and shellfish poisoning was not known before D. Hermann Sommer and
his colleagues at the University of California reported it in 1937.
Japanese investigators have recent evidence that a dinoflagellate
found in the South Pacific produces the poison found in certain
fish that causes the disease in humans known as ciguatera. Another,
Exuviaella marie-lebouriae occurring in areas around Japan has
caused short-necked clams to become poisonous. People eating them
came down with a disease resulting in fatty degeneration of liver
and kidney tissue. Other dinoflagellates, *Gymnodinium breve* and
Gonyaulax monilata, found in the Gulf of Mexico produce poisons
that kill fish and cause severe environmental problems due to
decaying fish and polluted waters in these areas.

The chemical structure of the paralytic poisons produced by *G.
catenella* and *G. tamarensis* have been worked out by investigators
at the University of Wisconsin. Saxitoxin is the major poison
produced by *G. catenella* and is found in California sea mussels
and Alaska butter clams, and 11-hydroxysaxitoxin sulfate is the

major poison produced by *G. tamarensis* and found in clams and
scallops along the New England coast and the Bay of Fundy. These
extremely poisonous substances cause paralysis by blocking sodium
channels in nerve and muscle cell membranes. They are water soluble,
slightly soluble in methanol and ethanol but insoluble in lipid solvents. They are quite heat stable at ordinary cooking temperatures,
making the food poisoning problem more acute. Some of the other
dinoflagellate poisons have been partially characterized by several
investigators in Japan, Hawaii and the continental United States.

TEXT

    As long as medical records have been kept, cases of food
poisoning have been recorded from eating mussels, clams and fish
in certain areas and at certain times of the year. The unpredictable
and sporadic occurrence of the poison, particularly in shellfish was
a great dilemma to people consuming shellfish and to public health
authorities. Shellfish have always made excellent food for people
along sea coasts. However, at times and with no apparent reason
the shellfish would become poisonous causing sickness and death of
the people that ate them. After a period of about 3 weeks the
shellfish would destroy or excrete the poison and would be safe
again to eat. In humans the first symptoms of paralytic shellfish
poisoning (PSP) include a tingling sensation and numbness in the
lips, tongue and finger tips and may be apparent within a few
minutes after eating poisonous shellfish. As the illness progresses
respiratory distress and muscular paralysis become severe and death,
apparently as a result of respiratory paralysis, occurs within 2 to
24 hours depending upon the size of the dose. If one survives 24
hours the prognosis is good.

    Through the years, from about 1780 onward, much speculation on
the cause of the poison in shellfish was carried on and attributed
to such things as high content of copper salts in the shellfish and
various types of putrefaction. It was not until about 1930 that
Dr. Hermann Sommer at the University of California Medical Center
observed the presence of a certain dinoflagellate blooming in the
water along the California coast when sea mussels were poisonous
and many people had become sick and died from eating them. On the
assumption that these organisms might be involved in the poisoning,
Dr. Sommer filtered a supply of the organisms from the water,
extracted them with a small amount of acidified water about pH 2
in the same manner as he extracted the poison from the mussels.
He found that the extract of the organisms killed mice in exactly
the same way as extracts of poisonous mussels. The organism was
identified as *Gonyaulax catenella* Kofoid (Sommer et al., 1937).
Dr. Sommer then placed nonpoisonous mussels in sea water containing
the organism in a bath in his laboratory and found that these
mussels soon became poisonous. When these mussels were placed in

a bath containing nonpoisonous organisms, the mussels lost the poison within a period of 2 to 3 weeks. His experiments duplicated the natural cycle of events and proved that the original source of the paralytic poison was from this particular dinoflagellate.

The discovery of the relationship of the poisonous dinoflagellate, *G. catenella*, to poisonous sea mussels led to the discovery of other poisonous dinoflagellates that cause shellfish to become poisonous. Koch (1939) found *Pyrodinium phoneus* to be responsible for the extreme toxicity of mussels in Belgium. Needler (1949) and Prakash (1963) established that the poison in scallops in the Bay of Fundy and in clams and mussels along the North Atlantic coast of the United States, the Maritime Provinces and the St. Lawrence estuary was caused by the dinoflagellate *Gonyaulax tamarensis* Lebour. This organism caused outbreaks of shellfish poisoning along the northeastern coast of England in the early summer of 1968 and sporadically has bloomed and caused shellfish to become poisonous in this area since then. Prakash and Taylor (1966) found poison in another species, *Gonyaulax acatenella* Whedon and Kofoid, which blooms along the coast of British Columbia and has caused shellfish in this area to become poisonous. The dinoflagellates *G. catenella*, *G. tamarensis* and *G. acatenella* are the only ones known to produce the paralytic poison and usually bloom at latitudes greater than 30° north or south in areas where the water temperature is around 13° to 18°C. However, in recent years reports from Venezuela state that shellfish poisoning similar to the paralytic type from eating clams and mussels has occurred. The areas involved were 10° to 11° north latitude and the organisms reported found were *G. tamarensis* and a species of *Cochlodinium* (Reys-Vasquez et al., 1979).

Other dinoflagellates have caused shellfish poisoning of different types. Severe cases of shellfish poisoning in humans from eating short-necked clams have occurred along the coast of Lake Hamana in Japan. The disease causes sickness and death in humans and animals by fatty degeneration of liver and kidney tissue and often results in death within a few days. Shellfish are assayed for the poison by injecting extracts of the shellfish into mice. Using 75% methanol, Akiba et al. (1950) extracted from the midgut of the toxic clams a toxic substance which he called venerupin. Later Nakajima (1968) found that the poison originated in the dinoflagellate *Exuviaella mariae-lebouriae* which is now known to be synonymous with *Prorocentrum minimum* var. *mariae-lebouriae* (Okaichi and Imatomi, 1979). Recently Yasumoto et al. (1979a) have found a new toxic dinoflagellate in mixtures of algae and detritus in ciguatera-endemic areas that appears to be the most likely cause of ciguatera. Ciguatera is a disease of humans occurring in tropical and subtropical areas from eating certain eels and reef fishes such as red snappers, sea basses, sharks, sturgeon fishes, groupers, barracudas, etc. Consumption of these fish at certain times results in nausea, vomiting, metallic taste,

dryness of the mouth, abdominal cramps, diarrhea, headache, prostration, chills, fever and general muscular weakness. Mortality rates
are low in this type of poisoning, but in several cases death has
resulted from various complications, mainly cardiovascular collapse.
Most of the poison in the fish is located in the liver and other
visceral organs. None of the fish appear to be direct plankton
feeders as adults and do not produce the poison in their bodies.
All evidence points to the fact that they acquire it through the
food chain by the consumption of smaller fish that feed on plankton.

Another important dinoflagellate, *Gymnodinium breve* Davis, has
caused devastating red tide blooms in the Gulf of Mexico, particularly along the West coast of Florida and produces a poison that is
toxic to fish, chicks and mice (Ray and Wilson, 1957). The tremen-
dous fish kills due to this organism along the Florida coast have
caused severe environmental problems due to the decaying fish and
polluted water in these areas. Associated with these red tides is
an ocean spray brought in by off-shore winds that causes throat and
nasal irritation in humans. Shellfish consuming the dinoflagellates
become toxic and when consumed by humans have caused intoxication
(McFarren et al., 1965). Yasumoto et al. (1979b) have described a
new type of shellfish intoxication in humans from eating mussels
and scallops collected from Tohoku District, Japan. The toxin is
lipid soluble and believed to originate from ingested plankton.
Other dinoflagellates have been reported to be poisonous but not
involved in shellfish or fish intoxications and these are listed in
Table 1 along with all of the known poisonous dinoflagellates, their
usual distribution and some properties of the poison they produce.

Paralytic Shellfish Poison

My studies on the purification and characterization of the
poison produced by *G. catenella* that causes the paralytic type
poisoning found in mussels and clams began with Dr. Hermann Sommer
at the University of California and Dr. Byron Riegel at Northwestern
University and their colleagues in 1945. Sea mussels were collected
along the California coast and Alaska butter clams were collected
in southeastern Alaska as a source of the poison. Purification of
the poison (now called saxitoxin) was accomplished, after several
years of research, by extracting the dark gland of the mussels or
the siphons of the clams with weak acid solutions of hydrochloric
or sulfuric acid at about pH 2. This extract was chromatographed
on carboxylic acid exchange resins (Amberlite XE-64 or CG-50) with
dilute acetic acid and finally on acid washed alumina in absolute
ethanol (Schantz et al., 1957; Mold et al., 1957). The poison was
assayed quantitatively with mice by a procedure developed by Dr.
Sommer (Sommer and Meyer, 1937). One mouse unit (MU) is defined
as the minimum amount of poison that will kill a 20-gram white
mouse in 15 minutes when one ml of the extract or serial dilution

Table 1. Known Poisonous Dinoflagellates

Dinoflagellate	Usual distribution	Poison
Gonyaulax catenella	North Pacific coasts, California to Japan, Chile, South Africa	Causes PSP[a] Structure determined
Gonyaulax tamarensis[b]	Coasts of New England, Canada, countries along North Sea	Causes PSP Structure determined
Gonyaulax acatenella	Coast of British Columbia	Causes PSP Poison not isolated
Pyrodinium phoneus	North Sea	Causes PSP Poison not isolated
Gonyaulax monilata[c]	Gulf of Mexico	Toxic to fish but not warm blooded animals Poison not isolated
Gonyaulax polyedra[d]	Coasts of southern California	Poison reported but not verified
Gymnodinium breve	Gulf of Mexico	Toxic to fish, chicks and mice, partially purified
Gymnodinium veneficum[e]	English Channel	Toxic to fish and mice
Exuviaella mariae-lebouriae[f]	Japan	Causes degeneration of liver and kidney tissue. Partially characterized
Dinoflagellate not identified	Torrid zone	Reported likely cause of ciguatera poisoning Partially characterized.

[a]PSP, paralytic shellfish poisoning.

[b]Called *G. excavata* in cases.

[c]See Aldrich et al., 1967.

[d]See Schradie and Bliss, 1962.

[e]See Abbott and Ballantine, 1957.

[f]Synonymous with *Prorocentrum minimum* var. *mariae-lebouriae*.

of it at pH about 4 is injected intraperitoneally. The specific
toxicity of the purified poison was found to be 5500 MU/mg solids
(Schantz et al., 1958). One MU is equivalent to 0.18 µg. Larger
amounts of poison will kill in shorter times. Death times of 4, 6,
8, and 15 minutes are equivalent to 2.5, 1.6, 1.3, and 1 MU, respectively. The dose may be calculated directly with the equation,
log dose = (145/t) - 0.2 where t is the time in seconds and death
occurs between 240 and 480 seconds. The purified poison (saxitoxin)
is a white solid, soluble in water, slightly soluble in methanol and
ethanol but insoluble in most organic solvents particularly those
immiscible with water such as ethyl and petroleum ether and chloroform. It has two titratable groups, pK_a 8.2 and 11.5. The molecular
formula is $C_{10}H_{17}N_7O_4$, molecular weight 300 as the free base (Fig.
1). It is usually kept as the hydrochloride salt (2 HCL, molecular
weight 372) and is very stable in this form and can be heated in
water solution at 120°C without loss of activity. The poison reacts
with the Benedict-Behre reagent (trinitrobenzoic acid) and the
Jaffe reagent (trinitrophenol) to produce blue and orange-red
derivatives, respectively (color reagents for creatinine) that are
approximately equivalent on a molar basis to the colors produced
when creatinine reacts with these reagents. Reduction of the poison
with hydrogen in the presence of a catalyst results in the uptake
of one mole of hydrogen per mole of poison at atmospheric pressure
and room temperature. This reaction destroys the toxicity and also
eliminates the color reaction with the Benedict-Behre and Jaffe
reagents (Schantz et al., 1961; Mold et al., 1957).

The original work on the chemical structure of the poison was
not completely successful because of the difficulties in obtaining
a suitable crystalline derivative for crystallographic studies. In
our laboratory at the University of Wisconsin, we finally obtained
a suitable crystalline derivative by reacting the poison with
p-bromobenzenesulfonic acid and in cooperation with crystallographers
at Iowa State University, Ames, Iowa, the structure was established
as illustrated in Figure 1 (Schantz et al., 1974, 1975) and later
verified by others (Bordner et al., 1975). Our studies showed that
the reduction, either catalytically or with sodium borohydride and
concurrent loss of toxicity referred to above, reduced one of the
hydroxyl groups of the hydrated ketone with the elimination of a
molecule of water. Treatment of saxitoxin with 7.5 M hydrochloric
acid at 100°C for three hours removed the carbamyl group leaving a
hydroxyl group in its place (Ghazarossian et al., 1976; Ghazarossian,
1977). This derivative had about 60% of the toxicity of saxitoxin
and established a means of making other derivatives of saxitoxin
with toxic activity or ones containing radioactive elements.

On the basis of our structure of saxitoxin, Tanino et al.
(1977) at Harbard University, synthesized saxitoxin and it blocked
sodium channels identical to saxitoxin isolated from the poisonous
dinoflagellate.

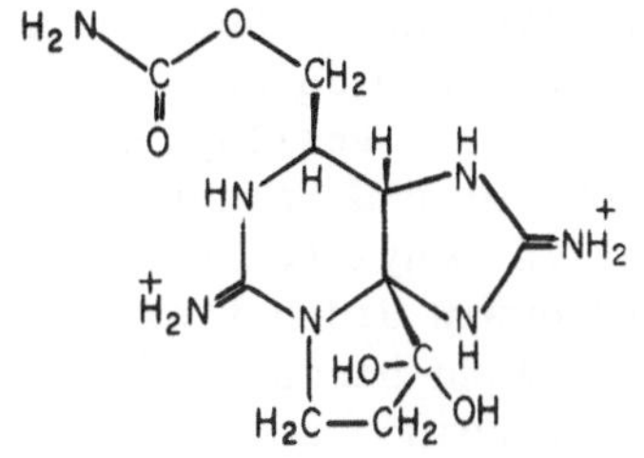

Saxitoxin (from G. catenella)

Fig. 1.   Structure of saxitoxin isolated from California sea
           mussels, Alaska butter clams and *G. catenella* (Schantz
           and Clardy et al., 1974, 1975).

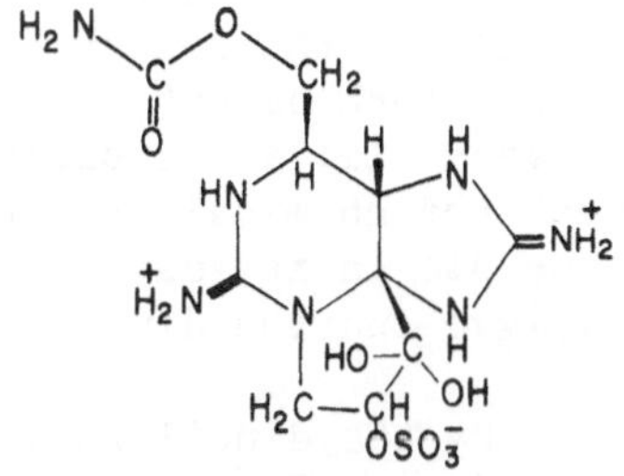

Il-Hydroxysaxitoxin Sulfate
(from G. tamarensis)

Fig. 2.   Structure of 11-hydroxysaxitoxin sulfate (gonyautoxin)
           isolated from Bay of Fundy scallops and *G. tamarensis*
           (Boyer et al., 1978).

Early investigations in our laboratory on the poison produced by *G. tamarensis* and found in scallops from the Bay of Fundy, showed the presence of at least two different poisons. One was a weakly basic poison accounting for 80 to 90 percent of the total toxicity and came off the carboxylic acid exchange resins from pH 7 down to 4, whereas the remainder came off from pH 3 down to 2, similar to the more basic saxitoxin (Schantz, 1960). Investigators at the University of Rhode Island (Shimizu et al., 1976) reported that the major poison from clams and scallops collected along the New England coast and from cultured *G. tamarensis* cells was 11-hydroxy-saxitoxin which they called gonyautoxin. Studies in our laboratory at the University of Wisconsin did not agree entirely with those at the University of Rhode Island because of the difference in the weakly basic nature of the poison compared with that expected of the proposed structure. We isolated the weakly basic poison from scallops and from cultured *G. tamarensis* and found, on the basis of NMR spectra, chemical analyses and properties, that the structure was the sulfonic acid ester of 11-hydroxysaxitoxin as shown in Figure 2 (Boyer et al., 1978). The strongly negative or acidic nature of the sulfate group offset the basic or strongly positive nature of the guanidinium group resulting in the weakly basic nature of the poison. When the sulfate group was hydrolyzed from the poison to 11-hydroxysaxitoxin, it had normal basic properties like saxitoxin. The structure of the poison, saxitoxin, from California sea mussels, Alaska butter clams and from cultured *G. catenella* from central California is identical (Schantz et al., 1966). The only detectable poison found in these *G. catenella* cells was saxi-toxin and the only poison found in the mussels was saxitoxin. However, some poisonous dinoflagellates found along the coast of southeastern Alaska have been reported to produce several poisons, but only saxitoxin was found in the Alaska butter clams. Some freshwater blue-green algae such as *Aphanizomenon flos-aquae* produce a poison reported to be very similar to saxitoxin (Jackim and Gentile, 1968). Although the chemical structure of the poison has not been established, the similarity may be an important link between the blue-green algae and the dinoflagellates.

The poisons produced by the dinoflagellates *G. catenella*, *G. tamarensis* and *G. acatenella* produce their effects by blocking the passage of an impulse along a nerve axon or muscle fiber. The block is due to the poison (saxitoxin and related poisons) binding to the sodium channel in the nerve or muscle cell membrane thus preventing the passage of sodium ions into the cell. The action is similar to that of tetrodotoxin from the puffer fish. Although the chemical structure of tetrodotoxin is quite different from saxitoxin it does contain a guanidinium group like saxitoxin. The action of these compounds has been reviewed by Kao (1966).

Public Health Aspects of Dinoflagellate Poisons

The public health problem caused by poisons produced by certain dinoflagellates is, for the most part, food poisoning in humans. Shellfish poisoning in past years has been a local public health problem in areas where the poisonous dinoflagellate bloomed and people collected the shellfish for food. Now that commercially harvested shellfish and fish that could contain the toxins and poisons are shipped by air to various parts of the world, the food poisoning problem could become more widespread unless proper care and control are maintained. These poisons are difficult to control because of the unpredictable and sporadic occurrence of the organisms producing them. Shellfish that feed on toxic as well as nontoxic microorganisms usually show no distinguishing signs in their appearance when feeding on the toxic organisms because they have a mechanism in the dark gland or the siphon that binds the poison. After the poisonous dinoflagellate has disappeared in the water and a nontoxic species has bloomed, the shellfish excrete or destroy the poison in a couple of weeks and they are safe to eat again. A person collecting shellfish for food therefore has no way to distinguish poisonous shellfish from edible ones except by testing them with some type of animal assay for the poison as described above. A slight numbness of the lips or finger tips a few minutes after eating a few shellfish should be a good indication to stop eating them. Usually ciguatera toxin in the ciguateric fishes becomes apparent only after people begin to get sick from eating them. The very palatable Alaska butter clam presents another problem because 60 to 80 percent of the poison (saxitoxin) is bound in the siphon, an organ of low metabolic activity, and a year or more is required for the poison to be eliminated. If the clams were exposed to a source of the poison each summer they could remain poisonous for many years.

In certain areas, government agencies carry out assays for the paralytic poison on clams and mussels where they are collected for food. If shellfish become dangerously poisonous, warnings are posted and publicized by radio and newspaper. The most practical means of controlling shellfish poisoning is by direct sampling and assaying of the shellfish by mouse assay in areas where they are harvested commercially and where picnickers commonly collect them for food. Education of the public regarding the danger of the sporadic occurrence of the poison and its cause is important, especially in areas where shellfish poisoning is common.

Estimates of the amount of paralytic shellfish poison to cause sickness and death have been made from accidental poisonings. Information obtained along the California coast from deaths due to poisonous sea mussels that fed on *G. catenella*, indicates that death occurred with a dose of about 20,000 MU or between 3 and 4 mg. Data gathered in Canada along the St. Lawrence Estuary on

poisonous clams that fed on *G. tamarensis* indicate that death
occurred in some individuals with a dose of 5,000 MU or about one
mg. Death usually occurs within 2 to 24 hours after eating
poisonous shellfish depending upon the magnitude of the dose.

The United States Food and Drug Administration has set the
maximum acceptable level for paralytic poison in fresh, frozen, or
canned shellfish at no more than 400 MU or about 80 µg per 100 grams
of edible portion. This amount or less has not been known to cause
sickness.

There is no known antidote for the paralytic shellfish poison,
but artificial respiration, which should always be employed when
respiratory distress becomes apparent, is believed to have saved
lives, particularly in cases where a borderline dose was consumed.
The dose that causes human sickness with other dinoflagellate
poisons is not known at present.

Saxitoxin has become a valuable tool in medical research on
nerve transmission because of its specific action of blocking the
passage of sodium ions through nerve and muscle cell membranes. A
considerable amount of the saxitoxin that we have purified in our
laboratory has been donated for this type of research.

REFERENCES

Abbot, B. C., and D. Ballantine. 1957. The toxin from *Gymnodinium
    veneficum*. J. Mar. Biol. Assoc. U. K. *36*:169.
Akiba, et al. 1950. Studies on the poisoning caused by Asari.
    Quoted by Okaichi and Imatomi, 1979.
Aldrich, D. V., S. M. Ray, and W. B. Wilson. 1967. *Gonyaulax
    monilata*: population growth and development of toxicity in
    cultures. J. Protozol. *14*:636.
Bordner, J., W. E. Thiessen, H. A. Bates, and H. Rapoport. 1975.
    The structure of a crystalline derivative of saxitoxin. J.
    Am. Chem. Soc. *97*:6008-6012.
Boyer, G. L., E. J. Schantz, and H. K. Schnoes. 1978. Character-
    ization of 11-hydroxysaxitoxin sulfate, a major toxin in
    scallops exposed to blooms of the poisonous dinoflagellate
    Gonyaulax tamarensis. J. Chem. Soc. Chem. Comm. *20*:889-890.
Ghazarossian, V. E. 1977. Ph.D. Thesis, University of Wisconsin/
    Madison.
Ghazarossian, V. E., E. J. Schantz, H. K. Schnoes, and F. M. Strong.
    1976. A biologically active acid hydrolysis product of
    saxitoxin. Biochem. Biophys. Res. Commun. *68*:776-780.
Jackim, E., and J. Gentile. 1968. Toxins of a blue-green alga:
    similarity to saxitoxin. Science *162*:915.
Kao, C. Y. 1966. Tetrodotoxin, saxitoxin, and their significance
    in the study of excitation phenomena. Pharmacol. Rev. *18*:997.

Koch, H. J. 1939. La cause des empoisonnements paralytiques
    provoques par les moules. Assoc. Fr. Av. Sci. Paris 63rd
    Session, pp. 654.

McFarren, E. F., H. Tanabe, F. J. Silva, W. B. Wilson, J. E.
    Campbell, and K. H. Lewis. 1965. The occurrence of a
    ciguatera-like poison in oysters, clams, and *Gymnodinium
    breve* cultures. Toxicon *3*:111.

Mold, J. D., J. P. Bowden, D. W. Stanger, J. E. Maurer, J. M.
    Lynch, R. S. Wyler, E. J. Schantz, and B. Riegel. 1957.
    Evidence for the purity of the poison isolated from toxic
    clams and mussels. J. Am. Chem. Soc. *79:*5235-5238.

Nakazima, M. 1968. Studies on the source of shellfish poison in
    Lake Hamana. Bull. Jap. Soc. Sci. Fish. *34*:130.

Needler, A. B. 1949. Paralytic shellfish poisoning and *Gonyaulax
    tamarensis*. J. Fish. Res. Board Can. *7*:490-504.

Okaichi, T., and Y. Imatomi. 1979. Toxicity of *Prorocentrum
    minimum* var. *mariae-lebouriae* assumed to be a causative
    agent of short-necked clam poisoning. Pages 385-388 *in*
    D. L. Taylor and H. H. Seliger, eds., Toxic Dinoflagellate
    Blooms. Elsevier/North Holland, New York.

Prakash, A. 1963. Source of paralytic shellfish toxin in the Bay
    of Fundy. J. Fish. Res. Board Can. *20*:983.

Prakash, A., and F. J. R. Taylor. 1966. A red water bloom of
    Gonyaulax acatenella in the strait of Georgia and its
    relation to paralytic shellfish toxicity. J. Fish. Res.
    Board Can. *23*:1265.

Ray, S. M., and W. B. Wilson. 1957. U. S. Fish and Wildl. Ser.
    Fish. Bull. *128*:469-496.

Reyes-Vasquez, G., E. Ferraz-Reyes, and E. Vasquez. 1979. Toxic
    dinoflagellate blooms in northeastern Venezuela during 1977.
    Pages 191-194 *in* D. L. Taylor and H. H. Seliger, eds.,
    Toxic Dinoflagellate Blooms. Elsevier/North Holland, New
    York.

Schantz, E. J. 1960. Biochemical studies on paralytic shellfish
    poisons. Ann. N. Y. Acad. Sci. *90*:843-855.

Schantz, E. J., V. E. Ghazarossian, H. K. Schnoes, F. M. Strong,
    J. P. Springer, J. O. Pezzanite, and J. Clardy. 1974.
    Paralytic poisons from marine dinoflagellates. Proc. 1st
    Int. Conf. on Toxic Dinoflagellate Blooms. November.
    Boston, Mass.

Schantz, E. J., V. E. Ghazarossian, H. K. Schnoes, F. M. Strong,
    J. P. Springer, J. O. Pezzanite, and J. Clardy. 1975. The
    structure of saxitoxin. J. Am. Chem. Soc. *97*:1238-1239.

Schantz, E. J., J. M. Lynch, G. Vayvada, K. Matsumoto, and H.
    Rapoport. 1966. The purification and characterization of
    the poison produced by *Gonyaulax catenella* in axenic
    culture. Biochem. *5*:1191-1195.

Schantz, E. J., E. F. McFarren, M. L. Schafer, and K. H. Lewis.
    1958. Purified shellfish poison for bioassay standardiza-
    tion. J. Assoc. Off. Anal. Chem. *41*:160-168.
Schantz, E. J., J. D. Mold, W. L. Howard, J. P. Bowden, D. W.
    Stanger, J. M. Lynch, O. P. Wintersteiner, J. D. Dutcher,
    D. R. Walters, and B. Riegel. 1961. Some chemical and
    physical properties of purified clam and mussel poisons.
    Can. J. Chem. *39*:2117-2123.
Schantz, E. J., J. D. Mold, D. W. Stanger, J. Shavel, F. J. Riel,
    J. P. Bowden, J. M. Lynch, R. S. Wyler, B. Riegel, and H.
    Sommer. 1957. A procedure for the isolation and purifica-
    tion of the poison from toxic clam and mussel tissues. J.
    Am. Chem. Soc. *79*:5230-5235.
Schradie, J., and C. A. Bliss. 1962. The cultivation and toxicity
    of *Gonyaulax polyedra*. Lloydia *25*:214.
Shimizu, Y., L. Buckley, M. Alam, Y. Oshima, W. E. Fallon, H. Kasai,
    I. Miura, V. P. Gullo, and K. Nakanishi. 1976. Structures
    of gonyautoxin II and III from the east coast dinoflagellate
    Gonyaulax tamarensis. J. Am. Chem. Soc. *98*:5414-5416.
Sommer, H., W. F. Whedon, C. A. Kofoid, and R. Stohler. 1937.
    Relation of paralytic shellfish poison to certain plankton
    organisms of the genus *Gonyaulax*. A. M. A. Arch. Pathol.
    24:537.
Tanino, H., T. Nakata, T. Kaneko, and Y. Kishi. 1977. A stereo-
    specific total synthesis of d,1-saxitoxin. J. Am. Chem.
    Soc. *99*:2818-2819.
Yasumoto, T., I. Nakajima, O. Yasukatsu, and R. Bagnis. 1979a. A
    new toxic dinoflagellate found in association with ciguatera.
    Pages 65-70 *in* D. L. Taylor and H. H. Seliger, eds., Toxic
    Dinoflagellate Blooms. Elsevier/North Holland, New York.
Yasumoto, T., Y. Oshima, and M. Yamaguchi. 1979b. Occurrence of
    a new type of toxic shellfish in Japan and chemical properties
    of the toxin. Pages 395-398 *in* D. L. Taylor and H. H.
    Seliger, eds., Toxic Dinoflagellate Blooms. Elsevier/North
    Holland, New York.

THE TOXIC PRINCIPLES OF *PRYMNESIUM PARVUM*

M. Shilo

Division of Microbial and Molecular Ecology
Institute of Life Sciences
Hebrew University
Jerusalem, Israel

ABSTRACT

The euryhaline chrysophyte *Prymnesium parvum*, widespread in brackish and marine habitats, produces toxic principles which cause mass mortality of fish. The toxins of *Prymnesium* show a broad spectrum of activity including lethal effects on gill breathing animals, cytotoxic effects on erythrocytes, nucleated mammalian cells, protozoa and bacteria, and different pharmacological effects. The toxins, purified from conditions in nature as well as from axenic cultures of *Prymnesium*, were found to be acidic polar phospho-proteolipids. Due to their ampiphatic nature, they form micelles in the aquatic millieu above the critical micellar concentration. The mode of action of the *Prymnesium* toxin in many different biological systems seems to be based on the reversible effect of the toxin on biological membranes, causing leakage and increasing their permeability which allows penetration of macromolecules such as trypan blue or radio iodinated bovine albumine. This activity is expressed only under specific conditions including pH and presence of cationic activators. Analysis of biosynthesis, excretion, and extracellular accumulation of the toxin now explains why its activity in nature is restricted. Complete control of economic damage can now be obtained in small bodies of water such as fishponds by using a sensitive field bioassay to detect sublethal doses of the toxin and by addition of ammonia to the ponds to control the organisms responsible for toxin production in the ponds.

The killing of *Prymnesium* occurs by osmotic lysis under appropriate pH conditions due to selective intracellular accumulation of the weak electrolyte driven by the pH differential

between the cell interior and its external medium.   Free diffusion
of the neutral molecule and trapping of the charged ion cause
increase in turgor swelling and lysis.   The example of *Prymnesium
parvum* can serve as a model for the study of other toxigenic algae
and cyanobacteria helping to uncover common biological principles.

TEXT

   Many different algae and cyanobacteria produce potent toxins,
and some of these have been shown to have important ecological
economic implications causing mass mortality of fish and other
aquatic animals, and intoxication of man and terrestrial animals
(Shilo, 1972).   It has been suggested that certain mass mortalities
of marine animals in the geological history of the earth caused by
toxigenic algae may have played a role in the formation of geological strata (Brongersma-Sauders, 1948).   Yet only in very few cases
do we understand the mechanisms of toxin biosynthesis, toxin accumulation in the environment, and its mode of action on populations
of sensitive organisms.

   The scattered information on algal toxins is associated with
the difficulties inherent in associating a specific alga as causative agent for a certain toxic phenomenon.   This is further compli-
cated by the sporadic toxic occurrences removed both spatially and
temporally from the appearance of the alga.   It has often been
observed that toxic phenomena appear after the peak of the algal
blooms has passed.

   An additional limitation in the study of toxigenic algae is
the difficulty in many cases to obtain axenic cultures, which are
the only way to exclude any role of contaminating microorganisms
in the expression of toxic phenomena.

   A phenomenon common to most toxigenic algae, including dinoflagellates, blue-green algae, and *Prymnesium parvum*, is the lack
of correlation between the number of algal organisms and the toxin
levels in the water body or experimental culture.   Appearances of
mass algal growth, therefore, does not necessarily indicate lethal
toxicities, or vice versa.   The importance of environmental and
physiological factors, as well as differences in the genetic
potential of toxigenic species to produce the toxin, have been
demonstrated for blue-green algae, dinoflagellates, and chrysophytes
(Shilo, 1972).

   I would therefore like to present at this meeting the case of
the toxin forming phytoflagellate *Prymnesium parvum*, widespread in
many regions of the world and economically important in the intoxication of aquatic animals in brackish water ecosystems (Shilo,

1967, 1971), in order to raise some of the more general points of
interest connected with the formation, accumulation and action of
algal toxins.   The advantage of use of *Prymnesium parvum* as a model
lies in the ease of its cultivation in axenic culture in defined
media, both photoautotrophically and heterotrophically.   The
ability to clone *Prymnesium* and to study toxin formation in single
colonies of the organism make it possible to approach genetic problems of toxin formation.   Further advantages of this system are the
availability of rapid dependable assay methods and the ability to
carry out ecological studies in the brackish water fishponds in
which the organisms are found in "blooms" throughout autumn, winter,
and spring.

In this case, different from the situation known for some
toxigenic, red-tide dinoflagellates and many toxin forming cyanobacteria, the toxic principles are excreted by the cells into the
environment.   The damage caused by the toxins is not restricted to
conditions of release of the toxins upon death and decomposition
of the toxigenic organisms, or to ingestion of filtering organisms
such as bi-valves which concentrate and store the toxigenic algae
from large water volumes.   Some of the major problems common to
most toxigenic algae for which the *Prymnesium parvum* system may be
helpful as a model are the following:

1.   There is a lack of correlation between appearance of the toxi-
     genic organism or its quantity in the environment and the onset
     of intoxication or the severity of the toxic phenomenon in
     sensitive populations.   In the case of *Prymnesium* toxicity in
     the ponds and concommitant fish, mortality is found in less
     than 1% of the afflicted fishponds.   For *Prymnesium parvum*
     laboratory and field experiment have revealed the role of
     each of a complex sequence of steps for the toxic expression
     to occur.   These steps include the intracellular biosynthesis
     of the toxin, its excretion, its accumulation in the water,
     its activation by cationic cofactors in the water, and its
     action on the target organs of sensitive organisms for the
     intoxication phenomenon to occur.   Each and all steps are
     necessary.   Our experiment showed that the non-toxic conditions
     found in many of the fishponds containing *Prymnesium parvum*
     could be related to conditions incompatable with one or more
     of the different stages of the sequence described.   Figure 1
     shows a scheme describing the different stages necessary for
     Prymnesium parvum toxicity.

2.   A condition found in *Prymnesium parvum* and worth consideration
     in other algal and cyanobacterial toxin systems is the fact
     that the ichthyotoxin of *Prymnesium parvum* requires activation
     by cationic cofactors (Shilo, 1971).   Furthermore, the nature
     of the activator, Na^+, Mg^{++}, or Ca^{++}, or polycationic polyamines
     determines the quantitative expression of the toxic effect.
     Thus the level of toxicity observed under different environ-

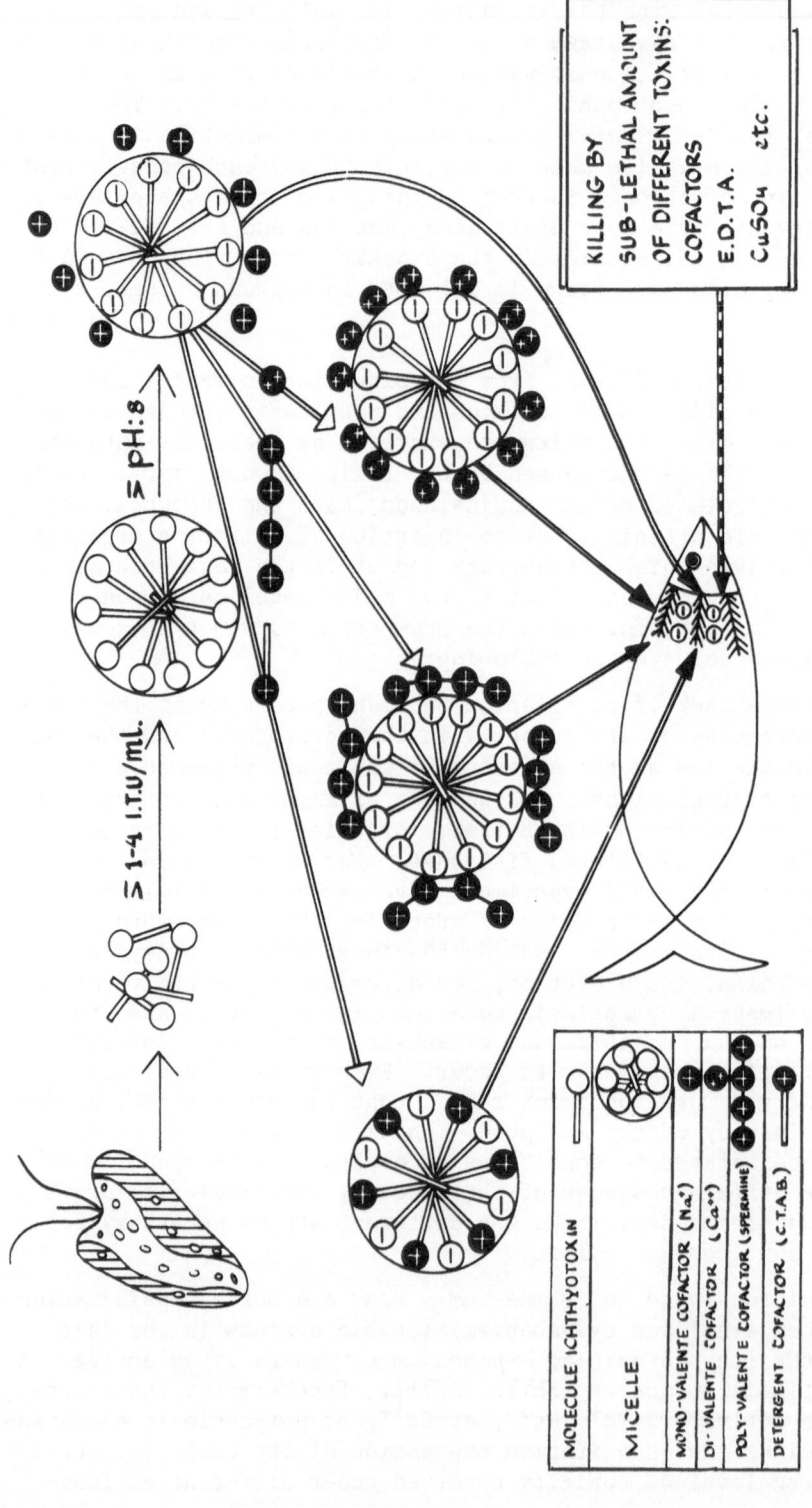

Fig. 1. Scheme for different stages necessary to induce *Prymnesium parvum* toxicity in fish.
I.T.U. = ichthyotoxin unit

mental conditions may vary greatly even when the amounts of the toxin involved remains the same.

3. The toxic principle of *Prymnesium parvum* was found to be composed of a family of closely resembling substances and not a single component (Shilo, 1971). The biological spectrum of activity of the *Prymnesium* toxins is extremely broad, including cytotoxic, hemolytic, neurotoxic, and lethal activity of isolated cells and organisms. The fact that the different components of the *Prymnesium* complex show differences in their biological or pharmacological activities may be helpful in finding the chemical nature responsible for the different activities. The question to what extent a similar situation may exist in other algal toxins remains open.

4. The question often arises as to the physiological or ecological function of the toxin in the life cycle of the toxigenic organism. In the case of *Prymnesium parvum* the hypothesis has been put forward that the toxin which is a proteo-phospho-lipid might be a membrane precursor. The excretion into the medium of this component reflects unbalanced oversynthesis of a normal membrane intermediates and its spilling out into the medium only under rare conditions which create unbalanced growth. The finding that toxin formation is greatly enhanced at low phosphate concentrations of the growth medium (Shilo, 1971) strengthens this notion. Toxin formation thus has no direct role or importance in the life of *Prymnesium parvum* and does not serve any evolutionary purpose. The uniqueness and rarity of conditions which allow this unbalanced synthesis may explain the rare appearance of toxicity in nature.

5. Wherever toxigenic algae or cyanobacteria appear and cause heavy economic losses, the problem of control of the organism or inactivation of the toxin becomes important. *Prymnesium* is the one case in which complete prevention of economic losses due to fish kills in fishponds is now possible in all brackish water regions. Control in this case is based on periodic testing of ponds for toxicity by a sensitive bioassay, including a potent cationic activator (Shilo, 1967, 1971, 1972). Wherever concentration sublethal to the fish are detected, aqueous ammonia is used as an effective *Prymnesium* lysing algicid. The mechanism of action of ammonia on *Prymnesium* was found to be the trapping and concentration of the protonated ammonia ion in the cell caused by the pH difference between the interior of the cell and its external millieu and the following entry of water swelling and lysis.

The *Prymnesium* Toxins

The toxic principles of *Prymnesium parvum* have been highly
purified and characterized by Ulitzur (1969; Ulitzur and Shilo,
1970b) from *Prymnesium* cells as well as from extracellular medium.
The toxin components have been isolated from axenic cultures as
well as from the water of fishponds containing "blooms" of
Prymnesium.

Extraction and purification of the toxin is based on the
differential solubility of the toxin in organic solvents and separation by a silicic acid column. The purification methods used
results in a 2000- to 3000-fold increase in specific activities of
Prymnesium hemolysin, cytotoxin, and ichthyotoxin with a yield of
50% of initial activity.

When purified or crude toxins were chromatographed in thin-
layer silica gel chromatographs (Ulitzur, 1969; Ulitzur and Shilo,
1970b), the toxins separated into six hemolytic spots. Purified
toxin of *Prymnesium* grown on medium containing ^{32}P was labeled
radioactively, and each of the six toxic fractions contained ^{32}P.
The chemical properties of the *Prymnesium* toxins have been found to
be similar to those of acidic polar lipids. Analysis of the purified toxin revealed 15 amino acids and a number of fatty acids,
0.47% phosphate (P_i), and 10–12% hexose sugars (Ulitzur and Shilo,
1970b). In its composition of fatty acids together with protein
and phosphate, the purified toxin resembles proteolipids.

A proteolipid structure of the toxin could explain various
effects on its biological activities. For instance, the observation
that the hemolytic, cytotoxic, and bacteriolytic activities are
destroyed by treatment of the toxin with monovalent cations at high
pH, while ichthyotoxic activity remained unaffected (Ulitzur and
Shilo, 1970b; Shilo and Rosenberger, 1960). A possible explanation
for this effect is that both hemolytic and ichthyotoxic activities
are expressed only when the lipid-protein complex is intact, and
that splitting of the complex inactivates hemolytic activity while
the lipid moiety alone retains its ichthyotoxic activity. The
sensitivity of the hemolysin, but not the ichthyotoxin, to enzymatic
inactivation by papain further supports the hypothesis that the
protein moiety may be involved in hemolytic activity (Ulitzur and
Shilo, 1970b).

By virtue of its content of a nonpolar lipid moiety and polar
components, the *Prymnesium* toxin would be expected to show amphi-
pathic properties and to form micelles or liquid crystalline
structures in water. Indeed, formation of micelles above the
critical micellar concentration is indicated by the concentration-
dependent dialyzability of the toxin through cellophane membranes
and by the filtrability through Millipore membranes of 0.1 μ mean

pore size which is similarly concentration-dependent. (Ulitzur, 1969).

The toxin micelles can be shown directly when examined by negative staining techniques in the electron microscope. The purified *Prymnesium* toxin appears to consist of a relatively homogeneous population of hollow spheres (Shilo, 1971).

Biological Activities of *Prymnesium* Toxin

The toxin of *Prymnesium parvum* has a broad spectrum of different biological activities in vivo and in vitro (Fig. 1). It seems that there are distinct toxic factors to a family of compounds having a similar composition rather than to a single factor with different activities.

Many gill-breathing aquatic animals show sensitivity to the toxin when immersed in toxin solutions. All teleost species tested were found to be sensitive (Ulitzur, 1965), including *Cyprinus carpio, Tilapia galilaea, Mugil cephalus,* and *Gambusia affinis* minnows; the latter fish is commonly used in laboratory and field tests for the assay of *Prymnesium* ichthyotoxicity. Many aquatic invertebrates are also killed by *Prymnesium* toxin; these include the bivales *Unio* sp. and *Dreissensia polymorpha* (Otterstrom and Steemann-Nielsen, 1939). The effect of *Prymnesium* toxin on amphibians is especially interesting since it demonstrates the relationship between gill-breathing and susceptibility to intoxication on submersion. In the amphibian species tested (*Rana pipiens* and *Bufo* sp.), the gill-breathing tadpoles only were found to be highly sensitive to the toxin, while the stages after metamorphosis were insensitive.

A unique property of *Prymnesium* toxin, is the enhancement of toxicity for immersed fish in the presence of Ca^{2+}, Mg^{2+}, or streptomycin. This activation is most dramatically expressed when *Prymnesium* preparations having little or no ichthyotoxicity of their own are rendered highly ichthyotoxic by cation addition (Ulitzur, 1965; Ulitzur and Shilo, 1964).

The ichthyotoxin and cation form an active complex, the toxicity of which is dependent on the nature of the cationic activator (Shilo, 1971).

A requirement for the cationic activation of *Prymnesium* ichthyotoxin is a suitable pH (<8). At pH 7 and below, little ichthyotoxicity is apparent of the toxin with cations, while above this value the ichthyotoxicity increases to a maximum at pH 9. This has been explained by the pH dependence of the complex for-

mation.   Once icthyotoxin is complexed with the cation activator at
high pH, the ichthyotoxicity is expressed even at lower pH (6-7)
(Ulitzur, 1965).

The first effects of the toxin on immersed fish can be observed
in less than 5 minutes. The fact that all gill-breathing animals
tested are sensitive to the toxin and that amphibians become
refractory upon metamorphosis strongly supports the hypothesis put
forward (Shilo, 1971) that the exposed gills are the primary target
of the toxin. When gambusias which have been immersed in lethal
toxin and cation mixtures for a short period are transferred into
a 1% trypan blue solution, their gills immediately become darkly
stained. Further direct evidence of damage to the permeability of
the gill epithelia is that minnows pretreated with toxin show
enhanced uptake of serum albumin ^{125}I and ^{131}I, and that such fish
are greatly (10-40 times) sensitized nonspecifically to various
fish toxicants (Ulitzur and Shilo, 1966). The increased permeability of the gill occurred only in conditions in which ichthyotoxin activity is expressed; it is cation activated, pH dependent,
and inhibited by sodium chloride. The damage to gill permeability
and the consequent sensitization to toxic agent is reversible;
pretreated fish transferred into toxin-free conditions for 4 to 8
hours gradually lost the enhanced gill permeability (Ulitzur and
Shilo, 1966).

It thus appears that the intoxication of gill-breathing
animals immersed in *Prymnesium* toxin consists of two stages. Initially, there is reversible damage to the gill tissues, resulting
in the loss of their selective permeability. This damage caused
by the ichthyotoxin is expressed only under the specific conditions
required for the ichthyotoxin activity. The second stage, leading
to death, is the response of the sensitized fish to any of a
number of toxicants which may be present in the milieu; this
includes the *Prymnesium* toxin itself, whose general toxicity to
the fish was indicated in the intraperitoneal injections mentioned
above. The nature of the toxin-cation complex determines the mode
of action of toxin on immersed fish.

Although the gill epithelium appears to be the only tissue of
the intact fish known to be affected by the toxin excreted by
Prymnesium in its natural milieu, gill cells are by no means unique
in their sensitivity to preparations of *Prymnesium* toxin. The wide
range of vertebrate cells shown sensitive in suspension to the
toxin includes nucleated fish and bird and nonnucleated mammalian
erythrocytes (Yariv and Hestrin, 1961; Bergmann and Kidron, 1966;
Reich, Bergmann and Kidron, 1965), normal human liver and amnion
cells, and tumor (Ehrlich ascites and HeLa) cells (Shilo and
Rosenberger, 1960; Dafni and Shilo, 1966). This activity has been
extended to bacterial protoplasts and spheroplasts and to *Mycoplasma*

(Ulitzur and Shilo, 1970a). The toxic action appears to be directed
toward the cytoplasmic membrane of these various cells, and its
final expression is their lysis.

The primary effect of *Prymnesium* cytotoxin seems to be its
interaction with the membrane cells.

In Ehrlich ascites cells *Prymnesium* toxin causes rapid fluxes
of the Na^+ and K^+. Within less than 30 seconds K^+ leaks out
rapidly followed by Na^+ entry, until a final equilibrium of intra-
and extracellular K^+ and Na^+ is reached (Dafni and Shilo, 1966;
Shilo, 1971). The total intracellular content of these ions in
the toxin treated cells increases within several minutes and may
well explain the phenomena of swelling and lysis as due to osmotic
imbalance. An attempt to approach the mechanism of action of the
Prymnesium toxin with membranes was made by studying the effect on
artificial thin lipid bileaflet membranes (Morau and Ilani, 1974).
The toxin caused marked increase in membrane conductance, and the
treated membranes became cation permselective.

The hypothesis was put forward (Morau and Ilani, 1974) that
aggregates of the toxin are intercalated in the membrane to form
negatively charged aqueous pores. The good correlation between the
increase in membrane conductance and the increase in membrane
permeability to urea after *Prymnesium* toxin treatment fitted well
with the hypothesis of formation of aqueous channels in the membrane.

Biosynthesis of *Prymnesium* Toxin

Growth of *Prymnesium* on simple mineral liquid media in the
light showed absolute requirements for vitamin B_{12} and thiamine
(Droop, 1954; McLaughlin, 1958; Rahat and Reich, 1963a, 1963b).
Heterotrophic growth of *Prymnesium* in the dark is accomplished in
glycerol-rich medium (Rahat and Jahn, 1965).

A comparison of the capacity to form toxin in different media
and under different environmental conditions clearly shows that
growth and toxin biosynthesis have different optimal requirements.
Two of the most striking examples of this difference are the
effects of light limitation and phosphate concentration on toxin
biosynthesis.

In the absence of light, a marked reduction of toxin production occurs in glycerol-enriched synthetic medium (Shilo, 1971),
although cell multiplication continues. In such dark-grown cultures there is a rapid dilution of intracellular toxin of the
inoculum and no detectable new toxin formation until late in the

stationary phase of growth. *Prymnesium* colonies grown on solid modified S50 medium in the dark exhibit hemolytic activity only after a 24-hour exposure to light.

Limitation of phosphate enhances toxin formation even before markedly affecting growth (Shilo, 1971). A ten- to twenty-fold increase in all the toxins (hemolysin, cytotoxin, and ichthyotoxin) is found in phosphate-starved cells, and even higher titers of these toxins are obtained in the culture fluid.

REFERENCES

Bergmann, F., and M. Kidron. 1966. Latent effects of haemolitic agents. J. Gen. Microbiol. *44*:233-240.

Brongersma-Sauders, M. 1948. The importance of upswelling water to vertebrate paleontology and oil geology. Verhandelingen der Koninklijke Nederlandsche Akademie van Wetternschaften, Afd Naturlennde Tweeds Sectie, Deel *45*(4):1-112.

Dafni, Z., and M. Shilo. 1966. The cytotoxic principle of the phytoflagellate *Prymnesium parvum*. J. Cell Biol. *28*:461-471.

Droop, M. 1954. A note on the isolation of small marine algae and flagellates in pure culture. J. Marine Biol. Assoc. U. K. *3*:511-514.

McLaughlin, J. J. A. 1958. Euryhalin chrysomonade: nutrition and toxigenesis in *Prymnesium parvum* with notes on *Isochrysis galbana* and *Monochrysis lutheri*. J. Protozool. *5*:75-81.

Morau A., and A. Ilani. 1974. The effect of *Prymnesium* on the electric conductivity of thin lipid membranes. J. Membrane Biol. *16*:237-256.

Otterstrom, C. V., and E. Steemann-Nielsen. 1939. Two cases of extensive mortality in fishes caused by the flagellate *Prymnesium parvum* Carter. Rep. Danish Biol. Sta. *44*:5-24.

Rahat, M., and L. T. Jahn. 1965. The growth of *Prymnesium parvum* in the dark and a note on its ichthyotoxin formation. J. Protozool. *12*:246-250.

Rahat, M., and K. Reich. 1963a. The B_{12} vitamins and growth of the flagellate *Prymnesium parvum*. J. Gen. Microbiol. *31*: 195-202.

Rahat, M., and K. Reich. 1963b. The B_{12} vitamins and methionine in the metabolism of *Prymnesium parvum*. J. Gen. Microbiol. *31*:203-208.

Reich, K., F. Bergmann, and M. Kidron. 1965. Studies on the homogeneity of *Prymnesium*, the toxin isolated from *Prymnesium parvum* Carter. Toxicon *3*:33-39.

Shilo, M. 1967. Formation and mode of action of algal toxins. Bacteriol. Rev. *31*:180-193.

Shilo, M. 1971. Toxins of Chrysophyceae. Pages 67-103 *in* S. Kadis, A. Ciegler, and S. J. Ajl, eds., Microbial Toxins. Vol. 7. Academic Press, New York.

Shilo, M. 1972. Toxigenic algae. Pages 233-265 *in* O. J. D. Hockenhull II, ed., Progress in Industrial Microbiology. Vol. 11. Churchill Livingstone Press, Edinburgh.

Shilo, M., and R. F. Rosenberger. 1960. Studies on the toxic principles formed by the chrysomonad *Prymnesium parvum*. Ann. N. Y. Acad. Sci. *90*:866-876.

Ulitzur, S. 1965. The mode of action of cofactors on the ichthyotoxic activity of *Prymnesium* toxin and other fish toxins. Ph.D. Thesis. Hebrew University, Jerusalem.

Ulitzur, S. 1969. Purification and separation of the toxins produced by the phytoflagellate *Prymnesium parvum*. Verh. Int. Verein. Limnol. *17*:771-777.

Ulitzur, S., and M. Shilo. 1964. A sensitive assay system for determination of the ichthyotoxicity of *Prymnesium parvum*. J. Gen. Microbiol. *36*:161-169.

Ulitzur, S., and M. Shilo. 1966. Mode of action of *Prymnesium parvum* ichthyotoxin. J. Protozool. *13*:332-336.

Ulitzur, S., and M. Shilo. 1970a. Effect of *Prymnesium parvum* toxin, cetyltrimethylammonium bromode and sodium dodecyl sulphate of bacteria. J. Gen. Microbiol. *62*:363-370.

Ulitzur, S., and M. Shilo. 1970b. Prodecure for purification and separation of *Prymnesium parvum* toxins. Biochem. Biophys. Acta *201*:350-363.

Yariv, J., and S. Hestrin. 1961. Toxicity of the extracellular phase of *Prymnesium parvum* cultures. J. Gen. Microbiol. *24*:165-175.

STUDIES ON THE ECOLOGY, GROWTH AND PHYSIOLOGY OF

TOXIC *MICROCYSTIS AERUGINOSA* IN SOUTH AFRICA[1]

William E. Scott, Deryl J. Barlow,[2]
and John H. Hauman

National Institute for Water Research
Council for Scientific and Industrial Research
P. O. Box 395
Pretoria, 0001, Republic of South Africa

ABSTRACT

Toxic blooms of *Microcystis aeruginosa* are a regular feature
of numerous impoundments throughout southern Africa. In the more
eutrophic impoundments, bloom formation can reach serious proportions, often resulting in cattle kills. The extent of the problem
in South Africa is taken into perspective. As a result of the
variability in appearance of the natural populations it is not
possible to identify the toxic strains from microscopical identification alone, but samples collected from various impoundments
over a period of four years indicated that toxicity, when present,
was always associated with a form described by Komárek (1958) as
Microcystis aeruginosa forma *aeruginosa* and never with the form
described as *M. aeruginosa* forma *flos-aquae*. Laboratory experiments
supplemented with field measurements indicated that *M. aeruginosa*
can grow well over a range of light intensities, giving it a competitive advantage over other algae. The alga adapts physiologically to different light intensities by changing its gas vacuole
and pigment composition. At high light intensities there is an

---

[1]Part of the work contained in this report was presented by the
first author under the title of "Observations on the ecology,
growth and physiology of *Microcystis aeruginosa* in the laboratory
and in the field" at the Symposium on "Health Aspects of Water
Supplies" Pretoria, November 15, 1979.
[2]Present address: Department of Botany, University of Natal,
P. O. Box 375, Pietermaritzburg, 3200, Republic of South Africa.

increase in gas vacuole content, while the reverse is true at low
light intensities. The pigment and gas vacuole changes enable *M.
aeruginosa* to grow rapidly under a variety of conditions and partly
explains the success of this alga in many turbid South African
impoundments.

INTRODUCTION

     Toxic blooms by strains of *Microcystis aeruginosa* have been
responsible for several instances of cattle kills in South Africa
(Steyn, 1945; Stephens, 1949; Toerien et al., 1976). It is therefore important to have a better understanding of factors which
promote growth and toxin production by these organisms. In this
paper we firstly report on the occurrence of toxic blooms in
southern Africa using both older records and our own more recent
observations. Secondly we also report on the results of experiments
where we investigated the response of *Microcystis* to a variety of
light intensities in the laboratory and in the field.

MATERIALS AND METHODS

Toxicity Tests

     Qualitative toxicity tests were routinely performed on samples
collected from the field by intraperitoneal injection of mice. The
field samples were then allowed to stand for a few hours to allow
the buoyant gas-vacuolate algae to collect at the surface of the
flask. One milliliter of the concentrated algae was frozen, thawed
twice and then subjected to mild sonication. Half of this mixture
(equivalent to approximately 4 to 6 mg dried algae) was then administered intraperitoneally to a mouse. The sample was regarded as
toxic if the animal died within 4 hours and if it had an enlarged
liver, which is symptomatic of *Microcystis* poisoning. Parallel
samples were microscopically examined to determine the species of
algae present, employing the key by Komárek (1958) to identify the
algae.

Culture Conditions

     The *Microcystis aeruginosa* culture used in laboratory experi-
ments originated from a toxic bloom which occurred in Hartbeespoort
Dam during 1974 (Toerien et al., 1976) and was isolated according
to the procedure described by Scott (1974). During cultivation
the alga changed from its colonial form to a unicellular form in
a similar fashion to that described for *Microcystis aeruginosa*

NRC-1 (Zehnder and Gorham, 1960). The culture has been given the number WR88 in the NIWR culture collection and was the source of the culture UV-006 in the collection of the University of the Orange Free State (Krüger and Eloff, 1977).

Cultivation of the alga was in 250 ml Erlenmeyer flasks equipped with a side arm allowing direct daily reading of the optical density (OD) with an EEL colorimeter (EEL Instruments, Ilford, U. K.). Each flask contained 150 ml of Volk and Phinney's medium (1968) with the following modifications: EDDHA-Fe was replaced by EDTA-Fe to give a concentration of 1.2 mg/dm^3 and the A5 (Stanier et al., 1971) trace element mixture replaced Hoagland's trace element solution.

The cultures were maintained at 25°C in a 16-hour light: 8-hour dark cycle at 60 to 70% relative humidity in a CONVIRON controlled environment chamber (Controlled Environments, Winnipeg, Canada). The light source in the CONVIRON consists of a mixture of fluorescent and incandescent lamps both manufactured by Sylvania Inc. Light intensity received by the cultures was manipulated either by changing the distance of the cultures from the light source or by covering the culture flasks with one or two layers of black silk netting. The light intensities received by the cultures were measured with a Lambda Li-Cor L-185 photometer in units of lux (lx). Light intensity was not uniform on the different shelves in the CONVIRON, showing a variation of about 20%. To ensure that replicate cultures received approximately the same total illumination while the experiments were in progress, the flasks were randomly placed on the shelves after the daily OD measurements.

Three light intensities (3600, 5700, and 18000 lx) were initially used. After 12 days of growth three replicates from 5700 lx were transferred to 720 lx and three to 230 lx, while three flasks grown at 18000 lx were transferred to 720 lx. The experiment was continued for 28 days.

Growth Measurements

Optical density values were obtained daily immediately after the 8-hour dark period in the 24-hour light:dark cycle, and growth rates (k) were calculated from

$$k = OD\ day\ (x + 1) - OD\ day\ (x)$$

Gas Vacuole Determinations

An estimate of the amount of gas vacuoles present in a *M. aeruginosa* suspension was obtained by comparing the OD before and

after gas vacuole collapse by a 6-atmosphere pressure application
(Walsby, 1969). The amount of gas vacuoles was expressed as a
percentage using the following formula:

$$\% \text{ Gas vacuoles} = \frac{(\text{OD before collapse}) - (\text{OD after collapse})}{(\text{OD before collapse})} \times 100$$

Gas vacuole content was determined on selected days during the
course of the experiment after aseptic removal of 10 ml subsamples
from each flask.

Pigment Determinations

    Pigment composition of the algal cultures was analysed by
comparison of absorption peaks at 440 nm for carotenoids and 662 nm
for chlorophyll after extraction in 80% acetone.

    In field material (see below) pigments were determined after
extraction of chlorophyll and carotenoids in 80% acetone. The
exact amount of pigments present in the extracts was calculated
from the extinction coefficients given by Allen (1968). The amount
of pigment is presented per unit dry-weight. It was not possible
to express the pigment analyses of laboratory experiments on a dry-
weight basis because the relatively small amounts of cultured
material available precluded accurate dry-weight determinations.

Electron Microscopy

    Samples for ultrastructural examination were filtered through
a 0.45 μm membrane filter and the resulting cell concentrate
embedded in 1.5% agar molten at 40°C. Agar blocks were prefixed
at 4°C for 16 hours in 2.5% glutaraldehyde in 0.05% phosphate buffer,
pH 7.2, rinsed twice in buffer and postfixed either in 1% osmium
tetroxide in buffer for 2 hours at 4°C or in 2% unbuffered potassium
permanganate (Mollenhauer, 1959) for 1 hour at ambient temperature.
The latter is a lipoprotein fixative and was used to enhance the
images of thylakoid arrangement. Postfixed samples were washed in
water, dehydrated through an ethanol series and embedded in Spurr's
soft resin (Spurr, 1969). Blocks were sectioned on a LKB Ultrotome
III using a diamond knife. Osmium-treated material was double-
stained in saturated uranyl acetate and lead citrate (Reynolds,
1963) but permanganate-fixed sections were stained with lead citrate
alone. Micrographs were taken on a Philips EM301 at an accelerating
voltage of 80 kV.

Field Incubation of *Microcystis*

A natural population of *Microcystis aeruginosa* was sampled from the surface of Hartbeespoort Dam in the late afternoon (1730h [5:30 pm]; March 7, 1979). Approximately 5 liters of this population, which appeared to be unialgal, was continuously stirred in a beaker and 100 cm^3 aliquots were pipetted into dialysis tubing. The tubes were sealed and suspended overnight in dam water. At dawn (0600h [6:00 am]; March 8, 1979) the tubes were suspended at selected depths from a float in the deepest part of the dam. To ensure that the tubes remained at the selected depths (0.3 and 28 meters) sinkers were attached to them. Samples were collected 24 and 48 hours after the start of incubation for dry-weight and pigment analyses.

Dry Weight Determinations

The dry weight of algae present in a dialysis bag was measured by filtering the total contents (100 cm^3) through a preweighed Nucleopore filter with a pore size of 8 μm. The filters were dried at room temperature for 2 days in a dissicator containing phosphorus pentoxide. Previous experiments had established that 2 days was sufficient to give a constant dry weight.

Other Field Measurements

Temperature, oxygen and pH values in Hartbeespoort Dam were measured with a Martek Instrument (Martek Inc., California). Light penetration into the water was measured with a Lambda Li-Cor L-185 photometer fitted with an underwater sensor which recorded photosynthetic active radiation in μEinsteins m^{-2} s^{-1}. The surface value was taken as 100% light penetration and the depths of 1% and 0.01% light penetration were calculated from the original measurements.

RESULTS

Occurrence of Toxic Blooms in South Africa

Earlier reports on the occurrence of toxic algae in South Africa were documented by Steyn (1949). In all cases the responsible alga was a *Microcystis* species, which has been described as a new species *M. toxica* by Stephens (1949). In a taxonomic revision of the planktonic blue-green algae, Komárek (1958) concluded that *M. toxica* was not a valid species, but a synonym of *M. aeruginosa* forma *aeruginosa*. Steyn found *M. toxica* in two provinces of South

Africa, the Transvaal and the Orange Free State. Most of the
occurrences were in the Vaal Dam, a man-made impoundment completed
in 1938 for supply of potable water to the greater Johannesburg
area and its catchment. Serious outbreaks of cattle poisoning
occurred in the Vaal Dam in 1940, 1942 and 1943. As a result of
the cattle kills, *Microcystis toxica* was proclaimed a weed in 1944.
Control of the algae with copper sulphate was initiated in 1942, a
practice that is still being continued. Figures supplied in Steÿn
(1973) compare the total cost of copper sulphate required for two
five-year periods, 1963 to 1967 and 1968 to 1972. In the first
the cost was R1000 and in the second period it escalated to
R37000 (1 SA Rand = 1.25 USA Dollar). Vaal Dam supplies more than
4 million people with water and is by far the most important water
body in South Africa. The latest cattle kills in the area were in
May 1980 and again the *Microcystis* infestation was treated with
copper sulphate.

Apart from the early reports associated with the Vaal Dam and
its catchment the alga was also described by Steyn (1949) to be
present in the Bon Accord Dam north of Pretoria (see also Louw,
1950). From these early reports it appears that the toxic algae
problem was really limited to relatively small and specific areas.
For a number of years no more reports of cattle kills or other
problems associated with toxic algae were published in southern
Africa.

Zillberg (1966) supplied circumstantial evidence that blooms
of *Anabaena flos-aquae* were responsible for gastroenteritis in
European children in Salisbury, Zimbabwe. The blooms occurred in
Lake McIlwaine which supplied water to the then European suburbs
in Salisbury. The non-European townships in Salisbury received
water from a different source and here the incidence of gastroenteritis was significantly lower.

Renewed interest in toxic algae was triggered by more cattle
kills in Hartbeespoort Dam, situated 40 km west of Pretoria, during
1973 and 1974 (Toerien et al., 1976) and by the observation that
a strain of *Microcystis*, present in the Erfenis Dam near Winburg
in the Orange Free State, was toxic (J. N. Eloff, personal communication, 1973). These events were partly responsible for stimulating
the current research activities on toxic algae in South Africa. We
now have direct evidence, based on mouse tests, of toxic algae from
a number of additional impoundments in the Transvaal and the Orange
Free State. The impoundments are Witbank Dam, Bronkhorstspruit
Dam, Roodeplaat Dam, Rietvlei Dam, Klipvoor Dam and Bospoort Dam
in the Transvaal, and Allemanskraal Dam (J. N. Eloff, personal
communication, 1980) and Stink Pan near Welkom in the Orange Free
State. Toxicity was associated with *Microcystis* in all cases but
one, i.e., Stink Pan, where toxicity was associated with *Nodularia
spumigena*. In order to obtain information on the presence of toxic

Table 1. Occurrence and Toxicity of Algal Blooms in Three Impoundments near Pretoria

Source and Date	Algal Species	Toxicity[a]
Hartbeesport Dam		
Sept. 1977 to Jan. 1978	No blooms (Lake covered by *E. crassipes*, the water hyacinth)	
April 1978	*Microcystis aeruginosa* forma *aeruginosa*	+
May 1978 to Sept. 1978	No large blooms	
Dec. 1978	*M. aeruginosa* f. *aeruginosa*	+
Feb. 1979	*M. aeruginosa* f. *aeruginosa*	+
March 1979	*M. aeruginosa* f. *aeruginosa*	+
April 1979	*M. aeruginosa* f. *aeruginosa*	+
May 1979	*M. aeruginosa* f. *aeruginosa*	−
June 1979	*M. aeruginosa* f. *aeruginosa*	−
Rietvlei Dam		
Oct. 1977 to Nov. 1977	*M. aeruginosa* f. *flos-aquae*	−
Nov. 1977	*M. aeruginosa* f. *aeruginosa*	+
Dec. 1977	*Anabaena circinalis*	−
Dec. 1977 to Jan. 1978	*M. aeruginosa* f. *flos-aquae* (60%) and *M. wesenbergii* 40%	− −
Feb. 1978	*M. aeruginosa* f. *aeruginosa*	+
March 1978	*Anabaena* sp. (70%) *M. aeruginosa* f. *aeruginosa* (30%)	− +
April 1978	*M. aeruginosa* f. *aeruginosa*	−
April 1978	*M. aeruginosa* f. *aeruginosa*	+
June 1978	*M. aeruginosa* f. *flos-aquae*	−
July 1978	*A. circinalis*	−
August 1978	*A. flos-aquae*	−
Sept. 1978	No blooms	

[a]Toxic = +; Non-toxic = −

(Table 1 continued)

Source and Date	Algal Species	Toxicity[a]
Roodeplaat Dam		
Nov. 1977 to Dec. 1977	Diatom bloom (*Melosira sp.*)	
Jan. 1978	*M. aeruginosa* f. *aeruginosa* (80%)	+
	M. aeruginosa f. *flos-aquae* (5%)	−
	M. wesenbergii (5%)	−
	A. circinalis	
	A. flos-aquae } (10%)	−
	A. spiroides	
	A. solitaria	
Feb. 1978	*M. aeruginosa* f. *aeruginosa*	−
March 1978	*M. aeruginosa* f. *aeruginosa*	−
April 1978	*M. aeruginosa* f. *aeruginosa*	+
May 1978 to Sept. 1978	No blooms	

[a]Toxic = + and Non-toxic = −

algae on a year-round basis, three impoundments were selected and sampled regularly for toxic algae. The results are presented in Table 1.

In all cases where positive toxicity was found it was associated with the presence of *M. aeruginosa* forma *aeruginosa* (or *M. toxica*). However, *M. aeruginosa* forma *aeruginosa* blooms were not always toxic. Toxicity could not be associated with *M. aeruginosa* forma *flos-aquae* or *M. wesenbergii*. In addition we could find no evidence that *A. circinalis* or *A. flos-aquae* were toxic.

The identification of the algae were strictly according to Komárek (1958), but in some cases it was not possible to decide from the colony form or shape to which of the two forms, forma *aeruginosa* or forma *flos-aquae*, the specimen should belong. In such cases the cell diameter decided the allocation.

Growth of *Microcystis* at Different Light Intensities

Similar sigmoid growth curves were observed when *Microcystis aeruginosa* was grown at 3600, 5700 and 18000 lx (curves (a), (b), and (e) in Figure 1). The initial lag phase decreased as light

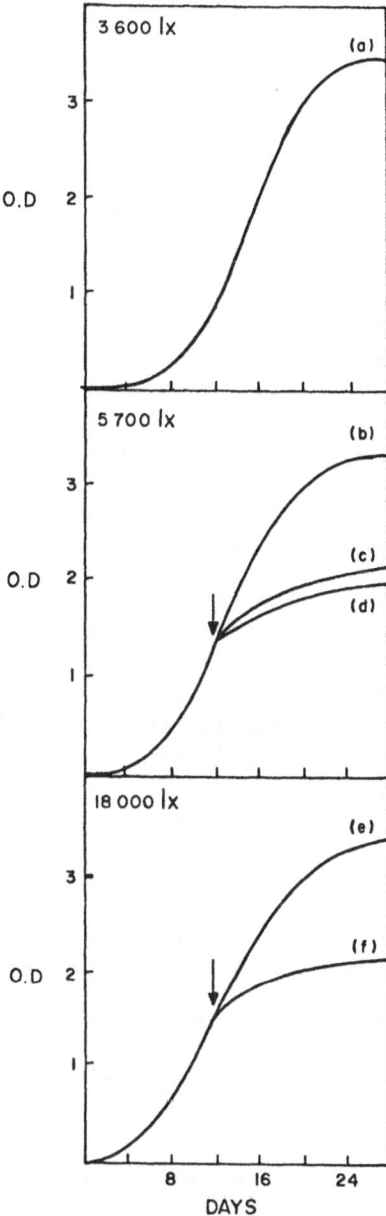

Fig. 1.   Growth curves of *M. aeruginosa* at different light intensities, (a) 3 600 lx, (b) 5 700 lx, (c) 5 700 → 720 lx,
(d) 5 700 → 230 lx, (e) 18 000 lx, (f) 18 000 lx → 720 lx.
Arrows indicate transfer to lower light intensities on the
12th day.

Table 2.   Optimum Growth Rates (k) of *Microcystis aeruginosa* at
           Different Light Intensities

Light Intensity	k
3600 lx	0.37
5700 lx	0.395
18000 lx	0.385

intensity increased.   This is clearly illustrated if the first 8
days of the three curves are compared.   It appears, therefore, that
once *Microcystis* has adapted to a particular light intensity, growth
can continue for several days at a rate independent of the actual
amount of light received by the algae.   The optimum growth rates
recorded at the three different light intensities are given in Table
2.   Although the growth rates were very similar, the physiological
state of the algae changed markedly.

   The cultures grown at 3600 lx had a normal green appearance
while the cultures at 18000 lx turned yellow after a few days.   The
cultures at 5700 lx had a yellow-green pigmentation which was intermediate between the two other treatments.   On the 12th day of the
experiment, flasks grown at 5700 and 18000 lx were transferred to
lower light intensities of 720 and 230 lx respectively (Figure 1).
These cultures turned green in a few days (Table 3) but were unable
to resume growth at high rates.   It appears, therefore, that light
intensities lower than 720 lx are not able to support vigorous
growth of *Microcystis*.

   In addition to the changes in pigmentation there were also
changes in the gas vacuolation of the algae (Table 4).

   The following conclusions can be made from the data in Table 4:

(a)   Gas vacuole content increases with increasing light intensities
      and the converse is true when cultures are transferred from
      higher to lower light intensities.

(b)   Gas vacuole content increases as the cultures age and become
      senescent.

Fine Structural Observations

   The results of the fine structural observations are summarized
in Table 5 and micrographs illustrating the *Microcystis* cells at
different light intensities and ages are given in Figures 2 to 7.

Table 3.   Chlorophyll:Carotenoid Ratios Estimated from Absorption
Peaks for *Microcystis* Grown at Different Light Intensities.

Light Intensity (lx)*	Chlorophyll:Carotenoid Ratio on Day		
	12	19	28
3600	1.11	0.9	0.68
5700	0.4	0.35	0.33
5700 → 720	0.4	0.81	1.19
5700 → 230	0.4	0.63	0.91
18000	0.19	0.30	0.19
18000 → 720	0.19	0.56	1.01

*Arrows indicate transfer from a higher to a lower light intensity
on the 12th day.

Table 4.   Gas Vacuole Content of *Microcystis* Grown at Different
Light Intensities.

Light Intensity (lx)*	Gas Vacuole Content on Day		
	12	19	28
3600	41.7	44.5	54.8
5700	58.6	62.0	64.8
5700 → 720	58.6	48.0	46.7
5700 → 230	58.6	50.0	47.0
18000	61.5	64.2	62.9
18000 → 720	61.5	48.6	42.7

*Arrows indicate transfer from a higher to a lower light intensity
on the 12th day.

Table 5. Fine Structural Observations on *Microcystis* Cultured at Different Light Intensities*

	3600			5700			18000			5700 → 720		5700 → 230		18000 → 720	
Day	12	19	28	12	19	28	12	19	28	19	28	19	28	19	28
Color of culture	G	G	G	YG	YG	YG	Y	Y	O	G	G	G	G	G	G
No. thylakoids per cross section	31	26	32	30	16	15	24	5	8	14	39	19	15	18	36
Thylakoid arrangement	gp	gp	gp	s	s	s	s	s	s	sp	gp	sg ip	gp	gp	gp
Polyglucoside arrangement	t	x	tx	tx	x	x	x	x	tx	tx	t	t	t	t	t
No. cyanophycin granules	0	0	2	1	1	2	0	0	3	0	0	0	0	0	0
No. polyhedral bodies	3	1	0	1	2	1	0	1	0	2	3	4	2	1	3
No. lipid droplets	1	0	6	6	1	6	5	3	8	3	1	5	2	3	3
No. polyphosphate granules	9	5	7	10	7	14	8	5	7	7	9	9	14	11	11

Light Intensity (lx)

Key: G = green; YG = yellow-green; Y = yellow; O = orange; g = grouped; s = single; p = parallel to cell wall; i = perpendicular to cell wall; t = associated with thylakoids; x = scattered through cytoplasm

*Values represent the mean of five median cross sections of cells at each light intensity.

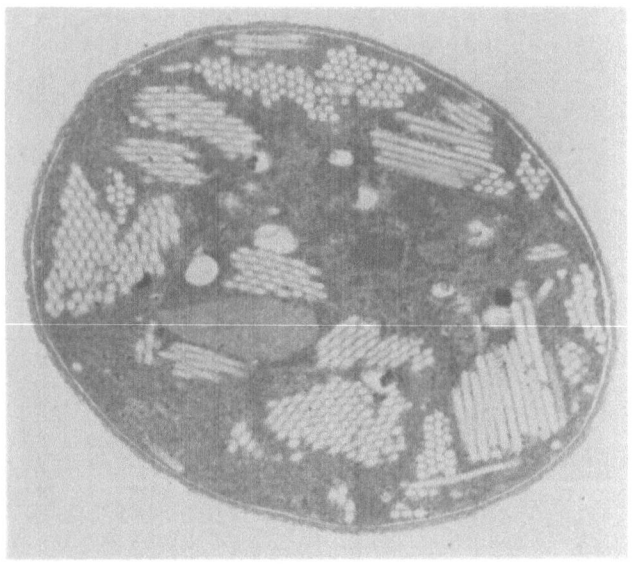

Fig. 2.   A *Microcystis aeruginosa* cell incubated for 12 days at
          3600 lx. Postfixed with osmium tetroxide, x 28500.

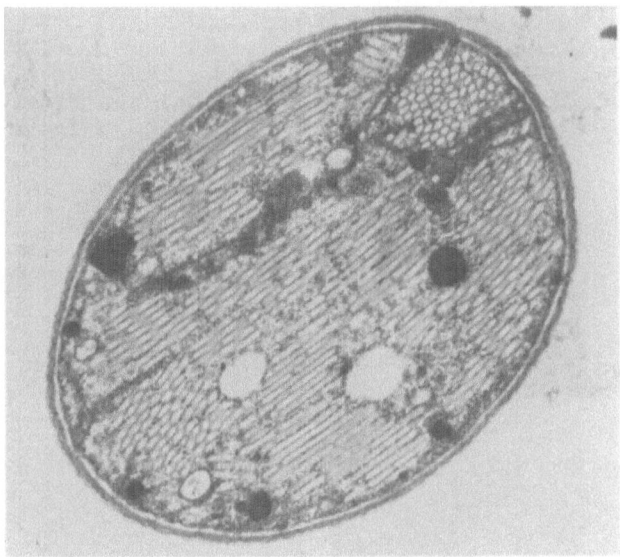

Fig. 3.   A *M. aeruginosa* cell incubated for 28 days at 18000 lx.
          Note tremendous increase in number of gas vacuoles. Post-
          fixed with osmium tetroxide, x 28500.

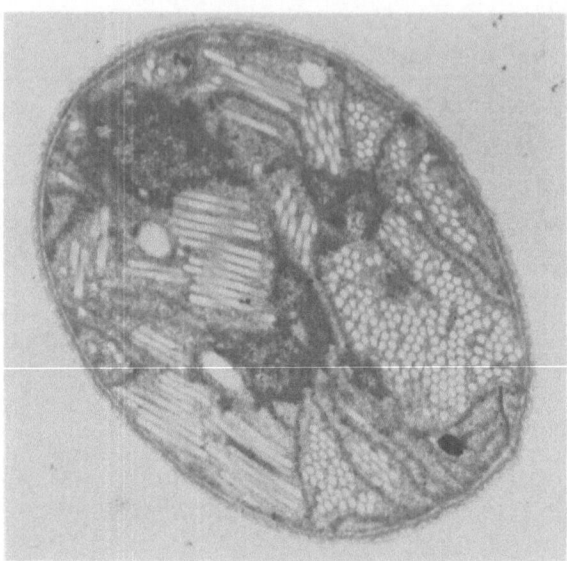

Fig. 4.   A *M. aeruginosa* cell incubated for 12 days at 5700 lx.
          Note single thylakoids perpendicular to the cell wall and
          polyglucosides arranged with the thylakoids. Compare the
          gas vacuole content with Fig. 2 and 3. Postfixed with
          osmium tetroxide, x 28500.

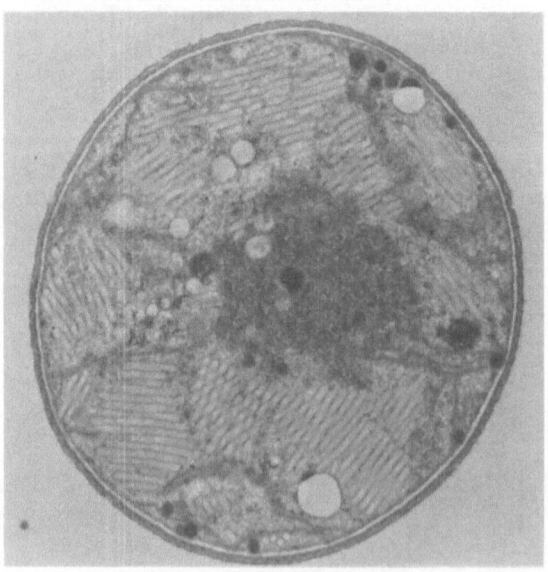

Fig. 5.   As Fig. 4, but incubated for 28 days. Note increase in
          gas vesicles with senescence. Postfixed with osmium
          tetroxide, x 28500.

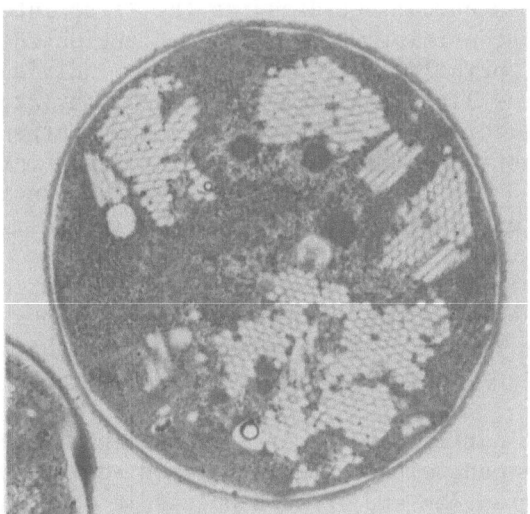

Fig. 6.   A *M. aeruginosa* cell first incubated for 12 days at 5700 lx
          (compare Fig. 4) and then transferred to a light intensity
          of 720 lx until the 28th day of incubation. Note thylakoids
          group together and parallel with the cell wall. Postfixed
          with osmium tetroxide, x 22500.

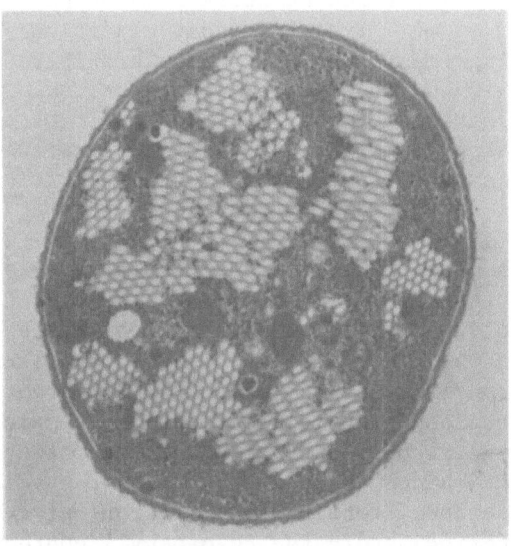

Fig. 7.   A *M. aeruginosa* cell first incubated for 12 days at 18000 lx
          and then transferred to a light intensity of 720 lx until
          the 28th day of incubation. Compare with Fig. 3. Postfixed
          with osmium tetroxide, x 28500.

The thylakoid number decreased as the light intensity increased. At lower light intensities units of 4 to 6 thylakoids were concentrically stacked parallel to the cell wall. Individual thylakoids were long. As the light intensity increased, thylakoids became more fragmentary and reorientated themselves singly perpendicular to the cell wall. Transfer from a high to a lower light intensity resulted in an increase in thylakoid numbers and a reorientation and regrouping of the membrane units parallel to the cell wall.

The conclusions derived from the direct measurement of gas vacuoles (Table 4) were confirmed by the micrographs. The gas vacuole contents in the cells increase as the light intensity increases and also increase as the cells become more senescent.

No definite pattern of the effects of light intensity on storage granules emerged, although aging seemed to increase the number of cyanophycin granules and lipid droplets in the cells.

Field Measurements

The laboratory observations on pigmentation of *Microcystis aeruginosa* were supplemented with field observations in Hartbeespoort Dam, where the *Microcystis* colonies were trapped in dialysis tubes kept at a fixed depth for periods up to 48 hours. The experiments were performed at a time when there was a thick *Microcystis* bloom in the center of the dam, reaching surface chlorophyll a concentrations in excess of 500 mg/m^3. Typical midday conditions observed in the dam while the experiment was in progress are illustrated in Fig. 8. The dam showed high surface pH and oxygen values. The oxygen concentration at 1 meter depth was less than half of the surface concentration and at 12 meters depth the water was anaerobic. The thick *Microcystis* scum reduced the light penetration to virtually zero at 2 meters depth.

The results of pigment analyses of *Microcystis* cells maintained at different depths are illustrated in Figure 9. Cells maintained at the surface of the dam had a linear drop in the chlorophyll: carotenoid ratio which was a direct result of a large decrease in the total chlorophyll content. Chlorophyll content of cells maintained at 3 meters depth and at the bottom (28 meters) both increased over the incubation period. The carotenoid content of the cells did not change as much as the chlorophyll content of the cells. At the surface, total carotenoid content decreased over the incubation period as did the chlorophyll content of these samples. At 3 meters depth and at the bottom, the carotenoid content of the cells increased with the net result that the chlorophyll:carotenoid ratios were slightly lower than at the beginning of the experiment.

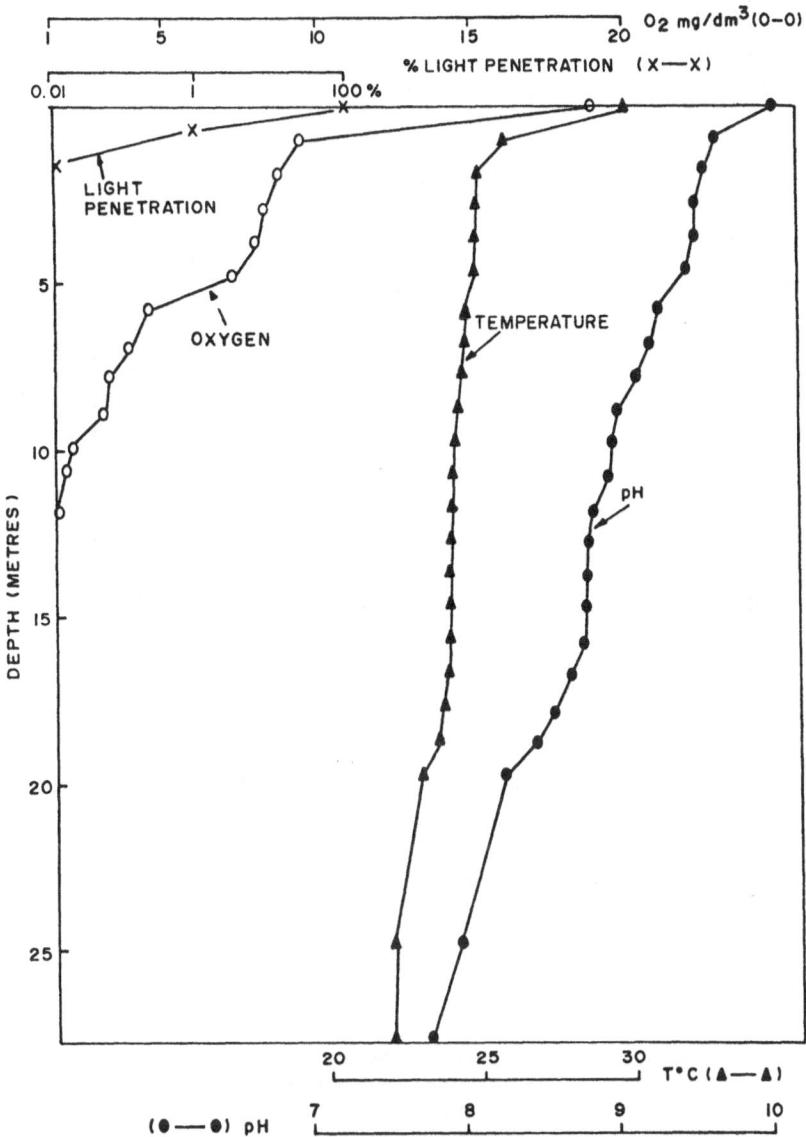

Fig. 8. Profiles of percentage light penetration, oxygen, temperature and pH in Hartbeespoort Dam (noon, March 7, 1979).

No direct measurement of gas vacuoles was made in the field
experiments. Cells maintained at the surface of Hartbeespoort Dam
for 48 hours were examined ultrastructurally (Figure 10). The
largest number of individual gas vesicles confirm the laboratory
experiments that exposure of *Microcystis* to high light intensities
increase gas vacuolation.

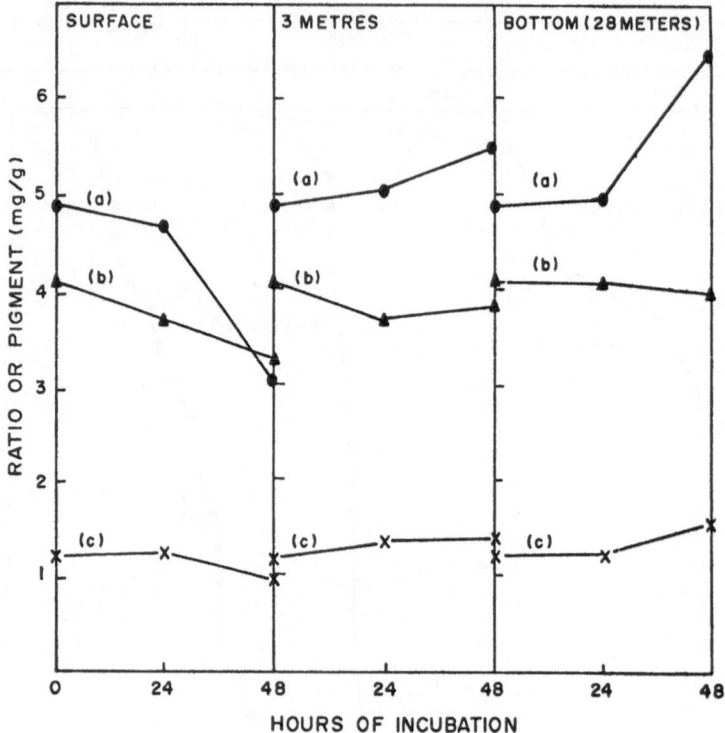

Fig. 9. Changes in the pigment content and chlorophyll:carotenoid
          ratio of *Microcystis* cells incubated at fixed depths in
          Hartbeespoort Dam. (a) ●———● chlorophyll a content;
          (b) ▲———▲ chlorophyll:carotenoid ratio; (c) x———x
          carotenoid content.

DISCUSSION

      Toxic blooms of *Microcystis* in South Africa are limited to
the Transvaal and Orange Free State provinces. There seems to be
no apparent reason to prevent the toxic alga from appearing in the
other regions of the country. Non-toxic *Microcystis* blooms are
found in both the Natal and Cape provinces of South Africa. These
blooms should be examined to establish whether they are the forma
aeruginosa or the forma *flos-aquae* of *M. aeruginosa*. Unfortunately,
toxic *Microcystis* is found in South Africa's most important water
body, the Vaal Dam. Large sums of money are spent yearly to combat
the alga with copper sulphate. Observations on toxic blooms in
three impoundments near Pretoria confirmed earlier reports on the ·
variability in toxicity in natural blooms (Gorham, 1964). No
explanation for this phenomenon is readily at hand.

      The similar optimum growth rates of *Microcystis aeruginosa* at
light intensities ranging from 3600 lx to 18000 lx indicate that

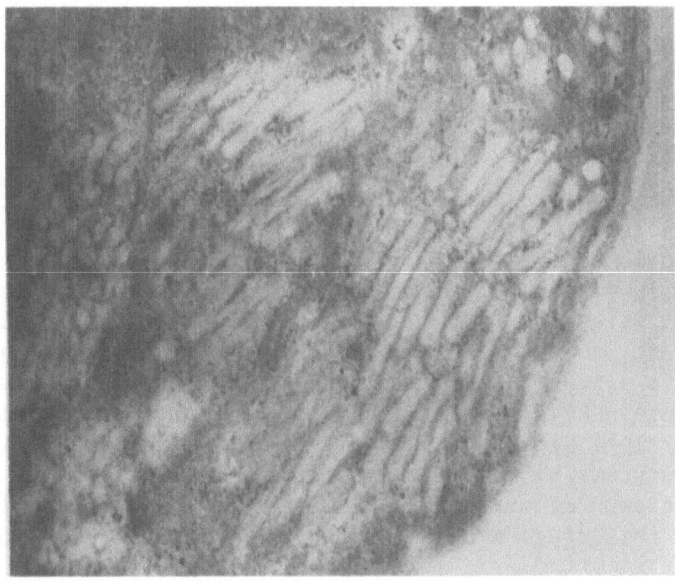

Fig. 10. Portion of a *M. aeruginosa* cell exposed to full daylight
          for 48 hours in a dialysis bag in Hartbeespoort Dam.
          Note tremendous increase in gas vesicles (similar to
          Figure 3). Postfixed with osmium tetroxide, x 67500.

light saturation for *Microcystis* growth is reached at relatively
low light intensities. Provided that there is no extensive shading
by a surface bloom, the intensity of 3600 lx will correspond to the
light intensity normally received at a depth of approximately 4 to
5 meters in a typical non-silty Transvaal impoundment such as
Hartbeespoort or Rietvlei Dams. The ability of *Microcystis aeruginosa* to grow fast at relatively low light intensities will give
it a competitive advantage over other algae in these impoundments
and partly explains why such large numbers of this alga frequently
develop in these impoundments.

     A further competitive advantage is the presence of gas vacuoles
in the cells of *Microcystis aeruginosa* which renders the algae more
buoyant thus preventing them from sinking to depths where the light
intensity may be limiting. Waaland et al. (1971) presented evidence
that gas vacuoles reduced the amount of light absorbed by photosynthetic pigments in *Nostoc muscorum* and concluded that these
structures acted as light shields. Walsby (1972) and Shear and
Walsby (1975) questioned the validity of these results and suggested that the main function of gas vacuoles was buoyancy regulation. Dinsdale and Walsby (1972) showed that transfer of *Anabaena
flos-aquae* from low (50 lx) to high (10000 lx) light intensities

resulted in a decrease in gas vacuole content. We were unable to
confirm these observations with *Microcystis aeruginosa* (Table 4)
and the present results together with the results of Porter and
Jost (1976) support the idea that gas vacuoles may act as light
shields. In addition, senescent cultures were found to have a
higher gas vacuole content.

When *Microcystis aeruginosa* is continuously exposed to high
light intensities the alga adapts by changing its pigment composition. This is achieved by reducing the chlorophyll content of the
cells. At lower light intensities and in complete darkness more
chlorophyll is synthesized (see Figure 9). *Microcystis aeruginosa*
is, therefore, able to adapt to changes in the light climate by
changing its pigment composition and gas vacuole content so that
it can maintain high growth rates. Suggestions to control unwanted
blue-green algal blooms, and by implication potential toxicity, in
eutrophic systems by managing conditions in such a way as to stimulate the growth of more desirable green algae, appear unlikely to
succeed at this stage because of the adaptability of the blue-green
algae.

ACKNOWLEDGMENTS

The authors wish to thank S. I. Funke for technical assistance.
The Department of Health partly funds a research program on toxic
blue-green algae at the NIWR [National Institute for Water Research].
Permission has been granted by the Department and by the Director
of the NIWR to publish this paper.

REFERENCES

Allen, M. M. 1968. Photosynthetic membrane system in *Anacystis
        nidulans*. J. Bacteriol. *96*:836-841.
Dinsdale, M. T., and A. E. Walsby. 1972. The interrelations of
        cell turgor pressure, gas-vacuolation and buoyancy in a
        blue-green alga. J. Exp. Bot. *23*:561-570.
Gorham, P. R. 1964. Toxic algae. Pages 307-336 *in* D. F. Jackson,
        ed., Algae and Man. Plenum Press, New York.
Komárek, J. 1958. Die taxonomische revision der planktischen
        blaualgen der Tschechoslowakei. Pages 10-206 *in* J. Komárek
        and H. Ettl, eds., Algologische Studien. Czechoslovak
        Academy of Science, Prague.
Krüger, G. H. J., and J. N. Eloff. 1977. The influence of light
        intensity on the growth of different *Microcystis* isolates.
        J. Limnol. Soc. S. Afr. *3*:21-25.
Louw, P. G. J. 1950. The active constituent of the poisonous
        algae, *Microcystis toxica* Stephens. S. Afr. Ind. Chemist
        4:62-66.

Mollenhauer, H. H.   1959.   Permanganate fixation of plant cells.
      J. Biophys. Biochem. Cytol. *6*:431-438.
Porter, J., and M. Jost.   1976.   Physiological effects of the
      presence and absence of gas vacuoles in the blue-green alga,
      Microcystis aeruginosa Kütz. emend. Elenkin.   Arch. Microbiol.
      110:225-231.
Reynolds, E. S.   1963.   The use of lead citrate at high pH as an
      electron opaque stain in electron microscopy.   J. Cell. Biol.
      17:208-212.
Scott, W. E.   1974.   The isolation of *Microcystis*.   S. Afr. J. Sci.
      70:179.
Shear, H., and A. E. Walsby.   1975.   An investigation into the
      possible light-shielding role of gas vacuoles in a plank-
      tonic blue-green alga.   Brit. Phyc. J. *10*:241-251.
Spurr, A. R.   1969.   A low-viscosity epoxy resin embedding medium
      for electron microscopy.   J. Ultrastruct. Res. *26*:31-43.
Stanier, R. Y., R. Kunisawa, M. Mandel, and G. Cohen-Bazire.   1971.
      Purification and properties of unicellular blue-green algae
      (order Chroococcales).   Bact. Rev. *35*:171-205.
Stephens, E. L.   1949.   *Microcystis toxica* sp. nov.:   a poisonous
      alga from the Transvaal and the Orange Free State.   Trans.
      R. Soc. S. Afr. *32*(1):105-112.
Steyn, D. G.   1945.   Poisoning of animals by algae (scum and water-
      bloom) in dams and pans.   Union of South Africa, Department
      of Agriculture and Forestry, Government Printer, Pretoria.
      9 pp.
Steyn, D. G.   1949.   Vergiftiging van mens en dier met gifplante,
      voedsel en drinkwater.   J. L. van Schaik, Pretoria.   264 pp.
Steyn, D. J.   1973.   Die eutrofikasiepeile van vier Transvaalse
      Damme.   M.Sc. Thesis.   University of Pretoria.   Pretoria.
      169 pp.
Toerien, D. F., W. E. Scott, and M. J. Pitout.   1976.   *Microcystis*
      toxins: isolation, identification, implications.   Water
      S. A. (Pretoria) *2*:160-162.
Volk, S. L., and H. K. Phinney.   1968.   Mineral requirements for
      the growth of *Anabaena spiroides*.   Can. J. Bot. *46*:619-630.
Waaland, J. R., S. D. Waaland, and D. Branton.   1971.   Gas vacuoles.
      Light shielding in blue-green algae.   J. Cell. Biol. *48*:
      212-215.
Walsby, A. E.   1969.   The permeability of blue-green algal gas-
      vacuole membranes to gas.   Proc. R. Soc. B. *173*:235-255.
Walsby, A. E.   1972.   Structure and function of gas vacuoles.
      Bact. Rev. *36*:1-32.
Zehnder, A., and P. R. Gorham.   1960.   Factors influencing the
      growth of *Microcystis aeruginosa* Kütz. emend. Elenkin.   Can.
      J. Microbiol. *6*:645-660.
Zillberg, B.   1966.   Gastro-enteritis in Salisbury European chil-
      dren -- a five year study.   Centr. Afr. J. Med. *12*:164-168.

AUTECOLOGICAL STUDIES ON *MICROCYSTIS*[1]

J. N. Eloff

University of the Orange Free State
Bloemfontein, South Africa

ABSTRACT

The influence of growth parameters such as light intensity,
light quality, light-dark cycles, temperature, pH, CO_2 and O_2
concentration, N and P nutrition and turbulence was determined on
the growth of toxic and non-toxic axenic *Microcystis* isolates under
laboratory conditions. The wide differences observed between
laboratory and natural cultures especially with regard to light
sensitivity were partially explained when laboratory cultures were
grown in the presence of bacteria and when experiments were carried
out under natural conditions over a long period. Equations describing the relationship between growth rate and growth parameters were
deduced. These equations may be of value in predicting growth of
Microcystis under natural conditions.

TEXT

Introduction

Microcystis is probably the most common bloom-forming organism

[1]This paper is a synthesis of some of the results of students and
colleagues in my Department during the past few years. The coop-
eration of the following persons is gratefully acknowledged:
Dr. G. H. J. Krüger, Mr. J. A. Pretorius, Mr. J. S. de Wet,
Mrs. M. M. van Vuren (nee Eksteen), Mr. P. W. J. van Wyk, Dr. R. L.
Verhoeven, Dr. W. L. J. van Rensburg, Mr. E. G. Groenewald, and
Mrs. H. E. Kriel.

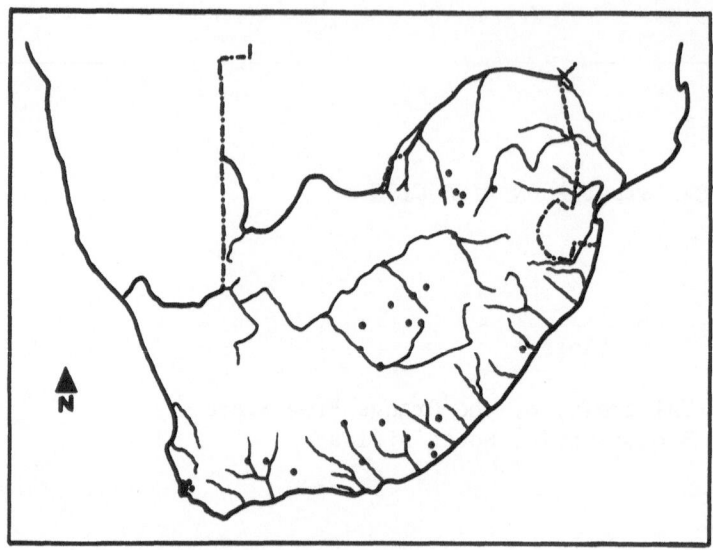

Fig. 1. Distribution of *Microcystis* in water bodies in South
         Africa. (Drawn from Bruwer, 1979).

in South African impoundments. From the distribution of lakes in
which *Microcystis* forms blooms (Figure 1 - compiled from Bruwer,
1979) it is clear that *Microcystis* occurs all over South Africa.

     The problems caused by blooms of *Microcystis* may be summarized
as follows: (a) increased water purification costs, (b) odor and
taste of drinking water, (c) aesthetic problems influencing the
recreational use of water bodies due to the smell of decaying
algae, (d) die-off of blooms leading to the development of anoxic
conditions and fish death, (e) cattle, sheep and even game deaths
regularly occurring due to ingestion of toxic *Microcystis*, (f)
Microcystis blooms implicated in an outbreak of gastroenteritis
and sublethal doses of *Microcystis* toxin leading to irreparable
liver damage in primates (Tustin et al., 1973).

     Consequently it was considered important to study various
aspects of *Microcystis* biology extensively in laboratory and later
field studies.

Purification and Growth of *Microcystis* Cultures

     Cells from the surface of a bloom were generally not as viable
as cells deeper in the water layer. Problems were frequently
experienced when efforts were made to grow cells originating from
the surface in the laboratory. This may be due to photooxidative
effects on the cells (Eloff et al., 1976). Transportation of

Fig. 2. General organization of growth room indicating Grolux
         fluorescent tubes, agar plates in petri dishes, culture
         flasks with cellulose stopper and paper hat to prevent
         dust falling on stopper, culture flasks with side arms to
         determine growth rate, Klett-Summerson photocolorimeter,
         test tubes containing single colonies isolated from agar
         plates.

Microcystis cells in stoppered containers in light frequently leads
to a complete loss of viability which may also be due to a photooxidative effect.

    The following procedure generally led to the growth of cells
from a natural bloom: The cell suspension was centrifuged at ca.
1000 g for 5 minutes and then at > 10000 g for 10 minutes, the
floating cells were resuspended in growth medium (BG 11, Stanier
et al., 1971) and recentrifuged. This procedure is very effective
in reducing the number of eucaryotic contaminants and is preferable
to the use of actidione. A small volume (4.5 ml) serial dilution

of the floating cells was made and grown in the growth room. The highest dilutions at which *Microcystis* cells grew was subsequently used. The growth conditions used (28°C, 10 to 20 $\mu E./m^2/s$) and the general growth room organization is apparent from Figure 2.

Various methods were tried in an effort to clean the cultures from contaminating bacteria. These included antibiotics (Vance, 1966), filtration through 5 and 8-nm pore size filters, washing with dilute chlorine water (Fogg, 1942) and with detergent and phenol (McDaniel et al., 1962), differential temperature treatments (Wieringa, 1968) and ultraviolet irradiation. None of these methods were effective in removing the bacteria associated with *Microcystis*, but unialgal cultures were readily obtained.

Obviously the best method would be to grow in dilute suspensions on agar and use standard microbiological techniques. The isolation of single colonies also would have the advantage that a clonal culture with homogeneous genetic properties would be obtained.

Unfortunately, small inocula of *Microcystis* do not grow well on agar and it is very difficult to remove bacteria sticking to the mucilage from colonies so that individual colonies that do grow on agar are always contaminated with bacteria. Before optimizing growth conditions of single cells of *Microcystis*, a procedure for changing colonial growth of *Microcystis* to single cells was developed. Under our laboratory conditions, cultures lose their colony habit within three to four months. This change was speeded up to ca. three weeks by shaking cultures vigorously in 1% Tween. Indications have been found that colony habit is associated with iron nutrition in some way and the decrease of iron content may lead to breaking up of colonies. Up until now it has not been possible to get a unicellular culture to start forming colonies again although laboratory cells kept under natural conditions have started clumping within a few weeks and cells grown in sodium acetate have started clumping (Figure 3) (van Wyk, 1980). Once unicellular cultures were obtained, the influence of environmental parameters on agar growth was determined.

In preliminary experiments it has been found that agar concentration and the type of agar plays an important role in the growth of single cells from small inocula (Pretorius, 1978). Generally the lower the agar concentration the better the growth and, surprisingly, purer agars gave less satisfactory results than impure agars. Separate sterilization of agar and medium (Allen, 1968) had a beneficial effect but very poor growth was obtained when plates were not taped. The medium concentration (0.5 strength vs. full strength) had no effect on either percentage colony recovery or colony diameter and in contrast to the results found with

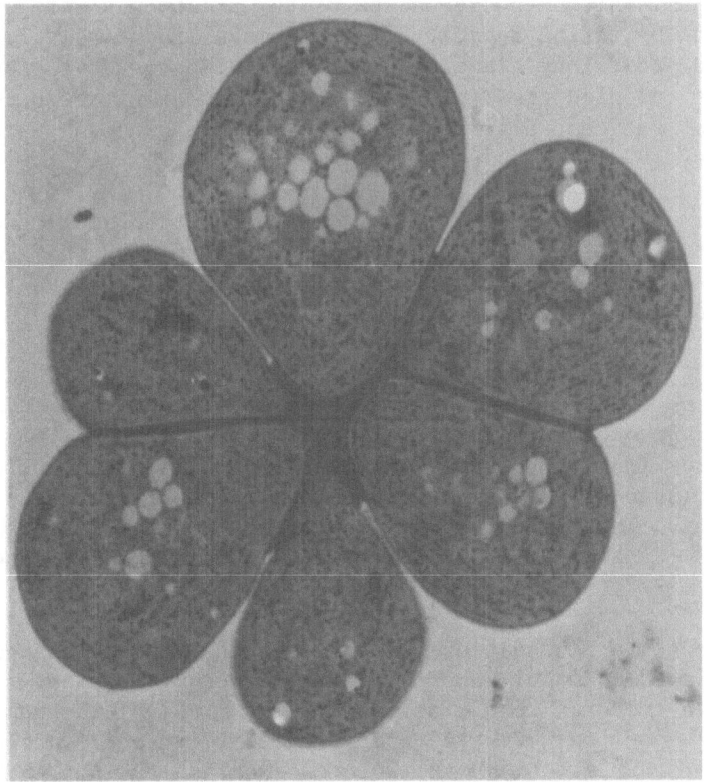

Fig. 3.   Electron micrograph demonstrating clumping of *M.* UV-007
          cells grown in 0.01 M sodium acetate.

liquid cultures (Krüger and Eloff, 1978a), the higher the light
intensity (up to 40 μE./m^2/s) the larger was the colony diameter.

It has been postulated that H_2O_2 forming in a reaction between
Mn^{2+} and citrate at high temperatures is responsible for agar
growth inhibition (Marler and van Baalen, 1965). By applying
various treatments and determining the influence of these treatments on agar growth H_2O_2 concentration (Pretorius, 1978) it was
deduced that the formation of peroxide in the medium does not completely inhibit the growth of *Microcystis* on agar but retards
growth. The superior growth of *Microcystis* on impure agars may be
ascribed to impurities inhibiting H_2O_2 formation.

It has also been stated that the pH of the agar plays a very
important role in determining recovery of a unicellular blue-green
alga on agar (Herdman et al., 1973). To examine this possibility,
experiments were conducted using a low Tris buffer concentration

(0.001 M), as higher concentrations were toxic and 1/4 strength
BG-11 medium. Although the end pH of the treatments were close due
to the weak buffering capacity of the Tris a linear relationship
between pH and colony diameter was found. The higher the pH, the
better the colonies grew (probably indicating an effect of CO_2 provision) but the control treatments yielded higher colony diameters
(Pretorius, 1978).

Because large inocula always grow better than small inocula,
the possibility of growth stimulants being formed by the growth
medium was considered. An axenic *Microcystis* culture was filtered
through an 0.22 μm filter (Nucleopore) and some of the spent filtered medium was added to liquid cultures. No effect whatsoever on
the growth rate was observed. When the filtered medium was added to
dilutions of a *Microcystis* culture before plating, instead of the
expected 100 colonies, 340 colonies grew out on the plate. Repeating the experiment gave the same results both when the cultures were
filtered or centrifuged before addition of spent media to dilutions
for plating. These results were explained by the formation of
nannocytes (Canabaeus, 1979) under certain conditions (Pretorius
and Eloff, 1977; Pretorius et al., 1977).

Under certain conditions one can see small protuberances in
Microcystis cells under Nomarski interference microscopy. When
Microcystis cultures were examined by transmission electron microscopy small spherical membrane-bound bodies (Figure 4) were observed
in all *Microcystis aeruginosa* isolates examined but not in *M. incerta*
or in *Synechococcus* isolates (van Wyk et al., 1979; van Wyk, 1980).
At first it was thought that these membrane-bound bodies represented
nannocytes, but no development of these bodies were observed. Due
to the ultrastructural similarity it was subsequently thought that
they were poly-β-hydroxybutyric acid (PHB) bodies but isolation and
chemical characterization showed that this was not the case. The
membrane-bound bodies probably function as a carbon reservoir formed
during photosynthesis because they are metabolized under stress conditions but their chemical identity has not been identified yet.

By applying the results obtained in the agar growth experiments
a technique was developed that gave satisfactory results as far as
isolation purposes are concerned. No problems were ever encountered
in growing colonies taken from agar plates in liquid medium if the
light intensity was not too high. The method still does not yield
good results for viable count purposes because frequently the yields
of replicate treatments are highly variable which may be due to
impurities present in the disposable petri dishes. The method has,
however, been applied to give satisfactory viable count data (Krüger
and Eloff, 1978a).

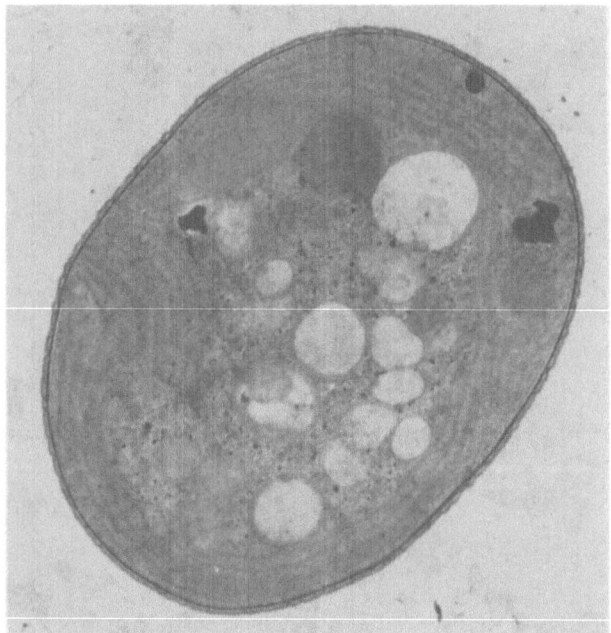

Fig. 4. Electron micrograph of *M.* UV-007 cell showing unidentified
membrane-bound bodies that were originally thought to be
poly-β-hydroxybutyric acid granules. (Photographed by
P. W. J. van Wyk).

Application of this method led to the isolation of clones of
a number of *Microcystis* isolates and to purification of some isolates as shown in Table 1.

The Influence of Nutrients on Growth of *Microcystis*

Phosphate nutrition. The influence of phosphate concentration
on the growth of a toxic (*M.* UV-006) and non-toxic (*M.* UV-007) isolate of *Microcystis* was compared. The half saturation constant of
the toxic isolate was much higher than that of the non-toxic isolate, 0.6 vs. 0.35 mg/liter, indicating that the non-toxic isolate
had a higher affinity for phosphate than the toxic isolate. The
yield coefficient of the non-toxic isolate was also higher than
that of the toxic isolate, 540 vs. 350 mg/liter, indicating that
M. UV-007 would produce much more dry matter per mg P than *M.* UV-006
(van Rensburg, 1980, unpublished information).

Contradictory results were obtained as to what extent *Microcystis* is able to make use of phosphate reserves for growth under

Table 1. Description of Cultures Growing in Culture Collection at University of the Orange Free State. (All *M. aeruginosa* unless otherwise stated) (AC = axenic culture, UC = unicellular (clonal) culture, MC = mixed (colonial) culture).

Culture No	Origin		Axenic	Toxic	Gas vacuoles	Size (S.D.) nm	Comments
UV-001	(UC) P.R. Gorham NRC-1, Little Rideau Lake, Ontario	1954	✓	✓	✓	4,97 (0,34)	
UV-002	(UC) Cambridge CC LB 1450/1, Wisconsin		-	-	-	-	
UV-003	(UC) Cambridge CC LB 1450/2		-	-	-	-	*M. incerta*
UV-004	Bethulie, South Afr. Hendrik Verwoerd Dam	1973	-	-	-	4,54 (0,47)	
UV-006	(MC) W.E. Scott, Pretoria, Hartbeespoort Dam	1976	-	✓	✓	6,15 (0,78)	
UV-007	(AC) R.Y. Stanier, Berkley 7005 (Gott-1450-1) G.C. Gerloff pre 1969		✓	-	-	4,99 (0,29)	
UV-008	(UC) F. Jüttner, Tübingen		-	-	-	4,91 (0,26)	
UV-009	(MC) J.F. Talling, Windermere		-	-	-	6,64 (0,64)	
UV-010	(MC) W.E. Scott, Pretoria, Witbank Dan	1977	-	✓	✓	7,27 (0,49)	
UV-011	(MC) W.E. Scott, Pretoria, Lindleyspoort Dam	1977	-	-	✓	4,89 (0,33)	
UV-017	(MC) W.E. Scott, Pretoria, Rietvlei Dam	1977	-	✓	✓	5,39 (0,98)	
UV-018	(UC) Gött-1450-1 G.C. Gerloff pre 1969		-	-	-	5,23 (0,56)	Similar to UV-007 different history
UV-019	Allemanskraal Dam, Winburg, S. Afr.	1979	-	✓	✓	8,46 (0,96)	
UV-020	(UC) C. Serruya B6 Kinneret		-	-	-	7,65 (0,76)	
UV-021	(MC) C.A. Bruwer, Allemanskraal Dam	1979	-	-	✓	5,72 (0,77)	
UV-022	(MC) C.A. Bruwer, Nooitgedacht Dam	1979	-	-	✓	5,25 (0,52)	
UV-023	(MC) C.A. Bruwer, van Ryneveldtspas Dam	1979	-	✓	✓	5,09 (0,37)	
UV-024	(MC) D.F. Toerien, Vaal Dam	1979	-	-	✓	4,88 (0,39)	
UV-027	Fish Pond Nir David, Israel	1980	-	✓	✓	6,21 (0,82)	
UV-028	Fish Pond Nir David, Israel	1980	-	✓	✓	8,57 (1,04)	
UV-029	(MC) C.A. Bruwer, Lake Arthur	1980	-	✓	✓	5,94 (0,83)	

phosphate-limiting conditions. In one set of experiments practically no growth was obtained when cells were grown in phosphate-
free medium whereas in another set of experiments *Microcystis* grew
reasonably well for 8 days before growth stopped (Barlow et al.,
1979). In the latter case, which was only carried out on a toxic
isolate, the disappearance of polyphosphate bodies and reappearance
of polyphosphate granules after cells were grown in phosphate-free
medium and later in phosphate-containing medium was monitored.
Within 24 hours of phosphate starvation, stored polyphosphate was
used whereas subsequent addition of phosphate led to redeposition
of polyphosphate in the centroplasm, DNA fibrils and between the
thylakoids (Barlow et al., 1979).

Nitrogen nutrition. The influence of various nitrogen sources
on the growth of axenic cultures of *M.* UV-006 and *M.* UV-007 was
investigated (de Wet, 1979). The results obtained are summarized
in Table 2.

From Table 2 it is clear that as far as the N-metabolism is
concerned there is a very big difference between the toxic and

Table 2.   Comparison of the Effects of Nitrogen Sources on Toxic
           and Non-toxic *Microcystis* Isolates (de Wet, 1979)

Effects	UV-007 (non-toxic)	UV-006 (toxic)
Inhibited by NH_4Cl conc. higher than	7.4 mg/l	250 mg/l
Inhibited by NH_4NO_3 conc. higher than	14.7 mg/l	250 mg/l
Inhibited by urea conc. higher than	25 mg/l	75 mg/l
Inhibited by amino acid nitrogen conc. higher than	247 mg/l	25 mg/l
Yield coefficient with NO_3^-	16.1	–
Yield coefficient with urea	15.5	–
Amino acids used as sole nitrogen source (in order of preference)	L-arginine	L-alanine
	L-citrulline	L-proline
	L-asparagine	L-arginine
	L-ornithine	L-asparagine
	L-proline	L-glycine
	L-leucine	
	L-isoleucine	

non-toxic isolates.

     The cells were centrifuged and washed with N-free medium to
remove any traces of N present in the inoculum. From all the preceding results it was clear that no N-reserves could be used by
Microcystis inoculated into N-free medium. When *M.* UV-007 was
grown in L-arginine, L-citrulline and L-asparagine as sole nitrogen
source however an increase in cyanophycin granule content of the
cells was observed (Figure 5). When *M.* UV-007 which had grown on
L-citrullin as sole nitrogen source was inoculated into a medium
with no nitrogen, the cyanophycin granules were used as nitrogen
source by the cells.

     Uptake and metabolism of various metabolites. To determine
to what extent various metabolites could be taken up and metabolized,
[14]C-labelled glucose, fructose, galactose, glycine, glycolic acid,
glyoxalate, glutamate, fumarate, malate and carbonate were added to
axenic cultures of a toxic (*M.* UV-001) and a non-toxic (*M.* UV-007)
isolate. The flasks were placed in the growth room at 28°C under
a light intensity of 15 μE./m^2/s. The percentage of the ^{14}C taken
up was determined after 1, 2, 4, and 8 days by centrifuging aliquots
and determining the radioactivity of the cells and the supernatant
with liquid sintillation counting. The percentages of total activity in the cells are shown in Table 3. It is clear that a very low
percentage of galactose, glycerol, glycolate and glyoxalate was
taken up within one day. The largest difference as far as uptake
is concerned between the toxic and non-toxic cells was with fructose where UV-001 took up more than 20 times as much fructose as
UV-007.

     To determine to what extent these metabolites were metabolized
or leached out of the cells the acitivity in the cells after 8 days
was deducted from the activity after the 1st day. From Table 3 it
is clear that whereas glucose, glutamate and malate were metabolized or leached to a large extent, glycerol, glycolate and glyoxalate were hardly leached or metabolized at all. Large differences were observed between the toxic and non-toxic isolates as
far as the leaching or metabolism of fructose, galactose and
fumarate were concerned. Unfortunately, the cultures were not
checked for the presence of bacteria after 8 days so that it is
possible that the differences may have been due to contamination
of *M.* UV-001 cultures (Eloff and van Vuren, 1978). It does seem,
however, as if differences do exist between toxic and non-toxic
isolates as far as uptake and possibly the rate of metabolism or
leaching of the tested carbon compounds is concerned.

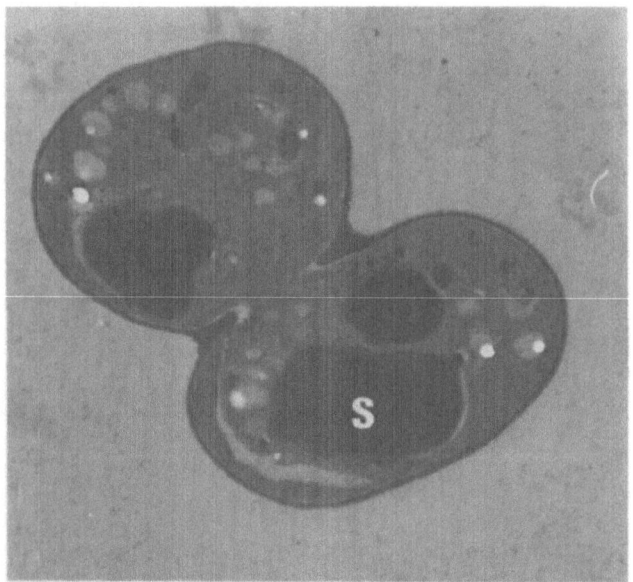

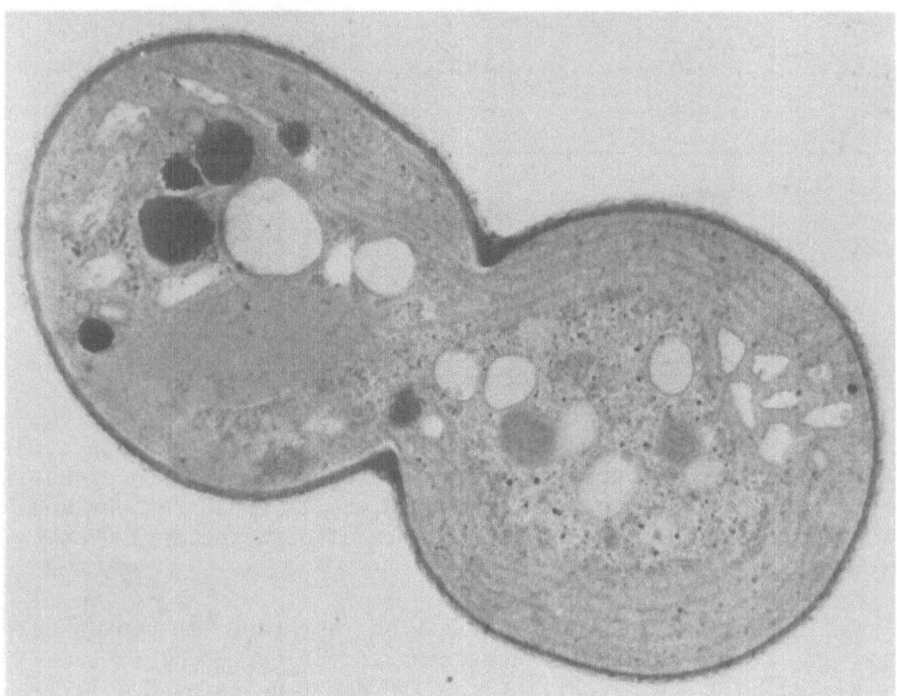

Fig. 5.   Electron micrographs of *M*. UV-007 cells that were grown on
L-citrulline as sole nitrogen source exhibiting cyanophycin
granules (S) (de Wet, 1979).

Table 3.   Uptake[a] and Metabolism or Leaching[b] of Various ^{14}C-
labelled Metabolites by Toxic and Non-toxic *Microcystis*
Cells

| Metabolite | M. UV-001 (Toxic) | | M. UV-007 (Non-toxic) | |
	% in Cells Day 1	Loss, Day 8-2	% in Cells Day 1	Loss, Day 8-2
Glucose	29	482	44	390
Fructose	46	518	2	-59
Galactose	4	242	2	-91
Glycerol	12	48	4	-10
Glycolate	5	32	4	-40
Glyoxalate	3	29	6	-69
Glutamate	46	485	51	613
Fumarate	24	562	14	35
Malate	34	313	56	686
Na_2CO_3	84	-48	78	-140

[a](% in cells at day 1).

[b](Activity day 8 - activity day 2, in d.p.m x 10^3).

Influence of CuSO₄ and Algicides on Growth

Copper sulphate treatment has been used extensively for com-
bating blooms of *Microcystis*. When $CuSO_4$ was added to exponentially
growing axenic *Microcystis* cells, growth was soon influenced but if
the Cu^{2+} concentration was not too high, cells could recover and
continue growing at the same growth rate after a few days (Verhoeven, 1978).

Cell concentration had an effect on resistance to copper toxicity. *M.* UV-006 cells at a concentration of ca. 80 Klett units
were killed by 0.3 mg/liter, at ca. 150 Klett units by 0.6 mg/liter,
and at ca. 300 Klett units were not killed by 0.8 mg/liter. *M.*
UV-010 (also a toxic isolate) had the same order of sensitivity,
but *M.* UV-007, the non-toxic isolate, was more resistant in that
cells at a concentration of ca. 150 Klett units could withstand
0.6 mg/liter Cu^{2+} but not 1.0 mg/liter Cu^{2+} (Verhoeven, 1978).

The color of cells that were affected changed to yellow-green. The chlorophyll a to pheophytin a ratio changed from 1.60 for healthy cells to 1.34 for cells that were influenced by Cu^{2+} (Verhoeven, 1978)

Ultrastructurally, copper decreased the electron-density of the nucleoplasm and caused aggregation of DNA fibrils. Furthermore, membrane-bounded inclusions and polyphosphate bodies disappeared and the well-defined thylakoids of normal cells were present as short membrane structures. In some cases the cytoplasm appeared loose from the cell wall. Carboxysomes were, however, generally present in the cells (Verhoeven and Eloff, 1979).

The influence of various herbicides on the growth of M. UV-006 was also determined (Groenewald, 1979). The results obtained are summarized in Table 4. Problems were encountered in dissolving some of the herbicides. When some solvents were used it was found that solvents such as methanol, ethanol, and ethyl acetate were even more toxic to *Microcystis* than many herbicides. Of the herbicides used in this preliminary test, Chloroquine and Vantoc seemed to be the most promising for controlling growth of *Microcystis* in nature.

Influence of Environmental Parameters

Turbulence. Some indications have been found that increasing turbulence decreased the standing crop of algae, including *Microcystis* (Ridley, 1979), but when laboratory cultures of *Microcystis* were grown at linear current velocities of 0 to 300 cm/sec, growth rate was not changed but cell viability increased with increasing turbulence (Krüger and Eloff, 1978a) (see also paper by Krüger and Eloff, this Conference Proceedings).

Temperature. The influence of temperature on the growth of four *Microcystis* isolates (M. UV-001 (toxic), M. UV-003, M. UV-006 (toxic) and M. UV-007) and a *Synechococcus* isolate (UV-005) was determined with a temperature-gradient incubator.

From the results obtained at 25 temperatures (10 to 40°C), the lower temperature limit (10.5 to 13.5°C), the optimal temperature (28.8 to 30.5°C) and upper temperature limit (35.0 to 40.0°C) were determined. Arrhenius equations, Q_{10} values and activation energies were calculated. An inflection in the gradient of the Arrhenius plot, indicating a sudden decrease in activation energy and an increase in temperature beyond 17.5°C, coincided with the temperature value for the onset of natural blooms of *Microcystis* (Krüger and Eloff, 1978b) (see also paper by Krüger and Eloff, this Conference Proceedings).

Table 4.   The Influence of Various Herbicides and Solvents on
           Growth of *M.* UV-006.

Herbicide or Solvent	Conc. which Had Effect, mg/l	Highest Conc. which Had No Effect, mg/l
φ Chloroquine	1.2	
φ Aurintricarboxylic acid	–	40
φ Vantoc C. L.	5.0	
φ p-Fluorophenylalanine	20	
* 2,3,6-Trichlorobenzoic acid	1000	
* 2,4-D	1000	
* 2,4,5-T	1000	
* 2-Methyl-chlorophenoxyacetic acid	1000	
* Isopropyl-N-phenylcarbamate	100	
φ Ethyl-N,N-diN-propylthiolcarbamate	200	
φ α,α-Dichloropropionic acid	66	
φ β-Chloroethyltrimethyl ammonia chloride	–	177.3
EtOH	0.8	
Acetone	–	23.7
DMSO	–	33
Ethyl acetate	0.89	
Chloroform	1.48	
Methanol	0.79	
Ethylene glycol monomethyl ether	–	28.8

*Compounds were dissolved in ethylene glycol monomethyl ether.

φCompounds were dissolved in water.

Light intensity. When the influence of light intensity on the
growth of seven isolates of *Microcystis* was determined it was found
that in all cases light intensities higher than ca. 15 μE./m^2/s led
to bleaching of cultures after 5 to 7 days (Krüger and Eloff, 1977).
This was surprising because *Microcystis* growing under natural conditions can apparently survive 100 times higher light intensities.
It also had important implications with regard to the growth of

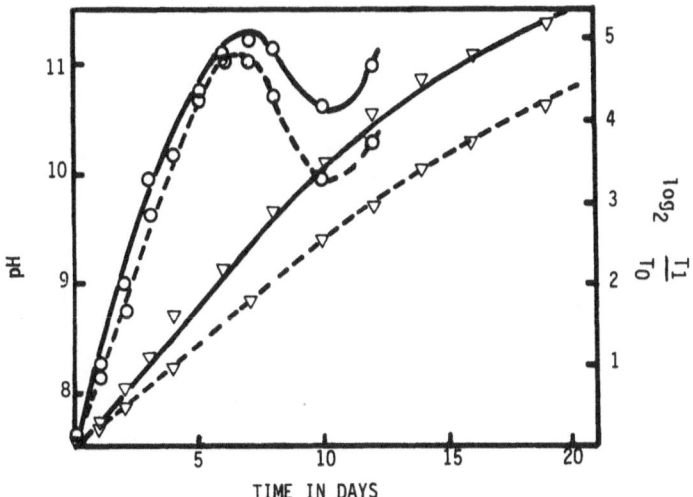

Fig. 6. The change in pH (-----) and turbidity (———) of cultures of *M*. UV-007 grown at 27 (○) and 9.5 (▽) µE./m²/s.

axenic *Microcystis* in a mass culture system as will be discussed in another paper (Krüger and Eloff, these Conference Proceedings).

pH

When the change of pH and turbidity was measured at low (9.5 µE./m²/s) and higher (27 µE./m²/s) light intensities it was found that the growth, expressed as $\log_2 T_1/T_0$, correlated very well with pH as shown in Figure 6 (redrawn from Krüger, 1978). From these results it was deduced that pH regulated the growth of *Microcystis* and it was decided to investigate this aspect.

Investigation into the influence of pH on the growth of uni-algal cultures of *Microcystis* strain 1036 (Gerloff et al., 1952) and *Microcystis* strain NRC-1 (McLachlan and Gorham, 1962) has been carried out by adjusting the pH with dilute HCl or NaOH every 12 hours and cropping after 7 days. Similar experiments were carried out with axenic *M*. UV-006 but the pH was monitored with a gel-filled electrode in the growth medium and growth was determined by measuring turbidity every 12 hours and resetting pH. Because the pH changed rapidly (ca. 0.5 pH units per hour from pH 6 with cells at a light intensity of 40 µE./m²/s, and ca. 0.3 pH units per hour with cells at a light intensity of 20 µE./m²/s) the experiment was repeated with the pH of the medium being adjusted every 2 hours for 3 1/2 days. It was found that some growth took place at a pH as low as 5.0 but that the optimal pH was 10.0, with no growth whatsoever at pH 11 (Figure 7). The growth inhibition at pH 5.0

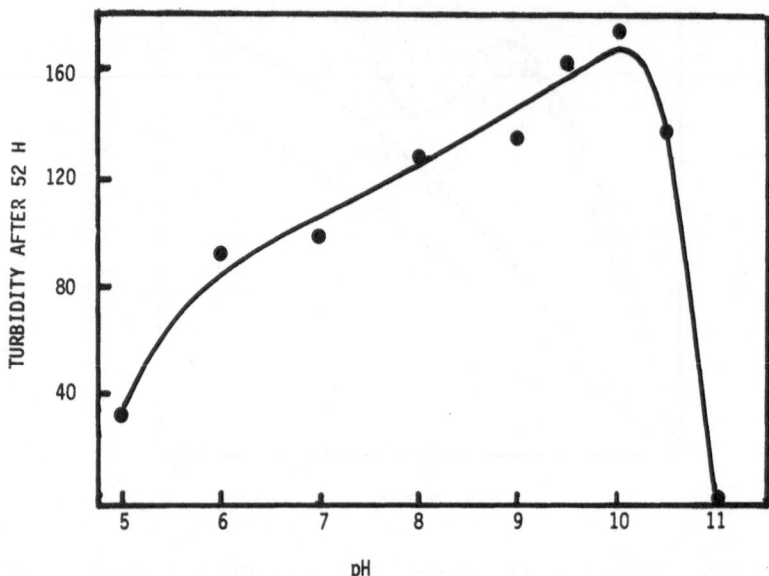

Fig. 7. The influence of pH adjusted every two hours on growth of
         M. UV-006.

was not due to Cl⁻ toxicity because when the pH was raised to 10 at
the end of the experiment the cells grew well. The growth inhibition at pH 11 was also not due to Na⁺ toxicity because the cells at
pH 10.5 were still growing well even after they had received more
NaOH than the pH 11.0 cells within the first few hours. The inhibition by high light intensities of growth of *Microcystis* may indeed
be ascribed to pH although contradictory results have been found
(Krüger, 1978).

In an effort to find a buffer that one could use in experiments
with *Microcystis*, the growth of M. UV-006 and M. UV-007 at six concentrations (0.1 M, 0.05 M, 0.01 M, 0.005 M, 0.001 M, and 0.0005 M)
of various "Good" buffers at pH values corresponding to respective
pKa values, were determined. Some of the results are summarized in
Table 5. It would appear that glycylglycine, Tricine, TES, HEPPS,
Bicine, CAPS, HEPES and CHEZ may be useful as buffers in the study
of *Microcystis* biology.

Sensitivity of Laboratory Cultures Relative to Natural Cultures

Adaptation to low light intensities. In a comparison it could
be shown that laboratory cultures generally have a much lower super-
oxide dismutase (SOD) activity than natural cultures (Eloff et al.,
1976). This could mean that laboratory cultures have become

Table 5. The Influence of Various Buffers on the Growth of *M.* UV-006 and *M.* UV-007.

Buffer	pKa	*M.* UV-007 (Non-toxic)		*M.* UV-006 (Toxic)	
		Conc. that Repressed Growth by > 50%	Order of Suitability	Conc. that Repressed Growth by > 50%	Order of Suitability
2 amino-2-ethyl 1,3 propanediol	9.4	0.001 M	11	0.001 M	9
2 amino-2-methyl 1,3 propanediol	9.4	0.005 M	8	0.0005 M	10
2 amino-2-methyl propanol	9.6	0.0005 M	12	0.0005 M	11
Bicine	8.4	0.05 M	3	0.01 M	6
CAPS	10.4	0.05 M	2	0.01 M	8
CHEZ	9.5	0.005 M	9	0.05 M	7
Glycinamide, HCl	8.2	0.005 M	10	0.0005 M	12
Glycylglycine	8.4	> 0.1 M	1	0.05 M	1
HEPES	7.5	0.05 M	7	0.1 M	5
HEPPS	8.0	0.05 M	5	0.05 M	4
TES	7.5	0.05 M	6	0.05 M	3
Tricine	8.2	0.1 M	4	0.05 M	2

adapted to low light intensities and have partly lost the ability
to synthesize SOD.

A change induced in growth medium by high light intensities.
When the chemical changes in the medium of cells growing under
stress and non-stress light intensities were monitored (Krüger and
Eloff, 1979a) indications were found that the CO_2 or O_2 concentrations and/or the ratio of CO_2 to O_2 played an important role in
the onset of light-stress conditions.

Removal of O_2. To determine whether removal of O_2 from
cultures would protect cultures of *Microcystis* from light-induced
stress conditions a system was evolved that removed O_2 from a
rubber-stoppered reaction flask without markedly influencing the
pH (Eloff, 1977). With two *Microcystis* isolates (UV-001 and
UV-007) and a *Synechococcus* isolate, removal of O_2 led to increased
yields.

Addition of CO_2. Although no additional effect of CO_2 addition could be observed when O_2 was removed from rubber-stoppered
flasks, addition of CO_2 had a dramatic effect on the resistance of
Microcystis culture to high light intensities (Krüger, 1978). When
sufficient CO_2 was added to the system even the highest light
intensity could be endured by the *Microcystis* cells. To investigate this problem the influence of very high light intensities and
light quality on the growth of *Microcystis* was examined.

Influence of high light intensities and light quality. When
different cell concentrations of *M.* UV-007 were illuminated with
different light intensities (50 to 1500 $\mu E./m^2/s$) in rubber-
stoppered containers for periods of up to 8 hours it was clear
that the higher the cell concentration, the higher the lethal
effect (see also Krüger and Eloff, 1979b). At light intensities
> 100 $\mu E./m^2/s$ the thylakoids were arranged perpendicular to the
cell wall, but at these higher light intensities there were fewer
thylakoids and the perpendicular arrangement suggested a protective
mechanism against high light intensity.

In rubber-stoppered containers the light intensity, x
($\mu E./m /s$) and death rate, y (per day) were exponentially related
as follows:

$$y = 0.372 \; x \; e^{0.0006x}$$

When the viability of cells was determined it was found that even
at 2000 $\mu E./m^2/s$ some cells remained viable after a 20 hour exposure. The lower the light intensity the longer was the time needed
to kill all cells. Cells kept at a light intensity of 50 $\mu E./m^2/s$
only lost viability after 256 hours (Figure 8). When bleaching
occurred under these conditions it was generally found that cells

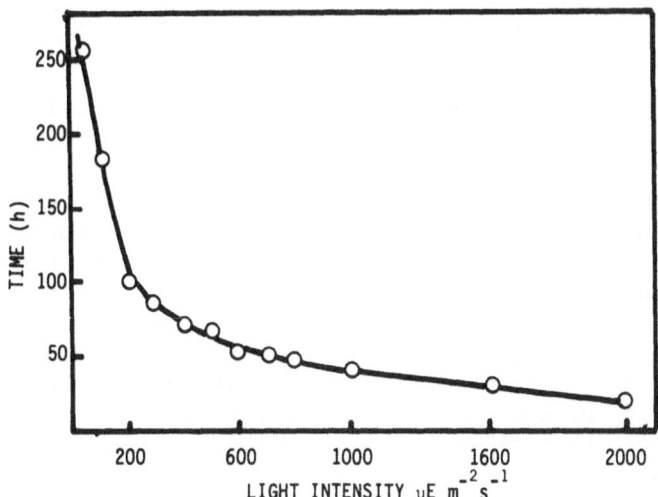

Fig. 8.   The influence of light intensity on the time lapse before
          M. UV-007 loses viability (van Vuren, 1979).

were not viable when the turbidity decreased to < 40% of the initial turbidity value (van Vuren, 1979).

Bubbling CO_2 through the cell suspension protected cells
against the lethal effect of high light intensity. The rate at
which 0.5% CO_2 was bubbled through was also critical. The protective effect of CO_2 could be abolished by increasing the O_2 content
of the gas bubbling through the system; once more indicating that
the CO_2:O_2 ratio plays a decisive role in the photoinhibition.

When action spectra for the growth of *M.* UV-007 and *M.* UV-006
(light intensities 5.5 to 7 µE./m^2/s) were determined using narrow
band interference filters it was clear that phycocyanin was the
most important receiver of light in both isolates. No evidence of
any chromatic adaptation could be found in any of the two isolates,
which may possibly be due to the absence of phycoerythrin.

When the same experiment was repeated at a higher light intensity (ca. 50 µE./m^2/s) to determine the action spectrum for photooxidation it was found that with *M.* UV-007 an unknown phycobilin
pigment absorbing green light at 574 nm plays the most important
role in the photoinhibition effect. In the case of *M.* UV-006
phycocyanin apparently plays an important role in photoinhibition
because 625 nm light had the highest inhibitory effect (van Vuren,
1979).

Influence of bacterial contamination on photoinhibition. When
we succeeded in purifying one of our toxic isolates (M. UV-010) we
noticed that this culture was much more sensitive to light intensity
than the M. UV-010 cultures still infected with bacteria. This suggested that bacteria may protect Microcystis against photoinhibition.

The growth of three M. isolates (UV-006, UV-007, and UV-010) at
different light intensities was compared to that of these isolates
reinfected with the bacteria originally present in each case. The
growth obtained after 21 days is represented in Table 6. From these
results it seemed as if bacterial contamination had very little
influence on the photoinhibition of M. UV-007, no influence on M.
UV-006 and a very big influence on M. UV-010. This could have had
something to do with the different bacteria present in these cases.
Long Gram-negative bacilli (3.6 x 0.6 µm) were present in M. UV-006
and 0.6 µm diameter Gram-negative cocci were present in M. UV-007.

To determine whether the protective influence of the bacteria
on the photoinhibition of Microcystis is due to a growth stimulant
being excreted or possibly to a growth inhibitor being inactivated
or decomposed by the associated bacteria an experiment was carried
out in which the bacteria were present in a nutrient broth in a
separate container. The results of such an experiment with M.
UV-010 are shown in Figure 9.

From the results it is clear that the bacteria present in the
separate container had a protective and/or a stimulating influence
on the growth of Microcystis at all the light intensities. The
logical explanation of this phenomenon is that the bacteria protect
Microcystis against photoinhibition by increasing the $CO_2:O_2$ ratio
through bacteria respiratory O_2 uptake and CO_2 release.

Table 6.   The Influence of Bacterial Contamination on the 21-Day
           Growth of Three Microcystis Isolates at Different Light
           Intensities.   (Growth Measured as Turbidity in Klett Units)

Light Intensity µE./m²/s	M. UV-006		M. UV-007		M. UV-010	
	Axenic	Bacterized	Axenic	Bacterized	Axenic	Bacterized
9	338	296	253	259	362	359
20	349	342	123	125	209	339
40	320	319	90	98	79	480
90	269	174	85	86	144	459

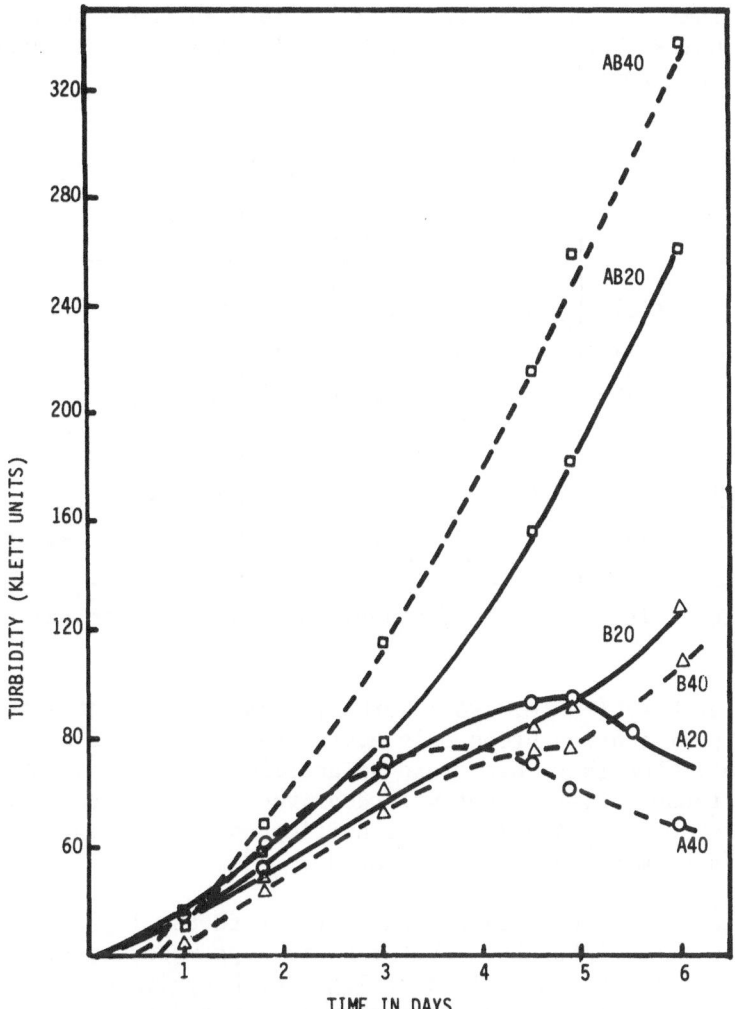

Fig. 9. The influence of bacteria on growth of *M*. UV-006 at two
light intensities, 40 μE./m²/s (-----) and 20 μE./m²/s
(———). A = axenic culture, B = bacterized culture,
AB = axenic culture with bacteria in a separate container.

This probably means that at least with one isolate of *Microcystis* the light sensitivity could be abolished to a large extent
by the action of bacteria present in the culture. This indicates
that the difference between light sensitivity of laboratory and
natural cultures may at least partially be ascribed to the presence of bacteria in natural cultures.

Table 7.   Increase in Turbidity within 5 (and 10) Days of *M*. UV-010
           Cells Grown at Different Light Intensities with Different
           Light/Dark Cycles[a]

Light/ Dark Cycle	Light Intensity, $\mu E./m^2/s$				
	32	20	8.5	4.5	2.1
24:0	-9 (16)	6 (18)	130 (197)	57 (106)	10 (14)
16:8	-16 (61)	91 (48)	99 (169)	36 (74)	5 (10)
12:12	4 (12)	120 (83)	91 (159)	30 (54)	7 (12)
8:16	116 (96)	115 (204)	66 (90)	22 (36)	6 (8)
4:20	117 (184)	54 (105)	15 (28)	8 (16)	3 (4)
0:24	-1 (-3)	-19 (-39)	-24 (-45)	-7 (-34)	-16 (-34)

[a]Average turbidity at start of experiment was 130 Klett units.

        Some authors (Vance, 1966; Gorham, 1964) found that after a
number of purification steps only contaminated cultures of *Microcystis* grew. Possibly this could be ascribed to a light inhibition
effect on those isolates that were deprived of the bacteria that
protected them from high light intensities and not to an obligatory
symbiotic system as has been suggested.

        Influence of photoperiod.   In contrast to the light/dark
cycles in nature, laboratory cultures are kept under continuous
light.   The influence of different light/dark cycles on the growth
of *M*. UV-010 at different light intensities was determined by
enclosing growth flasks with aluminum foil at predetermined times.
The results are shown in Table 7.

        It is clear that the length of the photoperiod had a large
influence on the high light intensity effect. With an 8-hour light
cycle the cells of this very sensitive *M*. isolate grew well at the
highest light intensity tested.

        These results may mean that the cells occurring in nature may
not differ all that much from laboratory cultures, but that their
increased resistance to light may be ascribed to the natural light/
dark cycle, to the presence of bacteria in natural populations and
to the migration of the cells in the water column.   To examine these
possibilities it was decided to compare the light sensitivity of
laboratory and pond cultures over a longer period than has been
done earlier.

Table 8. The Growth of *Microcystis* Freshly Isolated from a Pond, *M.* UV-006, and *M.* UV-010 After 1, 5, and 12 Days in Dialysis Tubing as % of Original Turbidity Value.

Depth cm	Light Int., $\mu E./m^2/s$	Pond *Microcystis*			*M.* UV-006			*M.* UV-010		
		1	5	12	1	5	12	1	5	12
0	580	81	114	170	91	111	243	70	96	287
10	340	144	132	174	110	119	–	117	134	–
20	190	150	133	171	98	129	263	119	145	–
40	52	129	117	125	90	108	206	109	–	219
100	0	90	115	101	81	101	140	53	125	117

Comparison of laboratory and pond cultures in field experiments. These experiments were carried out in cooperation with Moshe Shilo in fish ponds in the Beat Shean valley in Israel. Cells of *Microcystis* freshly isolated from a nearby pond and of *M.* UV-006 and *M.* UV-010 were suspended in dialysis tubing at 0, 10, 20, 40, and 100 cm depth as described elsewhere (Eloff et al., 1976). After 1, 5, and 12 days cells were collected and the turbidity was determined. Some of the results are given in Table 8.

The most important aspect of these results is that even under the cloudy conditions (light intensity at noon 720 $\mu E./m^2/s$) the growth of the pond *Microcystis* was inhibited at the surface and that the best growth took place at a depth of ca. 15 cm. The best growth of the *M.* UV-006 and *M.* UV-010 was also at a depth of ca. 20 cm. If the experiment was carried out in summer under sunny conditions the pond *Microcystis* would probably have died at the surface which accentuates the important role of algal movement and gas vacuoles in the biology of *Microcystis*.

CONCLUSION

From all these results it seems as if the apparent difference in light sensitivity between laboratory cultures and *Microcystis* growing in nature is not due to an adaptation of growth under low light intensities but rather an artifact of the laboratory growth conditions. Three aspects may be identified: (1) the bacterial influence as far as $CO_2:O_2$ ratios are concerned, (2) the fact that cells may recover to a certain extent in darkness, and (3) the importance of algal buoyance regulation in nature.

ACKNOWLEDGMENTS

    The financial assistance of the NIWR (National Institute for
Water Research) of the CSIR (Council for Scientific and Industrial
Research) and of the UOFS (University of the Orange Free State) is
gratefully acknowledged.

REFERENCES

Allen, M. M. 1968. Simple conditions for growth of unicellular
        blue-green algae on plates. J. Phycol. 4:1-4.
Barlow, D. J., W. L. J. van Rensburg, A. J. H. Pieterse, and J. N.
        Eloff. 1979. Effect of phosphate concentration on the
        fine structure of the cyanobacterium *Microcystis aeruginosa*
        Kütz. emend. Elenkin. J. Limnol. Soc. S. Afr. 5:79-83.
Bruwer, C. A. 1979. The Economic Impact of Eutrophication in
        South Africa. Republic of South Africa Department of Water
        Affairs, Tech. Rep. TR94. 48 pp.
Canabaeus, L. 1929. Über die heterocysten und gasvakuolen der
        blaualgen und ihre Beziehung zueinander. Pages 1-48 *in*
        R. Kolkowitz, ed., Pflanzenforschung, Vol. 13.
De Wet, J. S. 1979. Die groei van *Microcystis* met aminosure as
        stikstofbron. M.Sc. Thesis. UOVS. Bloemfontein, S. Afr.
Eloff, J. N. 1977. Absorption of oxygen from liquid cultures of
        blue-green algae by alkaline pyrogallol. J. Limnol. Soc.
        S. Afr. 3:13-16.
Eloff, J. N., Y. Steinitz, and M. Shilo. 1976. Photooxidation of
        Cyanobacteria in natural conditions. Appl. Environ. Micro-
        biol. 31:119-126.
Eloff, J. N., and M. M. van Vuren. 1978. (Unpublished results).
Fogg, G. E. 1942. Studies on nitrogen fixation by blue-green
        algae. I. Nitrogen fixation by *Anabaena cylindrica* Lemm.
        J. Exp. Biol. 19:78-87.
Gerloff, G. C., G. P. Fitzgerald, and F. Skoog. 1952. The mineral
        nutrition of *Microcystis aeruginosa*. Am. J. Bot. 39:26-32.
Groenewald, E. G. 1979. (Unpublished results).
Herdman, M., S. F. Delaney, and N. G. Carr. 1973. A new medium
        for the isolation and growth of auxotrophic mutants of the
        blue-green alga *Anacystis nidulans*. J. Gen. Microbiol. 79:
        233-237.
Krüger, G. H. J. 1978. The effect of physico-chemical factors on
        growth relevant to the mass culture of *Microcystis* under
        sterile conditions. Ph.D. Thesis. University of the Orange
        Free State. Bloemfontein, South Africa.
Krüger, G. H. J., and J. N. Eloff. 1977. Influence of light
        intensity on the growth of different *Microcystis* isolates.
        J. Limnol. Soc. S. Afr. 3:21-25.

Krüger, G. H. J., and J. N. Eloff. 1978a. The effect of agitation and turbulence of the growth medium on the growth and viability of *Microcystis*. J. Limnol. Soc. S. Afr. 4:69-74.

Krüger, G. H. J., and J. N. Eloff. 1978b. The effect of temperature on specific growth rate and activation energy of *Microcystis* and *Synechococcus* isolates relevant to the onset of natural blooms. J. Limnol. Soc. S. Afr. 4:9-20.

Krüger, G. H. J., and J. N. Eloff. 1979a. Chemical changes in growth medium of *Microcystis* under stress and non-stress light intensities. J. Limnol. Soc. S. Afr. 5:11-16.

Krüger, G. H. J., and J. N. Eloff. 1979b. The interaction between cell density of *Microcystis* batch cultures and light induced stress conditions. Z. Pflanzenphysiol. 95:441-447.

Marler, J. E., and C. van Baalen. 1965. Role of H_2O_2 in single-cell growth of the blue-green alga, *Anacystis nidulans*. J. Phycol. 1:180-185.

McDaniel, H. R., J. B. Middlebrook, and R. O. Bowman. 1962. Isolation of pure cultures of algae from contaminated cultures. Appl. Microbiol. 10:223.

McLachlan, J., and P. R. Gorham. 1962. Effects of pH and nitrogen sources on growth of *Microcystis aeruginosa* Kütz. Can. J. Microbiol. 8:1-11.

Pretorius, J. A. 1978. Groei van *Microcystis aeruginosa* op agar-medium. M.Sc. Thesis. UOVS. Bloemfontein, South Africa.

Pretorius, J. A., and J. N. Eloff. 1977. Occurrence of nannocytes in *Microcystis*. S. Afr. J. Sci. 73:245-246.

Pretorius, J. A., G. H. J. Krüger, and J. N. Eloff. 1977. The release of nannocytes during the growth cycle of *Microcystis*. J. Limnol. Soc. S. Afr. 3:17-20.

Ridley, J. E. 1969. Artificial destratification of waterworks impoundments for the control of algal blooms. Brit. Phycol. J. 4:215.

Stanier, R. Y., R. Kunisawa, M. Mandel, and G. Cohen-Bazire. 1971. Purification and properties of unicellular blue-green algae (order Chroococcales). Bacteriol. Rev. 35:171-205.

Tustin, R. C., S. J. van Rensburg, and J. N. Eloff. 1973. Hepatic damage in the primate following ingestion of toxic algae. Pages 383-385 *in* S. J. Saunders and J. Terblanche, eds., Liver: Proceedings International Liver Congress. Pitman, London.

Vance, B. D. 1966. Sensitivity of *Microcystis aeruginosa* and other blue-green algae and associated bacteria to selected antibiotics. J. Phycol. 2:125-128.

Van Vuren, M. M. 1979. Die invloed van ligintensiteit en ligkwaliteit op die groei van *Microcystis*. M.Sc. Thesis. UOVS. Bloemfontein, South Africa.

Van Wyk, P. W. J., J. N. Eloff, and R. L. Verhoeven. 1979. Investigation of membrane bound inclusion in *Microcystis*. Proc. Electron Microsc. Soc. S. Afr. 9:157-158.

Van Wyk, P. W. J. 1980. Ultrastrukturele karakterisering van
    sekere sellulêre insluitsel van *Microcystis*. M.Sc. Thesis.
    UOVS. Bloemfontein, South Africa.
Verhoeven, R. L. 1978. (Unpublished results).
Verhoeven, R. L., and J. N. Eloff. 1979. Effect of lethal con-
    centrations of copper on the ultrastructure and growth of
    Microcystis. Proc. Electron. Microsc. Soc. S. Afr. *9*:161-
    162.
Wieringa, K. T. 1968. A new method for obtaining bacteria-free
    cultures of blue-green algae. Antonie van Leeuwenhoek *34*:
    54-56.

IS A PLASMID(S) INVOLVED IN THE TOXICITY

OF *MICROCYSTIS AERUGINOSA*?

John H. Hauman

National Institute for Water Research
Council for Scientific and Industrial Research
P. O. Box 395
Pretoria, 0001, Republic of South Africa

ABSTRACT

Toxicity of *Microcystis aeruginosa* has been observed in various
impoundments in South Africa. The symptoms of toxicity (i.e.
swollen livers and short survival times) as well as the polypeptide
nature of the toxin are similar to those reported elsewhere. In
nature, the level, as well as the presence/absence of toxicity, is
highly variable. Plasmids have been demonstrated in a number of
blue-green algal species. The origin of toxicity in *M. aeruginosa*
is as yet poorly understood. Plasmids are known to be involved in
toxin production in certain bacteria, therefore the possibility of
plasmid involvement in the toxicity of *M. aeruginosa* is under
investigation. Laboratory cultures of a toxic strain of *M. aeruginosa* (WR 70) were supplemented by various concentrations of agents
known either to eliminate plasmids (e.g. acridine orange) or to
select for plasmid-free cells (e.g. sodium dodecyl sulphate) in
bacteria. Toxicity of the cultures was monitored by intraperitoneal injection of disrupted cells into mice. It was found that
cultures of toxic *M. aeruginosa* became non-toxic after growth in
suitable concentrations of acridine orange, streptomycin, sodium
dodecyl sulphate and chloramphenicol. These results indicate a
possibility of plasmid involvement in the toxicity of *M. aeruginosa* (WR 70).

INTRODUCTION

Microcystis aeruginosa, a planktonic blue-green alga (cyanobacterium) has been reported toxic to mammals in most countries

97

(Schwimmer and Schwimmer, 1968), including South Africa (Toerien
et al., 1976). Differences occur in the reported composition of
the toxin but, in each case, the pathological symptoms have been
the same (Elleman et al., 1978).

The genetics of toxin production in *Microcystis* have thus far
received little attention. Gorham (1964) reported that toxic and
non-toxic clones were isolated from the toxic strain NRC-1, which
was taken to indicate genetic heterogeneity of toxin production.
Vance (1977) suggested that toxin production in strain NRC-1 may
be the result of lysogenic conversion, as a prophage was induced,
using mitocin C, from toxic strains only.

Plasmids have been reported in various strains of filamentous
(Simon, 1978) and unicellular (Friedberg and Seijffers, 1979; Lau
and Doolittle, 1979; van den Hondel et al., 1979) cyanobacteria
but their function is, as yet, not well understood. Van den Hondel
et al. (1979) have suggested that plasmids may be involved in toxin
production in some of the cyanobacteria. On the other hand Kumar
and Gorham (1975), working with *Anabaena flos-aquae* NRC-44-1, concluded that its toxicity was not controlled by extrachromosomal
elements as toxicity was not lost after treatment with acridine
dyes.

The above results indicate a need for further information on
the genetics of toxin production in cyanobacteria. The effect of
various chemicals on the toxicity of *M. aeruginosa* culture was
therefore studied.

Agents tested for an effect on toxicity were acridine orange,
which eliminates sex factors from *Escherichia coli* (Hirota, 1960)
and cures toxigenic cultures of *Clostridium botulinum* of their
prophages and toxicity (Eklund and Poysky, 1973); sodium dodecyl
sulphate (SDS), which selects for plasmid-free cells in *E. coli*
(Tomoeda et al., 1968) and *Staphylococcus aereus* (Sonstein and
Baldwin, 1972); chloramphenicol, which eliminates the plasmid-coded
haemolysin of *E. coli* (Mitchell and Kenworthy, 1977) and streptomycin, which acts as a mutagen towards the non-chromosomal genes
of *Chlamydomonas* (Sager, 1962).

MATERIALS AND METHODS

A toxic, unialgal, non-axenic culture of *M. aeruginosa* (NIWR
strain WR 70), grown in modified Volk and Phinney's liquid medium
(Barlow et al., 1977) was used. The test agents, dissolved in
sterile distilled water, were added to the growing cultures at the
final concentrations given in Table 1. The cultures were kept at
20°C (± 1°C) in daylight with no aeration. Growth in the presence

of the agents tested was approximately 2 to 3 times slower than normal so that incubation of the test and control cultures was continued under the above conditions for 3 months before the cultures were harvested by centrifugation. As all the test agents were toxic at high doses, the cultures were preadapted by growth at concentrations which did not significantly affect the growth rate (Table 1). The harvested cells were disrupted by sonication and 0.5 ml of the resulting suspension, containing approximately double the normal LD_{100} in cell mass (Elleman et al., 1978), was injected intraperitoneally into mice. Two mice were used per culture, which was considered to be toxic if the mice died within 4 hours after injection and had enlarged livers, symptomatic of *Microcystis* poisoning (Ashworth and Mason, 1946).

RESULTS AND DISCUSSION

The toxicity of *M. aeruginosa* WR 70 has been a stable phenotype of this strain since its original isolation in 1976. This phenotypic stability was again demonstrated in the present experiments, where control cultures were still toxic at the end of the 3-month incubation period.

At suitable doses all the agents tested eliminated the toxicity of *M. aeruginosa* WR 70 cultures (Table 1). Of particular interest is the relatively low doses at which this was achieved compared with those curing plasmids in bacteria (Table 2). No explanation for this observation is readily at hand.

Table 1. Concentrations of Agents Tested for the Elimination of the Toxicity of the Toxic *M. aeruginosa* Strain WR 70

Agent	Concentrations Tested $(\mu g.ml^{-1})$ [a]							
	10^4	10^3	10^2	10	1	0.5	0.1	0.01
Acridine orange					-	-	+	+
SDS		-	-	-				
Streptomycin					-	-	-	+
Chloramphenicol					-	-	+	+

[a] Preadaption concentrations were the lowest indicated for each agent

- Indicates non-toxic ("cured") cultures

+ Indicates toxic cultures

Table 2.   Doses of Agents Curing Toxicity in *M. aeruginosa* and
           Doses of these Agents Curing Plasmids in Bacteria

Agent	Effective dose for *M. aeruginosa* ($\mu g.ml^{-1}$)	Effective Dose in Bacteria ($\mu g.ml^{-1}$)
Acridine orange	0.5	10-50 (Hirota, 1960) 20 (Eklund and Poysky, 1973)
SDS	10	20 (Sonstein and Baldwin, 1972) 10^5 (Tomoeda et al., 1968)
Chloram- phenicol	0.1	2 (Mitchell and Kenworthy, 1977)

In bacteria the agents tested vary in their mode of action on plasmid-bearing cells. Acridine orange actively causes the loss of plasmids (Hirota, 1960) by interefering with plasmid replication whilst chromosomal replication is unaffected (Yamagata and Uchida, 1969). SDS on the other hand appears to select for plasmid-free cells (Tomoeda et al., 1968) with the sex pili coded for by these plasmids playing a role in the higher susceptibility of plasmid-bearing cells to lysis by SDS (Adachi et al., 1972). Sager (1962) found that streptomycin acts as a mutagen towards the non-chromosomal genes in *Chlamydomonas*.

In bacteria acridine orange is only slightly effective in curing lysogenic strains of their prophages (Hirota, 1960; Jacob and Wollman, 1961), with Eklund and Poysky (1973) finding that, in one experiment, only 2 out of 69 colonies of a toxic *C. botulinum* strain were cured of their prophages and toxicity. Acridine orange is ineffective against the Hfr sex factor of *E. coli* (Hirota, 1960). Both of these factors are stably integrated into the host chromosome. On the other hand, acridine orange eliminates the F sex factor of *E. coli* (Hirota, 1960), which is cytoplasmic.

The ease of loss of toxicity, as evidenced by the action of various agents, together with the effectiveness of acridine orange at low doses, suggests that the genetic determinant of toxicity of this *M. aeruginosa* strain could be a non-integrated plasmid rather than a prophage, as has been suggested by Vance (1977) for *M. aeruginosa* NRC-1. More recently Sam et al. (1980) was able to cure *Agmenellum quadriplicatum* BG-1 of a 5.9 million dalton plasmid by growth of the cyanobacteria in 80 $\mu g.ml^{-1}$ SDS.

Further work on the isolation of the genetic determinant of toxicity in *M. aeruginosa* WR 70 is in progress but is being hampered by the resistance of this strain to lysis methods commonly used in plasmid isolation.

ACKNOWLEDGMENTS

Mrs. H. Krüger tested the cultures for toxicity.

This work was partly funded by a grant from the Department of Health and is published with the approval of the Director of the National Institute for Water Research.

REFERENCES

Adachi, H., M. Nakano, M. Inuzuka, and M. Tomoeda. 1972. Specific role of sex pili in the effective eliminatory action of sodium dodecyl sulfate on sex and drug resistance factors in *Escherichia coli*. J. Bacteriol. *109*:1114-1124.

Ashworth, C. T., and M. F. Mason. 1946. Observations on the pathological changes produced by a toxic substance present in blue-green algae (*Microcystis aeruginosa*). Am. J. Pathol. *22*:369-383.

Barlow, D. J., W. E. Scott, and S. I. Funke. 1977. Fine structural changes in response to different light intensities in cultured *Microcystis aeruginosa*. Proc. E. M. Soc. S. Afr. *7*:105-106.

Eklund, M. W., and F. T. Poysky, 1973. Bacteriophages and toxigenicity of *Clostridium botulinum*. Pages 31-39 *in* B. C. Hobbs and J. H. B. Christian, eds., The Microbiological Safety of Food. Academic Press, London and New York.

Elleman, T. C., I. R. Falconer, A. R. B. Jackson, and M. T. Runnegar. 1978. Isolation, characterization and pathology of the toxin from a *Microcystis aeruginosa* (= *Anacystis cyanea*) bloom. Aust. J. Biol. Sci. *31*:209-218.

Friedberg, D., and J. Seijffers. 1979. Plasmids in two cyanobacterial strains. FEBS Letters *107*:165-168.

Gorham, P. R. 1964. Toxic algae. Pages 307-336 *in* D. F. Jackson, ed., Algae and Man. Plenum Press, New York.

Hirota, Y. 1960. The effect of acridine dyes on mating type factors in *Escherichia coli*. Proc. Nat. Acad. Sci. U. S. A. *46*:57-64.

Jacob, F., and E. L. Wollman. 1961. Sexuality and Genetics of Bacteria. Academic Press, New York and London.

Kumar, H. D., and P. R. Gorham. 1975. Effects of acridine dyes and other substances on growth, lysis and toxicity of *Anabaena flos-aquae* NRC-44-1. Biochem. Physiol. Pflanzen. *167*:473-487.

Lau, R. H., and W. F. Doolittle. 1979. Covalently closed circular
    DNAs in closely related unicellular cyanobacteria. J.
    Bacteriol. *137*:648–652.

Mitchell, I. de G., and R. Kenworthy. 1977. Attempted elimination
    of plasmid-determined haemolysin, K 88 antigen and entero-
    toxin from *Escherichia coli* pathogenic for pigs. J. Appl.
    Bacteriol. *42*:207–212.

Sager, R. 1962. Streptomycin as a mutagen for nonchromosomal
    genes. Proc. Nat. Acad. Sci. U. S. A. *48*:2018–2026.

Sam, R. H., C. Sapienza, and W. F. Doolittle. 1980. Cyanobacterial
    plasmids: their widespread occurrence, and the existence of
    regions of homology between plasmids in the same and different
    species. Molec. Gen. Genetics *178*:203–211.

Schwimmer, M., and D. Schwimmer. 1969. Medical aspects of phycol-
    ogy. Pages 297–358 *in* D. F. Jackson, ed., Algae, Man and
    the Environment. Syracuse University Press, New York.

Simon, R. D. 1978. Survey of extrachromosomal DNA found in the
    filamentous cyanobacteria. J. Bacteriol. *136*:414–418.

Sonstein, S. A., and J. N. Baldwin. 1972. Loss of the penicillinase
    plasmid after treatment of *Staphylococcus aurous* with sodium
    dodecyl sulfate. J. Bacteriol. *109*:262–265.

Toerien, D. F., W. E. Scott, and M. J. Pitout. 1976. *Microcystis*
    toxins: isolation, identification, implications. Water
    S. A. *2*:160–162.

Tomoeda, M., M. Inuzuka, N. Kubo, and S. Nakamura. 1968. Effective
    elimination of drug resistance and sex factors in *Escherichia*
    coli by sodium dodecyl sulfate. J. Bacteriol. *95*:1078–1089.

Van den Hondel, C. A. M. J. J., W. Keegstra, W. E. Borrias, and
    G. A. Van Arkel. 1979. Homology of plasmids in strains of
    unicellular cyanobacteria. Plasmid *2*:323–333.

Vance, B. D. 1977. Prophage induction in toxic *Microcystis*
    aeruginosa NRC-1. J. Phycol. *13*(Suppl.):70.

Yamagata, H., and H. Uchida. 1969. Effect of acridine orange on
    sex factor multiplication in *Escherichia coli*. J. Mol. Biol.
    46:73–84.

A TOXIC BLOOM OF *ANABAENA FLOS-AQUAE*

IN HEBGEN RESERVOIR MONTANA IN 1977

Richard E. Juday, Edward J. Keller, Abe Horpestad[1],
Loren L. Bahls[1], and Stephen Glasser[2]

Department of Chemistry
University of Montana
Missoula, Montana

ABSTRACT

Because of the volcanic nature of the Yellowstone Park area,
waters draining it are highly mineralized. Consequently, Hebgen
Lake is also mineralized, with a high concentration of nutrients.
While the main lake is not highly productive, the Grayling Arm
commonly produces algal blooms both early and late in the season.
Predominating species in the early blooms are usually *Anabaena
spiroides* and *Aphanizomenon flos-aquae*. In June of 1977 a heavy
bloom of *Anabaena flos-aquae* occurred which developed toxic characteristics. A number of cattle and dogs died as a result, and it
was necessary to close the beaches and camp sites on the Grayling
Arm. Stomach contents of the dead animals were toxic to mice, and
anatoxin-a was isolated from algae samples by Wayne Carmichael,
Wright State University, Dayton, Ohio. The bloom declined during
July and was non-toxic by the end of the month.

Possible explanations for the bloom include early filling of
the reservoir because of expected poor runoff, and early warming.
Also, since cattle have unrestricted access to the lake, fertilization from that source could have been a factor.

---

[1]With Montana Water Quality Bureau, Helena, Montana.

[2]With U. S. Forest Service, Gallatin Natl. Forest, Bozeman, Montana.

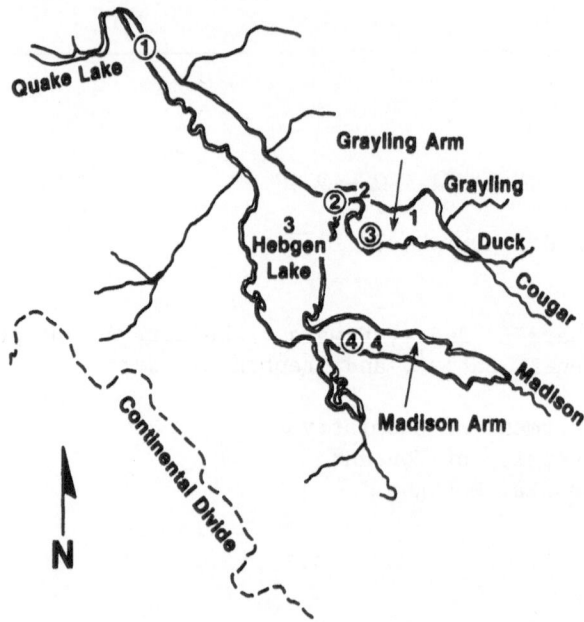

Fig. 1. Map of the Hebgen Reservoir area (U. S. G. S.). Circled
         numbers are National Environmental Survey (NES) sampling
         sites. Uncircled numbers are Water Quality Bureau and
         University of Montana sites.

TEXT

     Early in the summer of 1977, a heavy bloom of *Anabaena flos-
aquae* developed in the Grayling Arm of Hebgen Reservoir. Its toxic
nature was first noted on June 21 when a dog was reported to have
died after drinking water from the reservoir. Ultimately about
8 dogs and 30 head of cattle died from it. Algal material was
collected and tested for toxicity by Wayne Carmichael at Wright
State University and Jim Cutler at Montana State University. The
alga was identified by, among others, Loren Bahls (Montana Water
Quality Bureau, Helena), and Gerald Prescott (Flathead Lake
Research Station, Montana). The bloom gradually declined during
July and had disappeared by early August. This is the first incident of this nature ever documented in Montana.

     Hebgen Reservoir has not received a great deal of study up to
now. It was the subject of a Ph.D. thesis project during the summers of 1964 and 1965 (Martin, 1967). It was included in the
National Eutrophication Survey (NES) in 1975 and was studied by
Dr. Abe Horpestad of the Montana State Water Quality Bureau in
1976. As a result of the bloom, further analyses were performed
by the Montana Water Quality Bureau and the water chemistry laboratory of the University of Montana.

Table 1. Lake and Drainage Basin Characteristics (NES, 1977)

Morphometry	Whole Reservoir	Grayling Arm
Surface Area	51.27 km^2	7.8 km^2
Mean Depth	8.2 m	3.0 m (est.)
Maximum Depth	37.5 m	5.0 m
Volume	420 x 10^6 m^3	37 x 10^6 m^3
Retention Time	172 days	---

Tributaries and Outlet	Drainage Area (km^2)	Mean Flow (m^3/sec.)
Tributaries		
Madison River	1,157.7	13.940
Watkins Cr.	29.5	0.098
Madison R. (S. Fork)	211.6	3.920
Duck Cr.	305.4	1.350
Minor Tributaries and Immediate Drainage	653.2	9.545
	2,357.4	28.853
Outlet		
Madison River	2,408.7	28.830

Precipitation		
1975	89.7 cm	
Mean Annual	70.6 cm	

The reservoir is located in Montana adjacent to the northwest corner of Yellowstone Park. This area is located over a hot spot in the earth's crust similar to Iceland and Hawaii. As a result, there is much geothermal activity and local waters are highly mineralized, containing relatively much sodium and potassium but little calcium and magnesium. Phosphate levels are also relatively high. Figure 1 is a map of the area, and Table 1 contains morphometric data.

Except for the Grayling Arm, Hebgen Reservoir shares traits common to reservoirs compared with lakes:

1. It is drained through the hypolimnion and nutrient concentrations are orthograde, therefore it tends to become depleted in nutrients.

Table 2. Algal Taxa Identified

BACILLARIOPHYCEAE

Achnanthes sp.
Amphora sp.
 Asterionella formosa
Cocconeis sp.
 Cyclotella sp.
 Fragilaria crotonensis
Gomphonema sp.
 Melosira granulata
 Meridion sp.
 Navicula sp.
Pinnularia sp.
 Synedra sp.

CHRYSOPHYCEAE

Synura uvella
 Chrysococcus sp.

CRYPTOPHYCEAE

 Rhodomonas lacustris
 Cryptomonas ovata

DINOPHYCEAE

 Ceratium hirundinella

CHLOROPHYCEAE

 Ankistrodesmus falcatus
 Chlamydomonas sp.
 Chlorella ellipsoidea
 Elakatothrix gelatinosa
 Glycocystis sp.
Kirchneriella sp.
 Pediastrum sp.
 Pandorina morum
 Oocystis sp.
Scenedesmus sp.
 Schroederia setigera
 Staurastrum sp.

EUGLENOPHYCEAE

Trachelomonas sp.

MYXOPHYCEAE

 Anabaena spiroides
 Aphanizomenon flos-aquae

---

*These species were found rarely or in small numbers.

2. Since the water lost is colder, water temperatures of the
    reservoir tends to be higher than lakes.

3. The water level of Hebgen Reservoir undergoes a vertical fluc-
    tuation of about 7 m during the year, draining the Grayling
    Arm.

This extensive drawdown largely precludes the development of ben-
thic or littoral biotic communities along the margin of the lake
(Martin, 1967). Algae species identified from phytoplankton samples by Martin in 1965 are listed in Table 2, and the population
sequence of the five most common species is shown in Figure 2. It
will be noted the dominant *Anabaena* species was *A. spiroides*, and
that *A. flos-aquae* was absent. Since the plankton samples were
pooled and only one came from the Grayling Arm, it is not possible
to decide how much of this data can be ascribed to the Grayling
Arm.

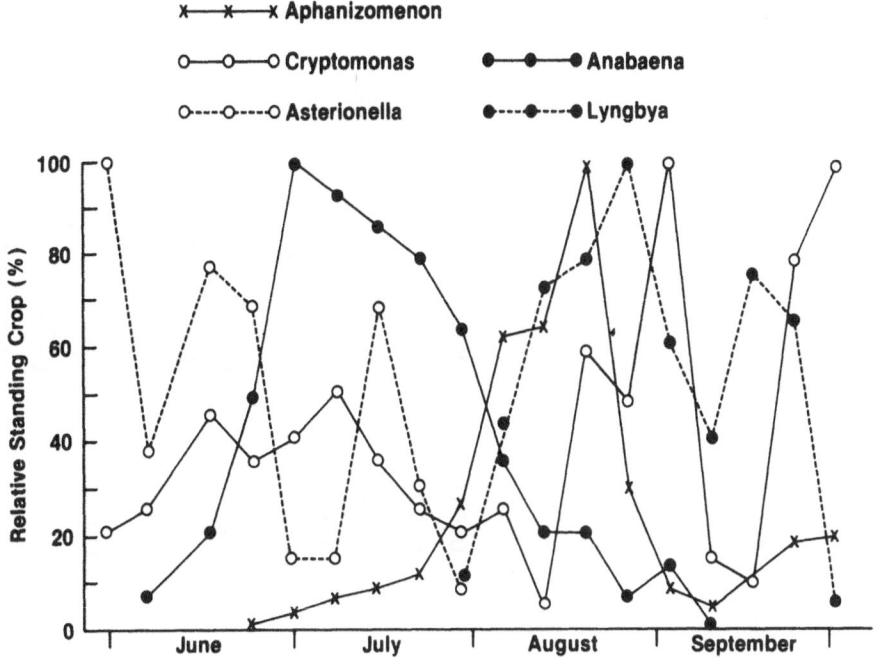

Fig. 2. Seasonal standing crops of the five most abundant phytoplankton species during 1965 (Martin, 1967).

During 1975 the National Environmental Survey (NES) included
15 Montana lakes and reservoirs in its study, in which Hebgen
Reservoir was included. The overall trophic quality was determined using six water quality parameters. Four sampling stations
were established on the reservoir. They are the circled numbers
in Figure 1, number 3 being in the Grayling Arm. Pertinent analytic data from the NES report is summarized in Table 3. The NES
sampled the reservoir three times. The maximum chlorophyll level
was in the Grayling Arm in the September sampling and was 11.5
µg/liter. The NES and Horpestad found that plankton growth was
nitrogen-limited in the early sampling. Horpestad found that it
was phosphate-limited in September. The NES rated the reservoir
tenth in productivity and classified it meso-eutrophic.

Analytical data from the Montana Water Quality Laboratory for
the reservoir in 1976 and 1977 are included in Tables 4 and 5,
respectively. Data from samples collected July 3 and 4, 1977 are
in Table 6.

Decreasing ammonia and nitrate levels (Table 5), with an
accompanying increase in o-phosphate show that the bloom was nitrogen-limited. Data from Tables 4 and 6 clearly show the difference
between the Grayling Arm and the rest of the reservoir in cation

Table 3. Analytic Data from Hebgen Reservoir for July 30, 1975 (NES).

Depth (m)	Secchi Depth (m)	NH_3-N Total (ppm)	Total Kjel. Nitrogen (ppm)	NO_2-NO_3 N Total (ppm)	Soluble O-Phos. (ppm)	Total Phos. (ppm)	Chlorophyll (µg/l)
Station 1							
0	4.0	0.03	0.2	< 0.02	0.022	0.024	
1.5		0.03	0.3	< 0.02	0.028	0.019	
8		0.02	0.2	< 0.02	0.021	0.014	2.0
13		0.04	0.2	< 0.02	0.031	0.018	
Station 2							
0	4.3	0.02	0.2	< 0.02	0.019	0.016	
1.5		0.02	0.2	< 0.02	0.021	0.017	
7		0.02	0.2	< 0.02	0.013	0.015	3.5
13		0.05	0.3	< 0.02	0.017	0.019	
Station 3							
0	—	0.03	0.5	< 0.02	0.008	0.037	
5		0.04	0.4	< 0.02	0.011	0.011	4.4
Station 4							
0	—	0.03	0.3	< 0.02	0.031	0.019	
5.5		0.04	0.3	< 0.02	0.027	0.024	3.2

Table 4. Montana Water Quality Laboratory Results From June 23, 1976 for Surface Samples From Three Sites[a]

Site	NH_3-N Total	NO_2-NO_3 N Total	Total Kjel. Nitrogen	Soluble O-Phos.	Total Phos.	Ca	Mg	Na	Cl	SO_4
1	< 0.01	< 0.01	0.08	0.006	0.019	11.2	1.5	5.0	1.9	2.5
2	< 0.01	< 0.01	0.05	0.008	0.014	10.5	2.4	19.0	9.8	9.5
4	< 0.01	0.01	0.24	0.002	0.018	6.8	0.0	48.0	22.	11.

[a]See Figure 1 for site locations. All values are in ppm.

Table 5. Montana Water Quality Laboratory Results for 1977 for Surface Samples From Site 2[a]

Date	NH_3-N Total	NO_2-NO_3 N Total	Total Kjel. Nitrogen	Soluble O-Phos.	Total Phos.
7/05	--	0.02	0.53	0.030	0.068
7/11	0.02	< 0.01	0.67	0.020	0.058
7/14	0.02	0.03	1.3	0.026	0.050
7/18	0.02	< 0.01	0.67	0.025	0.044
7/21	< 0.01	< 0.01	0.62	0.033	0.064
7/25	< 0.01	< 0.01	0.61	0.049	0.058

[a]All values are in ppm.

Table 6. Results from July 3, 4, 1977 Samples Collected by Juday and Keller. Units are ppm Unless Stated Otherwise

Site	Depth (m)	Secchi Depth (m)	T°C	DO	pH	Color[a]	Humic Acid (est)[b]	NO2-NO3 N total	Soluble O-Phos.	Total Phos.	Chl-a ug/l	Ca	Mg	Na	K
1	0	0.8	17	10.6	11.0	3.95	3.8	0.015	0.013	0.042	66.1	9.9	3.6	24.6	3.1
1	2		17	10.6		3.95	3.7	0.015	0.016	0.044	89.7	9.9	3.6	24.6	2.8
1	4		17	10.2		4.26	3.5	0.015	0.021	0.046	59.8	10.4	3.6	24.6	3.0
2	0	0.7	18	9.7	11.0	4.30	3.4	0.012	0.014	0.036	116.5	9.5	3.5	26.1	3.1
2	3		17	9.5		2.50	1.8	0.011	0.027	0.020	97.6	4.0	1.9	54.9	5.8
2	6		16	8.6	9.8	2.45	1.7	0.010	0.029	0.017	7.4	4.0	1.8	55.0	5.9
3	0	8.0	16	8.7	7.2	2.79	1.7	0.011	0.022	0.016	5.4	3.9	1.8	55.4	5.8
3	5		16	8.9		2.79	1.7	0.011	0.019	0.014	7.0	3.8	1.8	55.4	5.8
3	10		13	8.1		2.93	2.0	---	0.026	0.018	3.5	3.8	1.8	55.2	5.6
3	13		10	8.0		2.93	2.0	---	0.043	0.018	---	4.1	1.9	54.9	5.8
Duck Cr.					7.3	13.88	6.4	0.017	0.015	0.024		7.8	2.8	6.3	1.8
Cougar Cr.					7.5	2.78	1.8	0.015	0.012	0.029		7.7	1.9	4.9	3.7
Grayling Cr.					7.5	2.78	2.6	0.010	0.006	0.012		14.2	3.9	2.1	1.1
Madison R.					8.1	4.25	2.0	0.012	0.019	0.010		1.2	0.3	97.9	9.2
Quake Lake					7.4	2.21	1.9	0.009	0.009	0.021		8.2	3.4	32.3	3.7

[a] Color determined by method of Juday and Keller, 1980.
[b] Estimated from fluorescence values (Juday and Keller, 1980; Smart et al., 1976).

and anion content. These differences are also reflected in the
Grayling Arm inlets compared with the Madison River which dominates
the rest of the reservoir. For that reason it seems inappropriate
to try to assign a single trophic state to the reservoir. We clas-
sify the main reservoir mesotrophic, and the Grayling Arm eutrophic.
It is also clear that the main reservoir could not support the
Anabaena bloom. The material floated out of the Grayling Arm for
about 1 km in the main lake and then suddenly disappeared. The
data in Table 6 on the channel shows that while water from the
main lake was penetrating into the Grayling Arm, the algae did not
thrive in that water. The principal species of algae noted in the
main lake was *Aphanizomenon flos-aquae*. It was interesting to note
that there was no pheophytin present in extracts from the main
lake. In all of our other samples pheophytin levels amounted to
5 to 10% of the chlorophyll values.

A comparison of the 1975-1976 data with that of 1977 does not
indicate any decisive differences in nutrient levels except for
the elevated Kjeldahl nitrogen and total phosphorus levels in the
Grayling Arm. The differences in Secchi depths between Tables 3
and 6 are due to the fact that NES sampling was done from a heli-
copter and that the rotors are kept turning, while ours were made
in a boat, a more optimum condition. This was clearly shown in
the Seeley Lake study where NES reported values of 4.0 to 4.4 m
at a time when we were obtaining values of over 7.0 m.

The principal climatological difference between 1977 and previous years was the below normal precipitation of 1977. This
actually resulted in reduced productivity in the Clearwater Lakes
near Missoula (Juday and Keller, 1980), and the main part of the
reservoir was not highly productive. Perhaps the principal cause
of the bloom was the early filling of the reservoir in February
instead of April because of the anticipated low water. This
allowed three extra months for the bottom sediments of Grayling
Arm to be extracted for nutrients. Since cattle are allowed free
access to the Grayling Arm, this could also contribute to the
nutrient load. Early filling also allowed the water to warm more
rapidly and may also have shifted the cation-anion composition
further away from the Madison River levels.

In 1978 a light bloom of *Anabaena flos-aquae* occurred in the
Grayling Arm. Samples sent to Dr. Wayne Carmichael, Wright State
University, Dayton, Ohio were toxic to mice. Clonal isolates made
from the sample were toxic and produced a toxin like anatoxin-a
(Carmichael, personal communication).

REFERENCES

Juday, R. E., and E. J. Keller. 1980. A Water Quality Study of
      Placid Lake and its Drainages. Office of Water Research and
      Technology (OWRT).
Martin, D. B. 1967. Limnological Studies on Hebgen Lake, Montana.
      Ph.D. Thesis. University of Montana. Helena.
Smart, P. L., B. L. Finlayson, W. D. Rylands, and C. M. Ball.
      1976. The relation of fluorescence to dissolved organic
      carbon in surface waters. Water Res. *10*:805-812.
U. S. Environmental Protection Agency, Corvallis, Oregon Labora-
      tory. 1977. National Eutrophication Survey (NES) Working
      Paper 794 for Hebgen Reservoir. n.p.

MORPHOLOGY OF TOXIC VERSUS NON-TOXIC STRAINS

OF *APHANIZOMENON FLOS-AQUAE*

Marilyn M. Ecker, Thomas L. Foxall,
and John J. Sasner, Jr.

Department of Zoology
Spaulding Life Science Building
University of New Hampshire
Durham, New Hampshire 03824

ABSTRACT

In an attempt to distinguish toxic from non-toxic forms of the blue-green alga *Aphanizomenon flos-aquae*, observations were made of several morphological features characteristic of these cells. Both field collections and laboratory cultures were examined by light microscopy and scanning electron microscopy.

The following parameters were examined: sheath diameter, cell size, bacterial associations, and heterocyst morphology. All of these features were variable and were influenced by environmental conditions. Manipulation of culture conditions directly influenced cell morphology. Toxicity was determined by the standard mouse bioassay method. Culturing techniques induced an increase in the bacterial population associated with the nitrogen-fixing cell, the heterocyst.

Cellular features, including associated bacteria, have been suspected of being related to toxicity. Several of these features may be utilized for predicting toxicity, but information from a broader sampling base must be gathered before definitive conclusions can be made.

INTRODUCTION

The blue-green algae, *Aphanizomenon flos-aquae*, *Anabaena flos-aquae*, and *Microcystis aeruginosa* (Figures 1, 2, and 3) bloom

113

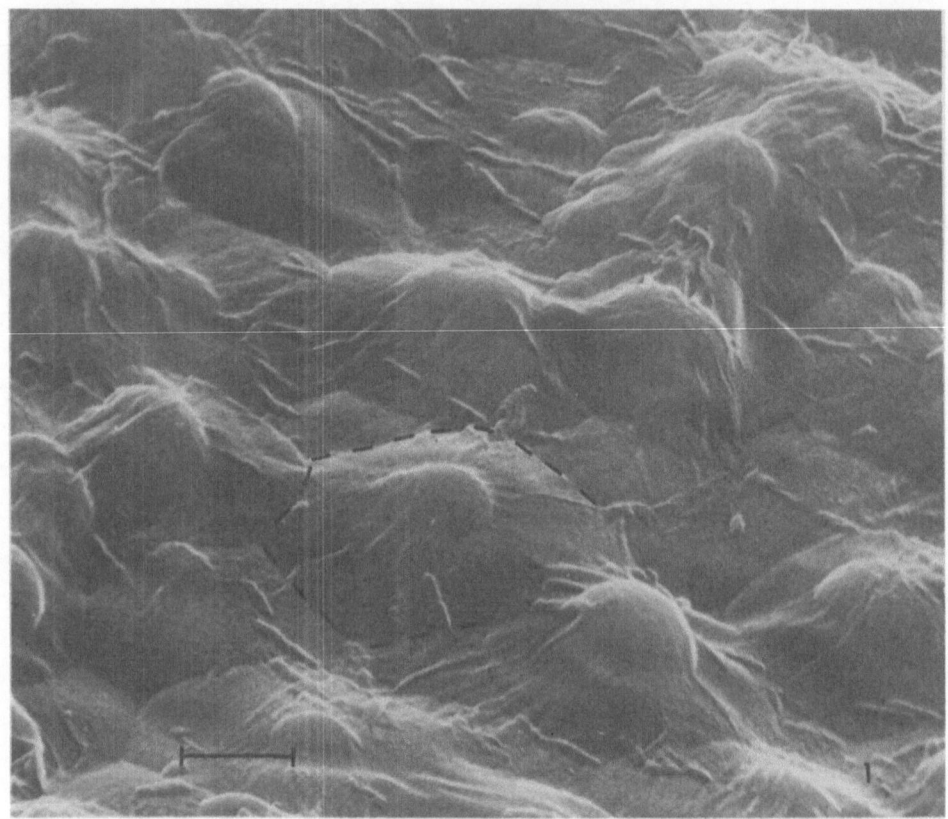

Fig. 1. *Microcystis aeruginosa* collected during bloom conditions.
Microflora appear to be associated with the sheath rather
than with the individual cells. Note consistency in
amount of sheath produced by each cell. This bloom was
non-toxic. Marker = 5 μm.

intermittently in several of New Hampshire's lakes and ponds.
These species occur in both toxic and non-toxic forms. While
sequentially occurring algal and zooplankton fluctuations are
expected in natural eutrophic systems, the proliferation of blue-
green algae may present significant ecological and economic consequences.

The ability of blue-green algae to rise in the water using
gas vacuoles, may selectively enhance their growth to the exclusion of other forms. When dense blooms occur, the oxygen and
light levels vary with depth in the water column. When the bloom
is toxic, fish kills, livestock and aquatic invertebrate mortalities, and water that is unsafe for human consumption may result.

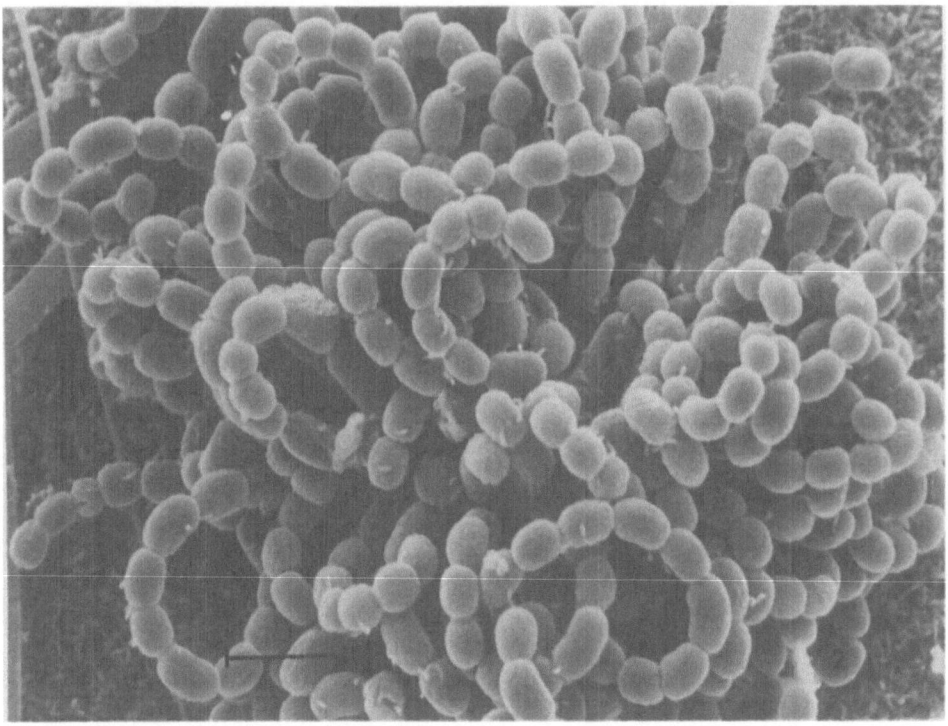

Fig. 2. *Anabaena circinalis* collected in plankton tow. Bacteria
         appear randomly on both vegetative cells and heterocysts.
         Sheath is detectable only at cellular junctions. Marker
         = 5 µm.

The primary treatment of blooms is by the application of an
algicide, usually copper sulfate ($CuSO_4$). This treatment results
in lysis of the algal cells and subsequent release of their cellular products, including endotoxins, into the water column. It is
important that water quality management officials know the nature
of the bloom before treatment.

Factors implicated in the variability of toxicity of blooms
include species differences, strains or subspecies, bacteria,
growth conditions, age of bloom and relative proportions of toxic
and non-toxic filaments within a sample (Carmichael and Gorham,
1977; Gentile and Maloney, 1969; and Vance, 1966). At the present
time, positive identification of a toxin-producing alga requires
time-consuming laboratory work in order to distinguish toxic from
non-toxic strains. The standard mouse bioassay for monitoring
saxitoxin levels during marine "red tide" outbreaks and, more

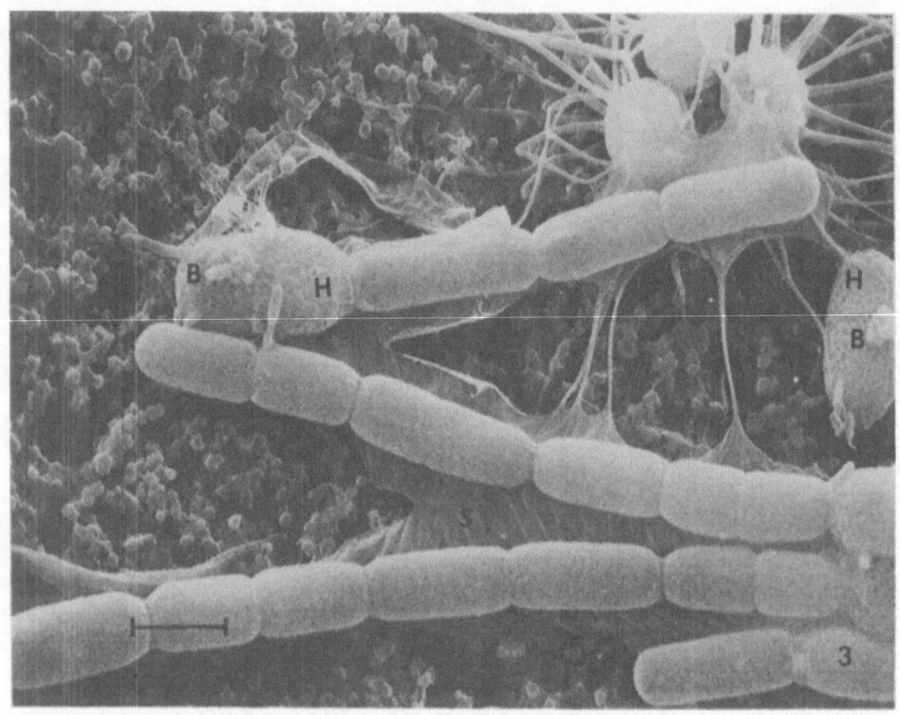

Fig. 3. *Aphanizomenon flos-aquae* from Kezar Lake, North Sutton,
         New Hampshire. Sheath is thin, firm and not in evidence
         at the heterocyst. This sample was positive for toxin,
         using the mouse bioassay. Heterocyst has irregular outer
         cell wall and associated microflora. Marker = 5 μm.

recently, a fluorometric assay are used (Shoptaugh, 1978). Both
of these methods are time-consuming, costly, require highly trained
personnel and extensive sampling.

     The object of this study was to closely examine the cellular
morphology of *Aphanizomenon* samples with the aim of differentiating
toxic from non-toxic strains.

MATERIALS AND METHODS

     Field collections were obtained from Kezar Lake, North Sutton,
New Hampshire, in the summer of 1977 and spring of 1978. All samples were monitored for toxicity using the mouse bioassay. After
cells were lysed by freezing and thawing, 1 ml of the extract was
added to 2 mls 0.1 N HCl and heated 5 minutes in a 100°C water
bath. Volume was then adjusted to 3 mls with water, H^+ concentra-

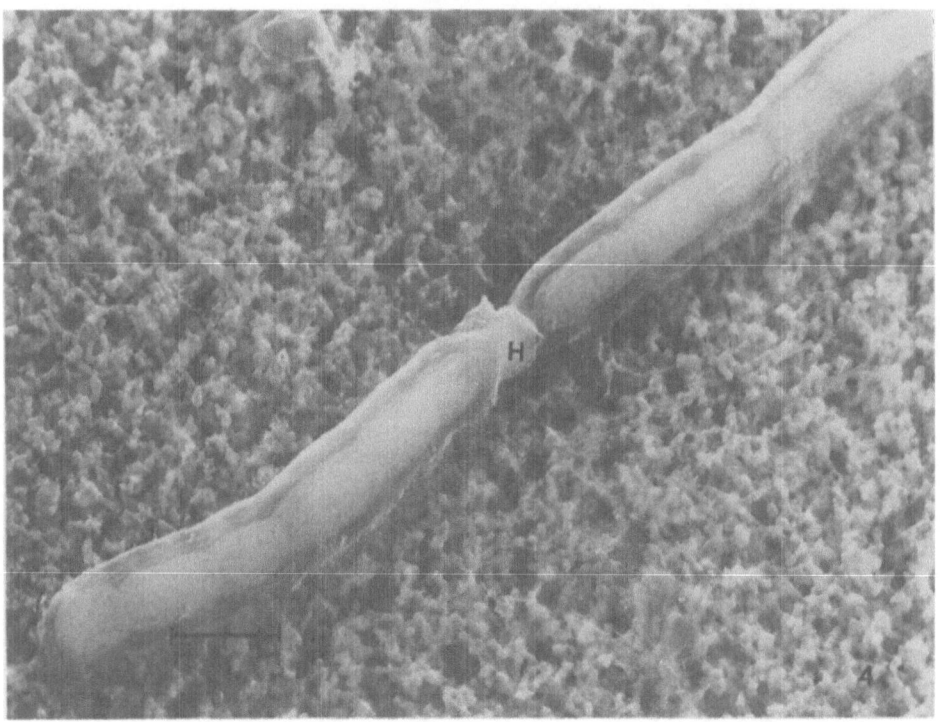

Fig. 4. *Aphanizomenon* filament from a freshly collected sample.
Filament has a moderate amount of firm sheath encapsu-
lating all cells. Amount of sheath varies at the heterocyst and no bacteria are seen. This sample was non-toxic.
Marker = 5 μm.

tion adjusted to pH 4.0 using either 0.1 N NaOH or 0.1 N HCl and
0.5 ml of the supernatant was injected (i.p.) into 20 g mice.

Wet mounts of all samples were examined with the light microscope. India ink and methylene blue were used for viewing sheath
diameters.

Representative samples were prepared for scanning electron
microscopy. Cells were collected on 0.45 μm millipore filters,
fixed in 1% E.M. grade glutaraldhyde, post-fixed in 1% OsO_4 and
dehydrated through an ethanol series. The preparations were criti-
cal-point-dried with CO_2, coated with gold/palladium and viewed in
an AMR1000A scanning electron microscope.

Laboratory cultures of toxic and non-toxic strains from Kezar
Lake, New Hampshire, were maintained in 20 liter carboys using

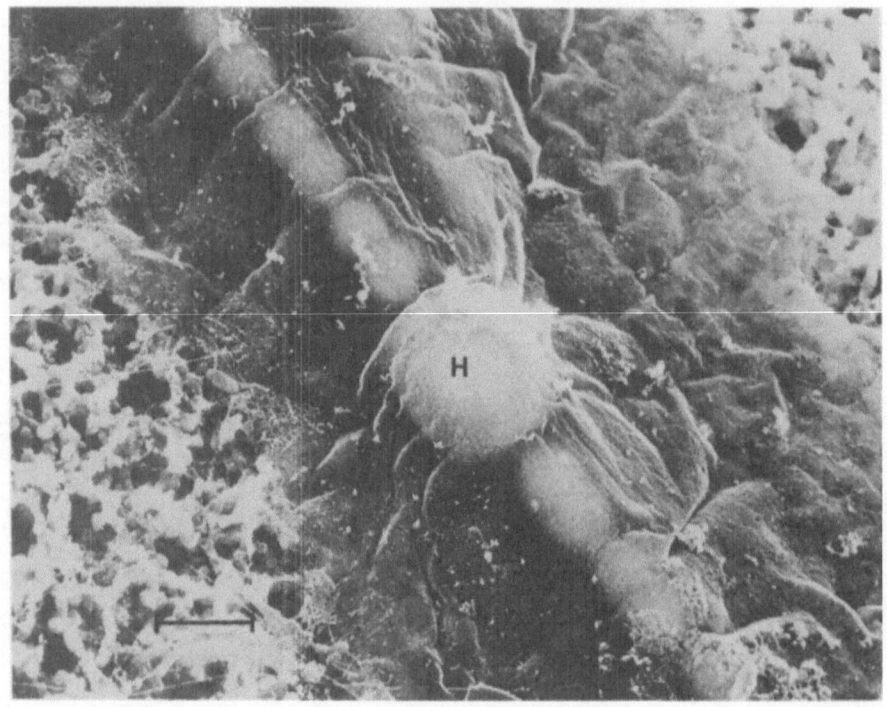

Fig. 5. *Aphanizomenon* from a culture grown in a salts medium
        under nitrogen gas. Filaments produced heavy, watery
        sheaths. This culture method produced many heterocysts
        (H) per filament, some juxtaposed. No bacteria are
        evident and the culture was non-toxic. Marker = 5 μm.

ASM-1 medium (Carmichael and Gorham, 1974), with aeration and constant light.

RESULTS

     Preliminary observations of *Aphanizomenon* filaments, in wet-
mount preparations, revealed three varying features: (a) the
amount of sheath material surrounding the trichome (Figures 3, 4,
5, and 9), (b) the number and position of heterocysts, and (c)
the microflora associated with the heterocyst. The extent of the
vacuolar system provided another variable among different samples,
some filaments appearing more homogeneous or dense than others.
These morphological parameters varied between toxic and non-toxic
samples and the question was raised as to the utility of these
features as indicators of toxicity. Would any one of these features hold as a reliable indicator? In order to be proven valid,

Table 1.   Salts Medium + Nitrogen[a]

To 1.0 liter of sterile distilled H_2O add:

$MgSO_4$. $7H_2O$ ------------- 0.2 g

K_2HPO_4 -------------------- 1.0 g

$FeSO_4$. $7H_2O$ ------------- 0.01 g

$CaCl_2$. $2H_2O$ ------------- 0.02 g

$MnCl_2$. $4H_2O$ ------------- 0.002 g

Na_2MoO_4 . $2H_2O$ ----------- 0.001 g

NaCl --------------------- 0.5 g

$NaHCO_3$ -------------------- 0.1 g

Trace Mineral Mix.* ------ 0.4 ml/liter

N_2 Gas ----------- Bubble for 5 minutes

*To 250 mls distilled H_2O add:

$CuSO_4$. $5H_2O$ ---- 2.5 mg
$ZnSO_4$. $7H_2O$ ---- 0.25 mg
$CoCl_2$. $6H_2O$ ---- 2.5 mg

---

[a]Courtesy of Prof. D. M. Green, Department of Biochemistry, University of New Hampshire.

a feature should be reproducible in laboratory cultures and exhibit a direct correlation with a fluctuation in toxicity.

Simple wet-mount staining techniques with india ink and methylene blue were effective for distinguishing sheath material. Prolonged examination in the presence of india ink caused fracture of the filaments. The fractures occurred at cellular junctions, leaving sheath material intact on the lateral portions of the cells. This could be an indication of a weakness or differential composition of sheath at cellular junctions. Initial observations revealed a sheath width of two cell diameters in a non-toxic sample.

Cells grown in an increased concentration of phosphate (2x) exhibited no visual differences from those in the standard ASM-1 medium.

Cells cultured in a salts medium (Table 1) under nitrogen gas, produced very extensive sheaths (Figure 5). The sheath was uniform in diameter around the filament regardless of the underlying cell type. The sheath was more extensive than any seen in field collec-

tions. The culture was not toxic. The volume of extracellular
polysaccharide produced has been attributed to age of culture, temperature and form of nitrogen available (Sanger and Dugan, 1972).
While the age and temperature of the culture were comparable with
control cultures, the medium contained no nitrogen source other
than the gas. Increased numbers of heterocysts resulted from this
technique, however, and individual cells were smaller than those
in the controls.

    Bacterial associations were easily observed at the light
microscope level. Toxic cultures displayed heterocysts that were
heavily populated with rod-shaped bacteria (Figure 6). Generally,
vegetative cells of toxic samples did not exhibit bacterial populations.

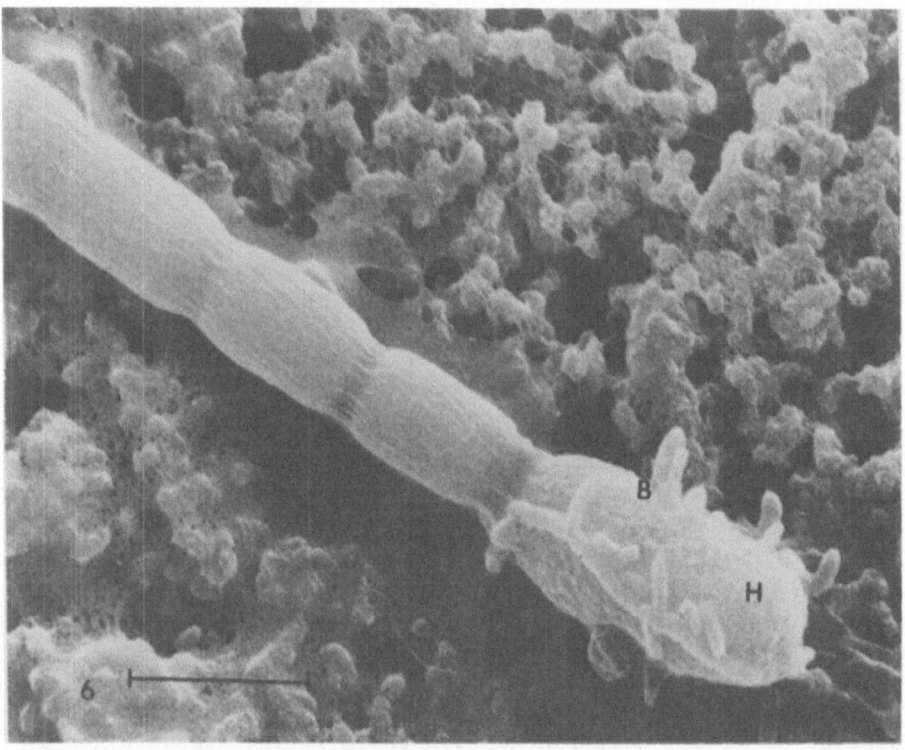

Fig. 6. *Aphanizomenon* filament from a freshly collected toxic
         sample. Outer cell wall of heterocyst (H) is textured
         and colonized by Gram-negative rod bacteria (B). Marker
         = 5 μm.

In non-toxic samples, large ciliate protozoans were commonly seen grazing on the surface of the blue-green algal filaments. This observation may suggest that the ciliates are feeding on epiphytic bacteria, thereby keeping the filament relatively free of microbes. If this is the case, high densities of bacteria would result during bloom conditions when the ciliate population cannot keep pace with the blue-green algal growth rate and the grazing pressure is reduced. We did not observe ciliates in toxic cultures, and no attempt was made to investigate their response to the toxin. Zooplankters were not seen feeding on *Aphanizomenon* filaments, however, some non-toxic samples were heavily populated with amoebae that ingested whole filaments.

In an effort to relate bacterial infestations and toxicity, toxic and non-toxic strains were cultured in ASM-1 medium without nitrate. The reduction in available nitrate-nitrogen led to an increase in the number of heterocysts as well as an increase in the number and types of bacteria associated with them (Figure 7). While no attempt was made to establish an LD_{100} for this sample, a

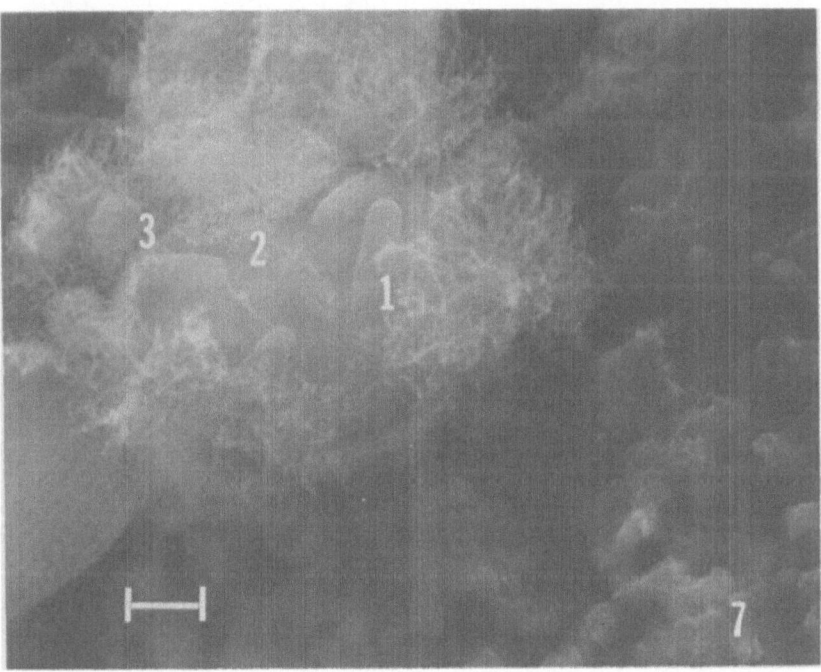

Fig. 7. *Aphanizomenon* cultured in ASM-1 medium without nitrates. This technique gave increased numbers of heterocysts as well as increased numbers and types of microflora (1, 2, 3) associated with them. Marker = 1 μm.

30 g mouse injected (i.p.) with 0.5 ml of extract died of respiratory failure in less than 1.25 minutes. The mouse displayed the
classic symptoms associated with the very fast death factor (VFDF)
(Gentile and Maloney, 1969; Gorham and Carmichael, 1980). These
symptoms are: convulsions, gasping respiration, death by respiratory arrest within 5 to 10 minutes.

     Scanning E.M. preparations of these samples revealed differ-
ences in the bacterial populations. Heterocysts from the ASM-1
cultures were colonized by Gram-negative rods, whereas heterocysts
from the ASM-1 minus nitrate supported rods as well as coccoid
forms. Both of these bacterial forms appeared to be surrounded by
a material similar in appearance to the algal sheath polysaccharide.

     Cell-fracture preparations were made to determine if the
microbial population was located on the surface or embedded within
the primary cell wall. Samples were fractured with sticky tape
after critical-point drying in the previously mentioned S.E.M.

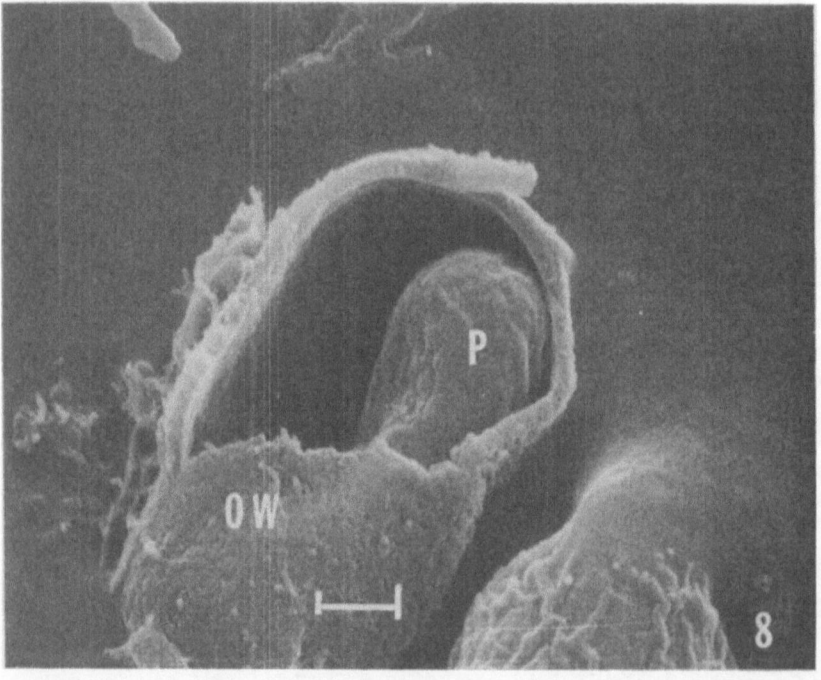

Fig. 8. Fractured heterocyst from toxic sample. Samples were
         readily fractured revealing an intact protoplast (P),
         whereas non-toxic samples were difficult to fracture and
         the protoplast was destroyed in the process. Marker = 1 μm.

preparation. The heterocyst from a toxic sample had a tendency to fracture through the thick outer wall leaving the protoplast intact (Figure 8). These cells were easily fractured and the bacteria were located only on the exterior of the outer wall. Non-toxic samples were very difficult to fracture and the percent yield of torn cells was small. When heterocysts did fracture, the proto-plast as well as the cell wall was torn. Mature heterocysts devel-oped a thick outer wall with a fibrillar appearance which may account for the intercellular space and the ease with which they fractured (Figure 10).

DISCUSSION

No one morphological feature proved to be positively corre-lated with toxicity. The limited number of preparations and the inability to identify toxicity of single filaments not-withstanding, some speculations are possible. The toxic cells appeared more vesiculated, more yellow-green and the sheath less than a cell diameter in width or not discernible at all with the light micro-

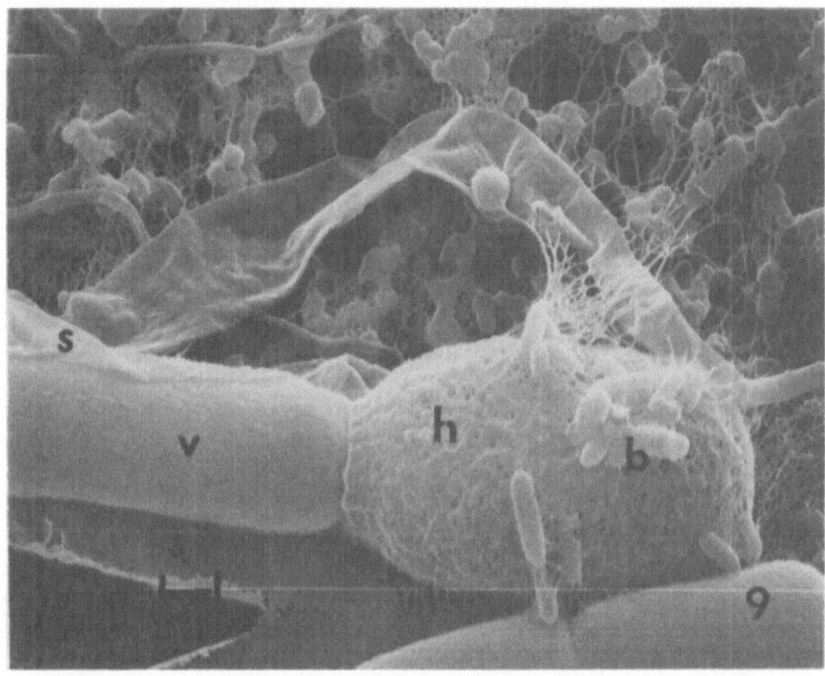

Fig. 9. Higher magnification of heterocyst (H) in Figure 3 showing sheath (S), vegetative cell (V) and bacteria (B). Marker = 1 μm.

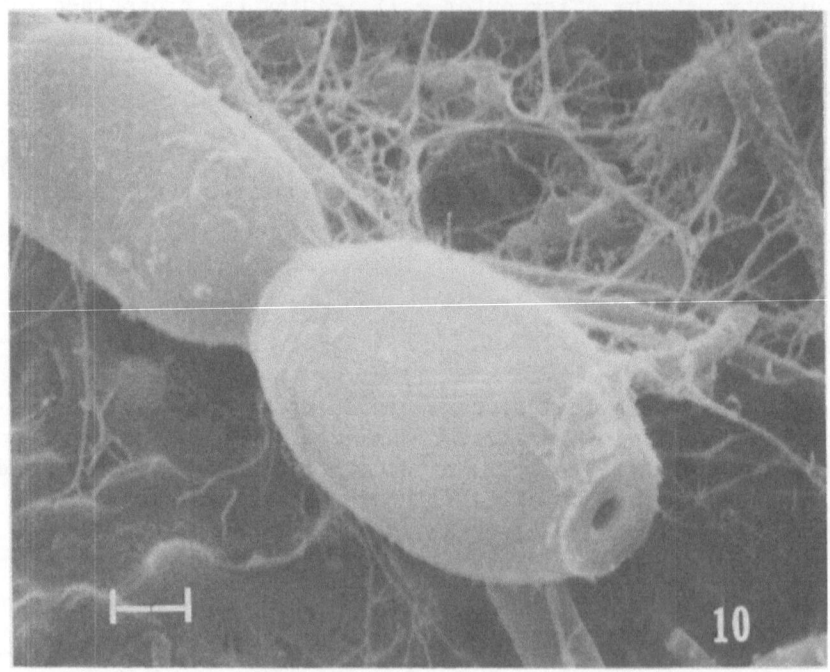

Fig. 10. Heterocyst from non-toxic field sample of *Aphanizomenon*.
Note smooth appearance of outer cell wall and the site
of the polar nodule. Marker = 1 μm.

scope. Filaments with high bacterial infestations on and around
the heterocyst were often toxic. Non-toxic cells appeared to have
fewer gas vacuoles and a more homogeneous color. The sheath
appeared uniform around vegetative cells and reduced or absent at
the heterocyst. Sheath width was equal to or larger than the diameter of a vegetative cell in non-toxic samples. Bacteria, when
present, were randomly distributed along the filament. Active
grazing by ciliates or predation by other protozoans may be indicative of a non-toxic form.

Observations made in this study; amount of sheath, heterocyst
cell wall appearance, bacterial associations and increased numbers
of gas vacuoles, can all be attributed to the age of the cells.

Sheath production is a major end product of photosynthesis
(Tischer and Davis, 1971) and cell metabolism. Actively growing
and metabolizing cells have a moderate to heavy sheath and in this
study were non-toxic.

Pearl (1976; Pearl and Kellar, 1978) has shown that the bacteria associated with blue-green algae assimilate amino acids and sugars as well as increase the efficiency of N_2-fixation. The mature heterocyst wall has been reported to contain more than 70% glucose as opposed to 35% in the vegetative cell (Dunn and Wolk, 1970). These two factors combined with the N_2-fixing ability of this cell may account for the concentration of bacteria. It is entirely conceivable that some blue-green algae and some bacteria are obligate symbionts. *Aphanizomenon* in the rafted form, collected from New England, has not been reported to be toxic. However, *Aphanizomenon* in the rafted form has been reported toxic from Oregon (Phinney and Peek, 1961) and other places (Gorham and Carmichael, 1980). The toxin from these blooms, however, was of the peptide type and not neurotoxic like the Kezar Lake strains.

Gentile and Maloney (1969) reported that toxicity increased as cultures became older and more dense. Our study supported their observations. In nature, algal blooms are very dense (> 10^6 cells/ml) and the morphological differences observed are expected as cells age and the bloom begins to die off. If decisions relative to the use of algicides in controlling noxious algal blooms must be made, it is important to find an indicator of toxicity that can be employed in the field.

ACKNOWLEDGMENTS

This research was supported through the New Hampshire Water Resources Research Center of the University of New Hampshire by grant #AO-47 NH from the Office of Water Research and Technology, United States Department of the Interior, as authorized under the Water Research and Development Act of 1978, Public Law 95-467.

REFERENCES

Carmichael, W. W., and P. R. Gorham. 1974. An improved method for obtaining axenic clones of planktonic blue-green algae. J. Phycol. *10*:238-240.

Carmichael, W. W., and P. R. Gorham. 1977. Factors influencing the toxicity and animal susceptibility of *Anabaena flos-aquae* (Cyanophyta) blooms. J. Phycol. *13*(2):97-101.

Dunn, J. H., and C. P. Wolk. 1970. Composition of the cellular envelopes of *Anabaena cylindrica*. J. Bacteriol. *103*:153-158.

Gentile, J. H., and T. E. Maloney. 1969. Toxicity and environmental requirements of a strain of *Aphanizomenon flos-aquae* (L.) Ralfs. Can. J. Microbiol. *15*:165-173.

Gorham, P. R., and W. W. Carmichael. 1980. Toxic substances from freshwater algae. Prog. Water Technol. *12*:189-198.

Pearl, H. W. 1976. Specific associations of the blue-green algae
    Anabaena and *Aphanizomenon* with bacteria in freshwater
    blooms. J. Phycol. *12*:431–435.

Pearl, H. W., and P. E. Kellar. 1978. Significance of bacterial-
    Anabaena (Cyanophyceae) associations with respect to N_2
    fixation in freshwater. J. Phycol. *14*:254–260.

Phinney, H. K., and C. A. Peek. 1961. Klamath Lake, an Instance
    of Natural Enrichment. Trans. Sem. on Algae and Metropolitan
    Wastes. April 27–29, 1960. Robert A. Taft Sanitary Engi-
    neering Center, Cincinnati, Ohio.

Sanger, V. K., and P. C. Dugan. 1972. Polysaccharide produced by
    Anacystis nidulans: its ecological implication. Appl.
    Microbiol. *24*:732–734.

Shoptaugh, N. H. 1978. Fluorometric Studies on the Toxins of
    Gonyaulax tamarensis and *Aphanizomenon flos-aquae*. Ph.D.
    Thesis. University of New Hampshire. Durham, New Hampshire.

Tischer, R. G., and E. B. Davis. 1971. The effect of various
    nitrogen sources upon the production of extracellular poly-
    saccharide by *Anabaena*. J. Exp. Bot. *22*:546.

Vance, B. D. 1966. Sensitivity of *Microcystis aeruginosa* and
    other blue-green algae and associated bacteria to selected
    anitbiotics. J. Phycol. *2*:125–218.

THE OCCURRENCE OF TOXIC CYANOPHYTE BLOOMS IN AUSTRALIA

Valerie May

National Herbarium of New South Wales
Royal Botanic Gardens
Sydney, New South Wales, 2000
Australia

ABSTRACT

Australian literature records stock losses due to excessive growths in water of the cyanophyte algae *Anacystis cyanea* (Küetz.) Dr. and Dail. (= *Microcystis aeruginosa* Küetz.), *Anabaina circinalis* Rabenh. and *Nodularia spumigena* Mert. (now *Nostoc spumigena* (Mert.) Drouet). These algal species were recorded in Australia at least sixty years ago, but their increasing occurrence, wide distribution and the resultant damage appear to be due to increasing pollution of the water.

Factors affecting the toxicity of these blue-green algae are discussed. Animals affected adversely include horses, cows, sheep, dogs, pigs, fowls, turkeys, laboratory guinea pigs, mice and probably various wild animals, including birds and fish.

Particular attention is given to *Anabaina circinalis*, its nomenclature, ease of distribution and factors affecting its establishment and growth. It appears that a series of physiological abilities in the group of blue-green algae allows these plants to take advantage of conditions prevailing under eutrophic or drought conditions, and so to outgrow green algae.

Control of the concentration of phosphorus appears to be the most practical method of controlling the occurrence of these blooms. My work indicated that blooms only developed when phosphorus levels reached a concentration of 0.5 ppm, and that at these times there was a high bacterial count which no doubt aided the freeing of phosphorus from the previously-enriched sediment.

Treatment of farm dams with alum and/or block ferric alum
(Alumina ferric R), applied before the summer rise in phosphorus
levels, limited this rise and reduced or inhibited cyanophyte bloom
formation.

TEXT

Certain blue-green algae, if growing sufficiently densely so
that they form a bloom in water, are known to be toxic to various
animals, possibly including man. Such toxic water blooms have been
reported in many parts of the world, including Australia. The
problem is of particular importance in my country since, generally,
we have limited water supplies. We have had these troubles in our
larger reservoirs and in very slow moving rivers, but most stock
losses have been caused by algal blooms occurring in small farm
dams. The first knowledge of the trouble is usually the death of
a few to hundreds of farm animals.

The toxic bloom occurs more densely in the surface layer of
the water, which develops a dense green paint-like appearance and
often produces a pungent smell reminiscent of the gamma isomer of
benzene hexachloride (BHC). The toxins are held within the live
cell and may be freed artificially by the use of ultrasonic vibra-
tions. Animals known to be affected are horses, cows, sheep,
fowls, turkeys, laboratory guinea pigs and mice. Probably various
wild animals, including birds and fish, are also susceptible.
Death is linked with liver damage and hemorrhage and can be quite
quick, in some cases occurring in minutes. Laboratory tests have
been usually carried out using intraperitoneal injections in test
animals. The only known treatment for affected animals is to
remove them from the source of infection and allow them to rest
comfortably. Little is known of the effect of repeated low dosage
or of any chronic damage caused; one of the few studies on this is
by McBarron et al. (1975), who treated some white Leghorn chickens
with *Anabaina circinalis* and found that, although no clinical
signs occurred with the given dosage, liver changes were pronounced,
with an overall picture suggesting acute degenerative change.

There is considerable circumstantial evidence suggesting that
the cyanophyte algae can be toxic also to man, but it is hard to
get firm evidence. Who volunteers for trials? I think a rather
neat proof of a spectacular result of toxicity to man may lie in
the following: There is a thriving market for dried *Spirulina*
(excellent as food), gathered by local people from a particular
lake in the Lake Chad area, Africa, but the material growing in a
nearby lake, which often contains a nearly pure growth of the same
species, is never used, and the reason appears to be that sometimes
a co-dominant cyanophyte occurs in that lake, and this is poten-

tially toxic (Léonard and Compère, 1967). Local memory could well
be long!

    The algae now known to cause these troubles in Australia are
the blue-green (cyanophyte) algae, *Anacystis cyanea* (Küetz.) Dr.
and Dail. (also known as *Microcystis aeruginosa* Küetz.), *Anabaina
circinalis* Rabenh. and, to a much lesser extent, *Nostoc spumigena*
(Mert.) Drouet (formerly *Nodularia spumigena* Mert.). Blooms of
these genera usually occur under somewhat similar conditions, often
in almost pure cultures, although *Anacystis* and *Anabaina* are frequently associated and can change relative proportions during a
single bloom (May, 1978). All these genera are widely distributed;
for instance I have records of *Anacystis* from all mainland states
except the Northern Territory, and the absence from there is possibly due to the absence of interested collectors! We have records
of the occurrence of these algae extending over sixty years, so the
increasing troubles they now cause appear to be due to changing
conditions, rather than to the introduction of new pests. Some
other algal genera, *Euglena* and *Phacus* in particular, we also sus-
pect of being toxic and work is continuing to test these genera.
There have been a number of stock deaths associated with dense
growths of these algae, but so far no confirmatory laboratory
results; it is possible that a very short-lived toxin is responsible. *Aphanizomenon*, recorded by Drouet (1973, p. 141) as an
ecophene of *Calothrix parietina* (Nageli) Thuret, is toxic in many
parts of the world, but so far has not been found in sufficient
density to cause problems in Australia; we do not know why.

    One of the problems in studying blooms of toxic algae has
been the variability and rapid fluctuation of both the bloom and
its toxicity. Animals die after drinking from a certain dam, but
sometimes others drink from it soon after with impunity. Part of
this reported variability is due to wind shifting the dense surface
blooms of these algae. If fences, for instance, limit the area of
a dam to which animals have access, a change in wind can rapidly
change the effective toxicity of the water available. Further,
even with a dense bloom blown into the bay from which animals
obtain their water, the depth and slope of the dam floor may affect
how far out beyond the wind-blown debris the animals can wade seek-
ing clear water. I have known sheep near shore to be poisoned and
killed, while deeper-wading cattle have survived satisfactorily.
On the other hand, I have known turkeys to peck at green blobs of
Anacystis in their water supply and die soon after.

    Serial studies of a stationary bloom, however, have demon-
strated changes in toxicity paralleling changes in growth stage
of the alga (at least in both *Anacystis* and *Anabaina*), and a bloom
can develop or disintegrate very quickly. When copper sulphate,
still regarded as the best and cheapest algicide, is used to rid

a dam of its bloom, the algal nuisance can, on occasion, disappear
and then regenerate within a week! Further, those conditions which
give maximum growth of the algal bloom are not necessarily those
which give it maximum toxin production, so that a bloom of a certain species can be called only "potentially" toxic until it has
been tested (Harris and Gorham, 1956, cited by Gorham, 1964).

Work in America has indicated, too, the variable genetic make-
up of blooms, and that both toxic and non-toxic strains can occur
in a single bloom (Gorham, 1964). Closely associated bacteria and
their toxins (Gorham, 1964) or toxin-depressants (Carmichael and
Gorham, 1977) further complicate the picture. It is only recently
that it has been possible to obtain reasonably constant toxicity
from a culture (Carmichael and Gorham, 1978). This has occurred
with a culture of *Anabaina flos-aquae* obtained from a single filament grown under standard conditions, free of all bacteria.

The susceptibility of different animals, the volume of infected
water ingested, and even the time spread in which a certain volume
of infected water is drunk may also all affect the degree of poisoning suffered.

In Australia, as elsewhere, most of these blooms develop during
late summer to autumn, with residual patches sometimes remaining,
or even flaring up again. Thus it appears that light and/or temperature are involved in the new season's growth. Also these algae
occur mostly in stationary or very slow moving water which is polluted by stock or drainage, so that the nutrient level is high.
In fact, increased pollution seems to be the cause of the increasing occurrence of these unwanted algal blooms.

Culture studies are very useful in determining the effects of
changing environmental factors on the growth of an alga, but I
think it is prudent always to relate any results from such studies
to field observations. It should, however, be noted that large
bodies of water show more rhythmic changes in their algal populations, while small bodies of water seem to show more evidence of
chance occurrences.

Of our two common toxic algae, it is more difficult to determine those field factors which affect the growth of *Anacystis
cyanea*, since this species can overwinter on the floor of dams
(Chernousova, 1968, cited by Whitton, 1973; May, 1973) and its
presence is not necessarily indicative of its activity.

Anabaina circinalis, on the other hand, is amenable to a
field study of its autecology, since it occurs either as growing
plants or as single cell reproductive structures (or spores).
Some notes on this species, including observations on its aute-

cology, especially as shown by field studies in Australia, are appended below.

Taxonomy of *A. circinalis*

Anabaina (Anabaena) circinalis Rabenhorst has recently (Drouet, 1978) been included as an ecophene of *Nostoc commune* Vauch. The even-more-cited name *Anabaina (Anabaena) flos-aquae* (L.) Bory has been excluded from the Nostocaceae by Drouet (1978). This taxon has previously been cited as *A. flos-aquae* (Lyngb.) de Bréb. by Desikachary (1959); it also was formerly listed among the synonyms of *Microcoleus vaginatus* (Vauch.) Gom. by Drouet (1968, p. 226).

I agree that there appears to be no vast discontinuity between *A. circinalis* and *Nostoc commune*; however I feel that the two forms, albeit probably only ecological expressions of the same super-species, should be considered separately. *A. circinalis* is easily recognized, is planktonic and widespread in distribution, differs in production of sheath material and may contain definite toxic chemicals so that it is a risk to stock, whereas the well-known, easily recognized and widely distributed *Nostoc commune* has a different habit and habitat, lacks any sign of toxicity and may safely be used as food for man or beast.

I, therefore, am in this paper still using the name which pervades the literature on toxic algae, viz. *A. circinalis*, but following Drouet in the spelling of *Anabaina*. I understand (Drouet, 1978, p. 83) this ecophene would include also what has been recorded in the literature of toxic algae as *A. flos-aquae*.

Distribution of *A. circinalis*

This species occurs widely distributed throughout the world and within my own country is apparently not restricted by geological or minor climatic considerations, or by the size of the water mass within which it occurs. *A. circinalis*, with its reproductive structures, can be recognized in drying mud, so that it appears that the species may be wind-, as well as water- and animal-distributed. As regards longevity, we have records of the regeneration of a *Nostoc*, a closely related taxon, after some 80 years of desiccation.

It seems that *A. circinalis* develops readily in any suitable new area of water. Ready dispersal of algae was well demonstrated by the interesting results of Macguire (1963, cited by Krebs, 1972), who showed that algae colonized bottles of artificial lake water up to 400 yards from small ponds within a few days. A further demonstration is that at a site in a river which I am studying

there was only an occasional occurrence of *A. circinalis* - consisting of three cases in 49 observations (each of 100 cc. of water)
over a period of two years. These were in November and December
1977 and in December 1978. However, a dam was built incorporating
this river site, so that water covered it in September 1979 and
thereafter the rate of flow of the water was altered. *A. circinalis*
developed with moderate frequency in February 1980, and occurred
in heavy concentration in March, in the first season of the dam's
construction during which the water depth of the whole dam only
ranged from 2 to 12 meters.

In a like manner, *Anacystis cyanea* was not recorded at this
site prior to the formation of the dam, but was recorded once in
October, was present in reasonable quantities in November, was
quite frequent in December and even more prolific in January. The
nearest source of these algae also showed a maximum of *Anacystis*
in January and of *Anabaina* in March.

This reported increase of blue-green algae in a newly constructed reservoir is apparently not exceptional. Mathew (1930)
describes an enormous undertaking to overcome excessive growth
of *Anabaina* in the just-completed and still filling Hume Dam (on
the New South Wales-Victoria border). The bloom must have been
quite extensive; approximately 22 tons of copper sulphate were
used to control it. This was a most unusual occurrence for its
time, and was no doubt due to the recency of construction, with
high soil disturbance and nutrient enrichment of the water. The
contrast between the algae of established water reservoirs of
some years ago and of the present time is shown better in comparing Yan Yean Reservoir, Victoria (West, 1909) with Burrinjuck Dam,
New South Wales (May, 1978). The former had a remarkable scarcity
of blue-green algae, while the latter has *A. cyanea* and/or *A.
circinalis* occurring frequently as unwanted summer blooms on surface water.

Effect of Light

Lund (1979) has stressed the effect of light in stimulating
the seasonal growth of *Asteroniella*. Ganf (1980) has demonstrated with in situ growth experiments in southern Australia that it
is light, rather than temperature, that limits winter growth of
a variety of genera. Possibly as a result of a light stimulus,
seasonal development of *A. circinalis* commences, and later blooms
of it are likely to occur over the summer to autumn period. *Anabaina* in the Australian dams studied can be found at a depth of
up to 50 meters, as well as in shallower waters.

Effect of Temperature

Although the growth of *A. circinalis* reaches a peak in summer, there is variation in the extent of the year in which the species may be found. In one reservoir studied (Carcoar Dam, Figure 1), the occurrence of this species was strictly seasonal, while in another reservoir some 280 km distant (Glenbawn Dam, Figure 2), studied in the same seasons (1978-1980), there was the same summer maximum in growth, but the species continued to occur throughout the year. The minimum temperature reached in these dams in each case (in the bottom water of the station nearest to the dam wall at a depth of approximately 30 m) was between 7° and 8°C. In Carcoar Dam this occurred in August, in Glenbawn Dam it occurred

Anabaina circinalis   OCTOBER 1977 TO MAY 1980   CARCOAR DAM   STATION 3

		OCT.	NOV.	DEC.	JAN.	FEB.	MAR.	APRIL	MAY	JUNE	JULY	AUG.	SEPT.
1977-78	TOP	−	−	−	−	□ □	■	COLLECTION MISSED	□ □	−	−	−	−
	MIDDLE	−	−	−	+	□ □	□		+ □	−	−	−	−
	BOTTOM	−	−	−	−	□ □	+		+ □	−	−	−	−
1978-79	TOP	−	−	−	■	✚	✚	□	−	−	−	−	−
	MIDDLE	−	−	−	✚	□	■	+	−	−	−	−	−
	BOTTOM	−	−	−	■	■	✚	+	−	−	−	−	−
1979-80	TOP	−	−	□	■ □	■	■	□	□				
	MIDDLE	−	−	□	✚ ■	■	■	□	□				
	BOTTOM	−	−	+	■ □	■	■	+	−				

Fig. 1.   Occurrence of *Anabaina circinalis* in Carcoar Dam, New
         South Wales, Australia, in water taken from near the sur-
         face (top), the mid-region (middle), and near the bottom
         (bottom) of the water column of Station 3, between October
         1977 and May 1980. One hundred ml of each water sample
         examined was filtered and the residue examined under the
         microscope. A rough measure of the relative frequency of
         occurrence was taken as the average number of trichomes
         present in a 0.5 mm diameter field of the same high power
         (H.P.) optical microscope. Key: ▬ = species missing;
         ✚ = quite rare occurrences; □ = 1 trichome/H.P. field;
         ■ = 10 trichome/H.P. field; ✚ = 20 trichome/H.P. field.
         No greater frequency was found in these samples at these
         times.

Anabaina circinalis   **OCTOBER 1978 TO MAY 1980**   **GLENBAWN DAM**   **STATION 4**

		OCT.	NOV.	DEC.	JAN.	FEB.	MAR.	APRIL	MAY	JUNE	JULY	AUG.	SEPT.
1977-78	TOP												
	MIDDLE												
	BOTTOM												
1978-79	TOP	□	□	■	■	✚	□	□	□	□	✚	■	□
	MIDDLE	□	□	■	■	✚	□	□	□	□	■	■	■
	BOTTOM	□	□	■	■	■	□	+	□	□	■	■	□
1979-80	TOP	□	—	+	■ ■	■	COLLECTION MISSED	■	□				
	MIDDLE	□	—	□	■ ■	■		■	□				
	BOTTOM	—	—	□	■ □	■		■	□				

Fig. 2.   Occurrence of *Anabaina circinalis* in Glenbawn Dam, New
South Wales, Australia, in water taken from near the surface (top), the mid-region (middle), and near the bottom
(bottom) of the water column of Station 4, between October
1978 and May 1980. Station No. 4 in Glenbawn Dam was in
a similar position to and of approximately the same depth
(ranging from 8 to 12 m) as Station No. 3 in Carcoar Dam.
One hundred ml of each water sample examined was filtered
and the residue examined under the microscope. A rough
measure of the relative frequency of occurrence was taken
as the average number of trichomes present in a 0.5 mm
diameter field of the same high power (H.P.) optical
microscope. Key: ▬ = species missing; ✚ = quite rare
occurrences; □ = 1 trichome/H.P. field; ■ = 10 trichome/
H.P. field; ✚ = 20 trichome/H.P. field. No greater frequency was found in these samples at these times.

in June. However, at Carcoar Dam, temperature readings of less
than 10°C continued from June to November, while at Glenbawn Dam
there was only the one month (June) when such low readings were
obtained. The corresponding date for the same minimum reading
(depth 50 m) in Burrinjuck Dam, in which *Anabaina* was found in
most months of the year, occurred in July, with temperatures of
less than 10°C continuing from June to August (May, 1978, Figure
2). It is possible that *Anabaina* can survive, in trichome form,
a low (7° to 8°C) winter temperature for a limited time, but that
the species cannot survive in the vegetative state over a prolonged
period of these low temperatures. Hence the duration of the vege-

tative state would depend on the duration rather than the minimum value of these low winter temperatures.

In shallower dams, such as Braidwood, which sometimes is rimmed with ice, *Anabaina* occurs strictly in summer only.

In Canada, Hammer (1964) recorded *A. flos-aquae* as appearing after the water temperature reached 5°C but bloom not appearing until after this temperature rose to 14°C.

Effects of Flood or Drought

Studies at Burrinjuck Dam (May, 1978) showed that *A. circinalis* did not reappear in the summer following a severe flood. Evidently the species was removed by flood more effectively than were, for instance, residual *Anacystis* growths. Lund (1979) also speaks of the wash-out of algae being important in the ecology of phytoplankton.

By contrast, excessive drought leads to: 1) less wash-out of algae, 2) higher concentrations of nutrients in the water because of higher evaporation, and 3) less aeration of the water. Resultant conditions are often associated with heavy developments of blue-green algae. Certainly it is during drought that most and heaviest blooms of blue-green algae occur, and, conversely, a wet year has very little, if any, such growth. May and Baker (1978) speak of a whole year's study of toxic algae from a series of dams being spoilt when there was "a very wet summer with much flooding and consequently only slight growth of toxic algae anywhere".

Conditions due to drought are similar to those which occur in deep anaerobic waters of a stratified dam which is eutrophic. Recent methods used for prevention or control of excessive growths of blue-green algae in such dams have included the pumping of air or oxygen into the anaerobic bottom layers of water to reverse some of the conditions described above. Possibly it is better to aerate only the hypolimnion, so that extra nutrients are not pumped into the photic zone.

Effect of Certain Nutrients

There are many records of correlation of bloom formation with nutrient enrichment, the suspected nutrient excess usually being phosphorus, nitrogen or organic matter.

In a small inland dam at Braidwood, New South Wales (May, 1972), algal blooms of *A. cyanea* and of *A. circinalis* developed only when the concentration of phosphorus reached 0.5 ppm. By contrast, it is known that surplus ("luxury") phosphorus can be held by the plant, and that sometimes heavy growths are recorded when the phosphorus concentration in the water is low. Moreover, it was found (May, 1974; May and Baker, 1978) that experimentally reducing the concentration of phosphorus in dams reduced or eliminated bloom formation.

In recent studies I have found a secondary increase in the density of *A. circinalis*, particularly in surface waters, occurring after the peak summer growth (Glenbawn Dam, Figure 2) which was correlated with an increase in the concentration of soluble phosphosus throughout the water column (Figure 3). It should be noted that this observation is based on a series of results, so it is not only due to short-term wind or rain effects. This observation, of course, counters the argument that the correlation between the concentration of phosphorus and bloom formation is only because phosphorus enrichment in summer coincides with other factors such as increased temperature or light intensity.

Nitrogen has been found to be in high concentrations (May, 1972) both following a heavy bloom and during a light bloom, so that it seems unlikely to be the factor determining bloom formation in the cases reported.

The effect on algal blooms of increased organic matter is likely to be indirect, through increasing bacterial growth, as described below.

It is of interest that at Burrinjuck Dam, *Anacystis* was found (May, 1978) resting on the sediment in bottom collections mainly in areas near the junctions of three different river tributaries of the dam, but the most frequent, dense, and persistent of such occurrences were at the junction of the two rivers which were subject to enrichment by sewerage and heavily polluted run-off. Here the bacterial count and phosphorus, nitrogen, iron and potassium levels were higher, and the oxygen and light lower than at the "control" clean-river junction.

Gorham et al. (1964) have shown in culture how nutrients, specifically relative concentrations of iron and manganese, affect the degree of coiling of *A. flos-aquae*.

Effect of Changes in pH

I have not found either of our common toxic algae occurring in the field with a pH of less than 6; often the reading is nearer 8 or 9. At these pH values blue-green algae are more efficient at using carbon dioxide (or bicarbonate) than are green algae (Shapiro, 1973).

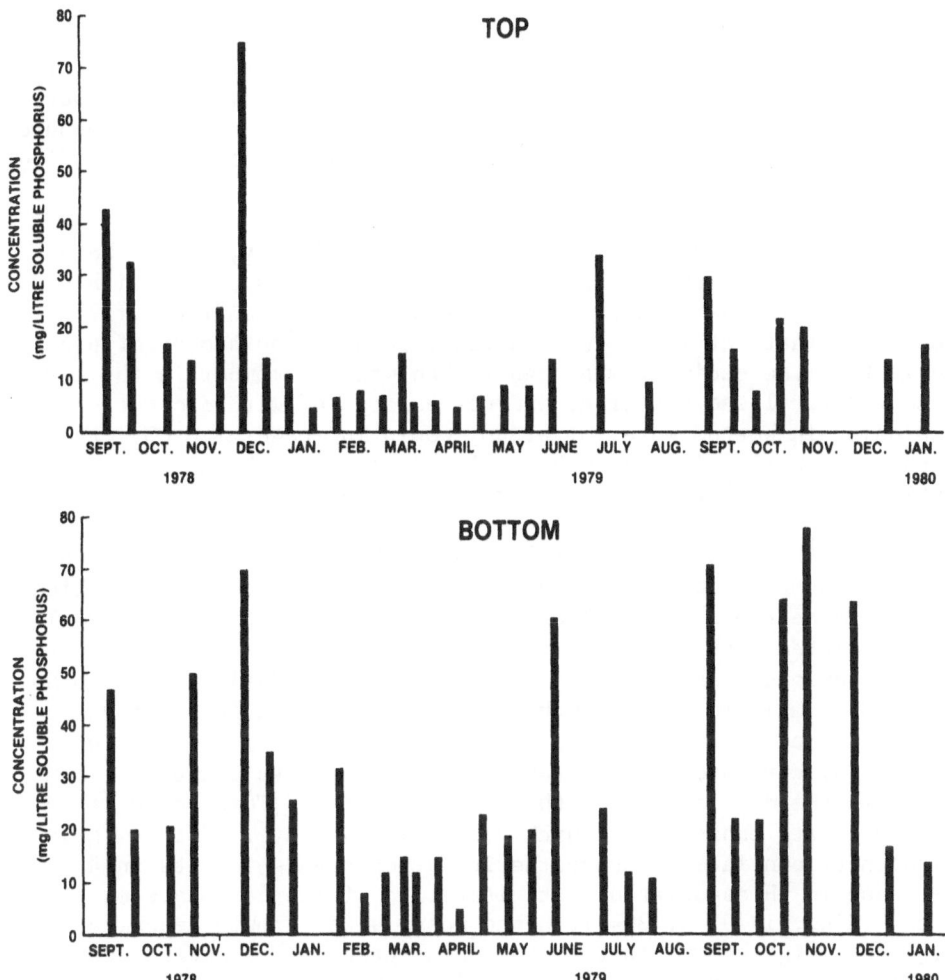

Fig. 3.   Levels of soluble phosphorus in water taken (1) near the
top, and (2) near the bottom of the water column of Glenbawn Dam, Station 4, between September 1978 and January
1980.   Figures provided by Water Resources Commission of
New South Wales.

Keenan (1973), reports that the photosynthesis of *Anabaina*
is dependent on the pH of the water within the range of pH 6.3 to
8.3, but not above this level.

Cumulative Factors

Various workers have stressed how the production of pseudo-
vacuoles in some blue-green algae may enable them to move vertically

and so to acquire both inorganic nutrients (at a lower water level
in a dam) and high light intensity (at a higher water level in the
dam). This faculty (Fogg, 1969) no doubt is of extreme benefit to
the algae concerned, as is the blue-green alga's ability to flour-
ish at low light intensity (Holm-Hansen, 1967), low oxygen pressures
(Stewart and Pearson, 1970), and low availability of carbon dioxide,
such as at high pH (King, 1970; Shapiro, 1973).

It has been shown (Silvey and Roach, 1964; May, 1972) that
peak periods of growth of blue-green algae are associated with
peaks in bacterial density in surface water, and Hendricks (1971)
has shown that, at least in a particular case, the number of bacteria from the sediment was higher than was the number in the surface water. These bacteria, whether associated with drought or
happening in bottom water of a dam, would indicate high CO_2 levels
conducive to all algal growth and low O_2 levels, which have been
shown (see above) to favor the growth of blue-green algae.

Sylvester and Anderson (1964) have shown that the freeing of
solutes and nutrients from the sediment occurs more in water which
is poorly aerated. These less aerobic conditions prevailing would
be associated with increased freeing of phosphorus from the underlying previously-enriched sediment (Patrick and Khalid, 1974), thus
further stimulating the growth of the blue-green algae. No doubt
this sequence of events (Kuentzel, 1969; May, 1972) explains why
additional carbohydrates in the water (= nutrients for bacteria)
increase the development of these algal blooms. These conditions
contrast with those under which green algae thrive. "At any given
alkalinity continued photosynthetic carbon uptake results in an
increase in pH associated with alteration of the carbonate equilibrium" (King, 1970, p. 2044), hence heavy algal growth leads to
high pH. King showed further that blue-green algae obtain adequate
supplies of carbon for photosynthesis at higher pH values than do
other algae. Hence high pH values further favor the growth of
blue-green algae. Especially in the case of a species, such as
A. circinalis, which can also fix atmospheric nitrogen, such a
series of physiological advantages can lead to a massive increasing
bias in favor of cyanophytes rather than green algae, where anaerobic conditions prevail in relatively stationary eutrophic water.
Hence the production of a bloom is likely.

Ganf (1980) recently has suggested that the seasonal "species
shift from green to blue-green algae is caused by .... a habitat
where their buoyancy mechanism gives them a competitive advantage
over other species present". This, no doubt, applies in the waters
he has studied, but of course is not the whole explanation of the
growth of a bloom of blue-green algae in a shallow farm dam, as
happens frequently.

As yet we know very little about the effect on each other of
various species in a seasonal succession, but there is a growing
literature on this subject. No doubt it is the interplay of various physiological abilities of a certain species that allows it
to achieve massive growths and form a bloom under a variety of conditions. No one factor is alone responsible. Certainly blooms do
not always happen only under exactly the same set of circumstances,
for instance at an exact time, temperature or location.

Control

Reverting again to the effect of nutrients in the water, much
work has indicated that variation in, among other factors, the
concentration of phosphorus, nitrogen, carbohydrates, carbon or
carbon dioxide, pH, iron, molybdenum or cobalt may affect or control the growth of these cyanophyte blooms. For instance, a recent
paper (Nicholas, 1980) suggests that sodium tungstate may be useful
to inhibit the growth of *Anabaina*, since this material is antago-
nistic to molybdenum, a trace element necessary for growth of the
alga.

A reduction in light, achieved artificially or by turbidity,
is certainly known to reduce all planktonic algae, but this is not
often practical. Since nitrogen and carbohydrates can be added to
water supplies by algae themselves, and since trace or minor elements are extremely difficult to eliminate completely, a reduction
in the concentration of phosphorus appears to be the most reasonable general method of controlling bloom formation. The obvious
methods to achieve this include avoidance of draining, storing, or
spraying phosphates near water storage or drainage areas as well
as leaving verges of waterways clear of stock, so reducing the
effect of their pollution. Similarly the use of drinking troughs
for stock keeps stored water cleaner.

My efforts towards the control of blooms of toxic algae were
directed towards reducing the phosphorus level in the water during
summer, the time of year when the level becomes highest. My hope
was that when phosphorus became available, the chemical "control"
material would combine with it before the unwanted alga could do
so. Hence we might obtain prevention rather than follow the
usual procedure of using an algicide after the bloom has developed.

First, alum and block ferric alum (Alumina ferric R) were
applied (200 mg/1) to a lake which had a history of toxic algal
blooms. The blocks were suspended in the water to prevent them
sinking into the underlying mud. Following treatment, no bloom
developed for the first time in five consecutive years (May, 1974).
Other small dams, with similar histories, were treated similarly
and again no blooms developed.

Next, a more ambitious effort (May and Baker, 1978) was applied to a series of farm dams all of which had suffered from toxic algae. Some dams were untreated and acted as controls, some had a single pre-summer treatment with ferric alum blocks, while other dams had repeated treatments. The gear, procedure and result-ant appearance of this alum treatment is described in May and Baker (1978). All dams showed a summer-time rise in concentration of total phosphorus, but the presence of alum definitely reduced the levels reached, and also decreased the incidence of algal blooms. The dosage (50 mg per liter) was evidently sufficient to reduce but not eliminate bloom occurrence. A somewhat higher dose is there-fore now recommended (100 mg per liter). The alum treatment not only reduced excessive growth of these unwanted algae; it also allowed an increase in the total number of algal species and in the average number of species per collection occurring in the treated dams, evidently leading to conditions more like those prevailing in pre-eutrophic times.

In large reservoirs individually determined rates of treatment would be preferable, or aeration may be more suitable, but at least the above-described treatment has provided a satisfactory way of coping with differing small dams, many of which have, in the past, presented a threat to our stock. The fact that this treatment is ecologically satisfactory is very pleasing and warrants its use in contrast to the use of algicides.

REFERENCES

Carmichael, W. W., and P. R. Gorham. 1977. Factors influencing the toxicity and animal susceptibility of *Anabaena flos-aquae* (Cyanophyta) blooms. J. Phycol. *13*(2):97-101.

Carmichael, W. W., and P. R. Gorham. 1978. Anatoxins from clones of *Anabaena flos-aquae* isolated from lakes of western Canada. Mitt. Int. Ver. Limnol. *21*:285-295.

Desikachary, T. V. 1959. Cyanophyta. Indian Council of Agricul-tural Research, New Delhi. 686 pp.

Drouet, F. 1968. Revision of the Classification of the Oscilla-toriaceae. Acad. Nat. Sci. Monogr. 15, Philadelphia, Pennsylvania. 370 pp.

Drouet, F. 1973. Revision of the Nostocaceae with Cylindrical Trichomes. Hafner Press, New York. 292 pp.

Drouet, F. 1978. Revision of the Nostocaceae with constricted trichomes. Nova Hedwigia *57*:1-258.

Fogg, G. E. 1969. The physiology of an algal nuisance. Proc. Roy. Soc. London. B. *173*:175-189.

Ganf, G. G. 1980. Factors controlling the growth of phytoplankton in Mount Bold Reservoir, South Australia. Pages 1-109 *in* Aust. Water Resources Council, Tech. Paper No. 48.

Gorham, P. R. 1964. Toxic algae. Pages 307-337 *in* D. F. Jackson, ed., Algae and Man. Plenum Press, New York.

Gorham, P. R., J. McLachlan, U. T. Hammer, and W. K. Kim. 1964. Isolation and culture of toxic strains of *Anabaena flos-aquae* (Lyngb.) de Bréb. Int. Assoc. Theor. Appl. Limnol. *15*:796-804.

Hammer, U. T. 1964. The succession of "bloom" species of blue-green algae and some causal factors. Verh. Int. Ver. Limnol. *15*:829-836.

Hendricks, C. W. 1971. Increased recovery rate of Salmonellae from stream bottom sediments versus surface waters. Appl. Microbiol. *21*(2):379-380.

Holm-Hansen, O. 1967. Recent advances in the physiology of blue-green algae. Pages 87-96 *in* Environmental Requirements of Blue-green Algae. Symposium. Proc. Water Poll. Control Fed. Administration. Univ. Wash., U. S. Dep. Interior.

Keenan, J. D. 1973. Response of *Anabaena* to pH, carbon and phosphorus. Am. Soc. Civil Eng., J. Environ. Eng. Div. *99*(EE5): 607-620.

King, D. L. 1970. The role of carbon in eutrophication. J. Water Poll. Control Fed. *42*:2035-2050.

Krebs, C. J. 1972. Ecology. Harper and Row, London. 694 pp.

Kuentzel, L. E. 1969. Bacteria, carbon dioxide and algal blooms. J. Water Poll. Control Fed. *41*(10):1737-1747.

Léonard, J., and D. Compère. 1967. *Spirulina platensis* (Gom.) Geitl. algae bleue de grande valeur alimentaire par sa richesse en protéines. Bull. Jard. Bot. Nat. Belgique *37*(1) Suppl.:1-25.

Lund, J. W. G. 1979. Changes in the phytoplankton of an English lake 1945-77. Hydrobiol. J. *14*(1):6-21.

Mathew, J. M. 1930. Hume Reservoir and algal infestation. Commonw. Eng. *17*:401-405.

May, V. 1972. Blue-green algal blooms at Braidwood, New South Wales (Australia). N. S. W. Dep. Agric. Sci. Bull. *82*:1-45.

May, V. 1974. Suppression of blue-green algal blooms in Braidwood Lagoon with alum. J. Aust. Inst. Agric. Sci. *40*:54-57.

May, V. 1978. Areas of recurrence of toxic algae within Burrinjuck Dam, New South Wales, Australia. Telopea *1*(5):295-313.

May, V., and H. Baker. 1978. Reduction of toxic algae in farm dams by ferric alum. N. S. W. Dep. Agric. Tech. Bull. *19*: 1-16.

McBarron, E. J., R. I. Walker, I. Gardner, and K. H. Walker. 1975. Toxicity to livestock of the blue-green alga *Anabaena circinalis*. Aust. Vet. J. *51*:587.

Nicholas, D. I. D. 1980. Mineral nutrient requirements and utilisation by algal flora of freshwater lakes. Pages 1-52 *in* Aust. Water Resources Council Tech. Paper No. 50. Aust. Govt. Publ., Canberra.

Patrick, W. H., Jr., and R. A. Khalid. 1974. Phosphate release
    and sorption by soils and sediments. Effect of aerobic and
    anaerobic conditions. Science *186*:53-57.
Shapiro, J. 1973. Blue-green algae: why they become dominant.
    Science *179*:382-384.
Silvey, J. K. G., and A. W. Roach. 1964. Studies in microbiotic
    cycles in surface waters. J. Am. Water Works Assoc. *56*:60-
    72.
Stewart, W. D. P., and H. W. Pearson. 1970. Effects of aerobic
    and anaerobic conditions on growth and metabolism of blue-
    green algae. Proc. Roy. Soc. Lond. B. *175*:293-311.
Sylvester, R. O., and G. E. Anderson. 1964. A lake's response to
    its environment. Am. Soc. Civil Eng., J. San. Eng. Div. *90*:
    1-22.
West, G. S. 1909. The algae of the Yan Yean Reservoir, Victoria:
    a biological and oecological study. J. Linn. Soc. Bot. *39*:
    1-88.
Whitton, B. A. 1973. Freshwater plankton. Pages 353-367 *in*
    N. G. Carr and B. A. Whitton, eds., The Biology of Blue-
    green Algae. Blackwell Scientific Publications, Oxford.

THE RECENT BLUE-GREEN ALGAL BLOOMS

OF LONG LAKE, WASHINGTON

Raymond A. Soltero and Donald G. Nichols

Department of Biology
Eastern Washington University
Cheney, Washington 99004

ABSTRACT

In 1976 and 1977, toxic blooms of *Anabaena flos-aquae* occurred in the upper end of Long Lake, Washington, an impoundment of the Spokane River downstream from the city of Spokane. A non-toxic bloom of *Microcystis aeruginosa* followed in 1978 despite significant reductions in phosphorus loading to the reservoir because of Spokane's new advanced wastewater treatment facility. *Microcystis aeruginosa*, *Anabaena circinalis* and *A. spiroides* pulsed in 1979, but standing crops were substantially less in magnitude.

It is postulated that abatement programs initiated in an upstream mining district may be the cause for the sudden appearance of cyanophytes in bloom proportions, which were previously minor constituents of the reservoir's phytoplankton community. Prior to 1976, the *A. flos-aquae* standing crop potential of the reservoir may not have been achievable because of heavy metal inhibition, regardless of nitrogen and phosphorus availability. However, following advanced wastewater treatment and decreased heavy metal suppression, higher N:P ratios gave a growth advantage to *M. aeruginosa* in 1978.

INTRODUCTION

In 1889, the city of Spokane, Washington installed its first combined sewer system without any form of treatment. The wastes were diluted and flushed out of sight by the large Spokane River. The Spokane population continued to grow, and discharges of wastes at several points along the river steadily increased. Unfortunately, this solution to their waste problem was indeed shortsighted.

The inability of the river to assimilate large volumes of wastes became apparent in the 1920's as sewage could be readily observed in the river, particularly following periods of heavy rainfall. By the mid-1950's millions of gallons of untreated industrial and municipal wastes were discharged daily.

Also, as the century progressed, several hydroelectric projects were completed on the river. The largest reservoir was created by Long Lake Dam in 1915, approximately 24 km northwest and downstream from Spokane. Strong public interest has been expressed over the years about the reservoir's deteriorated water quality with macrophyte growths and algal blooms being chief sources of concern.

The first major step toward correcting the pollution problems of the Spokane River system came in 1957 with the construction of the city's primary sewage treatment plant. In addition, state and federal regulatory efforts of the late 1960's and early 1970's resulted in industrial plants being in compliance with discharge standards. Although the situation had improved, there was still the problem of the city's combined sewer system. Bypasses of raw sewage and stormwater to the river were necessary at the treatment plant and 42 other outfall points because of hydraulic overloading of the collection system during periods of snowmelt or rainfall.

In response to directives from the State of Washington Department of Ecology (D.O.E.), Spokane commissioned a study which would outline a program to mitigate the sewer overflow problem and upgrade its sewage treatment plant. The city provided advanced wastewater treatment (AWT) for its sewage in December 1977. Advanced treatment was defined as 85 percent or better BOD_5 and phosphorus removal, and 90 percent or better suspended solids removal. Also, city officials are presently preparing to abate Spokane's combined sewer overflow problem.

Construction on the new AWT plant began in 1975, and in October of that year a bypass of raw sewage for three and one-half days occurred as part of the 55 million dollar project. Several Long Lake residents filed a 5.2 million dollar lawsuit[1] contending that the 110 million gallons of raw sewage released reduced their property values and resulted in toxic blue-green blooms in the late summer and fall of 1976, 1977 and 1978.

---

[1]Although the city of Spokane and the state of Washington have been found liable, the Superior Court of Spokane County on 20 July 1979 found ". . . the blooms in 1976, 1977, and 1978 were not caused by the October, 1975 bypass."

TEXT

Limnological Background

A synopsis of the investigations carried out on Long Lake preceding and just following AWT will be presented here before discussing its recent blue-green blooms. Morphometric data for the
reservoir are presented in Table 1.

Since 1959, the D.O.E. has frequently detected dissolved oxygen
concentrations less than 5.0 mg/liter at its water quality station
below Long Lake Dam. On September 1, 1964, the Washington Water
Pollution Control Commission (now D.O.E.) found oxygen levels lower
than 1.0 mg/liter just behind the dam from the nine meter level to
the bottom (Asselstine, W.W.P.C.C. unpublished data). In 1966,
during the week of September 12, Cunningham and Pine (1969) determined that approximately 40 percent of reservoir volume was anaerobic. The anaerobic condition was attributed primarily to volatile
solid stabilization occurring in the sediments with the assumption
that phytoplankton comprised the majority of the solids. It was
suggested that the nutrients supporting excessive algal growth came

Table 1. Morphometric Data for Long Lake at Maximum Capacity
          (Elevation 468.3 m)

Maximum Length	35.4 km
Maximum Effective Length	5.8 km
Maximum Width	1.1 km
Maximum Effective Width	1.1 km
Mean Width	571.8 m
Maximum Depth	54.9 m
Mean Depth	14.6 m
Area	208.4×10^5 m^2
Volume	304.9×10^6 m^3
Shoreline Length	74.3 km
Shoreline Development	4.6%
Bottom Grade	0.15%

from the reduced sediments and the Spokane primary sewage treatment
plant outfall via the Spokane River.

Haggarty (1970) pointed out that industrial waste from Spokane
was not a major pollution problem in the Spokane River but that the
major waste discharge to the river was the city's combined sewer
system. In addition, the survey showed that total coliform counts
exceeded the upper limit of the state class A standards of 240/100
ml in the reach from Upriver Dam to Long Lake Dam. Hence, the
contamination, apparently in the form of sewage, entered the river
as it passed through the city.

During the summer of 1971, another study of Long Lake was
conducted (Bishop and Lee, 1972) revealing that low dissolved oxygen
levels again existed below a thermocline and after fall turnover
surface levels dropped below 8.0 mg per liter. Condit (1972),
collaborating with Bishop and Lee, made two phytoplankton counts
and determined algal growth potential, via algal assay, from three
samples collected from behind Long Lake Dam.

The most comprehensive documentation of the deteriorated
condition of Long Lake was initiated in the spring of 1972 by
Eastern Washington University. In essence, these projects have
critically evaluated Long Lake and its major tributaries as a
dynamic system (Soltero et al., 1973; 1974; 1975; 1976; 1978; 1979;
and 1980).

A total of seven sampling stations were established on the
Spokane River and its major tributaries (Figure 1A). Five sampling
stations were established on Long Lake; station 0 was located
behind Long Lake Dam with the four remaining stations at approximately 8 km intervals from each other up the reservoir for 32 km
to station 4 (Figure 1B). The length of the sampling season was
approximately 7 months (June through December). Sampling frequency
was usually weekly during 1972 and 1973, and biweekly for all other
years.

Preceding AWT:

1. During the low flow periods and below the Spokane primary
    sewage treatment plant outflow, nutrient concentrations in
    the Spokane River (particularly phosphorus) increased sig-
    nificantly.

2. The Spokane treatment plant effluent was a major source of
    algal nutrients influent to Long Lake via the Spokane River.

3. The reservoir was a nutrient trap.

4. An internal density-current system was operative and attrib-
    uted to the complex interaction of density flow, thermal
    stratification and seasonal inflow and outflow. The river's

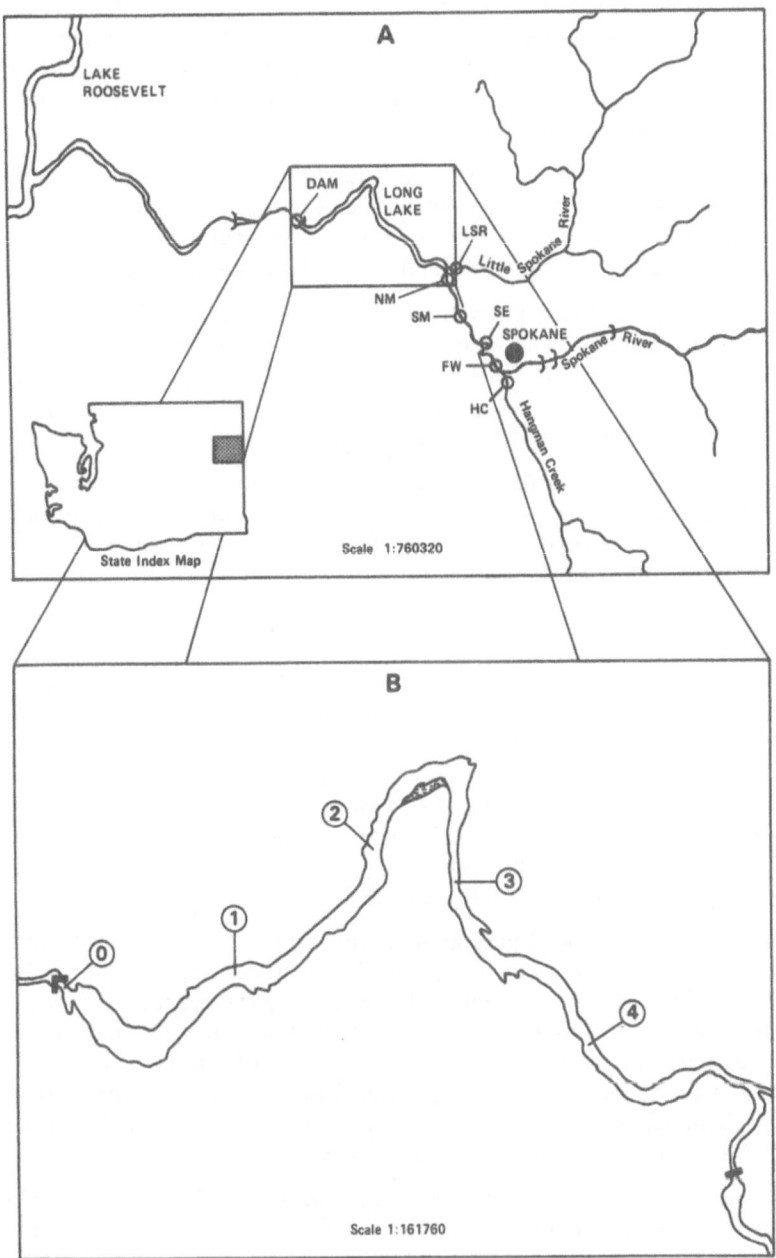

Fig. 1A and 1B. Map of the lower Spokane River system detailing the
study area. DAM = Base Long Lake Dam, LSR = Little
Spokane River, NM = Nine Mile, SM = Seven Mile,
SE = Sewage Effluent, FW = Fort Wright (bridge),
HC = Hangman Creek.

spring runoff of low salinity flowed throughout the entire depth
of the reservoir. In the summer with decreasing inflow and
increasing salinity, the down-reservoir flow of the river
occurred as an interflow at the depth of the power penstocks
(centerline elevation 457 m). In the fall, the lower temperature
of the inflowing river resulted in a bottom flow the length of
the reservoir and initiated fall homothermy.

5. The interflow of the river at the power penstocks elevation
   provided nutrients which promoted excessive algal growth,
   heightened hypolimnetic anoxia and increased hypolimnetic
   concentrations of orthophosphate and nitrate nitrogen.

6. Mean orthophosphate phosphorus, nitrate nitrogen, and silica
   concentrations of the euphotic zone were usually greater than
   0.01, 0.30, and 0.4 to 0.6 mg per liter, respectively. These
   levels are often considered limiting (Sawyer, 1947, 1952;
   Hutchinson, 1967; and Macan, 1970).

7. The reservoir was classified as eutrophic for all years of
   study on the basis of chlorophyll a concentrations, primary
   productivity, nitrogen and phosphorus concentrations, and
   hypolimnetic oxygen depletion.

8. Seasonal phytoplankton succession was similar for all years
   of study with a large part of the standing crop being made up
   of diatoms. Large *Melosire italica* standing crops were usually found in the spring. *Cryptomonas* sp. peaked in July
   shortly before a *Fragilaria crotonensis* pulse in August.
   Sphaerocystis schroeteri and *Oocystis* sp. pulses followed the
   F. crotonensis peak, and the greens were usually succeeded by
   another diatom pulse of *Melosira granulata* in the fall. An
   exception to the latter was the late summer and fall *Anabaena
   flos-aquae* pulses of 1976 and 1977.

9. The species of zooplankton present appeared to have been regulated by phytoplankton composition. However, the positive
   relationship between zooplankton and phytoplankton standing
   crops would indicate that grazing did not significantly limit
   phytoplankton production. Predation could have been a significant factor responsible for the lack of grazing pressure since
   both *Leptodora kindtii* and *Polyphemus pediculus* were abundant
   in the reservoir during much of the summer.

10. Through abatement of present high nutrient loadings to the
    reservoir and preservation of sediment integrity, it was pre-
    dicted that eutrophic Long Lake would experience a rapid
    recovery after the establishment of AWT with phosphorus removal
    (Thomas and Soltero, 1977). Sediment cores taken at station 1
    were zoned into annual layers according to profile concentra-
    tions of organic matter, nitrogen, phosphorus, and diatom
    frustules. The sedimentation rate was determined to be 26 mm

per year between 1958 and 1973. Below the 25 mm depth, phosphorus and diatom profiles were most obvious because clay particles entering the reservoir during spring runoff formed a
layer covering the biological material precipitated the previous year. This mechanism should effectively prevent internal
recycling of nutrients, particularly phosphorus, from the
bottom sediments.

Following AWT. Soltero et al. (1979; 1980) have concluded the
following:

1. Total and inorganic phosphorus loads from the sewage effluent
    have decreased approximately 89 and 92 percent, respectively.
    This reduction significantly lowered phosphate loads to Long
    Lake with a concomitant decline in reservoir concentrations.

2. Prior to AWT, the sewage effluent was the major contributor of
    phosphorus to the reservoir; however, the Spokane River above
    the treatment facility became the primary source following AWT.

3. The degree of hypolimnetic anoxia observed in 1978 and 1979
    was markedly less than that determined for 1973 and 1977, but
    similar to that observed in 1972, 1974 and 1975.

4. Phytoplankton biovolumes declined in all classes with the
    exception of the Cyanophyceae in 1978. An extensive *Microcystis
    aeruginosa* bloom occurred in the upper end of the reservoir
    during the fall and resulted in an overall increase in blue-
    green biovolume. *Microcystis* biovolumes were considerably lower
    in 1979.

5. Mean chlorophyll a concentrations and primary productivity
    values have significantly declined following AWT.

6. The overall decline in chlorophyll a concentrations was directly
    related to decreased phosphorus loading. Internal phosphorus
    loading did not appear to be a significant source as a propor-
    tionate decline in hypolimnetic concentrations occurred.

7. Zooplankton standing crops have also declined, possibly the
    result of decreased food availability. Algal toxicity in 1978,
    arising from the *Microcystis* pulse, may have also contributed
    to lower standing crops that year. The successional patterns
    for both years were similar to those observed prior to AWT.

8. Decreases in reservoir orthophosphate and chlorophyll a con-
    centrations, primary productivity, and hypolimnetic anoxia
    have indicated the reversion of eutrophic Long Lake to a more
    mesotrophic state.

Blue-green Bloom Toxicity
<u>   </u>

   Anabaena flos-aquae and *Microcystis aeruginosa* have been frequently observed in Long Lake, but have been minor constituents of
the phytoplankton community (Soltero et al., 1973; 1974; 1975; and
1976). Prior to the fall of 1976, neither organism had reached
bloom proportions.

   Reservoir sampling was curtailed in 1976, so limnological
conditions just prior to the toxic bloom of *A. flos-aquae* are
unknown. The bloom was restricted to the upper end of the reservoir and persisted for over a month. The deaths of four dogs were
attributed to drinking reservoir water containing *Anabaena*. In
addition, seven dogs, one horse and one cow were suspiciously sick
after drinking reservoir water. The reported deaths of two ducks
and a beaver were not verified.

   Table 2 shows the estimated *Anabaena* biovolumes and associated
toxicity of station 4 samples. Toxicity was established by the
death of white male mice (approximately 30 grams in weight) following a one ml intraperitoneal (i.p.) injection of a non-uniform
surface water grab sample. Controls were similarly injected with
distilled water and in all cases survived. Symptoms following a
lethal algal injection were prostration, piloerection, convulsion
and eventually respiratory failure. The toxinological properties
of *A. flos-aquae* are described by Gorham (1964), Carmichael et al.,
(1975) and Carmichael and Gorham (1978).

Table 2.   *Anabaena flos-aquae* Standing Crops (mm^3 per liter)[a] by
           Date and Station, Long Lake, Washington (1976)

Date[b]	Station				
	0	1	2	3	4
10/1	---[c]	---	---	---	216 (13,18)[d]
10/8	0.2	0.5	2.0	5.3	698 (8,10)
10/15	0.3	0	0.9	5.9	278 (4, 5)
10/22	---	---	---	---	1680 (within 24 hours)

[a]Samples were collected from the surface of the water in 0.5 liter
   sterilized bottles. Plankton volumes were determined by sedimentation and counting of 1-5 ml samples.
[b]Month/Day.
[c]--- = No data.
[d]The parenthetic figure is the number of minutes that transpired
   before death of mice injected i.p. with 1.0 ml of sample.

Table 3. Summer *Anabaena flos-aquae* Standing Crops (mm^3 per liter) by Date and Station, Long Lake, Washington (1977).

Date[a]	Station				
	0	1	2	3	4
6/28	0 (2.9)[b]	0 (9.2)	0.1 (7.9)	0 (3.6)	0 (2.7)
7/12	4.2 (13.7)	7.1 (26.7)	6.0 (21.7)	1.1 (15.6)	0 (5.0)
7/26	0.2 (3.6)	0.1 (3.8)	0.4 (2.1)	0 (2.6)	0 (4.1)
8/09	0 (2.4)	0 (4.8)	0.7 (14.8)	0 (13.1)	0.2 (5.1)
8/23	0 (4.3)	0 (17.9)	0 (5.1)	0.3 (4.4)	0.7 (13.5)
9/06	0.7 (15.8)	0 (36.4)	0 (28.0)	0 (31.0)	7.7 (37.6)
9/20	0 (3.8)	0 (7.4)	0 (6.2)	0 (7.4)	0 (7.5)

[a]Month/Day

[b]The parenthetic figure is the total phytoplankton biovolume for the sample.

Table 3 gives the *A. flos-aquae* standing crops that existed in 1977. Biovolumes were substantially less than those observed in 1976, and only one sample collected from a bay near mid-reservoir proved to be toxic. No animal deaths were reported in 1977.

Also in the summer of 1977, a blue-green bloom was noted in Lake Roosevelt, downstream from Long Lake. The major algal constituent of the bloom was later identified to be *A. flos-aquae*. This *Anabaena* pulse probably came about due to the growth of vegetative cells and/or akinetes washed out of Long Lake. No known animal deaths occurred as a result of the bloom. Intraperitoneal injection of mice from a collected sample showed no signs of toxicity. Carmichael (personal communication) freeze-dried a small sample of the bloom and following bioassay (i.p. injection) on mixed male/female mice estimated its LD_{50} at 30 mg/kg. Fresh sample (non-freeze-dried) also showed toxicity in his laboratory. All signs of poisoning pointed to a neuromuscular anatoxin-a type poison.

After the initiation of AWT and in contrast to the 1976 and 1977 findings, an extensive bloom of *Microcystis aeruginosa* occurred in the upper end of Long Lake during the late summer of 1978 (Table 4). *Anabaena flos-aquae* was also present, but contributed

Table 4. Summer *Microcystis aeruginosa* Standing Crops (mm³ per
          liter) by Date and Station, Long Lake, Washington (1978)

Date[a]	Station				
	0	1	2	3	4
7/05	0 (0.7)[b]	0 (1.2)	0 (0.6)	0 (1.0)	0 (1.1)
7/18	0 (0.3)	0 (0.6)	0 (0.4)	0 (1.3)	0 (1.7)
8/01	0 (5.1)	0 (5.5)	0 (10.9)	0 (3.6)	0 (1.1)
8/14	0 (2.6)	0 (4.9)	0 (3.7)	0 (6.3)	0 (3.4)
8/29	0 (5.1)	0 (3.4)	0.9 (9.9)	6.2 (14.8)	12.4 (16.3)
9/12	0 (1.5)	0 (0.8)	0 (3.3)	228 (229)	1512 (1522)

[a]Month/Day

[b]The parenthetic figure is the total phytoplankton biovolume for the
sample.

little to the overall standing crop. No animal deaths were reported
and mouse testing revealed no toxicity.

Toxicity testing was not carried out in 1979 due to lower blue-
green densities. *Microcystis aeruginosa* dominated the phytoplankton
standing crop at stations 2 (1.02 mm³ per liter) and 0 (2.31 mm³
per liter) in mid-August and early September, respectively (Soltero
et al., 1980). These biovolumes were considerably less than those
which occurred at stations 3 and 4 in 1978 (Table 4). *Anabaena
circinalis* and *Anabaena spiroides* were the major contributors to
the standing crop at station 4 during September. Biovolumes contributed by these two species were substantially less than those
determined for *A. flos-aquae* in 1976 (Table 2), but slightly higher
than the 1977 values (Table 3). *Anabaena flos-aquae* was observed
only once at station 4 during early September. By early October,
blue-green standing crop had declined throughout the reservoir
with the Bacillariophyceae becoming dominant.

Possible Cause of Blooms

A number of articles have pointed out the rather widespread occurrence of toxic blue-green blooms (e.g., Prescott, 1933; Ingram and Prescott, 1954; Gorham, 1964; Hammer, 1968), but why they occurred was not adequately addressed. A toxic bloom is a serious occurrence, but it is just as important to be able to delineate the sequence of events that gave rise to the bloom so that possible preventive measures could be taken in the future. The appearance and persistence of blue-green algal blooms in Long Lake, particularly following AWT, was perplexing, and their cause(s) need study and explanation.

We believe it is impossible to prove the 1976 and 1977 toxic blooms of *A. flos-aquae* were specifically and uniquely the result of the 1975 raw sewage bypass. The basis of this judgement, in part, is that nutrients associated with the bypass would have been essentially dispersed and discharged from the reservoir 11 months later. Our studies (Soltero et al., 1973; 1974; 1975; 1976; 1978; 1979; and 1980) have consistently shown that from mid-March to mid-June the volume of the reservoir can be exchanged as often as six times a month. The total number of exchanges a year has usually been thirty.

In addition, phosphorus and nitrogen loading to the reservoir was altered little during and following the 1975 bypass (Soltero, unpublished data). This was not unexpected since primary treatment removes little if any of these nutrients. The additional amount of phosphorus contributed because of the bypass equals about 0.05 percent of the annual load. This agrees with the fact that no significant differences in river or lake concentrations for nitrogen and phosphorus were determined during and following the bypass from those previously observed. Nutrient levels in Long Lake have been high enough to promote excessive algal growth for a number of years, and one would expect that a bloom, if indeed attributable to the 1975 bypass, would be more immediate following the bypass rather than a year later.

Also, materials sedimented during the bypass would have represented a very small fraction of the annual nutrient load (less than 0.01%) to Long Lake. The degradable bypass materials that probably settled out behind Nine Mile Dam (Figure 1 - NM) and in the upper end of Long Lake would have decomposed in a relatively short period, with the nutrients released to the overlying waters diluted and eventually dispersed following spring flows. In addition, visual observations and bottom dredging in the upper end of Long Lake have shown that the main channel is annually scoured of overlying sediment because of spring runoff. Thus, sedimented bypass materials would have been deposited farther into the reservoir and downstream from the major bloom site.

There are several hypotheses as to why blue-greens can dominate the plankton (cf., Fitzgerald, 1964; Hammer, 1964; Vance, 1965; Boyd, 1973; Reynolds, 1973; Shapiro, 1973; Lange, 1974; Morton and Lee, 1974; Schindler and Fee, 1974; and Keating, 1978). The physical chemical and/or biological mechanisms responsible may operate separately, in combination, directly and/or indirectly. The questions arise: Why hasn't Long Lake been more productive? Why did the *Anabaena* blooms of 1976 and 1977 occur suddenly and unexpectedly? A possible answer is that some form of suppression or inhibition has been operative prior to 1976. It would appear that factors such as light, temperature, turbidity, retention times, nutrient levels, etc., were not significantly different in 1976 and 1977 than in 1975.

We hypothesize that prior to 1976 the blue-green algal standing crop potential in Long Lake may not have been achievable even with excessive nutrient loading because of heavy metal inhibition. Furthermore, the federal regulatory efforts of the 1970's to clean up the mining wastes originating from the mining and smelter operation near Kellogg, Idaho, have resulted in reduced heavy metal concentrations in Long Lake (and therefore reduced inhibitory effect) permitting more blue-green growth.

Evidence to support our speculation comes from the fact that heavy metals (particularly zinc) have entered the reservoir at algicidal and/or algistatic levels (cf., Bartlett et al., 1974; Greene et al., 1975; Miller et al., 1975), thereby possibly influencing the quantitative and qualitative composition of the phytoplankton. In addition, trend analysis has shown that total mean annual zinc concentrations in the Spokane River at Post Falls, Idaho have decreased 50 percent (300 to 150 mg per liter) between 1973 and 1978 (Yake, 1979). Likewise, total mean annual zinc concentrations in Long Lake have declined from 155 to 110 mg per liter between 1970 and 1978. Thomas and Soltero (1977) have shown that the zinc content of the 1973 and 1974 zones of a Long Lake sediment core was about half of that determined for the 1968-1972 zones. The latter correlates well with improved collection and treatment implemented in 1973 for mining wastewaters.

Greene et al. (1978) report that a positive relationship existed between Spokane River flow and zinc concentrations influent to Long Lake. They also pointed out that zinc levels in the latter part of the growing season can fall to or below the lower range (30 to 100 mg per liter) of algal growth inhibition.

Algal assay (Greene et al., 1975; 1976) using Long Lake test waters has revealed a good relationship between maximum assay yields and the limiting nutrient after the removal of heavy metal inhibition (EDTA chelation) to *Selenastrum capricornutum*. Excellent correlation also existed between maximum algal assay yields following metal chelation and reservoir phytoplankton biovolumes and chlorophyll a concentrations. The above work demonstrated the ability of the assay to estimate indigenous phytoplankton standing crop of reservoir test water.

Additional assays using a Long Lake isolate, *Sphaerocystis schroeteri* showed that this organism produced maximal yield in control waters and that the addition of EDTA did not significantly change the final *Sphaerocystis* biomass (Greene et al., 1978). This data indicates that an indigenous phytoplankton was able to grow in an environment containing zinc levels shown to be inhibitory to other algae.

The growth of some algae, indigenous to Long Lake, may not be adversely affected by high concentrations of heavy metals; however, *A. flos-aquae* growth can be suppressed in the presence of heavy metals (Shiroyama et al., 1976). Shiroyama also showed that as Spokane River flows subside metal concentrations decline rendering Long Lake waters more suitable for *Anabaena* growth. Thus it is possible that with a decline of heavy metals entering the reservoir and concentrations dropping below inhibitory levels in the latter part of the 1976 growing season, *A. flos-aquae* was able to achieve maximal standing crop with other conditions optimal.

As previously pointed out, *Microcystis aeruginosa* dominated the plankton in the late summer of 1978 following AWT. The change in nutrient limitation from nitrogen to phosphorus and decreased heavy metal toxicity just prior to the *Microcystis* bloom, as determined by algal assay, may have been in part responsible for the predominance of *M. aeruginosa* (Soltero et al., 1979). *Microcystis* has also been shown to have a relatively low phosphorus requirement (Gerloff et al., 1952) and has demonstrated the ability to utilize sulfur as an alternative to phosphorus in its metabolism, giving preference to the element which is present in sufficient amount (Volodin, 1970). The requirement for sulfur has been shown to be considerably higher than for phosphorus (Gerloff et al., 1952).

The possible advantage afforded to *Microcystis* (e.g. over *Anabaena flos-aquae*) has only become evident since phosphorus limiting conditions. Studies of pure cultures of *M. aeruginosa* have also shown that the organism has a high requirement for nitrogen and produces maximum growth in cultures with high N:P ratios (Gerloff et al., 1952). In contrast, Schindler (1977) suggests that there is a direct relationship between the occurrence of nitrogen fixers such as *Anabaena* and low N:P ratios. Total inorganic nitrogen ($NO_3^- + NO_2^- + NH_3$) to orthophosphate concentration ratios (TIN:TIP) for the euphotic zone of Long Lake prior to AWT were approximately 5:1. However, since the initiation of phosphorus removal at the treatment plant, the TIN:TIP ratio has increased to approximately 20:1.

SUMMARY

The evidence indicates that as judged by available nutrients, the bloom potential of Long Lake has been high for a number of years and that blue-green blooms could have occurred previous to 1976. It is unlikely that the toxic *A. flos-aquae* blooms of 1976 and 1977 were the result of the October 1975 raw sewage bypass. A more plausible explanation may be that these blooms were the result of pollution abatement efforts in the Kellogg, Idaho mining district. Since the early 1970's, heavy metal concentrations in Long Lake (particularly zinc) have significantly decreased. This decline, in turn, possibly resulted in zinc concentrations dropping below inhibitory levels for *Anabaena* growth in the latter part of the 1976 growing season. Thus, our suggested explanation implies that the year of 1976 was a turning point for Long Lake.

In 1978, with AWT "on-line," declines in reservoir phytoplankton biovolumes occurred in all classes with the exception of the Cyanophyceae. Decreased metal inhibition probably was one factor allowing for the blue-green blooms. In addition, altered nutrient conditions (high N:P ratio) brought about by AWT may have selected for a blue-green, *M. aeruginosa*, whose metabolic requirements possibly gave it a competitive advantage over *A. flos-aquae*

Certainty cannot be ascribed to any cause(s) for either the *Anabaena* or *Microcystis* blooms. Although much is known of Long Lake and its tributaries, the recent blue-green blooms have demonstrated that our level of resolution was (and is) not sufficient to conclusively describe the circumstances giving rise to the blooms.

ACKNOWLEDGMENT

The work upon which this publication is based was supported in part by funds provided by the Office of Water Resources and Technology, and the State of Washington Department of Ecology.

REFERENCES

Bartlett, L., F. W. Rabe, and W. H. Funk. 1974. Effects of copper, zinc, and cadmium on *Selenastrum capricornutum*. Water Res. *8*:179-185.

Bishop, R. A., and R. A. Lee. 1972. Spokane River water quality study. D.O.E. Technical Report No. 72-001, Olympia, Washington. 72 pp.

Boyd, C. E. 1973. Biotic interactions between different species of algae. Weed Sci. 21:32-37.

Carmichael, W. W., D. F. Biggs, and P. R. Gorham. 1975. Toxicology and pharmacological action of *Anabaena flos-aquae* toxin. Science 187:542-544.

Carmichael, W. W., and P. R. Gorham. 1978. Anatoxins from clones of *Anabaena flos-aquae* isolated from lakes of western Canada. Mitt. Int. Ver. Limnol. 21:285-295.

Condit, R. J. 1972. Phosphorus and algal growth in the Spokane River. Northwest Sci. 46:190-193.

Cunningham, R. K., and R. E. Pine. 1969. Preliminary investigation of the low dissolved oxygen concentrations that exist in Long Lake located near Spokane, Washington. W.W.P.C.C. Technical Report No. 69-1, Olympia, Washington. 24 pp.

Fitzgerald, G. P. 1964. The biotic relations with water blooms. Pages 300-306 *in* D. F. Jackson, ed., Algae and Man. Plenum Press, New York.

Gerloff, G. C., G. P. Fitzgerald, and F. Skoog. 1952. The mineral nutrition of *Microcystis aeruginosa*. Am. J. Bot. 39:26-32.

Gorham, P. R. 1964. Toxic algae. Pages 307-336 *in* D. F. Jackson, ed., Algae and Man. Plenum Press, New York.

Greene, J. C., W. E. Miller, T. Shiroyama, R. A. Soltero, and K. Putnam. 1976. Use of algal assays to assess the effects of municipal and smelter wastes upon phytoplankton production. Pages 327-336 *in* R. D. Andrews, III et al., eds., Terrestrial and Aquatic Ecological Studies of the Northwest. Eastern Washington State College Press, Cheney, WA.

Greene, J. C., W. E. Miller, T. Shiroyama, R. A. Soltero, and K. Putnam. 1978. Use of laboratory cultures of *Selenastrum*, *Anabaena* and the indigenous isolate *Sphaerocystis* to predict effects of nutrient and zinc interactions upon phytoplankton growth in Long Lake, WA. Mitt. Int. Ver. Limnol. 21:372-384.

Greene, J. C., R. A. Soltero, W. E. Miller, A. F. Gasperino, and T. Shiroyama. 1975. The relationship of laboratory algal assays to measurements of indigenous phytoplankton in Long Lake, WA. Pages 93-126 *in* E. J. Middlebrooks et al., eds., Biostimulation and Nutrient Assessment Workshop. Ann Arbor Science, Michigan.

Haggarty, T. G. 1970. Water pollution in the Spokane River. W.W.P.C.C. Technical Report No. 69-1, Olympia, Washington. 8 pp.

Hammer, U. T. 1964. The succession of "bloom" species of blue-green algae and some causal factors. Verh. Int. Ver. Limnol. 15:829-836.

Hammer, U. T. 1968. Toxic blue-green algae in Saskatchewan. Can. Vet. J. 9:221-229.

Hutchinson, G. E.   1967.   A Treatise on Limnology.   II.   Introduc-
    tion to Lake Biology and the Limnoplankton.   John Wiley and
    Sons, Inc., New York.   1115 pp.
Ingram, W. M., and G. W. Prescott.   1954.   Toxic fresh-water algae.
    Am. Mid. Nat. 52:75-87.
Keating, K. I.   1978.   Blue-green algal inhibition of diatom growth:
    transition from mesotrophic to eutrophic community structure.
    Science 199:971-973.
Lange, W.   1974.   Competitive exclusion among three planktonic
    blue-green algal species.   J. Phycol. 10:411-414.
Macan, T. T.   1970.   Biological Studies of the English Lakes.
    American Elsevier, New York.   260 pp.
Miller, W. E., J. C. Greene, and T. Shiroyama.   1975.   Application
    of algal assays to define the effects of wastewater effluents
    upon algal growth in multiple use river systems.   Pages 77-92
    in E. J. Middlebrooks et al., eds., Biostimulation and
    Nutrient Assessment Workshop.   Ann Arbor Science, Michigan.
Morton, S. D., and T. H. Lee.   1974.   Algal blooms -- possible
    effects of iron.   Env. Sci. Tech. 8:673-674.
Prescott, G. W.   1933.   Some effects of the blue-green algae,
    Aphanizomenon flos-aquae, on lake fish.   Collect. Net 8:77-80.
Reynolds, C. S.   1973.   Growth and buoyancy of Microcystis aeruginosa
    Kütz. emend. Elenkin in a shallow eutrophic lake.   Proc. R.
    Soc. Lond. B. Biol. Sci. 184:29-50.
Sawyer, C. N.   1947.   Fertilization of lakes by agricultural and
    urban drainage.   J. New Eng. Water Works Assoc. 60:109-127.
Sawyer, C. N.   1952.   Some new aspects of phosphates in relation
    to lake fertilization.   Sewage Ind. Wastes J. 24:768-776.
Schindler, D. W.   1977.   Evolution of phosphorus limitation in
    lakes.   Science 195:260-262.
Schindler, D. W., and E. J. Fee.   1974.   Experimental lakes area:
    whole-lake experiments in eutrophication.   J. Fish. Res.
    Board Can. 31:937-953.
Shapiro, J.   1973.   Blue-green algae: why they become dominant.
    Science 179:382-384.
Shiroyama, T., W. E. Miller, J. C. Greene, and C. Shigihara.   1976.
    Growth response of Anabaena flos-aquae (Lyngb.) de Bréb. in
    waters collected from Long Lake Reservoir, Washington.   Pages
    267-275 in R. D. Andrews, III, et al., eds., Terrestrial and
    Aquatic Ecological Studies of the Northwest.   Eastern Wash-
    ington State College Press, Cheney, Washington.
Soltero, R. A., A. F. Gasperino, and W. G. Graham.   1973.   An
    investigation of the cause and effect of eutrophication in
    Long Lake, Washington.   Final Progress Report, O.W.R.R.
    Project 143-34-10E-3996-5501.   86 pp.
Soltero, R. A., A. F. Gasperino, and W. G. Graham.   1974.   Further
    investigations as to the cause and effect of eutrophication
    in Long Lake, Washington.   Completion Report, D.O.E. Project
    74-025A.   85 pp.

Soltero, R. A., A. F. Gasperino, P. H. Williams, and S. R. Thomas.
    1975. Response of the Spokane River periphyton community
    to primary sewage effluent and continued investigation of
    Long Lake. Completion Report, D.O.E. Project 74-144. 117 pp.

Soltero, R. A., D. M. Kruger, A. F. Gasperino, J. P. Griffin, S. R.
    Thomas, and P. H. Williams. 1976. Continued investigation
    of eutrophication in Long Lake, Washington: verification
    data for the Long Lake model. Project Completion Report,
    D.O.E. Project WF-6-75-081. 64 pp.

Soltero, R. A., D. G. Nichols, G. P. Burr, and L. R. Singleton.
    1979. The effect of continuous advanced wastewater treat-
    ment by the city of Spokane on the trophic status of Long
    Lake, WA. Completion Report, D.O.E. Project 77-108. 95 pp.

Soltero, R. A., D. G. Nichols, and J. M. Mires. 1980. The effect
    of continuous advanced wastewater treatment by the city of
    Spokane on the trophic status of Long Lake, WA during 1979.
    Completion Report, D.O.E. Project 80-019. (In press).

Soltero, R. A., D. G. Nichols, G. A. Pebles, and L. R. Singleton.
    1978. Limnological investigation of eutrophic Long Lake
    and its tributaries just prior to advanced wastewater treat-
    ment with phosphorus removal by Spokane, WA. Project
    Progress Report, D.O.E. Project 77-108. 67 pp.

Thomas, S. R., and R. A. Soltero. 1977. Recent sedimentary history
    of a eutrophic reservoir: Long Lake, Washington. J. Fish.
    Res. Board Can. *34*:669-676.

Vance, B. D. 1965. Composition and succession of cyanophycean
    water blooms. J. Phycol. *1*:81-86.

Volodin, B. B. 1970. Relationship between sulfur and phosphorus
    in the nutrition of *Microcystis aeruginosa* Kütz. Hydro-
    biologia *6*:47-54.

Yake, W. E. 1979. Water quality trend analysis -- the Spokane
    River basin. D.O.E. Report No. DOE-PR-6. 39 pp.

THE MOSAIC NATURE OF TOXIC BLOOMS OF CYANOBACTERIA

Wayne W. Carmichael[1] and Paul R. Gorham

Department of Botany
University of Alberta
Edmonton, Alberta T6G 2E9
Canada

ABSTRACT

The incidence of toxic populations of planktonic Cyanobacteria in hypertrophic Hastings Lake, Alberta (located 40 km east of Edmonton) was monitored during July and August of 1975, 1976 and 1977. Samples were collected at intervals from different stations, concentrated from 1 to 20 (exceptionally to 200 or 400) times with a plankton net, as required, and tested for toxicity by 1.0 ml intraperitoneal injection into 20 g mice. Species composition was determined microscopically and biomass densities were determined from dry weights. Strains of two of the three predominant species, *Microcystis aeruginosa* Kütz. emend. Elenkin and *Anabaena flos-aquae* (Lyngb.) de Bréb. were isolated, cultured and tested for toxicity. Based on signs of poisoning produced in mice and chicks by populations and isolates, *Microcystis aeruginosa* type-c toxin was found to be primarily responsible for the toxicity observed at different stations during all three seasons. Only one collection (June 27, 1976, concentrated 200 times) indicated the presence of anatoxin-b. On a single day, as well as on different days, population density, species composition and toxicity varied greatly from station to station. On some days, big differences occurred between stations that were only 10 m apart. In July, 1975, localized pulses of type-c toxin were detected for a few days which ranged in titer from 0.4 to 1.4 times the lethal oral dose of anatoxin-a for a 60

---

[1]Present address: Department of Biological Sciences, Wright State University, Dayton, Ohio 45435, U. S. A.

kg calf (estimated as 16,000 mouse units (MU) per 10 liters,
equivalent to 0.5 ml (i.p.) per 20 g mouse of the most dense population composed entirely of type-a). In 1976, one, and in 1977,
two localized pulses of type-c toxin of much lower titer were
detected. The use of toxicity as a marker has revealed that blooms
of Cyanobacteria tend to have a dynamic and mosaic structure.
Remarkably localized differences in strain as well as species composition can occur in response to morphometry and to prevailing
nutritional and other environmental factors that affect growth,
buoyancy, accumulation and dispersal. The dynamic and mosaic
nature of blooms contributes to the unpredictable occurrence of
lethal doses of cyanotoxins that may be ingested by livestock and
wildlife. Other contributing factors are differences in type, pro-
ductions, accumulation, release and inactivation of toxins and in
animal susceptibility.

INTRODUCTION

Unpredictable poisonings of livestock and wildlife by blooms
of freshwater Cyanobacteria (Cyanophytes) have been reported from
all continents. Toxic strains of a comparatively small number of
species are now known to be responsible for these poisonings.
Three toxigenic species are most commonly implicated. These are
Aphanizomenon flos-aquae (L.) Ralfs., *Microcystis aeruginosa* Kütz.
emend. Elenkin and *Anabaena flos-aquae* (Lyngb.) de Bréb. (or taxa
that are difficult to distinguish from them) (see Ref. end of paper).

Some 12 toxins, or toxin mixtures, have been tentatively
recognized or clearly identified to date from strains and/or blooms
of these toxigenic species. They are low-molecular-weight compounds
of two chemically distinct classes, alkaloids and polypeptides, and
they characteristically produce neuromuscular blocking effects
within minutes or hepatotoxic effects within hours, respectively.
So far, the structures of only one toxin (an alkaloid) is known and
confirmed by synthesis. This is anatoxin-a (antx-a) ($C_{10}H_{15}NO$)
from *Anabaena*. *Aphanizomenon* produces a neurotoxin that is thought
to be very similar if not the same as saxitoxin, the alkaloid isolated from *Gonyaulax catenella*. Microcystin (Mcyst) from *Microcystis*, partially characterized as a cyclic polypeptide, is a mixture
of two toxic peptides. Antx-c, from *Anabaena*, produces effects
that resemble those of Mycst, and are very similar to those of a
toxin from *Microcystis* called type-c. Type-c is a mixture of a
hepatotoxic peptide and a neurotoxin. Antx-b and -d are fast-acting
and are presumed to be alkaloids. There are a number of other
peptide-containing toxins, some of which may eventually prove to
have common components.

Poisoning episodes, besides being unpredictable, usually
occur in short-lived pulses. They sometimes do, but often do not,

recur in the same body of water in subsequent years. The principal factors involved in the rise and fall of a poisonous bloom can be summarized as follows:

1. Morphometry and environmental variables that affect growth, buoyancy, accumulation of populations of high density, and dispersal of toxigenic species.

2. Sufficient dominance of a bloom by one or more toxic strains to produce potentially hazardous titers of toxin(s).

3. The access that animals have to a bloom with a hazardous titer, differences in their susceptibilities, and dosage.

4. Senescence, leakage, lysis, grazing, microbial interactions, and decomposition of cells.

5. Detoxification mechanisms.

PURPOSE

The purpose was to provide information on the occurrence and dominance of toxic strains and the incidence, extent and recurrence of lethal or hazardous blooms in a hypertrophic lake by monitoring a number of stations during three successive growing seasons.

METHODS

Samples were collected from 3 to 25 stations in hypertrophic Hastings Lake, Alberta (40 km east of Edmonton) at intervals from late June to early September in 1975, 1976 and 1977. They were concentrated from 1 to 400 times, as necessary, with a plankton net.

After screening out debris, most of each sample was lyophilized, weighed and assayed for toxicity by intraperitoneal (i.p.) injection of 20 g mice. The remainder was used for microscopical determination of percent species composition. A few non-toxic samples were further concentrated by centrifugation and re-assayed. To check for depolarizing neuromuscular blockage, as produced by anatoxin-a some samples were also assayed with 40 to 60 g chicks.

At various times, from different stations, fairly large numbers of colony isolates of the three toxigenic species were made. Success in obtaining unispecific cultures was variable but as many as possible were grown and tested for toxicity (type and titer).

A lethal titer for livestock was assumed to be similar to the oral lethal dose of antx-a for a 60 kg calf. This is estimated as

16,000 mouse units (MU) per 10 liters, equivalent to 0.5 ml (i.p.) per 20 g mouse of a dense bloom that is 100% *Anabaena* type-a.

A potentially hazardous titer, which might cause sickness, was defined as 0.1 to 0.5 or more times a lethal titer.

RESULTS

Results of the study are presented in Figures 1, 2, and 3.

Maximum depth (west of station 19) is 8 m. Prevailing winds are from NW or SW. Morphometric features are given in Table 1. Samples were collected (and concentrated as necessary) from the upper 0.5 m of water at 25 stations. Density ranges are indicated by bar widths and percent species composition by hatching according to codes given in the figure. The dominant species were *Anabaena flos-aquae* (with traces of *A. circinalis* and *A. spiroides*), *Microcystis aeruginosa* and *Aphanizomenon flos-aquae*. "Other" species included *Coelosphaerium kutzingianum*, diatoms, dinoflagellates and green algae.

Densities were high and composition similar at stations 1, 2, 5, 6, and 7 among beds of *Myriophyllum*, *Potamogeton* and *Scirpus* in the NE end of the main basin, close to shore and to two of the low off-shore islands. Densities were low and composition varied greatly at all other stations. Nearby stations, such as 3, 4, and 5, differed significantly in density and composition. Stations 2, 2A and 2B (dominated by *Anabaena*) were toxic (T) (325 to 1130 MU/10 liters) but nearby station 7, which had a similar density and composition, was non-toxic (NT). Samples from low density stations 9, 17, 19, and 25, which had widely differing compositions, were non-toxic even when concentrated further by centrifugation.

Fig. 1.   Hastings Lake, Alberta, showing differences in spatial distribution of planktonic Cyanobacteria on a single day, July 9, 1976.

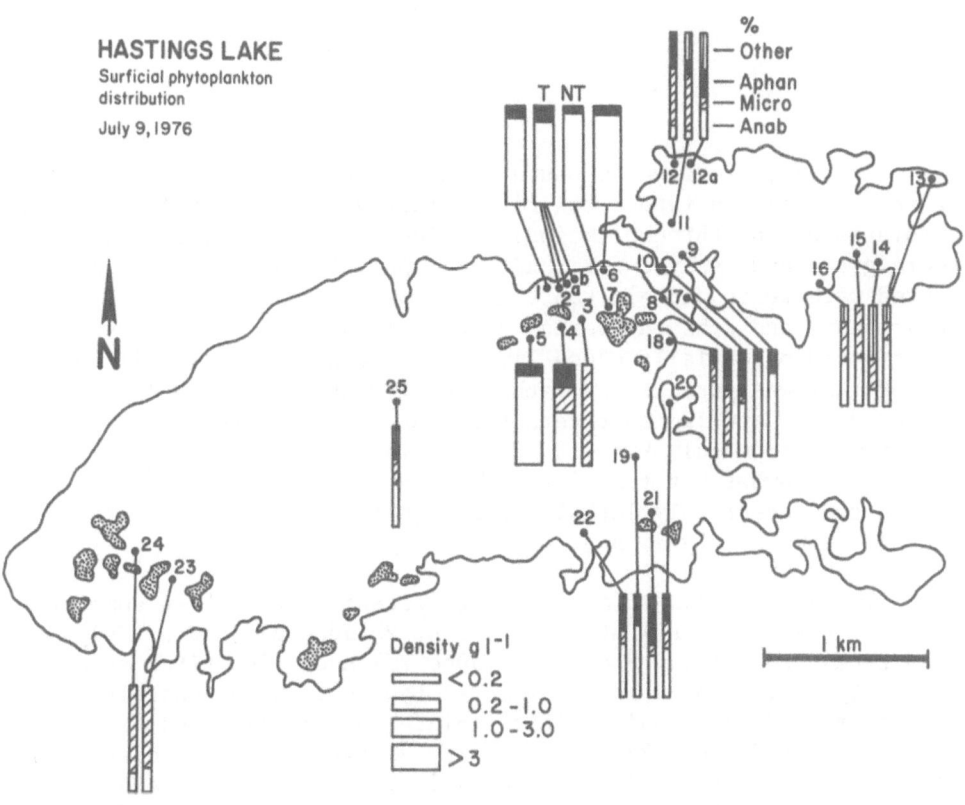

HASTINGS LAKE
Surficial phytoplankton
distribution

July 9, 1976

%
— Other
— Aphan
— Micro
— Anab

Density g l⁻¹
⊏⊐ <0.2
   0.2-1.0
   1.0-3.0
   >3

1 km

Fig. 2. Density, composition and toxicity of planktonic Cyanobac-
         teria at selected stations in Hastings Lake on different
         dates in 1975 and 1976. For toxicity assays, samples
         were concentrated 1 to 400 times (x) as necessary. Den-
         sities are indicated by bar widths, composition by hatch-
         ing, and toxicity by LD_{min} (intraperitoneal mouse minimum
         lethal dose for 100% death in mg/kg body weight) of
         assayed sample and titer (MU/10 liters = mouse units per
         10 liters water).

In 1975, *Microcystis* was a significant component at all stations and all had type-c toxicity. Thirty-four unispecific colony
isolates of the three principal species from the July 15 stations
were grown and assayed for toxicity. Only the 15 *Microcystis* isolates were toxic and all had a high type-c toxicity. On July 15,
stations 1 and 2 had different compositions but had toxicity titers
that would potentially cause livestock sickness; station 11 had a
titer that was potentially lethal for livestock while nearby stations 9, 10, and 12 had titers that were barely detectable. At
station 2, from July 9 to 15, density increased and composition
changed somewhat but toxicity increased from 0.02 times to 0.5
times the potentially lethal titer. There was not much further
change between July 15 and 18.

In 1976, *Microcystis* was a significant component at some but
not all stations. All stations except 2 (June 27) and 6 (July 4)
had type-c toxicity. Station 2 (June 27) had a low titer of antx-b
toxicity (which could have dominated type-c). Unfortunately attempts
to isolate and grow unispecific isolates of *Anabaena* and *Microcystis*
from station 2 (June 27) for toxicity assays were not successful.
At station 2, between June 27 and July 9, the density, composition,
titer and type of toxicity changed. Between July 9 and 197, composition and titer (but not type) changed again.

Stations 2, 2A and 2B, spaced 10 m apart, had differences in
composition and titer on both July 9 and July 17.

At station 6, between July 4 and 17, composition changed from
100% *Anabaena* which was non-toxic (at 1500 mg/kg) to 99% *Microcystis* which had a titer of 1450 MU/10 liters.

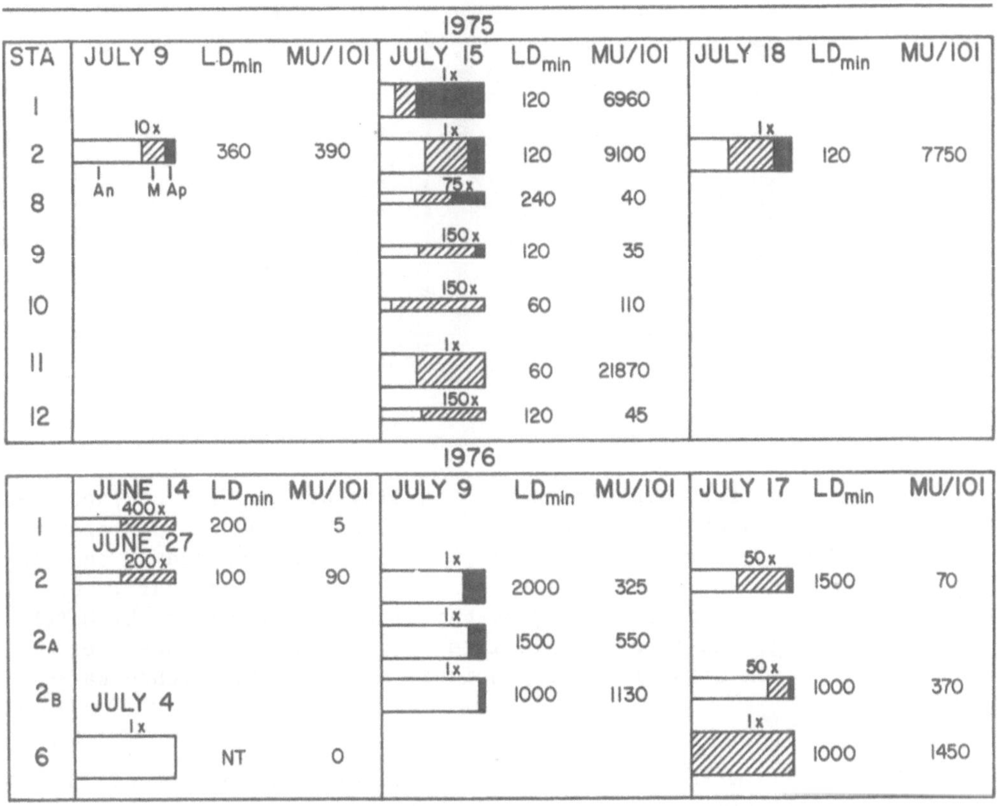

1975

STA	JULY 9	LD$_{min}$	MU/I0I	JULY 15	LD$_{min}$	MU/I0I	JULY 18	LD$_{min}$	MU/I0I
1				1x	120	6960			
2	10x	360	390	1x	120	9100	1x	120	7750
8	An M Ap			75x	240	40			
9				150x	120	35			
10				150x	60	110			
11				1x	60	21870			
12				150x	120	45			

1976

	JUNE 14	LD$_{min}$	MU/I0I	JULY 9	LD$_{min}$	MU/I0I	JULY 17	LD$_{min}$	MU/I0I
1	400x	200	5						
2	JUNE 27 200x	100	90	1x	2000	325	50x	1500	70
2A				1x	1500	550			
2B	JULY 4			1x	1000	1130	50x	1000	370
6	1x	NT	0				1x	1000	1450

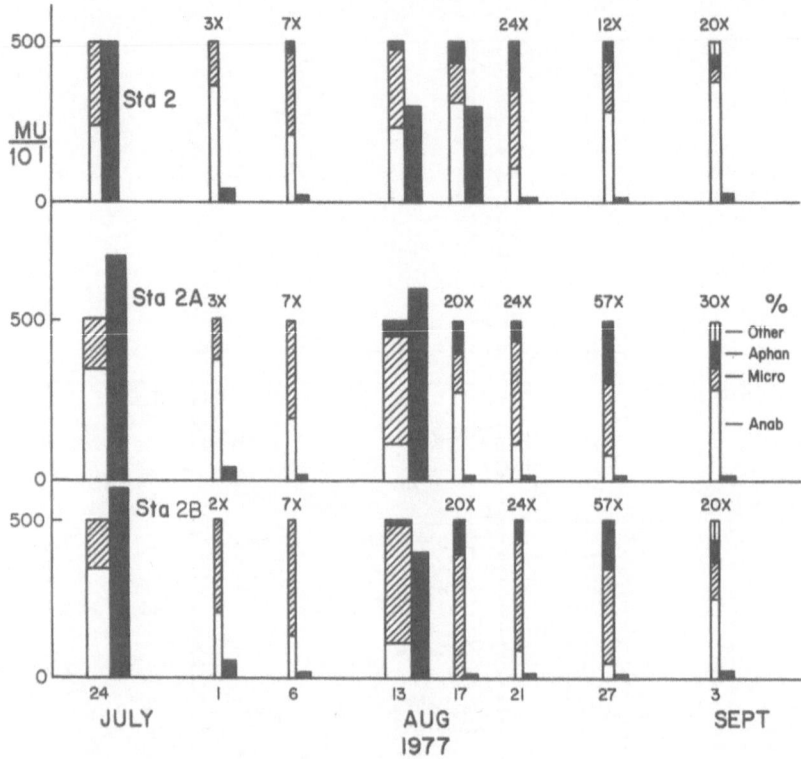

Fig. 3.   Density, composition and toxicity of planktonic Cyanobacteria at 3 stations in Hastings Lake, spaced 10 m apart,
on different dates in 1977. For toxicity assays, samples
were concentrated 1 to 57 times (x), as necessary. Densities are indicated by bar widths, composition by hatching, and titer (MU/10 liters) of toxicity (type-c) by
black bars of varying heights next to the hatched bars.

    In 1977, *Microcystis* was a significant component at all 3 sta-
tions at all times. Changes in density and composition throughout
the season were generally similar at all 3 stations but on 2 dates,
Aug. 17 and 25, there were noticeable differences in composition at
one station compared with the other two. Two pulses of blooming
occurred at all 3 stations on July 24 and Aug. 13 which were accompanied by 2 pulses of toxicity. The maximum titers observed were
only 0.02 to 0.04 times a potentially lethal titer. Except at station 2, which continued to have a significant titer between Aug. 13
and 17, titers at all 3 stations declined in a few days after a
pulse to barely detectable levels. This was correlated mainly with
changes in density rather than composition.

Table 1. Morphometric Features of Hastings Lake[a]

Parameter	Value
Elevation	735.8 m
Area (A)	8.44 km^2
Volume (V)	21.9 x 10^6 m^3
Maximum Length (L)	6156.0 m
Maximum Width (b)	2436.0 m
Maximum Depth (Z_m)	8.0 m
Mean Depth ($\bar{z}$)	2.5 m
Shoreline Length (L)	35.52 km
Shoreline Development (D_L)	3.4

[a]From M. Hickman and C. G. Jenkerson. 1978. Int. Revue Ges. Hydrobiol. *63*:1-24.

DISCUSSIONS AND CONCLUSIONS

Hastings Lake, in which this study was carried out, is typical of many other shallow, hypertrophic lakes in western Canada that regularly produce dense blooms of planktonic Cyanobacteria.

This study of populations of planktonic Cyanobacteria in surface waters has shown that toxicity, as well as density and species composition, tend to vary greatly from station to station on a single day, on different days throughout a growing season, and from one season to the next. Significant differences in toxicity and/or species composition can occur, with or without differences in density, within distances of 100 or even 10 meters.

Only the *Microcystis* cultures that were isolated from the lake were toxic and all were type-c. Probably strains of *Microcystis* were responsible for producing most of the type-c toxin observed at many stations during the three growing seasons.

None of the *Anabaena* strains that were isolated were toxic, but the numbers were small and, therefore, not adequately representative. There were some *Anabaena* blooms that contained no *Microcystis*. Since these had significant titers of type-c toxin,

the logical assumption is that antx-c producing strains of *Anabaena*
were present. Using the same argument, strains of antx-b must also
have been present in the one *Anabaena:Microcystis* bloom in which it
was detected.

By using concentrated samples, as necessary, toxins were
detected at many but not all stations at all times. One or two
pulses of toxin production occurred in each season at certain
stations. One that was potentially hazardous or lethal to livestock occurred in the first season, but pulses of a much lower
titer occurred in each of the following seasons. From this and
other evidence it can be concluded: 1. That toxic strains persist
in water bodies, often in very low amounts, for periods of years,
and 2. That localized, sub-lethal pulses of toxin production,
which would ordinarily escape detection, probably recur annually,
even though poisoning episodes do not.

The combined use of toxicity as a marker for changes in strain
composition and microscopical measurements for changes in species
composition has revealed that blooms of planktonic Cyanobacteria
are dynamic aggregations that have an ever-changing mosaic structure or pattern. Remarkably localized differences in strain and
species composition and density tend to occur in response to
morphometry and environmental variables that affect growth, buoyancy, accumulation and dispersal of the organisms.

This study is not the first to recognize the mosaic nature of
toxic Cyanobacteria blooms. Professor Theodore Olson of the University of Minnesota did studies similar to ours in the late
1940's to the early 1950's (Olson, 1951, ca. 1957 and ca. 1964).
He notes (Olson, ca. 1957) that "it is well known that the number
of organisms per unit volume of water may vary astronomically at
two points less than fifty feet apart along the same shore line".
In his collections eight species of Cyanobacteria were present, in
various mixtures, in the toxic blooms sampled. In all cases *Microcystis aeruginosa*, *Aphanizomenon flos-aquae* and *Anabaena Lemmer-
manni (flos-aquae)* were the dominant organisms. To give an example;
he collected samples, totaling 137 gallons of heavy surface bloom,
from 88 lakes in 1949. He found that 28 of the lakes were toxic
with *M. aeruginosa* and *Aph. flos-aquae* being the dominant organisms
from all the toxic blooms. His studies showed that toxic blooms
were far more common in occurrence than one would be led to believe
if the only index used were poisoning episodes involving animals.
He also noted how a toxic bloom can migrate from one side of the
water body to the other as wind conditions change. This results
in animal losses at a point in the lake one day, followed by losses
at another point the following day and no further losses at the
site where animals were originally lost (Olson, 1951).

The dynamic, mosaic nature of a bloom of planktonic Cyanobacteria is thus another, very important factor which must be added to the long list of other factors (summarized in the introduction) that work together to determine the unpredictable, often short-lived occurrence of poisonings of livestock and wildlife.

SUMMARY

1. A 3-year study of populations of planktonic Cyanobacteria in the surface waters of Hastings Lake, Alberta has been carried out.

2. Significant differences in toxicity and/or species composition can occur, with or without differences in density, between populations that are no more than 10 meters apart.

3. *Microcystis* type-c toxin was detected most widely and consistently during the three seasons and probably accounted for most of the toxicity observed.

4. Antx-b and antx-c were also detected.

5. Toxic strains apparently persist in water bodies for years, often in very low amounts.

6. In water bodies in which poisoning episodes have occurred, localized, sub-lethal pulses of toxicity probably recur in each growing season even though poisoning episodes do not.

7. Blooms of planktonic Cyanobacteria have a dynamic, mosaic structure which contributes significantly to the unpredictable, often short-lived occurrence of poisoning episodes.

REFERENCES

Carmichael, W. W., and P. R. Gorham. 1980. Freshwater cyanophyte toxins: types and their effects on the use of micro-algae biomass. (In press) *in* G. Shelef and C. J. Soeder, eds., The Production and Use of Micro-Algae Biomass. Elsevier/North Holland, New York.

Collins, M. 1978. Algal toxins. Microbiol. Rev. $42(4):725-746$.

Gentile, J. H. 1971. Blue-green and green algal toxins. Pages 27-66 *in* S. Kadis, A. Ciegler and S. J. Ajl, eds., Microbial Toxins, Vol. VII, Algal and Fungal Toxins. Academic Press, London. 401 pp.

Gorham, P. R., and W. W. Carmichael. 1979. Phycotoxins from blue-green algae. Pure Appl. Chem. $52(1):165-174$.

Gorham, P. R., and W. W. Carmichael. 1980. Toxic substances from
    freshwater algae. Prog. Water Technol. *12*:189-198.
Kirpenko, Y. A., L. A. Sirenko, V. M. Orlovskii, and L. F. Lukina.
    1977. Toxic blue-green algae and culture of the organism.
    In A. V. Topachevskiy and E. E. Kvetnetskia, eds., Naukogo
    Dumka. Ukrainian Scientific Publishing House, Kiev. 252 pp.
    (Russian).
Moore, R. E. 1977. Toxins from blue-green algae. Bioscience
    27(12):797-802.
Olson, T. A. 1951. Toxic plankton. Pages 86-95 *in* Proceedings
    of Inservice Training Course in Water Works Problems,
    February 15-16. University of Michigan, School of Public
    Health, Ann Arbor, Michigan.
Olson, T. A. ca. 1957. Water Poisoning - a Study of Poisonous
    Algae Blooms in Minnesota. Mimeo. report. 16 pp.
Olson, T. A. ca. 1964. Waterfowl and Algal Toxins. Mimeo.
    report. 18 pp.
Schwimmer, M., and D. Schwimmer. 1968. Medical aspects of phycol-
    ogy. Pages 279-358 *in* D. F. Jackson, ed., Algae, Man and
    the Environment. Syracuse University Press, New York.
    369 pp.
Shilo, M. 1972. Toxigenic algae. Pages 233-265 *in* O. J. D.
    Hockenhull, ed., Progress in Industrial Microbiology, Vol.
    11. Churchill Livingstone Press, Edinburgh.

COLLECTION, PURIFICATION, AND CULTURE

OF CYANOBACTERIA

Gary J. Court, J. Helen Kycia,
and Harold W. Siegelman

Biology Department
Brookhaven National Laboratory
Upton, New York 11973

ABSTRACT

The collection of Cyanobacteria from nature, isolation of pure
cultures, and large-scale culture techniques are prerequisites for
detailed studies. A plankton net is generally used to obtain concentrated collections from natural waters. Replicates of the concentrated collections are placed directly into several standard
liquid mineral media in the field. The mixed cultures are brought
to the laboratory and streaked on 0.75 to 1.00% agar plates of the
several mineral media. After a suitable incubation, individual
colonies are transferred to their respective liquid mineral media.
Alternation of the agar plating and liquid culture is continued
until axenic cultures are obtained. Other techniques are available
for obtaining axenic cultures if required. Purifications from lake
bottom sediments may provide a source of Cyanobacteria absent from
the waters. Culture collections can furnish important organisms
for study. Large-scale culture of Cyanobacteria is conveniently
achieved by simple continuous culture techniques in 20 liter glass
or polypropylene bottles.

Axenic Cyanobacteria rapidly growing in large-scale cultures
permit an assessment of nutritional and environmental factors on
their physiology, biochemistry, and toxin production.

TEXT

Detailed studies of the structure, physiology, biochemistry,
and toxin production of Cyanobacteria require a dependable and

reproducible supply of selected organisms. Axenic cloned strains
of Cyanobacteria should be used whenever possible. Several physical and chemical techniques are available to purify cultures to
the axenic condition. Cultural techniques are described here
which can be used to grow adequate amounts, greater than 100 g
fresh weight, of Cyanobacteria in about four weeks.

Many practical aspects of the purification and culture of
Cyanobacteria and phytoplankton are summarized by Castenholz, 1970;
Stanier et al., 1971; Siegelman and Guillard, 1971; Allen, 1973;
Guillard, 1975; Stanier and Cohen-Bazire, 1977; Siegelman and Kycia,
1979; Rippka et al., 1979; and Rippka et al., 1980. The Handbook
of Phycological Methods edited by J. R. Stein (1973a); The Biology
of Blue-green Algae edited by N. G. Carr and B. G. Whitton (1973b);
and The Blue-green Algae edited by G. E. Fogg, W. D. P. Stewart,
P. Fay, and A. E. Walsby (1973) also contains extensive information.

Methods summarized here for the collection, purification, and
culture of Cyanobacteria can be used to grow many kinds of freshwater and marine phytoplankton with only minor modifications.
Techniques adequately described in the cited literature are not
fully detailed in this report. It must be emphasized that there
are no definitive rules for purification and culture. Many suggestions are provided here, but in the final analysis there is no
substitute for the ingenuity and imagination of the individual
investigator.

Collection

Cyanobacteria not available from culture collections can be
obtained from natural sources. Government public health and water
pollution control agencies can often provide information on the
occurrence of blooms and particular locations for collection of
specific organisms.

A plankton net, about 80 μm mesh size, can be used to collect
and concentrate freshwater samples (available from Wildlife Supply
Co., Saginaw, Michigan; Ernest A. Case, Andover, New Jersey; or
V. W. R. Scientific Inc.). It is towed through the water from
shore or from a boat. Lake, pond, or stream bottom mud samples
may contain organisms not present in the waters during the collection period. Marine Cyanobacteria are obtained from sediments,
algal mats, rocks, or plankton tows. A tabulation of collection
sites of marine and freshwater Cyanobacteria is provided by Rippka
et al. (1979).

Collected liquid samples should be transferred into two to
four different liquid culture media in flasks at the collection

site. The initial cultures are held temporarily at room temperature (≈25°C), under diffuse daylight illumination. Mud and rock
samples are held in small bottles or vials; a moistened filter
paper inserted in the container lid, not in contact with the
sample, prevents water loss.

On return to the laboratory it is prudent to transfer portions
of the collections into the same two to four different liquid culture media. Observations of the initial cultures can provide suggestions for selection of a suitable culture medium. Samples
should be streaked on 0.75 to 1.00% agar plates as soon as possible
after collection to initiate purification. The agar plates are
prepared with the same two to four liquid culture media. Suspensions of the mud samples are placed in several liquid culture media
and streaked on agar plates as for liquid samples. Storage of mud
samples is probably satisfactory at 5°C. Laboratory cultures of
freshwater Cyanobacteria will generally grow at about 25°C. Illumination is provided with 500 to 1000 lux of continuous cool white
fluorescent light. Temperatures of about 18°C are required for
some marine forms and above 37°C for hot spring Cyanobacteria.

Cyanobacteria are often available from general culture collections and specialist collections. Lists of collections are provided by Stein (1973b) and Komárek (1973). The Indiana University
Culture Collection (IUCC) was transferred to the University of
Texas (UTEX), Austin, Texas (Starr, 1978).

The American Type Culture Collection, 12301 Parklawn Drive,
Rockville, Maryland, holds their collection of Cyanobacteria at
liquid nitrogen temperature. Storage of Cyanobacteria at low
temperature, when satisfactory, is an important method for maintaining genetic stability (Holm-Hansen, 1973).

Purification

The purification of Cyanobacteria is essential to avoid possible
complexities associated with contaminant microorganisms. Many forms
are easily purified, but others may require prolonged and dedicated
effort (Allen, 1973). Physical and chemical methods for cyanobacterial purification are listed by Allen (1973), Carr et al.
(1973a), and Carmichael and Gorham (1974). Additional useful
methods are described by Guillard (1973) and Hoshaw and Rosowski
(1973). Streaking on agar plates is probably the most widely used
purification method (Stanier et al., 1971; Allen, 1973; and Rippka
et al., 1979). Brief treatment, 1 to 2 minutes, of the inoculum
in a 50- to 100-watt ultrasonic bath, blending with glass beads
(McCurdy and Hodgson, 1974), or grinding in a glass homogenizer
(Allen, 1973) is used to break filamentous forms into short fila-

ments prior to streaking on agar. Washing filamentous forms in
sterile spot plates, depression slides (Guillard, 1973), or on 10
μm pore size polycarbonate membranes (Nuclepore Corp., Pleasanton,
California) can reduce contamination (J. B. Waterbury, personal
communication).

Alternation of streaking on agar and growth in liquid culture
media is continued until the culture is purified. Streaking of
the plates is simplified by placing three to four sterile 6 mm
glass beads on the agar plate, after the addition of inoculum, and
then horizontally shaking the plate. The beads are removed from
the plate prior to incubation. Agar plates are held in transparent
polystyrene boxes to reduce water loss and contamination (Stanier
et al., 1971). Colonies should be selected from agar plates as
soon as possible to avoid overgrowth by contaminants. The larger
the number of agar plates used and isolates selected, the better
the chances of obtaining an axenic isolate.

Purification of phototactically motile forms is assisted by
unilateral illumination (Stanier et al, 1971; Allen, 1973). Stock
cultures of Cyanobacteria are conveniently maintained on agar
slants. All sterile transfers are made in a laminar flow hood
equipped with a 0.3 μm air filter. Flaming of flasks and tubes is
avoided in laminar flow hoods because of disturbance of the laminar
air flow. If fungal contamination of agar plates occurs in the
laminar flow hood, a simple still-air hood may be substituted
(Guillard, 1973).

Media

A large number of liquid mineral media are available for culturing freshwater and marine Cyanobacteria. Their compositions
are given by Stanier et al. (1971); Siegelman and Guillard (1971);
Nichols (1973); McLachlan (1973); Carr et al. (1973a); Guillard
(1975); and Rippka et al. (1979). Selection of a satisfactory
freshwater medium is achieved by trying several media, e.g. the D
medium of Kratz and Myers (1955); ASM-1 medium of Gorham et al.
(1964); BG-11 or BG-11$_0$ medium of Allen (1973) and Rippka et al.
(1979); and WC medium of Guillard (1975). Marine media are similarly selected. Consultation of the literature provides initial
guidance, but there is no substitute for personal experience.

Equipment to prepare high quality water is commercially available. Distilled or reverse-osmosis waters are subjected to carbon
absorption, mixed-bed ion exchange, and 0.2 μm filtration to a
specific resistance of 18 megohms/cm. Complete purification units
are available from the Barnstead Co., Boston, Massachusetts, and
from Millipore Corp., Bedford, Maryland, or they can be assembled

from commercially available modules. Salts used for media preparation are reagent grade. For large volumes of artificial marine
media, noniodized food-grade NaCl and USP grade $MgSO_4.7H_2O$ can be
substituted for reagent-grade chemicals after preliminary comparative trials. Large volumes of culture medium are prepared in a
120 liter polyethylene drum with a bottom spigot. The drum is
mounted on a dolly about 50 cm high to permit filling of media
bottles. A heavy duty stirrer is mounted on a column attached to
the dolly with its stainless steel shaft and propeller in the drum.
The stirrer aids dissolution of the chemicals and pH adjustment.
A sealed plastic body combination pH electrode with a protected
glass tip is used for pH monitoring. The liquid culture medium is
dispensed into 10 liter or 20 liter autoclavable polypropylene
bottles, which are used instead of borosilicate glass bottles for
media sterilizing and storage. They are lighter, unbreakable and
have large handles for ease of lifting. A 20 cm^2 cover of two
layers of unbleached muslin with a double thickness of aluminum
foil between the muslin layers, is securely fastened over the
bottle mouth with a twist tie (available from garden stores). No
problems have been encountered using liquid culture media autoclaved in polypropylene bottles. Possible toxicity for sensitive
organisms should be tested (Blankley, 1973).

Silicone sponge closures (Bellco Glassware, Vineland, New
Jersey) are used to cover flasks and small bottles containing culture media. Bare aluminum foil covers should not be used on flasks
or bottles prior to autoclaving because they act as heat insulators
and barriers to the penetration of steam into the containers. Indicating autoclave tape is placed on all materials prior to sterilization.

Small Cultures

The general techniques used follow procedures carefully
described by Guillard (1975). The flasks are not aerated but are
shaken occasionally.

Large Cultures

Borosilicate glass or polypropylene bottle cultures, 8 to 20
liters, are aerated at about 350 ml per minute with 0.07 to 0.1%
CO_2 in air. The air-CO_2 mixtures are obtained by mixing 100% CO_2
from a cylinder with carbon-filtered air (Koby Junior Filters,
Koby Inc., Melrose, Maryland), using a back-pressure-compensated
gas proportioner (Aalborg Instruments, Munsey, New York).

The bottle cultures are illuminated with continuous cool
white fluorescent light at about 2000 lux. It may be desirable

to reduce the light intensity for three to four days after starting
a bottle culture by interposing one or two layers of window screen
between the culture and lamps.

The temperature selected for growing large cultures is generally a compromise between specific requirements and available
facilities. A large number of Cyanobacteria grow satisfactorily
at about 25°C. Forms demanding high or low temperatures are grown
in illuminated incubators, preferably with glass doors permitting
an unobstructed view of the cultures and external illumination if
desired. The doors are kept locked during culturing of toxic
forms.

Continuous Culture Apparatus

Cyanobacteria are cultured in 8 or 20 liter borosilicate glass
or polypropylene bottles. The system used is shown in Figure 1.
White or neoprene rubber stoppers are used, or silicone rubber
stoppers if rubber stoppers prove unsatisfactory. The stoppers
and associated tubing and fittings are wrapped in a double thickness of muslin and autoclaved. Nonwoven cellulose sterilization
wrap (American Hospital Supply, Evanston, Illinois) is currently
under trial and appears to be a satisfactory reusable substitute
for muslin. The stopper assembly and associated tubing are sterile-
transferred to the bottles in a laminar flow hood. A pair of stainless steel radiator hose clamps attached to the neck of borosilicate
glass bottles permits the rubber stopper to be securely fastened
with large rubber bands or string. All fittings are attached to
7 mm medium wall glass tubing with silicone rubber tubing. Aeration
inlet and venting outlets are through 30 or 37 mm membrane (0.3 μm)
air vents (Gelman Instruments, Ann Arbor, Michigan or Schleicher
and Schull, Keene, New Hampshire) which are reusable. Venting is
also accomplished with 25 mm or 47 mm polycarbonate Swin-Lok
holders (Nuclepore, Pleasanton, California) using 1 μm microfibre
filters (Whatman, Inc., Clifton, New Jersey). Glass filters are
preferred to other types in the venting outlets because they are
less subject to blockage on accumulation of the Cyanobacteria on
their surface. All tubing from the air-CO_2 supply to the aerator
tubes, and for all liquid transfers, except the silicone tubing in
the peristaltic pumps, are made with 1.2 m extension tubes having
luer lock end fittings on each end (Pharmaseal Laboratories,
Glendale, California). All glass tubing utilized for liquid transfer or aeration has a luer lock fitting attached with silicone
rubber tubing to allow attachment of the extension tubes. The luer
lock fittings are obtained from discarded extension tubes.

Aeration manifolds are constructed of polyvinylchloride
aquarium manifold pipe and plastic valves (Kenwill Co., Athens,

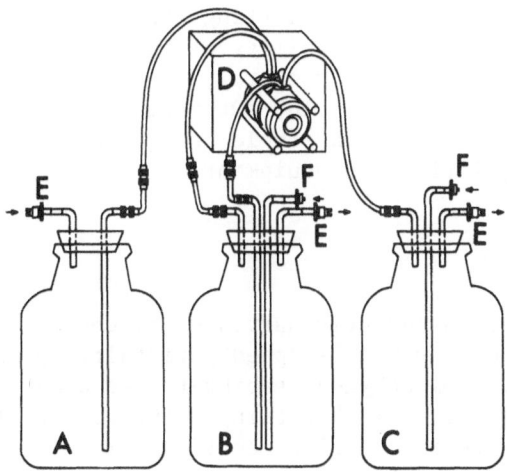

Fig. 1.   Continuous culture apparatus. The components are not
          drawn to scale and are as follows: A, sterile medium
          bottle;   B, culture bottle;   C, culture harvest bottle;
          D, synchronous peristaltic pumps;   E, glass microfiber
          filters;   and F, membrane air vent and air-CO_2 mixture
          inlet.

Georgia). A luer lock fitting is attached to each valve, adopting
a uniform convention for the direction of attachment of the fittings. A pair of peristaltic pump heads, mounted in tandem to
maintain synchrony, are driven by a variable speed motor for liquid
transfer (Cole Parmer, Chicago, Illinois). Flow rates of 1 to 10
ml/minute provide a useful range which may also be varied by changing the size of the pump head.

     The continuous culture apparatus simplified the growing of
Cyanobacteria. The following protocol is used for large scale
culture. A sterile borosilicate glass or polypropylene bottle is
inoculated with about 1 liter of an actively growing culture and 1
liter of sterile medium is added by pumping. The volume of the
culture is approximately doubled daily by pumping in sterile medium.
Sterile medium can also be added by siphoning. When the culture
reaches about mid-log phase, a sterile culture harvest bottle is
attached to the apparatus and filled to a volume of about 18 liters
by pumping the Cyanobacteria and medium from the culture bottle in
two to three days. Concurrently, the culture bottle is maintained
at a volume of about 18 liters by addition of sterile medium. The
newly filled culture harvest bottle is removed from the apparatus,
and allowed to reach the desired cell density. The procedure of
filling culture harvest bottles can be continued for four to six
weeks. Deterioration of the culture is indicated by excessive and
persistent foaming. Advantages of the peristaltic pump are: ease

of sterilizing, elimination of lifting sterile medium bottles, and
control by timers. The availability of inexpensive microprocessor-
controlled variable timers (Chrontrol, San Diego, California) permits the development of a simple, automated, continuous culture
system. Batch culturing of a single bottle can also be performed
with the same facilities and equipment.

Harvesting

Cultures are conveniently harvested by continuous centrifugation. The inlet of a variable speed peristaltic pump is connected
with 6.4 mm (O.D.) polyethylene tubing to the areation tube of a
culture bottle by a luer lock fitting. The outlet of the pump is
connected to the continuous flow centrifuge inlet. The maximum
flow rate is a function of rotor speed and ease of sedimentation.
Typically, a flow rate of 400 ml/min at 12,000 x g will provide
efficient harvesting of many Cyanobacteria. Following harvest, the
combined cell paste may need to be recentrifuged for 5 minutes at
18,000 x g to remove additional medium. Alternately, the harvested
cells are collected on sharkskin filter paper on a buchner funnel
to remove excess medium. Addition of $CaCl_2.2H_2O$ to a final concentration of 5 mM in the culture medium may assist the harvesting
of freshwater forms.

Special Methods

Herdman et al. (1970) synchronized growth of a cyanobacterium
for at least three divisions by deprivation of light and CO_2. Dialysis culture, a technique for examining the physiology and biochemistry of Cyanobacteria (see Jensen et al. 1972; Schultz and Gerhardt,
1969) permits culturing at constant nutrient conditions and studies
of the nature of autoinhibition processes. Membranes with a wide
range of molecular weight cutoffs are now available as a new inves-
tigative tool for dialysis culture.

Comments

The purification of Cyanobacteria is a critical prerequisite
for further studies. Agar plating is generally the procedure used
initially, but some recalcitrant Cyanobacteria are not freed easily
of contaminants and other techniques must be tried. Chemical and
physical treatments for purification must be adjusted in concentration and dosage, respectively, for each organism. Success in purification depends on persistence and patience. Treatments which can
induce mutations, such as ultraviolet irradiation, should be used
with caution.

ACKNOWLEDGMENTS

We thank Dr. John B. Waterbury, Woods Hole Oceanographic
Institution, Woods Hole, Massachusetts for many helpful suggestions.

This work was carried out under the auspices of the U. S.
Department of Energy and the National Institutes of Environmental
Health Sciences.

REFERENCES

Allen, M. M. 1973. Methods for Cyanophyceae. Pages 127-238 *in*
    J. R. Stein, ed., Handbook of Phycological Methods - Culture
    Methods and Growth Measurements. Cambridge University Press,
    London.
Blankley, W. F. 1973. Toxic and inhibitory materials associated
    with culturing. Pages 207-229 *in* J. R. Stein, ed., Handbook
    of Phycological Methods - Culture Methods and Growth
    Measurements. Cambridge University Press, London.
Carmichael, W. W., and P. R. Gorham. 1974. An improved method for
    obtaining axenic clones of planktonic blue-green algae. J.
    Phycol. *10*:238-240.
Carr, N. G., J. Komárek, and B. A. Whitton. 1973a. Notes on iso-
    lation and laboratory culture. Pages 525-530 *in* N. G. Carr
    and B. A. Whitton, eds., The Biology of Blue-green Algae.
    University of California Press, Berkeley and Los Angeles.
Carr, N. G., and B. A. Whitton, eds. 1973b. The Biology of Blue-
    green Algae. University of California Press, Berkeley and Los
    Angeles. 676 pp.
Castenholz, R. W. 1970. Laboratory culture of thermophilic cyano-
    phytes. Schweiz. Z. Hydrol. *32*:538-551.
Fogg, G. E., W. D. P. Stewart, P. Fay, and A. E. Walsby. 1973.
    The Blue-green Algae. Academic Press, London. 459 pp.
Gorham, P. R., J. McLachlan, U. T. Hammer, and W. K. Kim. 1964.
    Isolation and culture of toxic strains of *Anabaena flos-
    aquae* (Lyngb.) de Bréb. Int. Assoc. Theor. Appl. Limnol.
    15:796-804.
Guillard, R. R. L. 1973. Methods for microflagellates and nanno-
    plankton. Pages 69-85 *in* J. R. Stein, ed., Handbook of
    Phycological Methods - Culture Methods and Growth Measure-
    ments. Cambridge University Press, London.
Guillard, R. R. L. 1975. Culture of phytoplankton for feeding
    marine invertebrates. Pages 29-60 *in* W. L. Smith and M. H.
    Chaney, eds., Culture of Marine Invertebrates. Plenum
    Publishing Corp., New York.
Herdman, M., B. M. Faulkner, and N. G. Carr. 1970. Synchronous
    growth and genome replication in the blue-green algae *Ana-
    cystis nidulans*. Arch. Microbiol. *73*:238-249.

Holm-Hansen, O. 1973. Preservation by freezing and freeze-drying.
    Pages 195-205 *in* J. R. Stein, ed., Handbook of Phycological
    Methods - Culture Methods and Growth Measurements. Cambridge
    University Press, London.

Hoshaw, R. W., and J. R. Rosowski. 1973. Methods for microscopic
    algae. Pages 53-68 *in* J. R. Stein, ed., Handbook of Phyco-
    logical Methods - Culture Methods and Growth Measurements.
    Cambridge University Press, London.

Jensen, A., B. Rystad, and L. Skoglund. 1972. The use of dialysis
    culture in phytoplankton studies. J. Exp. Mar. Biol. Ecol.
    8:241-298.

Komárek, J. 1973. Culture collections. Pages 519-524 *in* N. G.
    Carr and B. A. Whitton, eds., The Biology of Blue-green
    Algae. University of California Press, Berkeley and Los
    Angeles.

Kratz, W. A., and J. Myers. 1955. Nutrition and growth of several
    blue-green algae. Am. J. Bot. *42*:282-287.

McCurdy, H. D., and W. Hodgson. 1974. The isolation of blue-green
    bacteria in pure culture. Can. J. Microbiol. *20*:272-273.

McLachlan, J. 1973. Growth media-marine. Pages 25-51 *in* J. R.
    Stein, ed., Handbook of Phycological Methods - Culture
    Methods and Growth Measurements. Cambridge University Press,
    London.

Nichols, H. W. 1973. Growth media-freshwater. Pages 7-24 *in*
    J. R. Stein, ed., Handbook of Phycological Methods - Culture
    Methods and Growth Measurements. Cambridge University Press,
    London.

Rippka, R., J. Deruelles, J. B. Waterbury, M. Herdman, and R. Y.
    Stanier. 1979. Generic assignments, strain histories, and
    properties of pure cultures of Cyanobacteria. J. Gen. Micro-
    biol. *111*:1-61.

Rippka, R., J. B. Waterbury, and R. Y. Stanier. 1980. (In press)
    in M. Starr, ed., The Procaryotes. Springer, New York.

Schultz, J. S., and P. Gerhardt. 1969. Dialysis culture of micro-
    organisms: design, theory, results. Bacteriol. Rev. *33*:2-47.

Siegelman, H. W., and R. R. L. Guillard. 1971. Methods Enzymol.
    23:110-115.

Siegelman, H. W., and J. H. Kycia. 1979. Large scale culture of
    dinoflagellate algae. Pages 115-120 *in* D. L. Taylor and
    H. W. Seliger, eds., Toxic Dinoflagellate Blooms. Elsevier/
    North Holland, New York.

Stanier, R. Y., and G. Cohen-Bazire. 1979. Phototrophic prokar-
    yotes: the Cyanobacteria. Ann. Rev. Microbiol. *35*:171-205.

Stanier, R. Y., M. Kunisawa, M. Mandel, and G. Cohen-Bazire. 1971.
    Purification and properties of unicellular blue-green algae
    (order Chroococcales). Bacteriol. Rev. *35*:171-205.

Starr, R. C. 1978. The culture collection of algae at the Univer-
    sity of Texas at Austin. J. Phycol. *14*(suppl):47-100.

Stein, J. R., ed. 1973a. Handbook of Phycological Methods – Culture
    Methods and Growth Measurements. Cambridge University Press,
    London. 448 pp.
Stein, J. R. 1973b. Introduction. Pages 1-4 *in* J. R. Stein, ed.,
    Handbook of Phycological Methods – Culture Methods and Growth
    Measurements. Cambridge University Press, London. 448 pp.

CULTURE OF PLANKTONIC CYANOPHYTES ON AGAR

E. A. Dale Allen and Paul R. Gorham

Department of Botany
University of Alberta
Edmonton, Alberta T6G 2E9
Canada

ABSTRACT

The difficulties experienced by phycologists in growing certain species of planktonic cyanophytes reliably on agar media have been traced to heat-induced chemical reactions in agar. Non-axenic clones or cultures of *Anabaena flos-aquae* (Lyngb.) de Bréb. NRC-44-1 and A-113-9-q-2, *Anabaena sub-cylindrica* Borge A 78-1, *Aphanizomenon flos-aquae* (L.) Ralfs. NRC-568 and *Microcystis aeruginosa* Kütz. emend. Elenkin NRC-1, SS-17 (NIVA collection) were tested on agars treated in various ways. Partial or complete lysis of filamentous forms was observed within minutes when they were streaked on washed Difco Bacto agar autoclaved separately from double-strength ASM-1 medium and combined at 35°C just prior to pouring. The lytic response could be simulated by surfactants such as Triton-X or Lux dishwashing detergent. Lysis of filaments was used to bioassay the titer of the lytic agent(s) in treated agars and extracts. Species and strains showed regular differences in susceptibility. Different brands and grades of agar, whether or not they were gelled and washed repeatedly before use, caused lysis after autoclaving. The titer of lytic agent(s) is a function of temperature and duration of heating. The agent(s) can be extracted from heat-treated agars by repeated leachings with distilled water. Substantial titers of lytic agent(s) are produced each time completely leached agar is autoclaved. Successful growth of streaks of the five test cyanophytes has been achieved on ASM-1 agar medium of pH 8.0 to 8.5 made with ·purified agar that is separately melted in a microwave oven at 80°C for 3 minutes, combined and sterilized by germicidal ultraviolet light for 15 minutes. Reliable growth of

colonies from single filaments or cells on agar medium that are
low in lytic agents will greatly facilitate purification and
cloning of planktonic cyanophytes.

INTRODUCTION

    The culturing of cyanophytes on agar media has been routinely
used by a number of phycologists for maintenance or purification
of certain species, but it has long been known that (certain) other
species could not be grown reliably on agar. Early work using agar
medium was carried out by M. B. Allen (1952) and M. M. Allen (1968).
The latter found that inhibition of the growth of *Anacystis nidulans*
could be relieved by autoclaving the agar and nutrients separately
and by reducing the agar concentration to 1.5%. Stanier et al.
(1971) followed M. M. Allen's 1968 procedure and successfully used
agar medium for isolation and growth of numerous cyanophytes.
Carmichael and Gorham (1974) reported inhibitors in gelled agar
that were removable by extensive washing. They were able to grow
single filaments of *Anabaena flos-aquae* and *Aphanizomenon flos-
aquae* in pour plates but noted that growth was often slow or inhibited during 2 to 3 days of incubation. They also noted little or
no growth occurred when streaked on surfaces of the washed agar
medium. Close examination of a number of species that would not
grow on agar medium revealed that with some, the inoculum disappeared
within 10 to 60 minutes while with others, it remained intact but
without growth for many days.

    Ley and Mueller (1946) found that a substance which was
extractable from agar by methanol and characterized by them as a
fatty acid, strongly inhibited the growth of *Neisseria gonorrhoeae*.
They were able to reduce inhibition by incorporating starch in
their medium and they presumed that it adsorbed the fatty acid.
Robbins and McVeigh (1951) and Yaphe (1959) have described the
inhibitory effects of acid-hydrolyzed agar on the fungus *Tricho-
phyton mentagrophytes* and the bacterium *Escherichia coli*, respectively. Yaphe found that a similar type of growth inhibition of
E. coli was produced by 5-hydroxymethyl-2-furaldehyde and 3,6-
anhydro-L-galactose, both of which are produced by acid hydrolysis
of agar.

TEXT

    When clonal isolates of two species of planktonic cyanophytes
were streaked on plates of sterile washed agar containing no added
nutrients either complete lysis of the cells or fragmentation of
the trichomes by lysis of one, two or three cells at irregular
intervals (clipping) occurred within an hour or so (Figure 1). By
placing a small block of this agar gel in a drop of nutrient medium

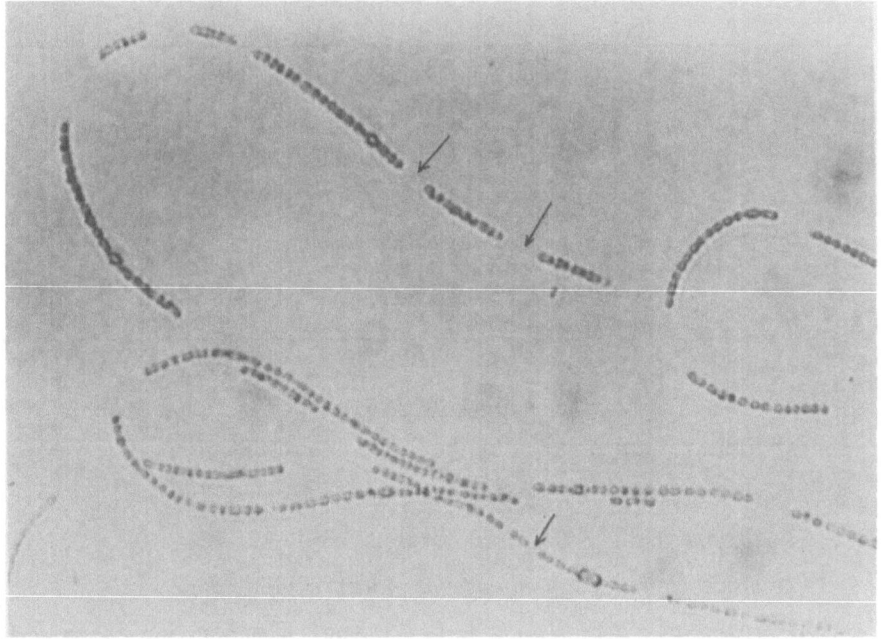

Fig. 1. Trichomes of *Anabaena sub-cylindrica* A 78-1 on heat-treated
agar medium shows lysis of groups of cells (clipping - at
arrows). Photographed through agar from below.

containing a few trichomes of the test species, it was found that
substances causing rapid lysis diffused out of the gel. Agar gels
which had been washed in several changes of distilled water but not
autoclaved did not produce lysis. Four commercial agars that represented a range of manufacturers and purity were made into gels
and tested for lysis with five non-axenic, clonal isolates of
planktonic cyanophytes (Table 1) using the block assay and streaked
plates. Autoclaved gels (1.5% w/v) of the four agars caused lysis
in varying degrees. Susceptibility to complete lysis or clipping
varied among the isolates tested.

Washing agar gels in several changes of distilled water
removed the material which caused cell damage, but autoclaving or
boiling to melt them for pouring made them lytic or inhibitory
again. The washing and heating cycle could be repeated several
times without diminution in the toxic effects. This indicated
that the lytic or inhibitory compounds are primarily degradation
products of the agars and not impurities that were present from
the outset.

It was found that washed agars could be heated to 80°C for a
short time and melted without lytic or inhibitory effects being

Table 1. Agars and Non-axenic Clonal Isolates of Cyanophytes Tested

---

Agars

B.D.H. shredded (Japanese)

Difco "Bacto"

Difco "Purified"

Oxoid No. 1

Cyanophytes

Anabaena flos-aquae (Lyngb.) de Bréb. NRC-44-1

Anabaena flos-aquae (Lyngb.) de Bréb. A-113-9-q-2

Anabaena sub-cylindrica Borge A 78-1

Aphanizomenon flos-aquae (L.) Ralfs. NRC-568

Microcystis aeruginosa Kütz. emend. Elenkin NRC-1,SS-17
   (NIVA collection)

---

observed. A microwave oven was found to be the best means for
controlling the heating. Washed gel could be raised to the melting
point quickly without producing localized over-heated areas near
the walls of the container as occurred when heating was done in an
autoclave or a hot plate.

Washing agar gel with distilled water passed through cartridges
for organic removal and deionization, caused the pH to drop to about
4. Addition of ASM-1 medium, (Carmichael and Gorham, 1974), which
is poorly buffered, to the melted gel raised the pH very little and
inhibition of cyanophyte growth still occurred. This was remedied
by adding solid sodium bicarbonate to the washed gel before melting
to give a concentration of 2 mM in the final ASM-1 agar medium.
Medium prepared in this way had a pH of 8.5 and supported good
growth of cyanophytes.

Procedure for Preparing Sterile Agar Medium for Cyanophytes

Agar preparation. With a good quality agar a concentration
of 0.75% gives a sufficiently stiff gel. Make 200 ml of double-
strength gel with charcoal-filtered, deionized distilled water and
pour into a 20 cm x 20 cm Pyrex baking dish to give a depth of 5 mm.

When the gel has set, cut it into 5 mm cubes with a sharp knife and transfer the cubes to a wide-mouth Erlenmeyer flask of sufficient capacity to contain several times the volume of the agar. Wash on a magnetic stirrer for about 18 hours. Five or six changes of the purified distilled water during the working day plus one overnight appear to be adequate. Cutting into small cubes increases the surface area and reduces the loss during the washing process.

Incorporation of nutrients. Put approximately 100 ml of drained washed agar cubes in a 250 ml Erlenmeyer flask, add 34 mg of sodium bicarbonate and cover with a glass beaker. Autoclave 100 ml of double-strength ASM-1 medium separately and cool to room temperature. Place the agar gel in the microwave oven and heat to 80°C until the cubes just melt (2 to 3 minutes). Remove the melted agar from the microwave oven, mix with the 100 ml of sterile double-strength ASM-1 medium and dispense into sterile plastic or glass Petri dishes immediately.

Sterilization. Medium prepared in the foregoing way is usually quite low in microbial contaminants. If glass Petri dishes are used, a second heating in the microwave oven after 24 hours may be carried out. This improves sterility but there is some increase in inhibition. Plastic Petri dishes warp if heated in a microwave oven. Exposure of opened poured plates to ultraviolet light for 10 to 15 minutes at a distance of 20 cm from a 15 watt germicidal lamp in a laminar-flow transfer hood provides an acceptable degree of sterility, without causing lysis or inhibition of cyanophyte growth.

RESULTS AND DISCUSSION

Sensitivity of cyanophytes to the lytic factor(s) in heated agar gels varies with the species. Of those used as test species *Anabaena* A 78-1 was the most resistant while *Microcystis aeruginosa* NRC-1,SS-17 was the most susceptible. Heated agar gels could cause lysis within a few minutes of contact. Using phase-contrast microscopy, the expulsion of cell contents through a perforation near one end could be seen. Shortly after this the remaining cell wall disappeared. Similar effects can be induced by exposing the cells to low concentrations of surfactants such as Tween-80, Triton-X or Lux liquid dishwashing detergents. In less extreme cases where heating was less or washing of the agar was incomplete, only a few cells in a trichome were damaged causing breakage to occur at these points. This sort of clipping damage appears as gaps in trichomes on an agar surface.

Acidic gels completely inhibited growth of all the test species but did not cause cell disruption. Trichomes placed on

Fig. 2.   Margin of actively growing *Anabaena sub-cylindrica* A 78-1
on ASM-1 washed agar medium, pH 8.5.

acidic plates remained visibly unchanged for over a week. Growth
of bacteria associated with the cyanophytes was also inhibited.
Good growth of *Anabaena sub-cylindrica* (Figure 2) and all of the
other test species was obtained on the agar medium at pH 8.5.

    Initiation of growth of filamentous species on agar medium as
outlined above occurs within a few hours and may be seen as undulations in the filament (Figure 3). The lysis of cyanophytes
appears to be linked to water-soluble agents, as yet unidentified,
generated in the agar by heating above 80°C. The titer is a
function of both temperature and duration of heating. The production of hydrogen peroxide in mineral media through the action of
heat and light with manganese as described by Marler and van Baalen
(1965) does not seem to be involved, as judged by experiments in
which we exposed test species to 0.6% H_2O_2 for 24 hours. Little
or no growth of cyanophytes in pour plates of washed agar medium
noted by Carmichael and Gorham (1974) was probably caused by
variation in the autoclaving and reheating procedures that were
used.

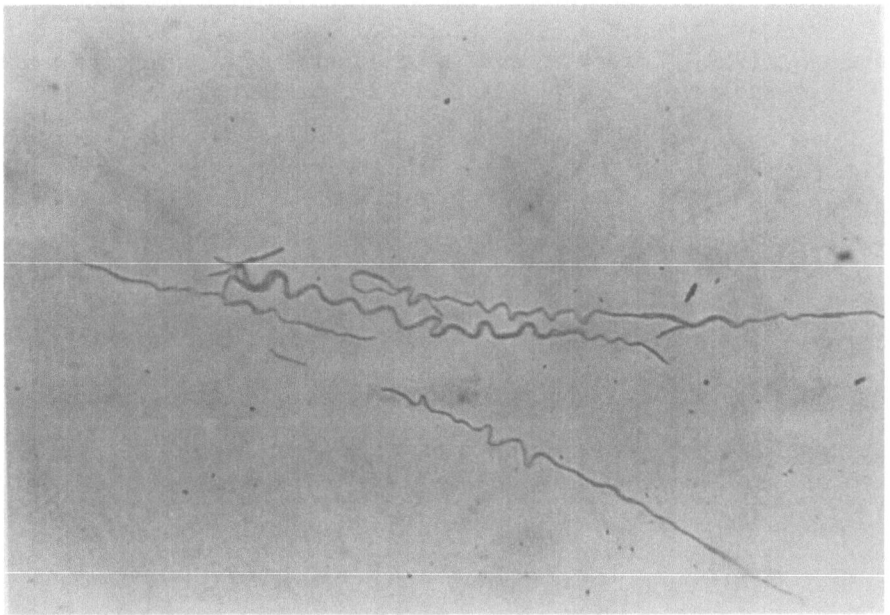

Fig. 3.   Undulations in trichomes of *Anabaena flos-aquae* NRC-44-1
          during early stages of growth on ASM-1 washed agar medium,
          pH 8.5.

REFERENCES

Allen, M. B.   1952.   The cultivation of Myxophyceae.   Arch. Mikro-
     biol. *17*:34-53.
Allen, M. M.   1968.   Simple conditions for growth of unicellular
     blue-green algae on plates.   J. Phycol. *4*:1-4.
Carmichael, W. W., and P. R. Gorham.   1974.   An improved method
     for obtaining axenic clones of planktonic blue-green algae.
     J. Phycol. *10*:238-240.
Ley, H. L., Jr., and J. H. Mueller.   1946.   On the isolation from
     agar of an inhibitor for *Neisseria gonorrhoeae*.   J. Bacteriol.
     52:453-460.
Marler, J. E., and C. van Baalen.   1965.   Role of H_2O_2 in single-
     cell growth of the blue-green alga, *Anacystis nidulans*.
     J. Phycol. *1*:180-185.
Robbins, W. J., and I. McVeigh.   1951.   Observations on the inhib-
     itory action of hydrolyzed agar.   Mycologia *43*:11-15.

Stanier, R. Y., R. Kunisawa, M. Mandel, and G. Cohen-Bazire. 1971.
    Purification and properties of unicellular blue-green algae
    (order Chroococcales). Bacteriol. Rev. *35*:171-205.
Yaphe, W. 1959. Effect of acid hydrolyzates of agar on the growth
    of *Escherichia coli*. Can. J. Microbiol..*5*:589-593.

THE EFFECT OF PHYSICO-CHEMICAL FACTORS ON GROWTH

RELEVANT TO THE MASS CULTURE OF AXENIC *MICROCYSTIS*

G. H. J. Krüger and J. N. Eloff

Department of Botany
University of the Orange Free State
Bloemfontein, Republic of South Africa

ABSTRACT

In order to produce relatively large quantities of axenic *Microcystis* cells on a routine basis for chemical and biological investigations of toxin biology, the influence of various factors on growth was investigated. Special attention was given to effects of light intensity.

By applying the data gained in the above-mentioned investigations, high yields of axenic *Microcystis* could be obtained in an all-glass mass culture system.

INTRODUCTION

Microcystis aeruginosa Kütz. emend. Elenkin seems to be a major constituent of many waterblooms (Vance, 1965; Gorham and Carmichael, 1979) especially in South Africa (Toerien, 1977; Bruwer, 1979) where a high proportion of *Microcystis* blooms are toxic. The occurrence of waterblooms of blue-green algae (Cyanobacteria), however, is usually sporadic; their appearances and disappearances are rather sudden and frequently unpredictable so, with these organisms, collection of material from a natural bloom poses problems. Natural blooms never consist of only the required organism. For studies of the chemical and biochemical characteristics of a particular alga, the use of axenic or at least unialgal cell material is essential. When isolating biologically active substances, relatively large quantities of cell material are usually required.

Furthermore, the production of certain cell constituents is highly dependent on environmental conditions. The toxicity of *Microcystis aeruginosa* cells in their natural environment and in laboratory culture has, for example, been shown to depend on growth conditions (Hughes et al., 1958; Gorham, 1960). Temperature, aeration rate (Gorham, 1964), nutrient composition (Brown, 1974) and pH (see paper by Eloff, this conference) are important variables.

One of the major obstacles in the elucidation of the structure and properties of toxins produced by Cyanobacteria, as well as the factors controlling their biosynthesis and those of other excretions, has, inter alia, been difficulties associated with the mass culture of algae in the laboratory under controlled and sterile conditions (Brown, 1974). The culture of *Microcystis*, in particular, has posed problems (Zehnder and Gorham, 1960) probably due to low growth rates (Hoogenhout and Amesz, 1965) and sensitivity to light intensity (Stanier et al., 1971; Jüttner, 1977; Krüger and Eloff, 1977). The successful culture of *Microcystis* in the laboratory requires a thorough knowledge of the factors controlling its growth. Very few in-depth studies on the factors involved in the laboratory culture of *Microcystis* have been reported, most probably because axenic material has not been available and because growth has frequently been associated with unexpected cell lysis.

In this paper, data are presented concerning the unexplained phenomenon of cell lysis and the effect of some physico-chemical factors, such as light intensity, population density, pH, CO_2 concentration, O_2 concentration, turbulence, and temperature on growth of *Microcystis*.

By applying the data gained in the above-mentioned investigations, high yields of axenic *Microcystis* were obtained on a routine basis in a 60 liter all-glass mass culture unit for chemical and biological investigation of toxin biology.

MATERIALS

Organisms Used

Microcystis aeruginosa:	NRC-1, a gift from P. R. Gorham, Univ. of Alberta, Edmonton, Alberta, Canada. Referred to as *M.* UV-001.
Microcystis incerta:	from Culture Center of Algae and Protozoa, Cambridge. Referred to as *M.* UV-003.
Microcystis aeruginosa:	isolated from Hendrik Verwoerd Dam near Bethulie, South Africa. Referred to as *M.* UV-004.

Microcystis aeruginosa: a gift from W. E. Scott, CSIR, Pretoria, isolated from Hartbeespoort Dam near Pretoria, South Africa. Referred to as *M.* UV-006.

Microcystis aeruginosa: Berkeley 7005, a gift from R. Y. Stanier, Institut Pasteur, Paris. (Originally from Göttingen Culture Collection). Referred to as *M.* UV-007.

Microcystis aeruginosa: a gift from F. Jüttner, Tübingen, Germany. (Originally from American Type Culture Collection -- strain 22663). Referred to as *M.* UV-008.

Microcystis aeruginosa: a gift from W. E. Scott, CSIR, Pretoria, isolated from Lindleyspoort Dam near Zwartruggens, South Africa. Referred to as *M.* UV-011.

Synechococcus sp.   previously named *Anacystis nidulans* (Berkeley strain 6301) a gift from R. Y. Stanier, Institut Pasteur, Paris. Referred to as *S.* UV-005.

Some properties of these isolates are summarized in Table 1.

Table 1.   Properties of Organisms Used

University O.F.S. Culture Collection	Toxic	Gas Vacuoles	Colonial Habit	Axenic
M. UV-001	√	√	−	−
M. UV-003	−	−	−	−
M. UV-004	−	−	−	−
M. UV-006	√	√	−	√[a]
M. UV-007	−	−	−	√
M. UV-008	−	−	−	−
M. UV-011	−	−	√	−
S. UV-005	−	−	−	−

[a]Not axenic when experiments discussed in § Effect of light intensity on growth, were carried out.

RESULTS

Effect of Light Intensity

Althouth blue-green algae appear to be able to withstand
light intensities in their natural environment, sudden and spontaneous die-off of natural blooms frequently occur (Abeliovich and
Shilo, 1971; 1972a). It is well known that laboratory cultures of
some blue-green algal strains (Stanier et al., 1971) and *Microcystis*
in particular (Zehnder and Gorham, 1960; Hughes et al., 1958;
Jüttner, 1977) are very sensitive to light and frequently lysed
under relatively low light intensities. Abeliovich and Shilo (1972b)
have attributed this phenomenon to photooxidation of the algal cells.

Effect of light intensity on growth. The influence of light
intensity on growth of seven *Microcystis* isolates (Table 1) was
investigated, using a batch culture system (Krüger and Eloff, 1977).
As a very good correlation existed between turbidity and cell number,
turbidity was used as the measure of growth (expressed as Log_2
Tt/To). The test organisms differed with regard to toxicity,
colonial habit, and possession of gas vacuoles. One (later two) of
the isolates were axenic and the others were unialgal.

Growth of these cultures was generally inhibited by light
intensities higher than 10 $\mu E./m^2/s$. The optimal light intensity,
i.e. the intensity at which cultures could grow for extended periods,
ranged from 7.5 to 22 $\mu E./m^2/s$ with *M.* UV-008 the most sensitive and
M. UV-003 the most resistant towards high light intensities. The
specific growth rate (μ = cell divisions/day) at the optimal light
intensity ranged from 0.14 for *M.* UV-008 to 1.32 for *M.* UV-003. The
maximum specific growth rates did not differ much ranging from 0.52
to 0.69, except for *M.* UV-003 which once again had the highest value
(μ = 1.32/day). Among the *Microcystis aeruginosa* isolates, *M.* UV-001
had the highest maximum specific growth rate (0.69/day). Sensitivity
towards light intensity was not related to toxicity, presence of
bacteria, colonial habit, cell size, or presence of gas vacuoles.

Typical growth curves obtained at different light intensities
with *Microcystis* UV-007 are depicted in Figure 1.

From this figure it is apparent that at the higher light
intensities (> 14 $\mu E./m^2/s$ in this case) a relatively high growth
rate was maintained only for a short period. This is followed by
a decline in turbidity which is not attributable to depletion of
mineral nutrients in the medium because the cultures at lower
light intensities grew well for prolonged periods and reached much
higher turbidity values.

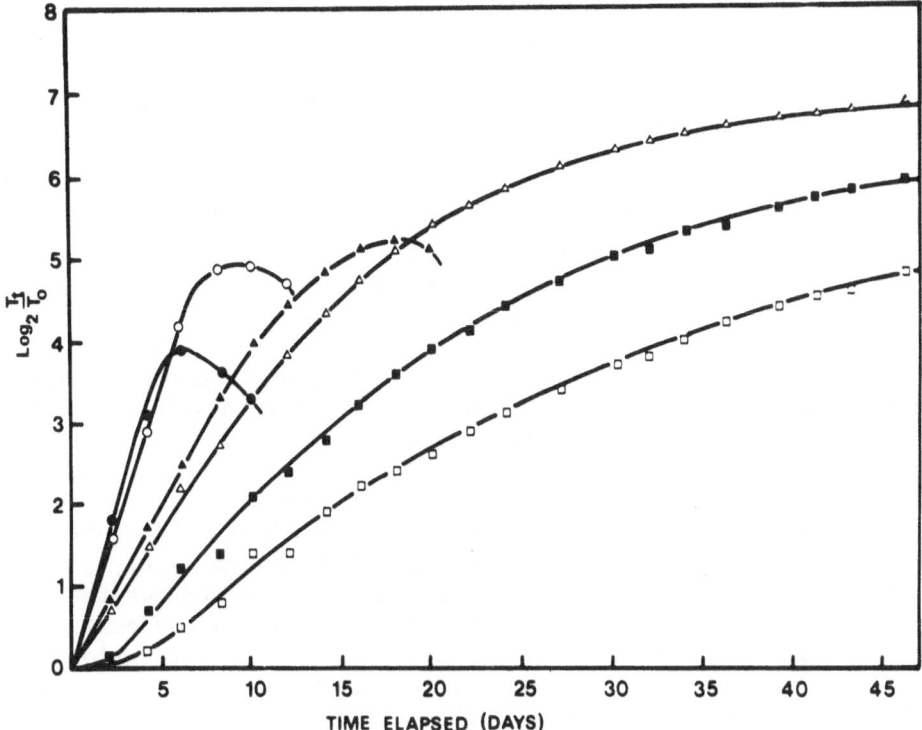

Fig. 1.   Typical growth curves of *M.* UV-007 at different light
intensities. Light intensity expressed as $\mu E./m^2/s$:
● = 36,   ○ = 21.7,   ▲ = 14.3,   △ = 9.5,   ■ = 6.8,
□ = 4.4

According to Stanier et al. (1971), *Microcystis incerta*
(UV-003) should not be described as *Microcystis* because it never
forms gas vacuoles and the small rod-shaped cells differ very much
from typical *Microcystis* cells. Our results showed that *M. incerta*
has a specific growth rate at least twice that of any of the others
and thus corroborates in physiological terms the existing morphological differences as a taxonomical criterium.

When similar experiments were carried out with a *Synechococcus*
isolate, no bleaching of cells took place at the highest light
intensity used (36 $\mu E./m^2/s$). This raises the question of the
origin of the difference in sensitivity towards light and also of
the mechanism of light inhibition involved. The fact that these
organisms grow in nature under much higher light intensities indicate that the light inhibition encountered in the laboratory may
be due to a complex interaction between light intensity and other
parameters and not to light per se.

The high sensitivity of laboratory cultures of *Microcystis* towards light may be attributed to one or more of the following:

(i)   Increase in pH.

(ii)   Decrease in CO_2 concentration or increase in O_2 concentration due to photosynethesis.

(iii)   Growth factors formed under stress light conditions.

(iv)   Some other parameters present in natural blooms leading to protection of cell pigments against photooxidative bleaching, e.g. presence of bacteria or a mucilage sheath.

Interaction between cell density of *Microcystis* batch cultures and light-induced stress conditions. Sensitivity towards light of *Microcystis* cultures having different cell concentrations was determined by measuring growth over a 20-day period (Krüger and Eloff, 1979b). It could be shown that despite the mutual shading effect encountered in *Microcystis* cultures with high cell densities, these cultures were much more sensitive towards light than cultures with a low cell density. From this it was deduced that the light-induced inhibition is closely correlated with a cell-density-mediated parameter such as carbon dioxide concentration, oxygen concentration, or pH, and not with light intensity per se. Inhibition of growth is caused by the accumulation or depletion of some substance in the culture medium, hence the enhancement of inhibition by an increase in cell concentration.

Morphological changes in *Microcystis* grown at different irradiance levels. Cell size of most algal species is influenced by environmental conditions (Ruzicka, 1971) such as nutritional status (Bursche, 1961) and temperature (Kullberg, 1977). To gain a better understanding of the phenomenon of die-off of *Microcystis* batch cultures at light intensities > 10 $\mu E./m^2/s$, the influence of aging and light intensity on the morphology of cells of a toxic (UV-006) and a non-toxic (UV-007) isolate of *Microcystis* were investigated. Both isolates were axenic.

Batch cultures of the *Microcystis* isolates were subjected to different light intensities between 4.5 and 40 $\mu E./m^2/s$. Cell counts and cell volume measurements were performed every 48 hours with a Coulter Counter, equipped with a 100-channel Coulter Channelizer.

Fig. 2.   Change in cell size distribution curves with time of *M*. UV-007 cultures grown at different irradiance levels. The age of the cultures (days) is shown on the curves. Irradiance levels are shown in right hand corners in $\mu E./m^2/s$.

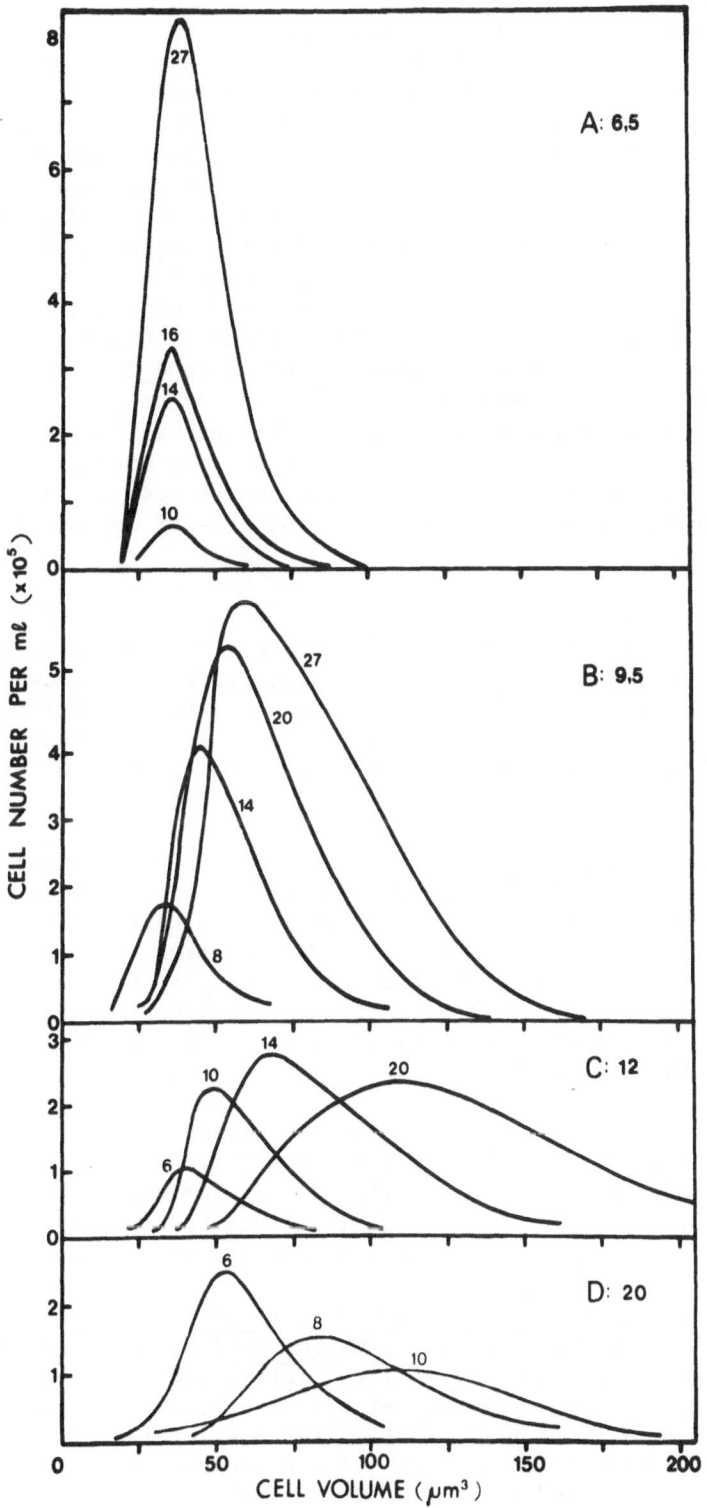

Size-range curves were plotted for the non-toxic *M*. UV-007
cell suspensions at different time intervals grown at different
light intensities. Some of these curves are shown in Figure 2.
Light intensity and culture age had a pronounced effect on cell
size and size range. Average cell volume varied from 33 to 140
μm^3, cells at higher light intensities being larger and having a
larger size range. At 6.5 $\mu E./m^2/s$, cell size remained relatively
small and constant at 20 to 70 μm^3 throughout the 27 days duration
of the experiment. However, at the slightly higher light intensity of 12 $\mu E./m^2/s$, cell size already ranged from 20 to > 200 μm^3
after 20 days. Cell-size distribution curves of cultures grown at
30 and 40 $\mu E./m^2/s$ are not included in Figure 2 because no clear
distribution curves were obtained. After 2 to 4 days of growth at
the latter light intensities there was a homogeneous distribution
of cell sizes over the entire range. This probably means that
after 4 days organized growth was seriously disrupted, with some
cells increasing in volume, some dividing and some disintegrating.
After 5 days, bleaching was observed and die-off commenced.

The effects on the toxic *Microcystis* isolate (*M*. UV-006) were
similar but less pronounced (distribution curves not presented).
Changes in maximum cell diameter (and, therefore, maximum cell
volume) were much less than with UV-007 (Figure 3). Furthermore
the cell-size distributions of the cultures remained fairly constant and had a fairly narrow range compared to those of UV-007 at
corresponding light intensities.

In general, cell size, and especially size variability, appear
to be sensitive indicators of physiological state with cells under
stress conditions being larger and associated with a larger size-
range. Cell size is apparently a more sensitive parameter for the
recognition of stress conditions than turbidity, chlorophyll content, or cell number, because long before any of these parameters
changed, dramatic changes occurred in the cell-size distribution
of *M*. UV-007 (Figure 2).

The wide range of cell diameters observed at different irradiance levels (3.4 to 7.2 μm for the non-toxic and 1.8 to 6.4 μm
for the toxic isolate), makes questionable the use of cell size as
a taxonomical character (Geitler, 1932; Desikachary, 1959; Komárek,
1958) without careful consideration of environmental conditions.

Changes in the growth medium with respect to pH, CO_2 concentration, O_2 concentration, and alkalinity of *Microcystis* cultures
grown at stress and non-stress light intensities. The growth
medium used for algal cultures of all experiments (including the
mass culture of *Microcystis*) was Medium BG-11 (Stanier et al.,
1971) modified as follows: the citrate concentration was doubled
and 0.006 g/liter $FeSO_4$ was used instead of ferrous citrate

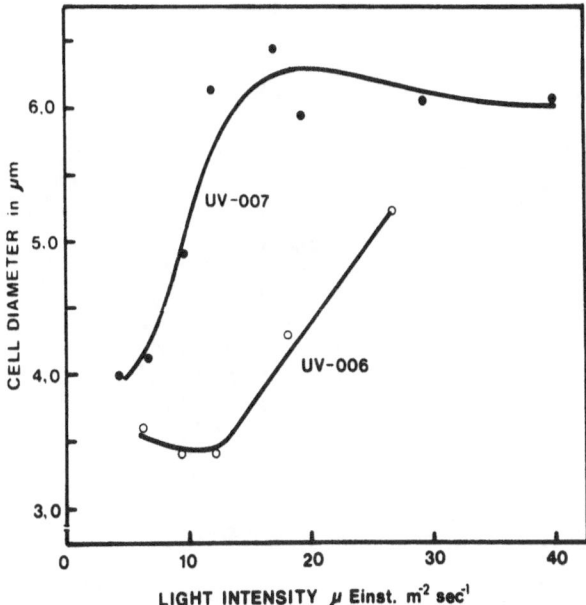

Fig. 3.   Influence of light intensity on the maximum cell size of
          M. UV-007 and *M*. UV-006 reached in a 27-day growth period.

(Herdman et al., 1973) and 0.058 g/liter $Na_2SiO_3.9H_2O$ was added
(Eloff et al., 1976). The trace-metal mix A5 (Arnon, 1935) was
used. The buffer capacity of this medium proved to be rather poor
(Kruger and Eloff, 1979a).

   It was assumed that the inhibition of growth in batch cultures
could not be caused by a deficiency of some mineral nutrients in
the medium, since it proved to contain a large excess of mineral
nutrients. Five-times-diluted medium yielded very good growth --
90% of control after 30 days of growth in batch culture. Furthermore, at higher light intensities, bleaching and lysis occurred at
relatively low cell concentrations, i.e. at a stage when only a
small proportion of the available nutrients could have been utilized by the algae (Krüger and Eloff, 1977). The chemical changes
of the growth medium due to growth of *Microcystis* UV-007 were thus
investigated in an attempt to explain growth inhibition of batch
cultures at higher light intensities.

   Figure 4A shows that the pH of the culture medium increased
at all irradiance levels. The cell suspension of the high-light-
intensity treatments (27 and 19 µE./m^2/s) reached pH values of
11.06 and 11.04 after 6.5 and 7 days respectively after which a
sharp decrease in pH occurred. These pH curves showed a close
resemblance to the corresponding growth curves (Figure 4B).

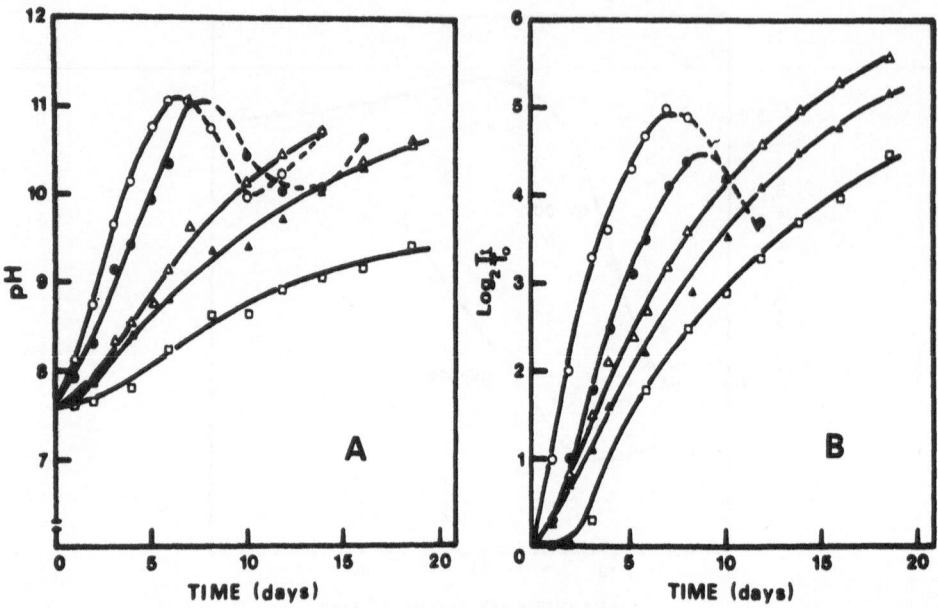

Fig. 4A and 4B.   (A) Changes in pH, and (B) turbidity (i.e. growth)
                  curves of M. UV-007 cultures grown at different
                  irradiance levels ($\circ$ = 27, $\bullet$ = 19, $\triangle$ = 12,
                  $\blacktriangle$ = 9.5, $\square$ = 6.0 µE./m /s).

The time when the highest turbidity values (Log_2 Tt/To) were
reached evidently corresponded with the maximum pH values reached
and the decrease in turbidity coincided with a decrease in pH.
The rise in pH accompanying growth indicated that CO_2 was assimi-
lated faster than it was supplied through diffusion and had to be
provided from carbonate and/or bicarbonate, with consequent formation of hydroxyl ions. (Carbonate alkalinity was replaced by non-
carbonate alkalinity.)

An experiment was therefore conducted to determine changes in
turbidity, total alkalinity, pH, oxygen and carbon dioxide concentrations with time on samples of cultures subjected to two irradiance levels over a 13-day period. At one irradiance level, *Microcystis* cultures can grow successfully for prolonged periods (9.5
µE./m²/s) and at the other, bleaching and reduction of cell numbers
usually commences after ca. 6 days (26 µE./m²/s) (Krüger and Eloff,
1979a). These results are presented in Figure 5.

It was evident that the time when die-off (Figure 5A) of cultures grown at the higher light intensity commenced almost coincided with the time when the pH value was maximal (Figure 5B), the
O_2 concentration was maximal (Figure 5C) and when the total inor-

ganic carbon (ΣCO_2) and the concentration ratio, $\Sigma CO_2 : O_2$ in the medium was minimal (Figures 5D and 5E).

Blue-green algae in general (King, 1970) and *Microcystis* in particular, (Talling, 1976) have a high capacity for the removal of CO_2 from alkaline and CO_2-depleted solutions, but from the results it appears that the cessation of growth in cultures grown at the stress light intensity, was caused by depletion of ΣCO_2 due to photosynthetic activity and insufficient replenishment of CO_2 by diffusion. At this low ΣCO_2 concentration and high pH value ribulose diphosphate carboxylase would not be able to assimilate CO_2 (Jackson and Volk, 1970; Owens and Esaias, 1976). On the other hand, conditions in the medium at the time cessation of growth commenced, i.e., low CO_2 concentration, high O_2 concentration (Pritchard et al., 1962) and high pH (Orth et al., 1966), were favorable for glycolic acid production and hence photorespiration.

It also appeared that pH per se was not responsible for cessation of growth at the higher light intensity (pH = 10.75) because the pH reached an even higher value (pH = 10.78) after 13 days at the lower light intensity where healthy growth lasted for prolonged periods. The same argument applies to O_2 concentration per se.

Interaction between light inhibition of cultures and CO_2 concentration of aerating gas. An investigation of the effect of the CO_2 concentration of the aerating gas on growth of *Microcystis* cultures grown at different stress light intensities revealed that bleaching and die-off of cells of cultures could be prevented by aeration with the appropriate CO_2 concentration. Growth curves of cultures aerated with different CO_2 concentrations (in air) grown at 98 $\mu E./m^2/s$ are presented in Figure 6. The aeration rate was 80 ml gas per minute. Batch cultures with a volume of 250 ml cell suspension were used.

It is evident (Figure 6) that the initial growth rate of the different CO_2 treatments (CO_2 concentrations ranged from 0.03 to 0.52 percent) did not differ largely. At all irradiance levels tested (25, 37 and 98 $\mu E./m^2/s$) however, the CO_2 concentration of the aerating gas had a marked effect on the final cell number obtained. In general there appeared to be a good correlation between the Co_2 concentration of the aerating gas and the light intensity the cultures could withstand as shown in Table 2.

At the highest light intensity (98 $\mu E./m^2/s$) good growth was maintained for the duration of the experiment (6.6 days) only at the treatment aerated with 0.52% CO_2 (Figure 6). The cell number of the latter treatment was 1.76×10^8 cell/ml after 6.6 days corresponding to 8.4 doublings (calculated from cell number).

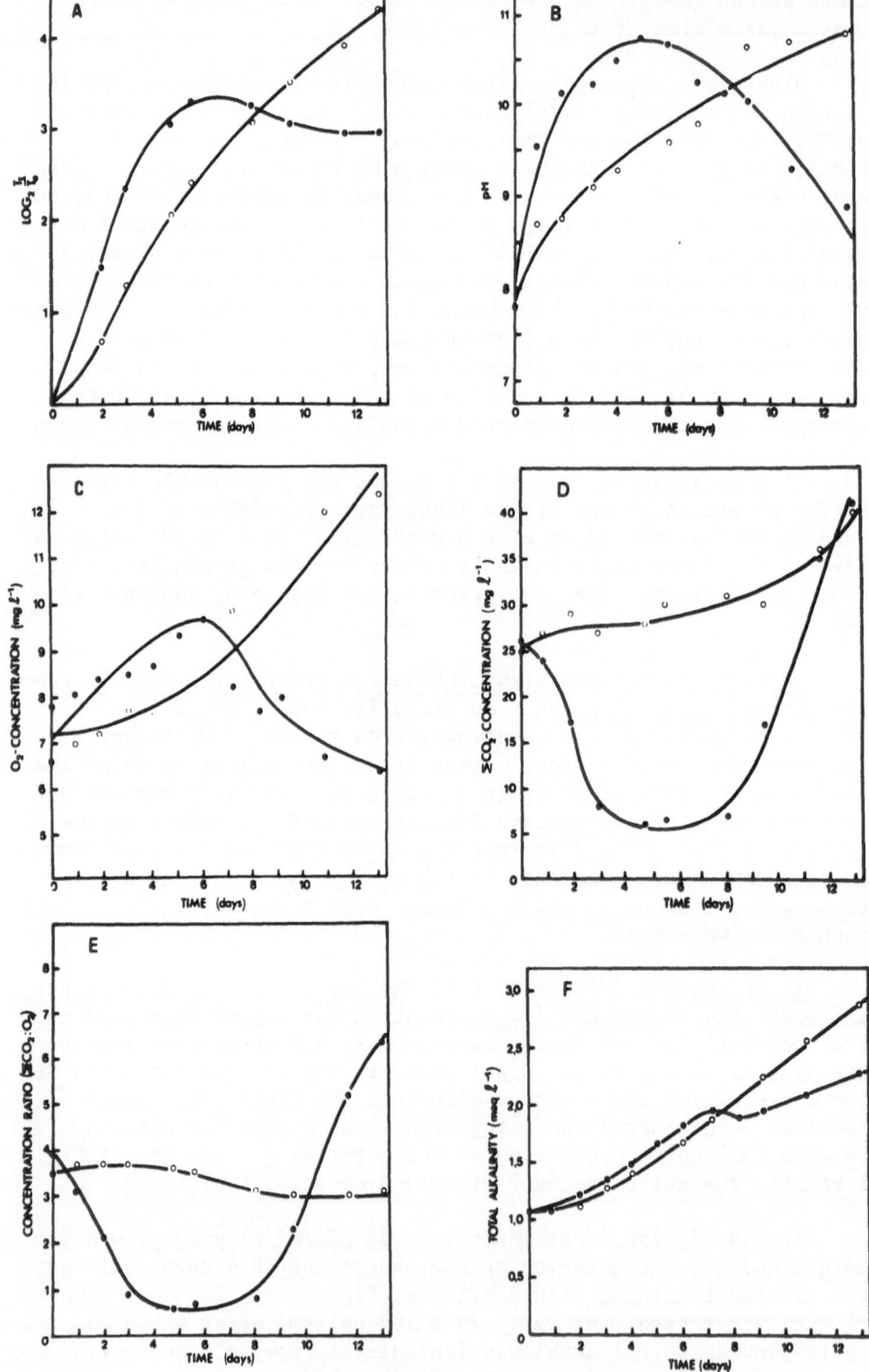

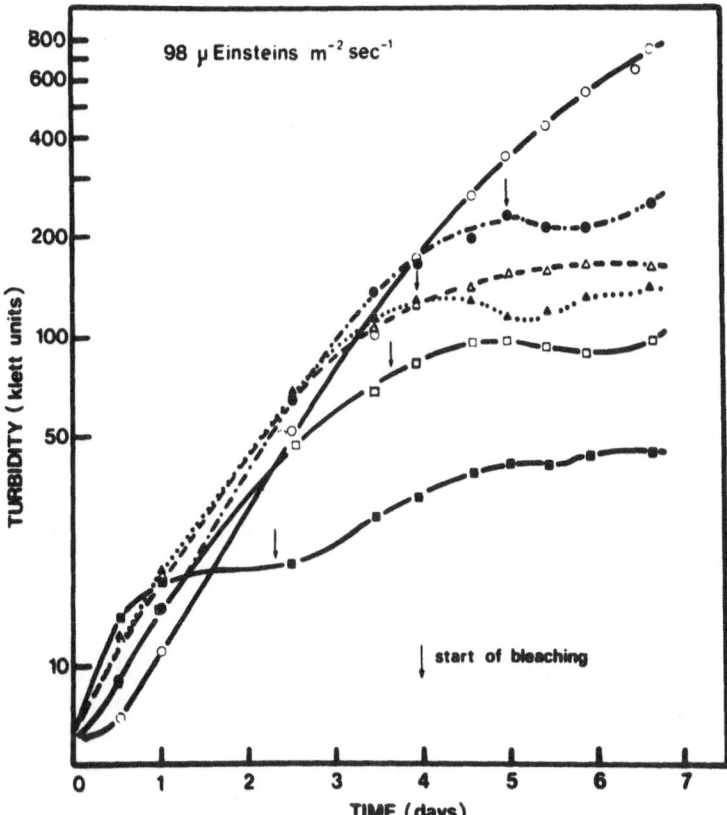

Fig. 6. Influence of CO_2 concentration of the aerating gas on the
growth of *M.* UV-007 in batch culture grown at a light
intensity of 98 μE./m^2/s (o = 0.52, ● = 0.23, △ = 0.13,
▲ = 0.05, □ = 0.03 (air) percent CO_2. ■ = No aeration).

Fig. 5. The change with time in various parameters of cultures
of *M.* UV-007 grown at non-stress, 9.5 (o) and stress, 26
(●) μE./m^2/s light intensities. A = Turbidity (growth
curve); B = pH; C = O_2 concentration; D = CO_2 concentration; E = Ratio of total inorganic carbon $\Sigma CO_2{:}O_2$
concentration; F = Total alkalinity.

Table 2.   Time in Days before Bleaching took place at Different
           CO_2 Concentrations in Aerating Gas at Three Light
           Intensities

Light Intensity (μE./m^2/s)	CO_2 Concentration (Percentage)				
	0.03	0.05	0.13	0.25	0.52
25	5	nb[a]	nb	nb	nb
37	4.3	5.4	nb	nb	nb
98	3.4	4	(4)[b]	(5)[b]	nb

[a] nb = No Bleaching.

[b] Cells did not bleach, but cessation of growth took place at this
stage.

        It has been calculated (Kuhl and Lorenzen, 1964) that ca. one
liter of CO_2, under laboratory conditions, is needed for the production of one gram of algal dry mass.   In the case of *Microcystis*
the CO_2 concentration gradient between the aerating gas and cell
suspension is apparently more important than the volume of gas
sparged per unit of time.   A relatively high external CO_2 concentration may be necessary to provide an adequate diffusion rate of
CO_2 to the carboxysomes of the *Microcystis* cells.   It has been
reported that CO_2 may play a catalytic role in the Hill reaction
and thus is a catalyst of photosystem II (Heise and Gaffron, 1963;
Metzner and Gerster, 1976).   Carbon dioxide may similarly protect
Microcystis cells against photoinhibition.

        The resistance of natural blooms of *Microcystis* may be
explained by the presence of high concentrations of bacteria and
other heterotrophic organisms which increases the $CO_2:O_2$ concentration ratio by respiration (Dor, 1974) (see also paper by Eloff,
these Conference Proceedings).

        Furthermore, the CO_2 concentration had a marked influence on
cell diameter; the lower the CO_2 concentration the bigger the cells
were at the end of the experiment (Figure 7).   These results indicate that the cell size of *Microcystis* cultures are not directly
influenced by light intensity but rather by the CO_2 concentration
of the medium which is dependent on the growth rate of the cells
in the culture (See § titled Morphological changes in *Microcystis*
shown at different irradiance levels).   These results also have
interesting implications with regard to taxonomical aspects.

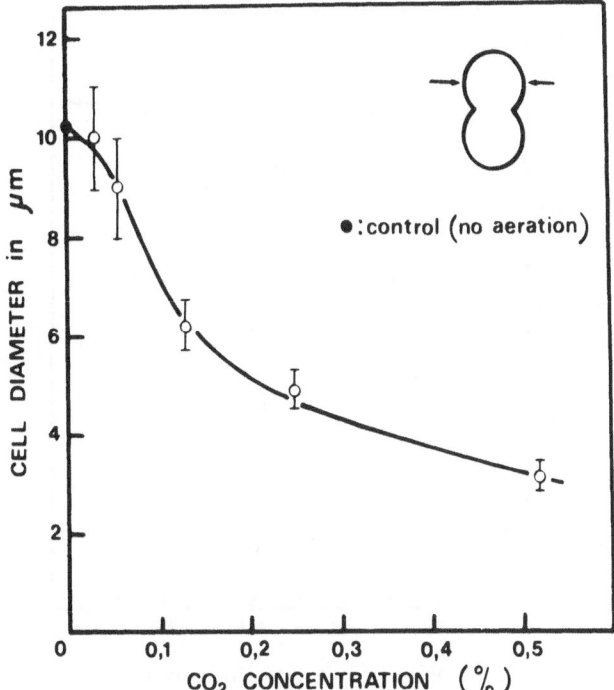

Fig. 7. Influence of CO_2 concentration of the aerating gas on the diameter of *M.* UV-007 cells after a growth period of 6.6 days at a light intensity of 98 $\mu E./m^2/s$. Diameter measured as shown.

Mathematical relationship between specific growth rate of *Microcystis* and light intensity. By using aerated cultures of *M.* UV-007, the relationship between specific growth rate (μ) and light intensity could be determined for light intensities up to 110 $\mu E./m^2/s$. These values were substituted in the Michaelis-Menten equation proposed by Shelef et al. (1968) as well as in the exponential equation proposed by Ruzicka and Simmer (1970). The latter equation proved to describe the relation best. The function was:

$$\mu = \mu max(1-e^{-I/S})$$

μ = specific growth rate, μmax = maximum specific growth rate, I = light intensity, S = a saturation constant whose value equals the illumination intensity for μ = 0.632 x μmax

When values predicted according to the above equation were compared with values obtained experimentally, the two sets of data correlated highly significantly (r = 0.98; n = 7) (Figure 8).

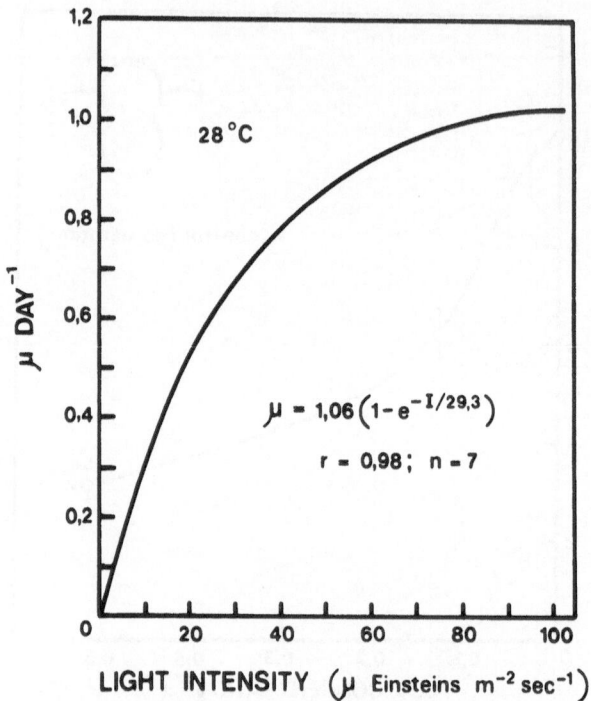

Fig. 8.   The influence of light intensity on the specific growth
rate of *M.* UV-007 in batch cultures. Mathematical
relation based on the exponential equation of Ruzicka
and Simmer (1970) shown on figure (I = light intensity,
1.06 = μmax).

Under the experimental conditions, light saturation was
obtained at ca. 90 μE./m²/s. Furthermore the highest growth rate
(μmax) for *M.* UV-007 proved to be 1.06/day at 28°C under conditions
where light and CO_2 were non-limiting. This corresponds to a doubling time of 15.6 hours.

The Effect of Temperature on Specific Growth Rate and Activation Energy

The effect of temperature on growth of four *Microcystis* isolates (two toxic and two non-toxic) and a *Synechococcus* isolate
was investigated (Krüger and Eloff, 1978c). The experiments were
carried out in a temperature-gradient incubator. The system was
illuminated continuously and turbidity (measured every 12 hours)
was used as the measure of growth.

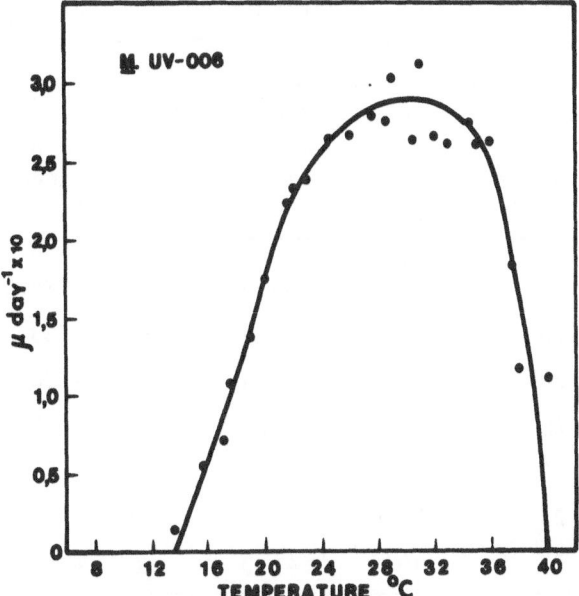

Fig. 9. Effect of incubation temperature on the specific growth
rate of *M*. UV-006.

A typical example of the effect of temperature on specific
growth rate (μ) is shown for *M*. UV-006 in Figure 9. As found previously (Sorokin and Krauss, 1962; Beljanin and Trenkensu, 1977),
the descending part of the curve was much steeper than the ascend-
ing part, indicating that a sudden decrease in growth rate occurred
when the upper temperature limit for the particular organism was
surpassed.

A comparison of the temperature requirements and temperature
tolerance of the organisms investigated is given in Table 3. When
evaluating the data it is important to bear in mind that the light
intensities employed were below saturating intensity.

The relatively broad optimal growth temperature ranges obtained
may be ascribed to the low light intensities employed. Therefore,
at this stage, it is not possible to state conclusively that the
Microcystis and *Synechococcus* isolates investigated have broad
optimal temperature ranges. To estimate the optimal temperature
value that would be valid under different environmental conditions,
it was decided to use the mean value of the optimal temperature
range (Table 3). Beljanin and Trenkensu (1977) found that the
position of the temperature optimum did not vary appreciably at
widely different irradiance levels when specific growth rate was
used as a measure of growth.

Table 3. Comparison of the Temperature Requirements and Temperature Tolerance of *Microcystis* and *Synechococcus* Isolates

Organism	Prevailing Light Intensity (µE./m²/s)	Lower Temp. Limit °C	Upper Temp. Limit °C	Optimal Growth Temp. Range °C	Thermal Growth Optimum °C
M. UV-001	20	12.0	36.0	24. - 34.0	29.0
M. UV-003	20	10.5	36.5	26.0 - 33.0	29.5
M. UV-006	20	13.5	40.0	26. - 34.5	30.3
M. UV-007	20	13.2	36.4	26. - 35.0	30.5
	33	12.0	35.0	25.5 - 32.0	28.8
S. UV-005	25	11.0	--		
	33	10.3	44.3	26. - 43.	34.3

Among the *Microcystis* species investigated, *M.* UV-003 had the lowest lower temperature limit (10.5°C). *M.* UV-006 (isolated from Hartbeespoort Dam, Pretoria) had the highest upper temperature value (40.0°C), which was 3.6 to 5°C higher than the value for the other *Microcystis* isolates. *M.* UV-006 could have diverged in this respect and become adapted to tolerate higher temperatures. The other cultures have been kept under laboratory conditions for 20 years compared to 3 years for *M.* UV-006.

The optimal growth temperatures did not differ much for the *Microcystis* isolates. The values ranged between 29.0 and 30.5°C and agreed well with the values reported for *M. aeruginosa* NRC-1 (identical to our *M.* UV-001) by Zehnder and Gorham (1960) and Gorham (1964) and for strain 7005 (Stanier et al., 1971) which is identical to our *M.* UV-007. The *Synechococcus* species investigated (*S.* UV-005) could tolerate relatively wide extremes of temperature (Table 3).

Van't Hoff Arrhenius plots: calculation of activation energy. When light intensity is held constant it is possible to describe the effect of temperature on growth rate of algae mathematically by applying the Arrhenius equation (Goldman and Carpenter, 1974):

$$\mu = Ae^{-E/RT}$$

A = Arrhenius frequency factor/day; E = activation energy (cal/mole); R = gas constant (cal/°K/mole) T = temperature (°K)

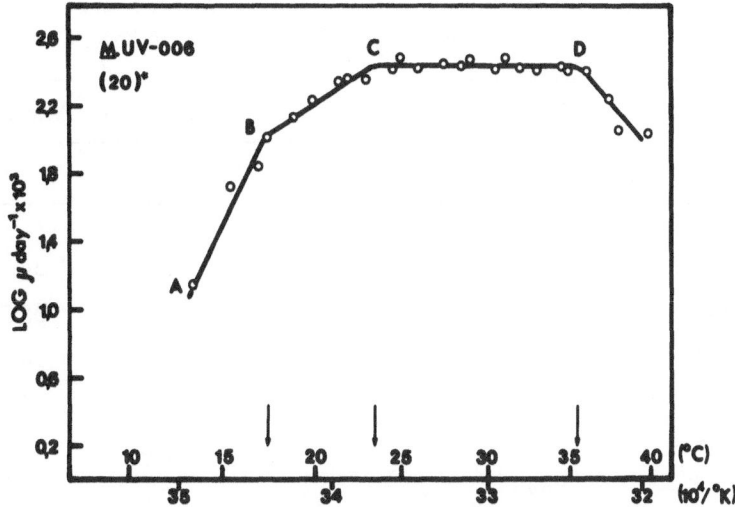

Fig. 10.   Arrhenius plot of the relationship between Log specific
growth rate and temperature for *M.* UV-006.   (*Light
intensity in µE./m²/s)

The activation energy is a valuable constant as it can be used to
predict the effect of temperature on growth rate over the normal
temperature range (Pirt, 1975).

Figure 10 depicts a typical Arrhenius plot (Log specific
growth rate as a function of absolute temperature).   In all cases
abrupt changes occurred in the slopes of the curves, demonstrating
that different Arrhenius relationships are applicable in different
temperature ranges.   This data agrees with results obtained for
Chlorella (Sorokin, 1960) and for the thermophilic *Synechococcus
lividus* (Meeks and Castenholz, 1971).   Thus the activation energy
or "temperature characteristic" for a particular organism is constant only over a short temperature range.   Similar results have
been reported for bacteria (Du Preez and Toerien, 1978).

These changes in the slopes of the logarithmic curves, i.e.,
changes in activation energy, can be explained in more than one
way.   Generally it can be attributed to the fact that different
reactions limit the rate of the overall growth process at different temperatures (Crozier, 1924; Sorokin, 1960; Pirt, 1975).

Three distinct inflexion points occurred (B, C and D) in all
cases.   The temperature values of these inflexion points are summarized in Table 4.   In all cases the most drastic changes in the
slopes of the curves occurred at inflexion point B.   Accordingly
there was a big difference in activation energy values below and
above this point (Krüger and Eloff, 1978c).   For all organisms

Table 4. Temperatures at Which Inflection Points Occurred in
          Arrhenius Plots of *Microcystis* and *Synechococcus*.

Organism	Light Intensity µE./m²/s	Inflection Point Temp.		
		B	C	D
M. UV–003	20	13	22	34
M. UV–006	20	17.5	24	35.5
M. UV–007	20	14	20	34
	33	14	25	34
S. UV–005	25	16	22	–
	33	17	23	44

investigated, inflexion point B may be regarded as the temperature
where, according to Sorokin (1960), a change from one master reaction to another took place. Inflection point B (at 17.5°C) in the
Arrhenius plot for *M.* UV–006 correlated very well with observations
that *Microcystis* blooms such as that from which *M.* UV–006 was isolated, frequently start forming when the water temperature in
Hartbeespoort Dam is between 16 and 17°C (Scott et al., 1977).
This could be due to the activation energy being higher (78.7 kcal/
mole) at temperatures below 17°C than at temperatures above 17°C
(22.3 kcal/mole) and explains in thermodynamic terms why the blooming of *Microcystis* in the Hartbeespoort Dam only takes place when
the temperature rises above 17°C.

The occurrence of the second inflection point C could probably
be considered to be an artifact. In this temperature range, light
rather than temperature could have been the limiting factor.

Arrhenius equations and Q_{10} values computed for the test organisms are summarized in Table 5. These equations may be used to predict specific growth rate at different temperatures within the
appropriate temperature range at the specified light intensity.
Implications of the results are discussed in detail in a previous
paper (Krüger and Eloff, 1978c).

Effect of Agitation and Turbulence of the Growth Medium on Growth and Viability

It has been reported (Fogg et al., 1973) that artificial circulation of water bodies suppress growth of blue-green algae.

Table 5. Arrhenius Equations and Q_{10} Values for the Growth of Different *Microcystis* and *Synechococcus* Isolates

Isolate	Light Intensity $\mu E./m^2/s$	Temperature Range °C	Frequency Factor A/Day	Arrhenius Equation (μ = per day)	Q_{10}
M. UV-003	20	13 - 22	5.938×10^7	$\mu = 5.938 \times 10^7\ e^{-5665/T}$	2.0
		Above 22	3.549	$\mu = 3.549\ e^{-746/T}$	1.08
M. UV-006	20	17.5 - 24	7.642×10^{15}	$\mu = 7.642 \times 10^{15}\ e^{-11248/T}$	3.6
		Above 24	0.073	$\mu = 0.073\ e^{+433/T}$	0.95
M. UV-007	20	14 - 20	2.075×10^{22}	$\mu = 2.075 \times 10^{22}\ e^{-15552/T}$	6.3
		Above 20	4.364×10^3	$\mu = 4.364 \times 10^3\ e^{-2983/T}$	1.3
	33	14 - 25	2.254×10^{11}	$\mu = 2.254 \times 10^{11}\ e^{-8155/T}$	2.6
		Above 25	3.821	$\mu = 3.821\ e^{-794/T}$	1.09
S. UV-005	25	16 - 22	5.322×10^7	$\mu = 5.322 \times 10^7\ e^{-5629/T}$	1.9
		Above 22	0.981	$\mu = 0.981\ e^{-373/T}$	1.0
	33	17 - 23	7.006×10^3	$\mu = 7.006 \times 10^3\ e^{-2983/T}$	1.4
		Above 23	0.485	$\mu = 0.485\ e^{-132/T}$	1.0

Similarly Soeder and Stengel (1974) pointed out that although
hydrostatic pressure does not appear to be an ecologically important factor for algae, it has an unfavorable effect on blue-green
algae which contain gas vacuoles.

In a mass culture unit built by us (Krüger and Eloff, 1978a)
the cell suspension is continuously circulated through a system of
glass tubes by a centrifugal pump. Each time the cells pass through
the pump they are briefly subjected to a complex of hydromechanical
factors such as high turbulence, hydrostatic pressure, cavitation
and accelerated current velocity.

The influence of turbulence and agitation on growth and viability of *Microcystis* in batch cultures was, therefore, determined.
Cultures were agitated by means of magnetic follower-stirrers at
speeds of 60, 180, 300, 500, and 750 rpm which correspond to linear
velocities of 25, 75, 126, 209, and 314 cm/sec respectively (maximum velocity calculated on the inner surface of flask, 1.5 cm from
the bottom). The duration of the experiment was 20 days. *Micro-*
cystis UV-007 was used as test organism (Krüger and Eloff, 1978b).

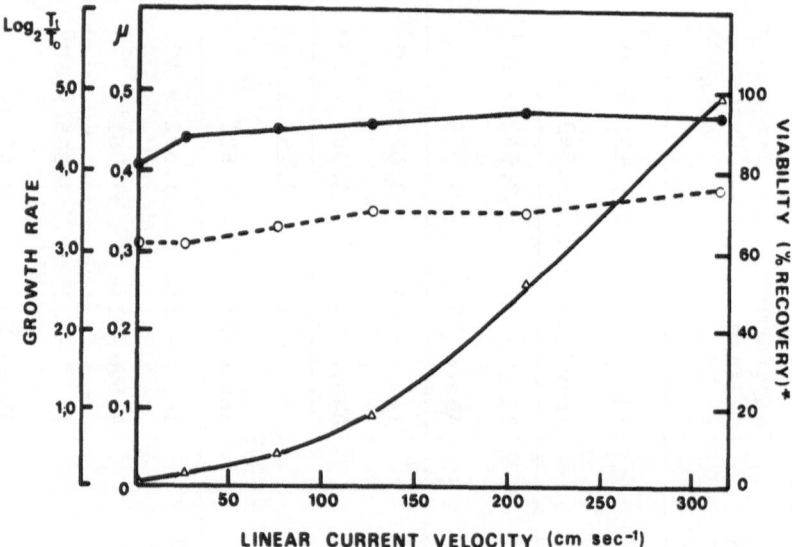

Fig. 11.   Influence of turbulence (linear current velocity) on
           growth and viability of *Microcystis*. Key: ● = Degree
           of multiplication (Log$_2$ Tt/To);   o = Specific growth
           rate (μ);   Δ = Viability (percentage recovery of ini-
           tial cell count after 20 days); Tt = Turbidity at time
           t (20 days); To = Turbidity at start of experiment;
           * = Average of two viability determination methods
           (Table 2).

The results proved that turbulence and agitation (as executed in this experiment) had no adverse effect on growth rate. A highly significant correlation existed between increasing linear current velocity and percentage viability as determined by a plating and serial dilution method (Figure 11).

Though the hydromechanical effects that may be caused by a centrifugal pump could not adequately be simulated by magnetic followers it seemed probable that no adverse effects should be encountered with *Microcystis* in the mass culture system.

Mass Culture of *Microcystis*

When a large volume of algae is cultured in vessels, illumination of the cells for optimal growth is prevented due to self shading. Under such conditions aeration of the culture, i.e., provision of CO_2 and removal of O_2 is also difficult to accomplish effectively. The main requirements of a satisfactory system for culturing large volumes of an algal suspension under controlled and axenic conditions would thus be:

(i)   Effective sterilization of the system prior to inoculation.

(ii)   Effective aeration while maintaining sterility.

(iii) Effective control of temperature and pH.

(iv)   Effective illumination of as large an area of cell suspension as possible.

To our mind a tubular all-glass system would fulfill these requirements best. Such a system with a capacity of 60 liters was consequently constructed on the same lines as the 110 liter culture system built by Jüttner et al. (1971) which was based on that of Tamiya et al (1953) and Setlik et al. (1966). A diagrammatic representation of the layout is given in Figure 12. The entire system was built from borosilicate glass components.

Twelve 2-meter-long glass tubes with an internal diameter (I.D.) of 50 mm were coupled by "U" bends to form a vertical flattened spiral. The upper end of the spiral was coupled to a gas exchange vessel (E) having an I.D. of 70 mm. The cycle was completed by connecting the inlet and outlet of an all-glass centrifugal pump to the bottom of the gas exchange vessel and the lower end of the spiral of tubes respectively. By use of three-way valves C and D, the flow of the suspension could be short-circuited, decreasing the effective volume to ca. 15 liters. The gas-exchange vessel was provided with the following side tubes:

(i)   I, for introduction of the air/CO_2 mixture.

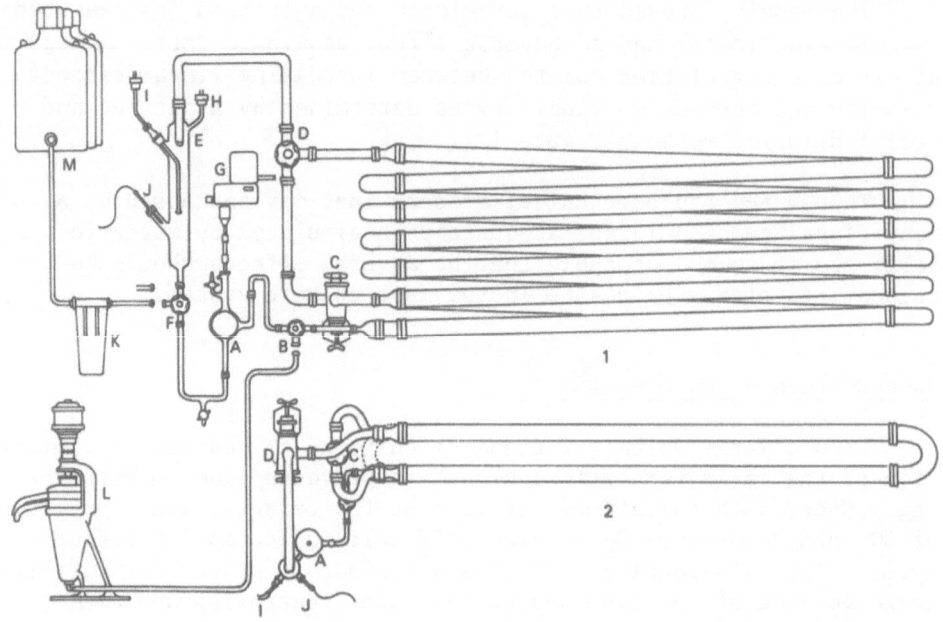

Fig. 12.   Diagrammatic representation of the construction of the
           mass culture system. 1: Side view; 2: Top view (see
           text for explanation).

(ii)   H, for exit of air to remain pressure equilibrium. This
        tube was coupled to an Allihn condenser which open end was
        shut off with a filter to maintain sterility.

(iii)   J, for introduction of an autoclaveable pH electrode.

The cell suspension was circulated through the system by the pump
(A) and dispersed at the top of the gas exchange vessel to ensure
efficient gas exchange. The CO_2 concentration of the aerating gas
was regulated by mixing pure CO_2 and air by means of flow meters.
The gases were filtered by 0.2 µm membrane filters at a rate of
5 liters per minute. The apparatus was sterilized by passing
steam, generated by an electrode boiler at a pressure of 70 kPa,
for 6 to 8 hours through the system.

     It was found that filtration of the growth medium (see §
titled: Morphological changes in *Microcystis* grown at different
irradiance levels) through a membrane filter (0.2 µm pore size)
led to a large reduction in Fe concentration (Krüger and Eloff,
1978a). Fe-free medium was thus sterilized by filtration and autoclaved $FeSO_4$ was subsequently introduced into the system separately.

Temperature of the growth room was controlled by means of an air conditioner (28°C). Cells were harvested by a continuous centrifuge.

By rigorously controlling environmental factors such as CO_2 concentration and pH, high yields of *Microcystis* cells could be obtained. Growth obtained with the non-toxic axenic *Microcystis* strain UV-007 in the mass culture system, is presented graphically in Figure 13.

The change in conditions regarding irradiance level, CO_2 concentration of the aerating gas and pH is also shown. When inspecting the growth curves (Figure 13A) it must be kept in mind that light intensity was not constant throughout. (In subsequent operations high light intensity was used throughout and had no adverse effect on growth.)

The pH was controlled between pH 8.5 and 10 by manipulation of the CO_2 concentration of the aerating gas (Figure 13B and 13C) by increasing the CO_2 concentration from 0.03% initially to a final concentration of 3.8%. It is clear that as long as the CO_2 concentration of the aerating gas mixture is sufficiently high, O_2 produced by photosynthesis, had no inhibitory effect on the cells even at high light intensities. This indicates that high CO_2 concentration plays a very important role in protecting cells against photooxidation and is of crucial importance in the successful mass culture of *Microcystis*.

After seven days of growth, the cell density was 6.7×10^7 cells per milliliter. At this stage no sign of bleaching was noticeable and the cells seemed to be healthy when investigated microscopically. The degree of multiplication was 4.6 doublings (calculated from cell number) or 4.9 doublings (calculated from turbidity) in 7.6 days. The cell yield obtained was 8.5 g per liter of fresh mass (total yield, 508 g) and 1.02 g per liter of dry mass (total yield, 60.9 g). It would be possible to increase the yield to 0.25 g dry mass per liter per day without increasing light intensity beyond 130 $\mu E./m^2/s$ by running the system on a semi-continuous basis, i.e., harvesting 50% of the cell suspension at a turbidity of ca. 500 Klett units every 2 days and replenishing with medium.

In subsequent operations a high irradiance level was used throughout and had no adverse effect on growth. Not only the non-toxic *Microcystis* UV-007 but also toxic *Microcystis* UV-006 and UV-010 (originally from Allemanskraal Dam) grew very well in our system if the CO_2 concentration (pH) was controlled. In the latter case the colonies appeared as flocks, but grew well. *Microcystis* UV-006 was grown for periods up to 12 weeks, with weekly harvesting

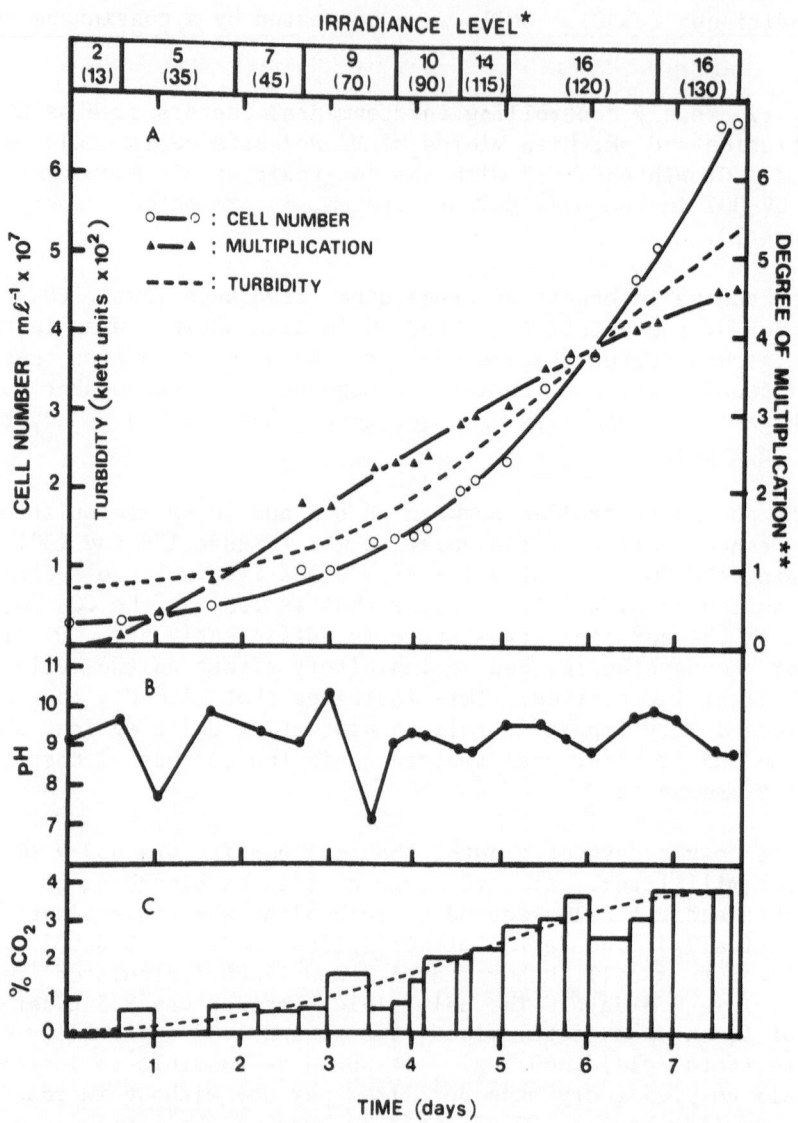

Fig. 13.   Representation of the growth of *M.* UV-007 in the mass
           culture system, indicating conditions regarding irra-
           diance level, CO_2 concentration of the aerating gas,
           and pH of the suspension during the course of the
           experiment.
             * Numbers on top of Figure A represent number of fluores-
               cent lamps employed at the corresponding times, and the
               numbers in parentheses give an indication of the cor-
               responding mean irradiance levels in $\mu E./m^2/s$.
           **Degree of multiplication calculated from cell counts
               (Log_2 Nt/No).

of ca. 95% of the suspension, without any deleterious effect in the mass culture system. This confirms once more that *Microcystis* can withstand severe turbulence for extended periods.

Because the CO_2 concentration is so important the system has subsequently been provided with a pH-stat unit. If the pH in the gas exchange vessel rises above the preset value a magnetic valve is actuated which introduces CO_2 in the system.

ACKNOWLEDGMENTS

The financial support by the University of the Orange Free State and the National Institute of Water Research of the Council for Scientific and Industrial Research is gratefully acknowledged.

REFERENCES

Abeliovich, A., and M. Shilo. 1971. CO_2 dependent viability of blue-green algae. Israel J. Med. Sci. *7*:9.

Abeliovich, A., and M. Shilo. 1972a. Photooxidative death in blue-green algae. J. Bacteriol. *3*:682-689.

Abeliovich, A., and M. Shilo. 1972b. Photooxidative reactions of c-phycocyanin. Biochem. Biophys. Acta *283*:483-491.

Arnon, D. I. 1935. Microelements in culture solution experiments with higher plants. Am. J. Botany *25*:322-325.

Beljanin, V. N., and A. P. Trenkensu. 1977. Growth and spectro-photometric characteristics of the blue-green alga *Synechococcus elongatus* under different temperature and light conditions. Arch. Hydrobiol. Suppl. *51*:46-66.

Brown, P. J. 1974. An Investigation of the Toxin of *Microcystis aeruginosa*: Effects of Varying the Concentration of Selected Salts upon Toxicity and the Effects of the Toxin on Some Physiological Parameters. Ph.D. Thesis. Texas A. and M. University, College Station, Texas. 72 pp.

Bruwer, C. A. 1979. The Economic Impact of Eutrophication in South Africa. Republic of South Africa Department of Water Affairs, Tech. Rep. TR94. 48 pp.

Bursche, E. M. 1961. Änderungen im chlorophyllgehalt und im zellvolumen bei planktonalgen, hervorgerufen durch unter-schiedliche lebensbedienungen. Int. Rev. Hydrobiol. *46*:610-652.

Crozier, W. J. 1924. On biological oxidations as function of temperature. J. Gen. Physiol. *7*:189.

Desikachary, T. V. 1959. Cyanophyta. Indian Council of Agricul-tural Research, New Delhi. 686 pp.

Dor, I. 1974. High density, dialysis culture of algae on sewage. Water Res. *9*:251-254.

Du Preez, J. C., and D. F. Toerien. 1978. The effect of tempera-
    ture on the growth of *Acinetobacter calcoaceticus*. Water
    S. A. *4*:10-13.
Eloff, J. N., Y. Steinitz, and M. Shilo. 1976. Photooxidation of
    Cyanobacteria in natural conditions. Appl. Environ. Micro-
    biol. *31*:119-126.
Fogg, G. E., W. D. P. Steward, P. Fay, and A. E. Walsby, eds. 1973.
    The Blue-green Algae. Academic Press Inc., London. 459 pp.
Geitler, L. 1932. Cyanophyceae. Pages 130-148 *in* Akademische
    Verlagsgesellschaft m.b.H. Leipzig.
Goldman, J. C., and E. J. Carpenter. 1974. A kinetic approach to
    the effect of temperature on algal growth. Limnol. Ocean-
    ography *19*:756-766.
Gorham, P. R. 1960. Toxic waterblooms of blue-green algae. Can.
    Vet. J. *1*:235-245.
Gorham, P. R. 1964. Toxic algae. Pages 307-336 *in* D. F. Jackson,
    ed., Algae and Man. Plenum Press, New York,
Gorham, P. R., and W. W. Carmichael. 1979. Phycotoxins from blue-
    green algae. Pure Appl. Chem. *52*(1):165-174.
Heise, J. J., and H. Gaffron. 1963. Catalytic effects of CO_2 in
    CO_2 assimilating cells. Plant Cell Physiol. (Tokyo) *4*:1-11.
Herdman, M., S. F. Delaney, and N. G. Carr. 1973. A new medium
    for the isolation and growth of auxotrophic mutants of the
    blue-green alga *Anacystis nidulans*. J. Gen. Microbiol. *79*:
    233-237.
Hoogenhout, H., and J. Amesz. 1965. Growth rates of photosynthetic
    microorganisms in laboratory cultures. Arch. Mikrobiol. *50*:
    10-25.
Hughes, E. O., P. R. Gorham, and A. Zehnder. 1958. Toxicity of a
    unialgal culture of *Microcystis aeruginosa*. Can. J. Micro-
    biol. *4*:225-236.
Jackson, W. A., and R. J. Volk. 1970. Photorespiration. Ann.
    Rev. Plant Physiol. *21*:385-432.
Jüttner, F. 1977. Thirty liter tower-type pilot plant for the
    mass cultivation of light and motion sensitive planktonic
    algae. Biotechnol. Bioeng. *19*:1679-1688.
Jüttner, F., H. Victor, and H. Metzner. 1971. Massenansucht
    phototropher organismen in einer automatischen kulturanlage.
    Arch. Mikrobiol. *77*:275-280.
King, D. L. 1970. The role of carbon in eutrophication. J. Water
    Poll. Control Fed. *42*:2035-2051.
Komárek, J. 1958. Die taxomische revision der planktischen
    blaualgen der Tschechoslawakei. *In* J. Komárek and H. Ettl,
    eds., Algologische Studien. Tschechoslawakishe Akademie
    der Wissenschaften, Prague. 358 pp.
Krüger, G. H. J., and J. N. Eloff. 1977. The influence of light
    intensity on the growth of different *Microcystis* isolates.
    J. Limnol. Soc. S. Afr. *3*:21-25.

Krüger, G. H. J., and J. N. Eloff. 1978a. Mass culture of *Micro-cystis* under sterile conditions. J. Limnol. Soc. S. Afr. 4(2):119-124.

Krüger, G. H. J., and J. N. Eloff. 1978b. The effect of agitation and turbulence of the growth medium on the growth and viability of *Microcystis*. J. Limnol. Soc. S. Afr. 4:69-74.

Krüger, G. H. J., and J. N. Eloff. 1978c. The effect of temperature on specific growth rate and activation energy of *Microcystis* and *Synechococcus* isolates relevant to the onset of natural blooms. J. Limnol. Soc. S. Afr. 4:9-20.

Krüger, G. H. J., and J. N. Eloff. 1979a. Chemical changes in the growth medium of *Microcystis* batch cultures grown at stress and non-stress light intensities. J. Limnol. Soc. S. Afr. 5:43-48.

Krüger, G. H. J., and J. N. Eloff. 1979b. The interaction between cell density of *Microcystis* batch cultures and light induced stress conditions. Z. Pflanzenphysiol. 95:441-447.

Kuhl, A., and H. Lorenzen. 1964. Handling and culturing of *Chlorella*. Pages 159-187 *in* D. M. Prescott, ed., Methods of Cell Physiology, Vol. 1. Academic Press, New York and London.

Kullberg, R. G. 1977. The effects of some ecological factors on cell size of the hot spring alga *Synechococcus lividus* (Cyanophyta). J. Phycol. 13:111-115.

Meeks, J. C., and R. W. Castenholz. 1971. Growth and photosynthesis in an extreme thermophile, *Synechococcus lividus* (Cyanophyta). Arch. Mikrobiol. 78:25-41.

Metzner, H., and R. Gerster. 1976. Energy conservation in photosynthesis models. III. Role of bicarbonate ions in oxygen evolution. Photosynthetica 10:302-306.

Orth, G. M., N. E. Tolbert, and E. Jimenez. 1966. Rate of glycolate formation during photosynthesis at high pH. Plant Physiol. 41:143-147.

Owens, O. V. H., and W. E. Esaias. 1976. Physiological responses of phytoplankton to major environmental factors. Ann. Rev. Plant Physiol. 27:461-483.

Pirt, S. J. 1975. Principles of Microbe and Cell Cultivation. Blackwell Scientific Publications, Oxford. 274 pp.

Pritchard, G. G., W. J. Griffin, and C. P. Whittingham. 1962. The effect of carbon dioxide concentration, light intensity and isonicotinyl hydrazide on the photosynthetic production of glygollic acid by *Chlorella*. J. Exp. Bot. 13:176-184.

Ruzicka, J. 1971. Morphologische variabilität der algen, hervorgerufen durch kultivierungsbedienungen. Arch. Hydrobiol. Suppl. 39:146-177.

Ruzicka, J., and J. Simmer. 1970. Measurement of productivity of algal strains by characteristic constants. Algol. Stud. (Trebon) 1:33-40.

Scott, W. E., M. T. Seaman, A. D. Connell, S. I. Kohlmeyer, and
    D. F. Toerien. 1977. The limnology of some South African
    impoundments. I. The physico-chemical limnology of Hart-
    beespoort Dam. J. Limnol. Soc. S. Afr. 3:43-58.
Setlík, I., J. Komárek, and B. Prokeš. 1966. Short account of
    the activities from 1960-1965. Ann. Rep. of the Laboratory
    of Experimental Algology. Laboratory of Algology, Trebon.
Shelef, G., W. J. Oswald, and C. G. Golueke. 1968. Kinetics of
    algal systems in waste treatment. Sanitary Engineering
    Research Laboratory College of Engineering and School of
    Public Health, Report No. 68-4. University of California,
    Berkeley, California. 183 pp.
Soeder, C., and E. Stengel. 1974. Physico-chemical factors
    affecting metabolism and growth rate. Pages 714-740 *in*
    W. D. P. Steward, ed., Algal Physiology and Biochemistry.
    Blackwell Scientific Publications, London.
Sorokin, C. 1960. Kinetic studies of temperature effects on the
    cellular level. Biochim. Biophys. Acta 38:197-204.
Sorokin, C., and R. W. Krauss. 1962. Effects of temperature and
    illuminance on *Chlorella* growth uncoupled from cell
    division. Plant Physiol. 37:37-42.
Stanier, R. Y., R. Kunisawa, M. Mandel, and G. Cohen-Bazire.
    1971. Purification and properties of unicellular blue-
    green algae (order Chroococcales). Bacteriol. Rev. 35:171-
    205.
Talling, J. F. 1976. The depletion of carbon dioxide from lake
    water by phytoplankton. J. Ecol. 64:79-121.
Tamiya, H., E. Hase, K. Shibata, A. Mituya, T. Iwamura, T. Nihei,
    and T. Sasa. 1953. Kinetics of growth of *Chlorella*, with
    special reference to its dependence on quantity of available
    light and on temperature. Pages 207-232 *in* J. S. Burlew,
    ed., Algal Culture from Laboratory to Pilot Plant. Carnegie
    Institution of Washington Publication 600, Washington, D. C.
Toerien, D. F. 1977. A review of eutrophication and guidelines
    for its control in South Africa. Pages 1-110 *in* CSIR
    Special Report WAT 48. Pretoria, Republic of South Africa.
Vance, B. D. 1965. Composition and succession of cyanophycean
    water blooms. J. Phycol. 1:81-86.
Zehnder, A., and P. R. Gorham. 1960. Factors influencing the
    growth of *Microcystis aeruginosa* Kütz. emend. Elenkin.
    Can. J. Microbiol. 6:645-660.

PRELIMINARY TESTS OF TOXICITY OF *SYNECHOCYSTIS* SP.

GROWN ON WASTEWATER MEDIUM

E. P. Lincoln and W. W. Carmichael[1]

Agricultural Engineering Department
University of Florida
Gainesville, Florida

ABSTRACT

A toxic, coccoid Cyanobacteria 1.4 to 1.6 μm in diameter, was found to be dominant during summer in an algal culture used to recycle swine wastes. It has been referred to the genus *Synechocystis*, without species designation. Electron micrographs and phase microscopy suggest its identity with strain 6714 of Stanier et al. (1971), a chemo- and photo-heterotrophic species of *Synechocystis* formerly included in *Aphanocapsa*. It most closely resembles *S. limnetica* as described by Komárek (1976). The alga typically occurs as a diplococcoid uniformly distributed in the water column at densities up to 0.3×10^9 cells/ml. The whole cell absorption spectrum shows a pronounced phycocyanin peak at 630 nm. Use of the dried alga in poultry feeding trials resulted in high mortality in one instance and very low mortality in another, indicating the variable presence of toxic strains. Injection of the toxic strain i.p. at 1000 mg/kg into mice was lethal in all cases. Survival time ranged from 5 to 7 hours, liver damage was evident, and lethargy and paleness of the extremities preceded collapse. Symptoms resembled those of anatoxin-c but were attenuated. A 30 day feeding trial with 18 grower swine on diets containing 16.5% algae of the toxic strain produced no sign of toxicity in any of the experimental animals. Swine appeared to be immune to the toxin at cumulative dosages of 0.14 kg/kg. The apparent widespread occurrence of this species of *Synechocystis* in wastewater cultures intended for the

---

[1]Address: Department of Biological Sciences, Wright State Univ., Dayton, Ohio.

production of animal feeds and water treatment renders its toxicological properties of special interest.

INTRODUCTION

     A minute diplococcoid Cyanobacteria that produces toxic strains
has been found to dominate a large outdoor algal culture used for
the production of protein and the renovation of wastewater. The
culture, located at the Swine Research Unit of the University of
Florida, uses the effluent from an anaerobic lagoon as the principal nutrient source, with additional medium being drawn from an
adjoining facultative lagoon. The Cyanobacteria, while present the
year round, becomes dominant only with the advent of hot weather,
after which it may constitute up to 95% of the algal biomass for
periods of three to six months. It has been referred to the genus
Synechocystis (Sauvageau, 1892), but no species designation has
been made. The alga will be referred to here as *Syn.* sp. to avoid
confusion with the generic term in a less specific context.

TEXT

Morphology

     The individual cells are either round or slightly elliptic,
and occur most frequently joined in pairs. Often two such pairs
will be united to form a tetrad. The diameter of the single cell
is 1.4 to 1.6 μm (microns) and the long axis seldom exceeds 2 μm.
Under phase optics, the color is brown to dark blue-green, and a
hint of central internal structure is evident. Suspended cells,
because of their small size, undergo pronounced Brownian motion
and tend to appear motile.

     An electron micrograph of a typical cell pair, magnified
27,500 diameters is shown in Figure 1. It clearly demonstrates
the major fine structural components of the Chroococcales. Most
evident are the polyphosphate granules forming a circular cluster
in the center of the cell, and which probably are the internal
structures visible under phase microscopy. The slightly elongated
polar region of these cells contrasts with their more globular or
oblate image under the light microscope and may be a distortion
incurred during fixation. The distinct peptidoglycan layers overlying the cell membrane in this specimen show a discontinuity in
the constricted region between the cells, and may indicate approach-
ing telophase. Whether or not such cytoplasmic bridges are typical
of the predominant diplococcoid form has not been determined.

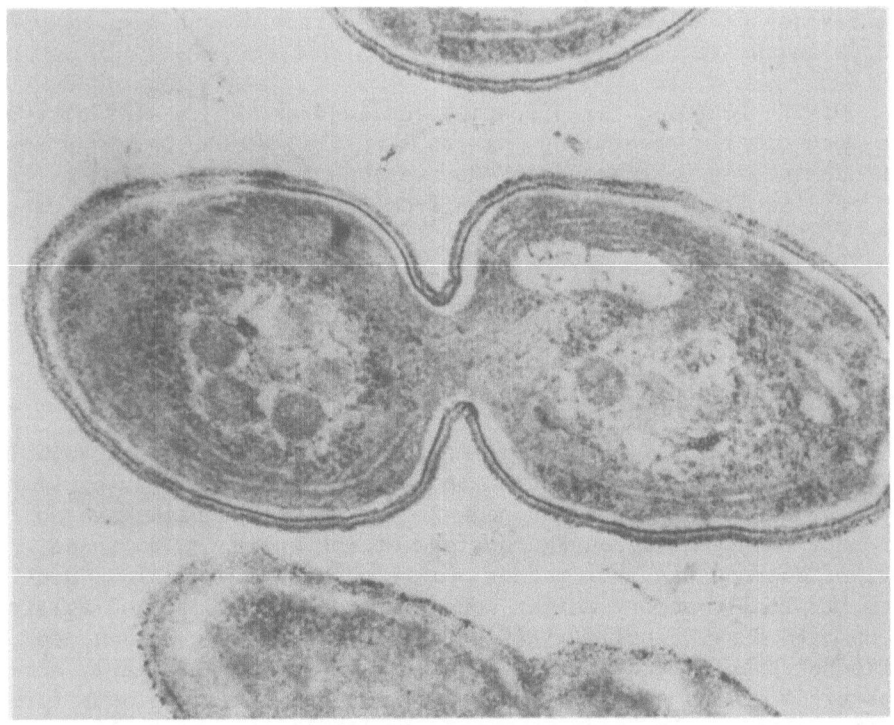

Fig. 1.   Electron micrograph of *Synechocystis* sp. magnified 27,500
          diameters. Polyphosphate granules appear as a cluster at
          center of cell.

Taxonomy

A recent review of the genus *Synechocystis* by Komárek (1976)
indicates that it is widely occurring in both marine and freshwater habitats, with forms ranging in size from less than 1 μm to
more than 20 μm in diameter. Spheroidal to broadly oval cells of
finite size, typically in pairs arranged in perpendicular planes,
are taken to identify a coherent group of species within the
Croococcales, distinct from *Synechococcus* which vary in shape and
size, and occur in chain-like colonies. The name *Synechocystis*,
which translates as "with an echo cell", is appropriate to this
morphological grouping which would seem to define a recognizable
genus, based ultimately on the geometry of cell division.

Stanier et al. (1971) after analyzing 11 strains of this
group (their typological group II) in pure cultures found them
homogeneous in DNA base-pair composition and temperature tolerance,
and included them in *Aphanocapsa*. They have since been reassigned

to *Synechocystis* (Stanier and Cohen-Bazire, 1977). This first
designation, and the occasional confusion with *Microcystis* or *Anacystis*, may have led to the relative obscurity of the name *Synechocystis* in the recent literature.

Of the fourteen named species reviewed by Komárek (1976) the
one most closely resembling *Syn.* sp. in size, morphology, and apparent habitat, is *S. limnetica*. However, no firm conclusions as to
its species identity can be drawn from the present data. It does
seem from the E.M. photographs and light microscopy to be essentially identical with strain 6714 of Stanier et al. (1971) and
Rippka (1972), also of undetermined species.

Ecology

Macroscopically, *Syn.* sp. imparts a brilliant lime-green color
to the culture. The absorption spectrum of whole cell suspensions
has a pronounced phycocyanin peak at 630 nm. The cells have no
tendency to aggregate on the surface of the water, as do those,
say, of *Microcystis*, but rather remain suspended at uniform density, typically at dry weight concentrations of 200 to 600 mg/liter,
or up to 0.3×10^9 cells/ml (Figure 2). Growth rate of *Syn.* sp.,
while not precisely quantified, is very rapid and noticeably exceeds
that of *Chlorella* sp. or *Monodus guttula*, the other dominant forms
in the culture. Its sudden ascendence leading to the replacement
of *Monodus* in the spring of 1979 is shown in Figure 3. The ordi-
nate in this case refers to *Syn.* sp. cells/ml. *Monodus* counts are
multiplied by a factor of 20 to give relative abundance in terms
of algal biomass.

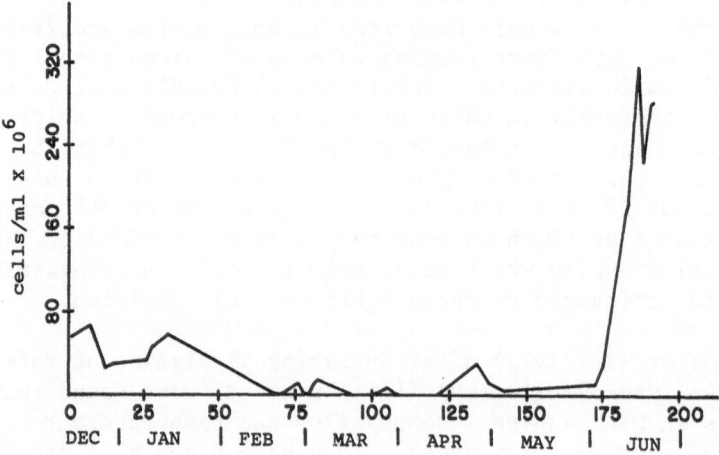

Fig. 2. Concentration of *Synechocystis* sp. in the algae culture.

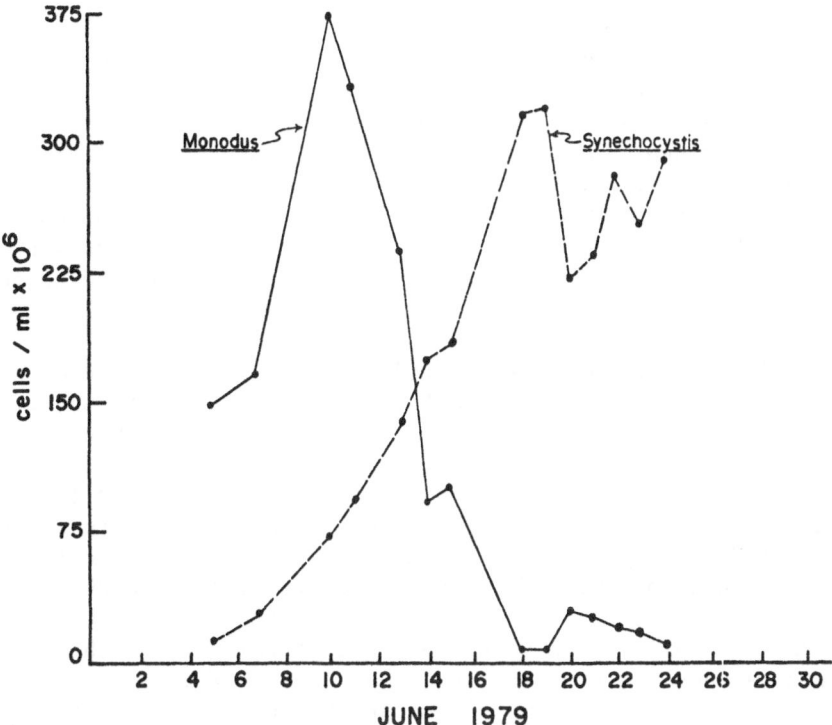

Fig. 3. Cells/ml of *Synechocystis* sp. and *Monodus* x 20 showing
relative abundance in terms of algal biomass.

     In outdoor cultures, *Syn.* sp. can maintain high densities
with little photosynthetic activity, as indicated by its uniform
distribution throughout the water column during thermal stratifi-
cation, and its often increasing concentrations over prolonged
periods of cloudy weather. It also thrives at extremely high
levels of irradiation. These characteristics are consistent with
the findings of Rippka (1972) for the above mentioned strain 6714,
which was found to be both photoheterotrophic and chemoheterotro-
phic.

     Syn. sp., while able to tolerate high temperatures (40°C),
may also remain the dominant alga at temperatures down to freezing.
Its relative insensitivity to radiation and temperature, variables
profoundly affecting most algae, must give *Syn.* sp. a broad competitive advantage over other algae in the culture. In fact, it
has thus far been impossible to determine precisely what causes
its eventual replacement by *Chlorella* sp., which may occur any
time from late summer to mid winter.

Toxicity

Relatively pure crops of *Syn.* sp. have been harvested in quantities of a ton or more (dry weight) for experimental use as a protein supplement in swine and poultry rations. Its toxic properties became evident in the second poultry feeding trial. Birds on diets containing 10% to 20% algae had close to 100% mortality, and those on 5% algae showed symptoms of paralysis after three days of feeding. These symptoms were reversed when the algal feed was withdrawn. The first feeding trial with *Syn.* sp. produced no toxic effects, indicating that only certain strains are toxic.

Samples of the toxic strain collected in October, December, and May from different points in the culture system were injected i.p. into mice (ICR-Swiss, 20 to 25 gram males) at dosages ranging from 250 to 2000 mg/kg (dry weight algae/kg body weight). The results are given in Table 1. Dosages of 1000 mg/kg were lethal in all cases and, from these data, the LD_{50} appears to lie in the range of 500 to 750 mg/kg. Liver damage of the anatoxin-c type (Carmichael and Gorham, 1978) consisting of swollen, reddened lobes and engorgement with blood in the hepatic lobules was noted for all three algal samples. The observed signs of poisoning such as lethargy and paleness of the ears and other body extremities also resemble anatoxin-c effects. Although the survival time of 5 to 7 hours was considerably longer than the 1 to 3 hours typical of anatoxin-c at the LD_{50} range, the general effects of *Syn.* sp. toxin are indicative of a peptide of the anatoxin-c type.

Because the primary use of the cultured algae is as a substitute for soybean meal in swine rations, a preliminary feeding trial was conducted on grower swine (23 to 55 kg) in which 16.5% of the diet, by weight, consisted of *Syn.* sp. The trial was run for 30 days, with 18 animals on the algae diet and 18 controls on standard rations. No mortality, nor any sign of toxicity was recorded for any of the animals on the algae diet. A total of 213 kg of *Syn.* sp. was consumed by the experimental group, a cumulative dosage of some 140 mg/kg. Since this algae meal was from the same batch that proved toxic to both mice and poultry, it must be concluded that swine are highly tolerant of the toxin, or entirely immune to it.

Species differences among vertebrates in susceptibility to algal toxins are apparently pronounced. As pointed out by Carmichael and Biggs (1978) such differences can have a fundamental physiological basis relating to the sensitivity of the neuromuscular junction rather than to degree of absorption and detoxification. The genetic difference between animal species may thus have significant bearing on the evaluation of a given anatoxin, and the actual danger it presents in a particular situation.

Table 1. Results of i.p. Injection of *Synechocystis* sp. into Mice at Varying Dosages; Deaths/N

Dosage mg/kg	Algal Sample		
	Oct.	Dec.	May
2000	-	4/4	-
1500	-	4/4	4/4
1200	5/5	-	-
1000	5/5	4/4	4/4
750	-	4/4	2/4
500	2/5	0/4	0/4
250	0/5	-	-

CONCLUSION

Confirmation of the above results is presently being undertaken. Aside from the necessity of verification from a toxinological standpoint, data on the relative toxicity of *Syn.* sp. is currently of considerable practical importance. A number of algal recycling systems are under development in various parts of the world, most of which have the production of feed for swine, poultry, or fish as a major objective. A small coccoid cyanobacterium, apparently identical to *Syn.* sp., has been found in stabilization and high-rate ponds in Israel, Singapore, Thailand, and the Philippines in both swine-waste and sewage media (Lincoln and Hall, 1980). As in the Florida project, it periodically attains dominance and may become a major component of the algal harvest. If swine, and perhaps fish, can be shown to be immune to its toxin, and poultry highly susceptible, then such information has obvious economic significance in that it can allow algal recycling systems to be put to best use in a particular cultural setting. Human sensitivity to *Syn.* sp. toxin is likewise a prime consideration if, as it appears, this alga is widely occurring in systems intended for wastewater treatment.

ACKNOWLEDGMENTS

This investigation was conducted under the auspices of the Institute of Food and Agricultural Sciences (IFAS) of the University

of Florida with additional funding from the U. S. National Science
Foundation, Grant No. DAR 78-23886.

REFERENCES

Carmichael, W. W., and D. F. Biggs. 1978. Muscle sensitivity
differences in two avian species to anatoxin-a produced by
the freshwater cyanophyte *Anabaena flos-aquae* NRC-44-1.
Can. J. Zool. *56*(3):510-512.

Carmichael, W. W., and P. R. Gorham. 1978. Anatoxins from clones
of *Anabaena flos-aquae* isolated from lakes of western
Canada. Mitt. Int. Ver. Limnol. *21*:285-295.

Komárek, J. 1976. Taxonomic review of the genera *Synechocystis*
Sauv. 1892, *Synechococcus* Nag. 1849, and *Cyanothece* gen.
nov. (Cyanophyceae). Arch. Protistenk. Bd. *118*:119-179.

Lincoln, E. P., and T. W. Hall. 1980. Critical factors in the
large scale production of microalgae. Proc. Workshop on
Waste Treatment and Nutrient Recovery from High-Rate Algae
Ponds, Feb. 27-29, Singapore.

Rippka, R., J. Waterbury, and G. Cohen-Bazire. 1974. A cyanobacterium which lacks thylakoids. Arch. Microbiol. *100*:
419-436.

Sauvageau. 1892. Bull. Soc. Bot. France 14, ses. extraord.
(CXV-CXVI).

Stanier, R. Y., and G. Cohen-Bazire. 1977. Phototrophic prokaryotes: the Cyanobacteria. Ann. Rev. Microbiol. *31*:225-274.

Stanier, R. Y., R. Kunisawa, M. Mandel, and G. Cohen-Bazire. 1971.
Purification and properties of unicellular blue-green algae
(order Chroococcales). Bacteriol. Rev. *35*(2):171-205.

CHEMICAL NATURE AND MODE OF ACTION

OF A TOXIN FROM *PANDORINA MORUM*

G. M. L. Patterson, D. O. Harris, and W. S. Cohen

School of Biological Sciences
University of Kentucky
Lexington, Kentucky 40506

ABSTRACT

Pandorina morum, a green alga, was shown in earlier work to produce a toxic substance which adversely affects the growth of other algae in mixed cultures or when grown in the presence of culture filtrates or extracts of *Pandorina* cells. This paper reports on current methods used to extract the toxin.

The toxin extracts were then tested and found to inhibit several types of algae and obligate aerobic bacteria. Facultative anaerobic bacteria were insensitive to the toxin. The toxin was also found to inhibit photosynthetic electron transport in isolated thylakoids as well as mitochondrial electron transport in isolated potato mitochondria.

INTRODUCTION

Pandorina morum (Bory) produces a toxic substance which adversely affects the growth of other algae in mixed cultures or when grown in the presence of culture filtrates or extracts of *Pandorina* cells (Lefevre, 1956; Rice, 1954; Harris, 1970).

Harris (1970b), in a study of the inhibitory effects of culture filtrates from several members of family Volvocaceae, found that *Pandorina morum* produced the most effective growth inhibitory substance, while *Volvox globator* was most sensitive to inhibition. These two organisms formed that basis of a model system for the study of algal-produced growth inhibitors.

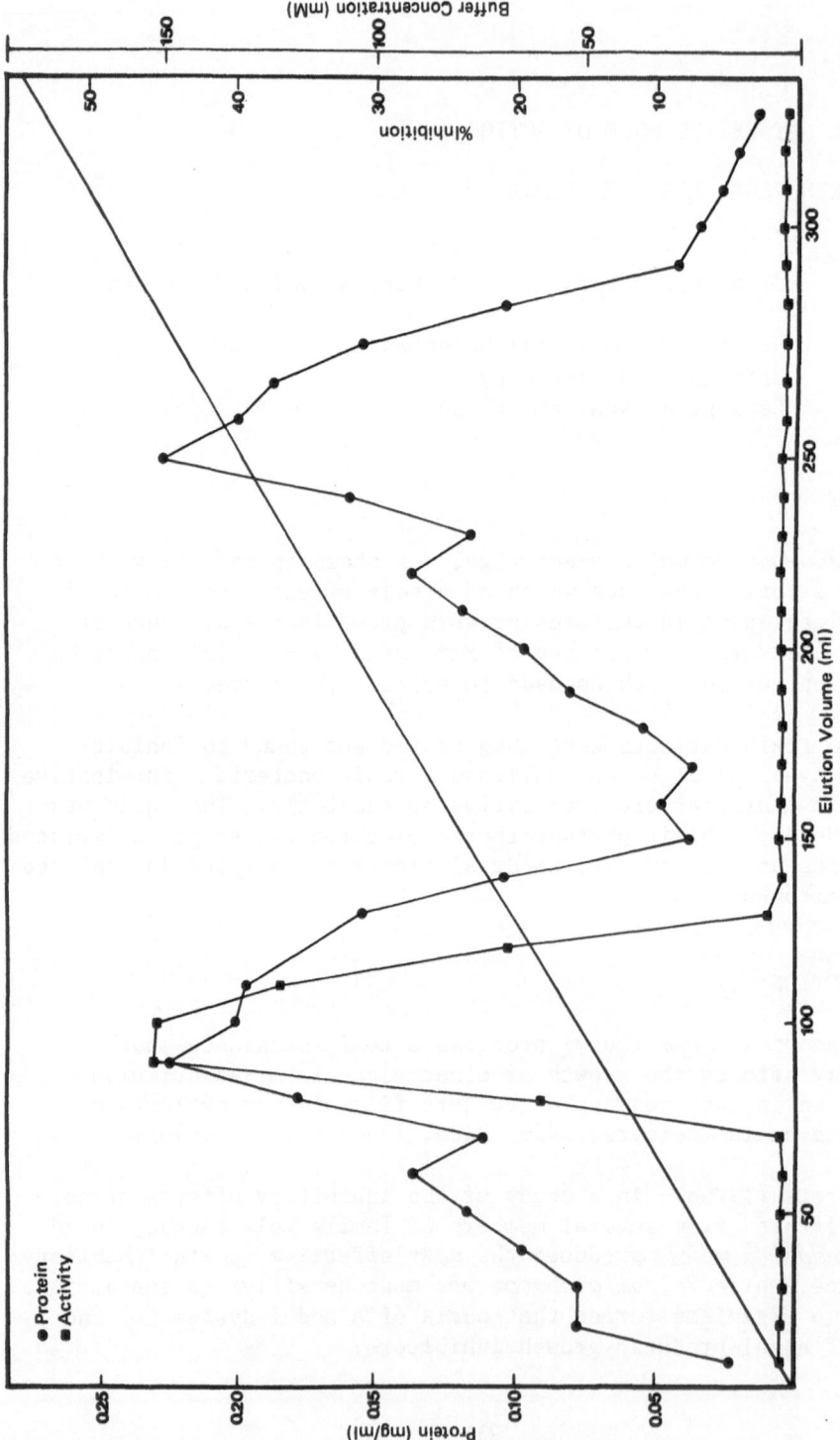

Fig. 1. Elution profile of the toxin from DEAE-cellulose. Protein (—●—) was estimated by the method of Lowry, 1951; % Inhibition (—■—) refers to inhibition of photosynthetic electron transport by aliquots of the eluate.

TEXT

Purification of the Toxin

Culture of *Pandorina morum* (UTEX 18) were grown in 10 liter Pyrex solution bottles filled with modified *Volvox* medium (Harris, 1970a). Light was supplied by banks of cool-white fluorescent tubes giving an intensity of 1000 foot-candles at the surface of the flasks. The cultures were grown under conditions of continuous illumination and aeration (supplemented with approximately 3% CO_2) at 24°C.

Colonies were harvested from the culture medium by centrifugation, and the pellet homogenized by sonication in 30 second bursts with cooling periods. Cell debris was removed by centrifugation, and the homogenate was dialyzed against 20 volumes of glass distilled water for 90 hours at 4°C. The external solution was changed at 12 hour intervals, pooled, and flash-evaporated in vacuo at 40°C.

The dialyzate was then buffered with 25 mM potassium phosphate buffer, pH 7.0, and applied to a 3.7 x 75 cm column of Sephadex G-25, and eluted with the same buffer.

Active fractions from the first Sephadex column were applied to a 3.7 x 75 cm column of DEAE-cellulose, precycled in accordance with the manufacturer's specifications, and eluted with a linear gradient of phosphate buffer. The elution profile from the DEAE-cellulose column is shown in Figure 1. Active fractions were pooled, concentrated in vacuo, and applied to a second G-25 column (1.6 x 100 cm), and eluted with 25 mM phosphate buffer.

Active fractions from the second G-25 step were concentrated in vacuo, desalted by passage through Bio-Gel P2 (toxin excluded) and further purified by descending paper chromatography on Whatman 3MM paper, using a solvent system of n-propanol, water, ammonium hydroxide (100:50:1). The region containing activity was cut from the chromatogram and the toxin was eluted with methanol:water (1:1). This material was again concentrated in vacuo and stored at -15°C. Final yields averaged 5 to 7% of the starting activity.

The toxin is heat-labile, acid-labile, and passes through a dialysis membrane. Estimates of molecular weight by the use of gel chromatography indicate that the toxin is in the range of 1000 to 1500 daltons. Purified toxin has a strong ninhydrin response, a peptide-like absorbance in the ultraviolet region, and yields free amino acids on hydrolysis, indicating that the toxin is a non-cyclic peptide.

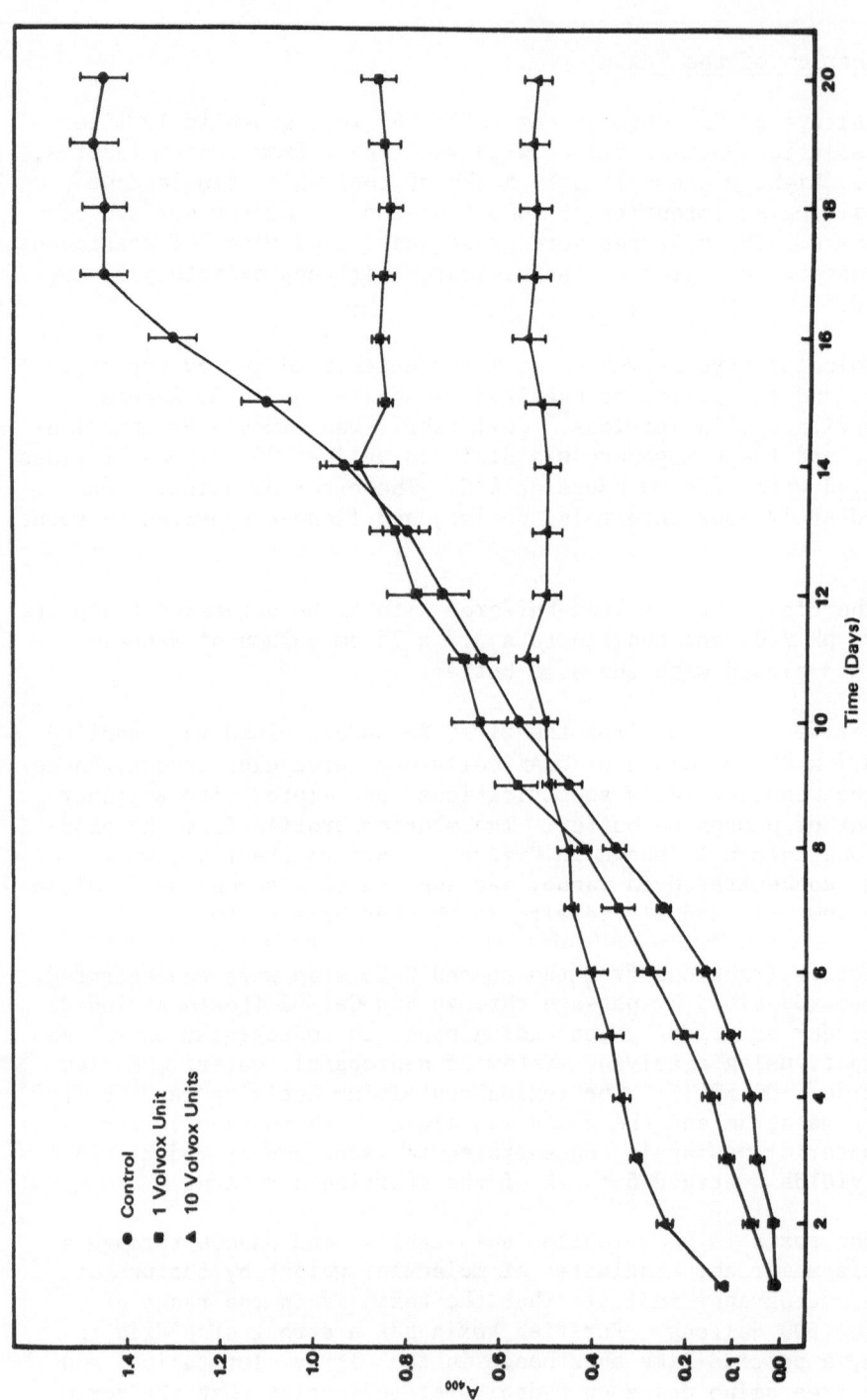

Fig. 2.   Growth of *Dysmorphococcus globosus* in the presence and absence of *Pandorina* toxin. Cultures were grown in 10 ml of modified *Volvox* medium in 18 x 150 mm culture tubes. Growth was estimated at 24 hour intervals by measuring absorbance at 400 nm in a Bausch and Lomb Spectronic 20.

Definition of Toxic Activity and Summary of Toxic Effects

Since an accurate, sensitive chemical assay has not yet been developed for the toxin, bioassay remains the only reliable method of detection and quantitation. Toxic activity is defined in terms of arbitrary *Volvox* units; one unit is the amount required to inhibit the growth of *Volvox globator* by 50% under standard conditions of assay. The effect of the toxin on growth of *Dysmorphococcus globosus*, a unicellular green alga, may be seen in Figure 2.

The toxin is inhibitory to a wide variety of algae, including members of the Chlorophyta, Cyanophyta, Rhodophyta, Euglenophyta, and Chrysophyta. Higher plants are also susceptible to the toxin, with growth of hydroponically cultured *Phaseolus vulgaris* being strongly inhibited by a toxin concentration of 5 *Volvox* units (Figure 3). Growth of the aquatic macrophyte *Lemna minor* is also inhibited by the toxin (data not shown).

The toxin also has an adverse effect on the growth of bacteria in liquid culture. Of the species tested, those which were incapable of anaerobic growth (obligate aerobes) were found to be sensitive to the toxin, while facultative anaerobes were insensitive (Table 1).

Table 1. Effect of *Pandorina* Toxin on the Growth of Various Bacteria[a]

Name	Anaerobic Growth	Sensitivity
Bacillus cereus	+	-
Mycobacterium smegmatis	-	+
Proteus vulgaris	+	-
Pseudomonas fluorescens	-	+
Pseudomonas aeruginosa	+	-
Escherichia coli	+	-

[a]Cultures were grown on Nutrient Broth (Difco) with 0.1% each of glucose and yeast extract. Growth was estimated by increase in absorbance at 400 nm. Insensitive species are those which showed no significant decrease in growth at 10 *Volvox* units.

Fig. 3.   Effect of *Pandorina* toxin on growth of *Phaseolus vulgaris*.
Hydroponically grown (Witham et al., 1971) plants with
Pandorina toxin added to the culture medium.

Inhibition of growth in bacteria and algae may also be
reversed. Incubation of bacterial cells for up to 96 hours in the
presence of the toxin, followed by centrifugation and resuspension
in medium without toxin, results in growth which approximates that
of control cultures. Since bacteria plated on agar containing the
toxin produce colonies of uniform small size, it appears that the
toxin acts as a growth depressant, rather than by selectively
killing some cells while allowing others to grow normally.

Effect of the Toxin on Photosynthetic Electron Transport in Isolated Thylakoids

Early experiments (Harris and Caldwell, 1974) had shown that
the toxin inhibited the ability of isolated thylakoids to photo-

Table 2.   Inhibition of Electron Transport by *Pandorina* Toxin[a]

Reaction Sequence	Toxin[b]	Rate (µequiv mg/chl/hr)	% Inhibiton
H_2O → Ferricyanide	-	974	-
	+	60	94
H_2O → Dimethylquinone	-	513	-
+ DBMIB[c] (II)	+	0	100
Diaminodurene/Ascorbate			
→ Methyl Viologen +	-	2490	-
DBMIB (I)	+	2485	0

[a]Reaction mixtures contained: 1.25 mM $K_3Fe(CN)_6$ (water to ferri-cyanide); 0.5 mM 2,5-dimethyl-p-benzoquinone, 1.5 mM $K_3Fe(CN)_6$, and 2 µM DBMIB (Dibromothymoquinone) (water to DMQ); and 1 mM DAD (Diaminodurene), 2.5 mM neutralized Ascorbate, 0.1 mM methyl viologen, 0.5 mM NaN_3, and 4 µM DCMU (Dichlorophenldimethyl urea) (DAD/Ascorbate to methyl viologen).   Chlorophyll concentration was 22 µg/ml, toxin was present at 100 µl/ml, and electron flow was uncoupled by 1 µM nigericin.   Thylakoids were isolated according to Cohen (1978).

[b]The + = Toxin present; and - = Toxin not present.

[c]DBMIB = Dibromothymoquinone.

synthetically reduce artificial electron acceptors.   To further localize the site of action, the effect of the toxin on partial electron transport reations was examined.   Electron transport was monitored polarographically by following oxygen concentration changes with a Clark-type electrode.

Table 2 shows the effect of the toxin on electron transport through both Photosystem I and II, and on each photosystem isolated by the use of the intersystem blocking agent dibromothymoquinone (DBMIB) and artificial electron donors and acceptors (Figure 4). While whole chain and Photosystem II electron flow are inhibited by the toxin, Photosystem I electron flow is unaffected.   The effect of the toxin on electron flow in Tris-treated[1] and control chloroplasts is summarized in Table 3.   This treatment eliminates

---

[1]Tris = N-Tris (Hydroxymethyl)-amino methane.

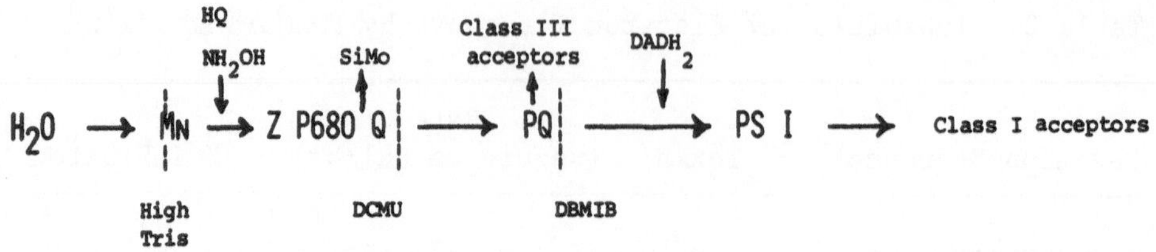

Fig. 4. Simplified scheme for photosynthetic electron transport
         showing sites of inhibition and main sites of action of
         artificial electron donors and acceptors.

water as a donor of electrons and allows the use of artificial
electron donors which supply electrons to Photosystem II closer to
the reaction center. Dichlorophenldimethyl urea (DCMU)-insensitive
Photosystem II electron flow from water to Silicomolybdate (accept-
ing electrons from unidentified primary acceptor of Photosystem II
(Q), the primary quencher of Photosystem II fluorescence) is also

Table 3. Effect of *Pandorina* Toxin on Electron Transport in Control
          and Tris-Treated Chloroplasts[a]

Reaction Sequence	Toxin[b]	Rate (μequiv mg/chl/hr)	% Inhibition
$H_2O \rightarrow$ Methyl Viologen	–	826	–
	+	267	68
Hydroquinone/Ascorbate	–	1056	–
$\rightarrow$ Methyl Viologen	+	273	74
Hydroquinone/Ascorbase			
$\rightarrow$ Methyl Viologen	–	437	–
(Tris-Treated)	+	129	70

[a] Reaction mixtures contained: 0.2 mM methyl viologen, 0.5 mM NaN_3
(water to methyl viologen); 0.33 mM hydroquinone, 0.5 mM neutralized ascorbate, 0.2 mM methyl viologen, 0.5 mM NaN_3 (Hydroquinone/
Ascorbate to methyl viologen). Electron flow was uncoupled with
0.75 μM nigericin. Toxin was present at 75 μl/ml. Tris-treated
chloroplasts were prepared by incubation in 0.8 M Tris-HCl (pH
8.2) for 15 minutes followed by washing and resuspension in medium
without Tris.
[b] The + = Toxin present; and – = Toxin not present.

Table 4. Effect of *Pandorina* Toxin on DCMU-Insensitive Silicomolybdate Reduction in Chloroplasts[a]

Reaction Sequence	Toxin	Rate (µequiv mg/chl/hr)	% Inhibition
$H_2O \rightarrow$ Ferricyanide	-	445	-
(I + II)	+	79	82
$H_2O \rightarrow$ Silicomolybdate	-	79	-
+ DCMU (II)	+	0	100

[a]Reaction mixtures contained: 1.25 mM $K_3Fe(CN)_6$, (water to ferri-
cyanide); 33 µM 12-molybdosilicic acid, 4 µM DCMU (water to silicomolybdate). Electron flow was uncoupled by 10 mM NH_4Cl. Chlorophyll concentration was 23 µg/ml.

markedly reduced by the toxin, as summarized in Table 4.

Further characterization of the site of action of the toxin
has been made by examination of changes in the fluorescence yield
of Photosystem II chlorophylls when dark adapted chloroplasts are
initially illuminated (fluorescence induction). Maximal levels of
variable fluorescences are markedly reduced in the presence of the
toxin, rather than increased, indicating that the site of action
of the toxin is not on the reducing side of Photosystem II. Inhibition on the reducing side of Photosystem II by herbicides like
DCMU increases the maximal level of fluorescence and alters the
kinetic patterns of induction transients.

The observation that electron flow at low light intensities
is more sensitive to the toxin than electron flow at saturating
light intensities provides additional evidence for an inhibitory
site close to the Photosystem II reaction center (data not shown).

Effect of the Toxin on Mitochondrial Electron Transport

Pandorina toxin also strongly inhibits mitochondrial electron
transport (Patterson et al., 1979). Table 5 shows the effect of
the toxin on State III respiration in isolated potato (*Solanum
tuberosum*) mitochondria (Laties, 1974). The similar pattern of
inhibition with both succinate and malate plus pyruvate as substrates indicates that the site of action is on the oxidizing side

Table 5. Effect of *Pandorina* Toxin on Electron Transport in Mito-
          chondria[a]

Reaction Sequence	Toxin	Rate (nmoles O_2/mg protein/min)	% Inhibition
Succinate → O_2	–	239	–
	+	125	48
Malate + Pyruvate	–	106	–
→ O_2	+	47	56
Diaminodurene/Ascorbate → O_2	–	216	–
+ Antimycin A	+	234	0

[a]Reaction mixtures contained: 400 mM Mannitol, 25 mM TES–NaOH (pH
7.4), 5 mM $MgCl_2$, 5mM K_2HPO_4, 1 mg/ml Bovine serum albumin, and
mitochondrial protein equivalent to 0.151 mg/ml. Toxin was present at 75 µl/ml. State III/State IV cycles were initiated by
addition of 300 nmoles of ADP. For DAD/Ascorbate to O_2, 0.5 mM
Diaminodurene and 2.5 mM neutralized ascorbate were added.

of Coenzyme Q (unidentified primary acceptor of Photosystem II)
(Figure 5). Electron transport from DAD/Ascorbate to oxygen in
the presence of Antimycin A is unaffected by the toxin, indicating
that the site of action is in the region of Site II of oxidative
phosphorylation.

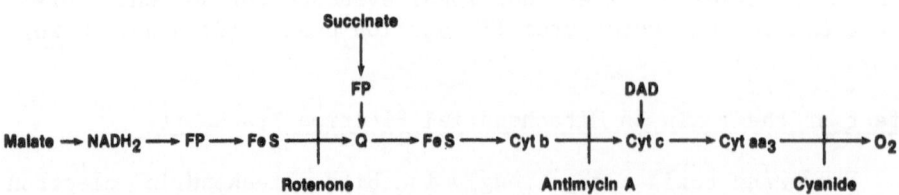

Fig. 5. Simplified scheme of mitochondrial electron transport
         showing sites of inhibition and main sites of action of
         artificial electron donors and acceptors.

CONCLUSIONS

The mode of action of the *Pandorina* toxin has been examined at both the cellular and subcellular level. The adverse effects on electron transport processes observed in toxin-treated organelles and cells may provide a basis for the toxic properties of this compound. The observation that bacteria which carry out neither process are immune to the toxin tends to support the above hypothesis.

REFERENCES

Cohen, W. S. 1978. The coupling of electron flow to ATP synthesis in pea chloroplasts stored in the presence of glycerol at -70°C. Plant Sci. Letters *11*:191-197.

Harris, D. O. 1970a. A model system for the study of algae growth inhibitors. Arch. Protistenkunde *113*:230-234.

Harris, D. O. 1970b. Inhibitors produced by green algae (Volvocales). Arch. Mikrobiologie *76*:47-50.

Harris, D. O. 1971. Inhibition of oxygen evolution in *Volvox globator* by culture filtrates from *Pandorina morum*. Microbios *3*:73-75.

Harris, D. O., and C. D. Caldwell. 1974. Possible mode of action of a photosynthetic inhibitor produced by *Pandorina morum*. Arch. Mikrobiologie *95*:193-204.

Laties, G. G. 1974. Isolation of mitochondria from plant material. Methods Enzymol. V *32a*:589-600.

Lefevre, M., and G. Farrugia. 1956. Sur quelqes proprietes des eaux de ruissellement contribuant au remplissage des mares et etangs. C. R. Acad. Sci. *235*:234-236.

Lowry, O. H., N. J. Rosebrough, A. L. Farr, and R. J. Randall. 1951. Protein measurement with the Folin phenol reagent. J. Biol. Chem. *193*:265-275.

Patterson, G. M. L., D. O. Harris, and W. S. Cohen. 1979. Inhibition of photosynthetic and mitochondrial electron transport by a toxic substance isolated from the alga *Pandorina morum*. Plant Sci. Letters *15*:293-300.

Rice, T. R. 1954. Biotic influences affecting population growth of planktonic algae. U. S. Dep. Int. Fish. Bull. *87*:227-245.

Witham, F. H., D. F. Blaydes, and R. M. Devlin. 1971. Experiments in Plant Physiology. Van Nostrand Reinhold, New York.

WATER-ASSOCIATED HUMAN ILLNESS IN NORTHEAST PENNSYLVANIA

AND ITS SUSPECTED ASSOCIATION WITH BLUE-GREEN ALGAE BLOOMS

Wayne H. Billings

Pennsylvania Department of Environmental Resources
Bureau of Community Environmental Control
Stroudsburg, Pennsylvania

ABSTRACT

On August 8, 1979 the Bureau of Community Environmental Control, Department of Environmental Resources, in collaboration with the Pennsylvania Department of Health, undertook the investigation of an outbreak involving 12 children and one adult at a lake-shore community located in northern Monroe County.

Within the next two weeks, three additional outbreaks occurred at another lake heavily used for recreational purposes in Pike County, approximately 20 miles away from the first site.

Though symptoms, referred to by local doctors as summer flu, varied widely, contact with the lake waters was found to be a common factor at both lakes. Symptoms which ranged from gastrointestinal involvement to hayfever-like symptoms came upon the affected individuals either during a period of water contact or within several hours thereafter.

Testing for standard bacterial contaminants in both the lake waters and consideration of possible viral contaminants provided no clues to the causitive agent or agents involved.

Investigation was then conducted of comments made by several of the individuals interviewed that the lakes were considerably dirtier this summer and that the larger lake was taking on a decidedly green color. This led to the discovery of very large numbers of the blue-green algae genus *Anabaena* in the waters of the larger of the two lakes.

As very little information was available through official
channels on any ties between blue-green algae and human sickness,
much time was lost in attempting to determine if the algae could
be the causitive agent. Although a search of the available literature eventually showed that at least three of these outbreaks had
a high probability of being caused by the blue-green algae genus
Anabaena, a more critical study is obviously needed. Recommendations have been submitted that an in-depth study be conducted with
the intent of developing maximum allowable levels for toxin-producing algae or the actual toxic substances produced for beaches and
other water-associated recreation sites.

INTRODUCTION

Although the ability of certain blue-green algae to produce
substances toxic to man is now well documented in the literature
of the scientific community, this relationship has not been widely
recognized by official public health agencies. Thus, when a series
of apparent water-borne outbreaks occurred in northeast Pennsylvania,
blue-green algae were not considered during the important initial
phases of this investigation. Only a chance comment by one of the
affected individuals led to the investigation of a possible correlation between a "natural" contaminant and the illnesses.

It is the purpose of this paper not only to discuss these suspected cases and problems experienced in this investigation, but
also some of the public health implications of a "new" etiologic
agent.

CASE REPORTS

Case I

The first of these cases was brought to the attention of the
Pennsylvania Department of Environmental Resources, Bureau of Community Environmental Control office on the afternoon of August 8,
1979. It involved Arrowhead Lake (designated Lake A), a lake-shore
vacation community in northern Monroe County. It was reported that
between twenty and thirty children as well as several adults had
become ill. The symptoms reported were gastrointestinal in nature
and occurred diring or shortly after swimming at the beaches of
this lake.

Responsibilities for epidemiological investigations in Pennsylvania are divided between the Pennsylvania Department of Health and
the Department of Environmental Resources. Arrangements were therefore made with the local Pennsylvania Department of Health office

for a public health nurse to interview those families and children
still vacationing at the lake.

Interviews by the public health nurse revealed that twelve
children between the ages of four and twelve years and one adult
had become ill. Although additional individuals were suspected to
have been involved, only these thirteen were still at the lake.
The onset of their symptoms which local doctors termed the "summer
flu", occurred between Saturday, August 5 and Thursday, August 9,
1979. Each of the group experienced symptoms during or within a
maximum of twelve hours of having had contact with the lake water.
The length of time which the individual symptoms persisted varied
greatly, though none lasted longer than five days. Most symptoms
subsided within seventy-two hours.

One case of rash occurred in a woman resident at the lake.
The rash developed on her arms and legs shortly after she waded
along the edge of the lake and splashed water on her arms to cool
off.

Table 1 lists the symptoms found and frequency of occurrence
in the Lake A group.

Table 1. Symptoms and Frequency of Occurrence at Lake A

Symptoms	Number of People
Headache	8
Stomach Cramps	9
Nausea	5
Vomiting	7
Diarrhea	11
Fever	5
Rash	1
Sore or Red Throat	3

As part of the Department of Health investigation, stool
specimens were taken from four children. These specimens were
tested for both *Salmonella* and *Shigella*. A throat swab was conducted of one young girl still exhibiting sore throat symptoms to
test for possible viral involvement. All these tests were subsequently found negative.

Physical investigation of the area and sampling of the water
of Lake A was conducted by the Department of Environmental Resources.

This lake has four swimming beaches located at various points
around the shoreline. Of these, beaches 1, 2, and 3 had been implicated as possible contact sites. The general layout of the lake
and location of the beaches are shown in Figure 1.

The weather that had been hot and dry for the past six weeks
changed quite suddenly on the evening of August 9th, the night
before the Community Environmental Control inspection. A heavy

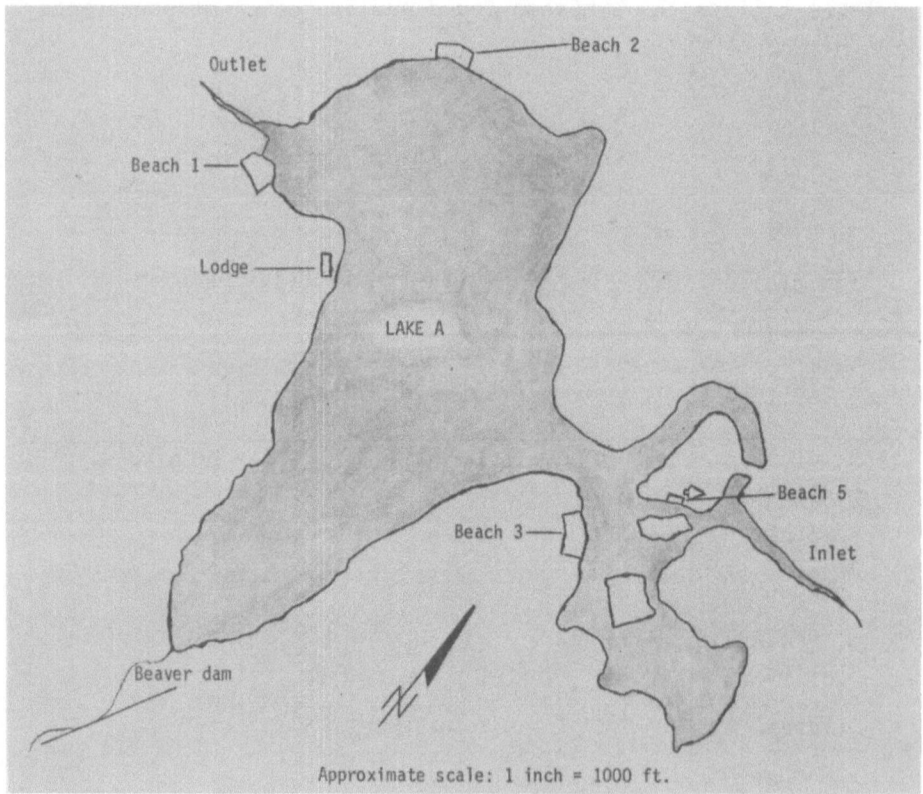

Fig. 1. General layout of Lake A and location of beaches.

rain fell and the temperature dropped markedly. This cool, over-
cast weather lasted for several days.

The lake water was brownish in color. This was typical of
lakes in the area fed by streams containing tannic acid, leached
from the surrounding forest, swamps and peat bogs. Under closer
inspection, the water appeared to contain a fine floculant-like
material suspended throughout. A small amount of brownish-foamy
scum was noted along the shoreline.

Several lake-shore properties where sewage malfunctions had
previously occurred were checked. None of these systems were mal-
functioning at the time of the inspection nor was evidence found
of recent discharge. One sewage malfunction was found discharging
effluent to the gutter in the second line of houses approximately
100 yards from the lake.

Pennsylvania Department of Environmental Resources' Regulations
require regular weekly sampling of all public beaches for bacteri-
ological quality. A history was therefore available of the quality
of the lake water for the preceding weeks of beach operation. The
samples taken prior to this outbreak showed fecal coliform levels
at or below 40 per 100 ml of lake water. Additional samples taken
the day after the rainstorm showed increased levels of fecal coliforms at beach number 2. This level then exceeded the allowable
limit of 1000 fecal coliforms per 100 ml of sample. Beach number
2, therefore, was closed for swimming for a two-day period. During
the time the beach was closed, the bacterial count rapidly fell to
pre-rainstorm level.

In discussing this situation with the lake manager, he mentioned another problem that occurred at the same time this series
of illnesses began. A beaver dam located on one of the small
feeder streams to the lake was torn out by a local peat mining
company. This water and a considerable amount of organic material
washed into the lake leaving residue on several properties.

Case II

On August 10, 1979 while still investigating this initial
outbreak, a second complaint was received. This case involved
Lake Wallenpaupack (designated Lake B), a larger man-made lake in
Pike County, approximately twenty miles northeast of Lake A. Here,
a father, son and daughter experienced very severe hayfever-like
symptoms while swimming in the lake. The mother also said that
she and the daughter's boyfriend, who was visiting, both had earaches the evening of the same day. The earaches occurred several
hours after swimming. The lake was noted to have taken on a distinct green color during the several days preceding the incident.

Case III

On August 28, 1979, a third complaint was received at the
Stroudsburg office of the Bureau of Community Environmental Control.
The complaintant was owner of a group of rental cottages at Lake B
in Pike County. He stated that fifteen of his vacationers had been
sick at varying times since the 19th of August. The symptoms initially reported were vomiting, nausea, and diarrhea which had
lasted between twenty-four and forty-eight hours. Upon further
questioning, the owner admitted that additional symptoms of irri-
tated eyes, sore throat, fever, and earaches had occurred. He had
not mentioned these as he did not feel they could be related to
the lake. As all those sick were swimmers, he was sure that the
source of the problem was the lake water. The owner also noted
that the symptoms occurred a short time after contact with the lake
water. The people involved had already ended their vacations and
returned home. No direct interviews were conducted.

Samples taken of the wells serving these cottages showed no
bacterial contamination. A check of the on-lot sewage disposal
systems showed no evidence of malfunctions.

Case IV

A fourth outbreak occurred during this August period that was
not officially reported to the Bureau of Community Environmental
Control, but was discovered during a later part of the investigation. As could best be determined, twenty to thirty of the participants in a summer aquatic program, held at Lake B by the local
high school, were involved. This group, which represented about
25% of the total class, showed various symptoms including eye irritation, sore throat, earache, sneezing, runny noses and swollen
lips. Onset of symptoms had been rapid, generally during or within
two to three hours of contact with the lake water. Symptoms in the
affected individuals had cleared within two to three days. Rein-
troduction to the lake water caused a reoccurrence of these symptoms. Because of these occurrences, the classes in the lake were
discontinued.

The relative location of Cases II, III, and IV are shown in
Figure 2.

Other Incidents

On August 30 the aquatic biologist from the Department of
Environmental Resources joined in the investigation. Checks were
made of the water at both involved lakes. These tests confirmed

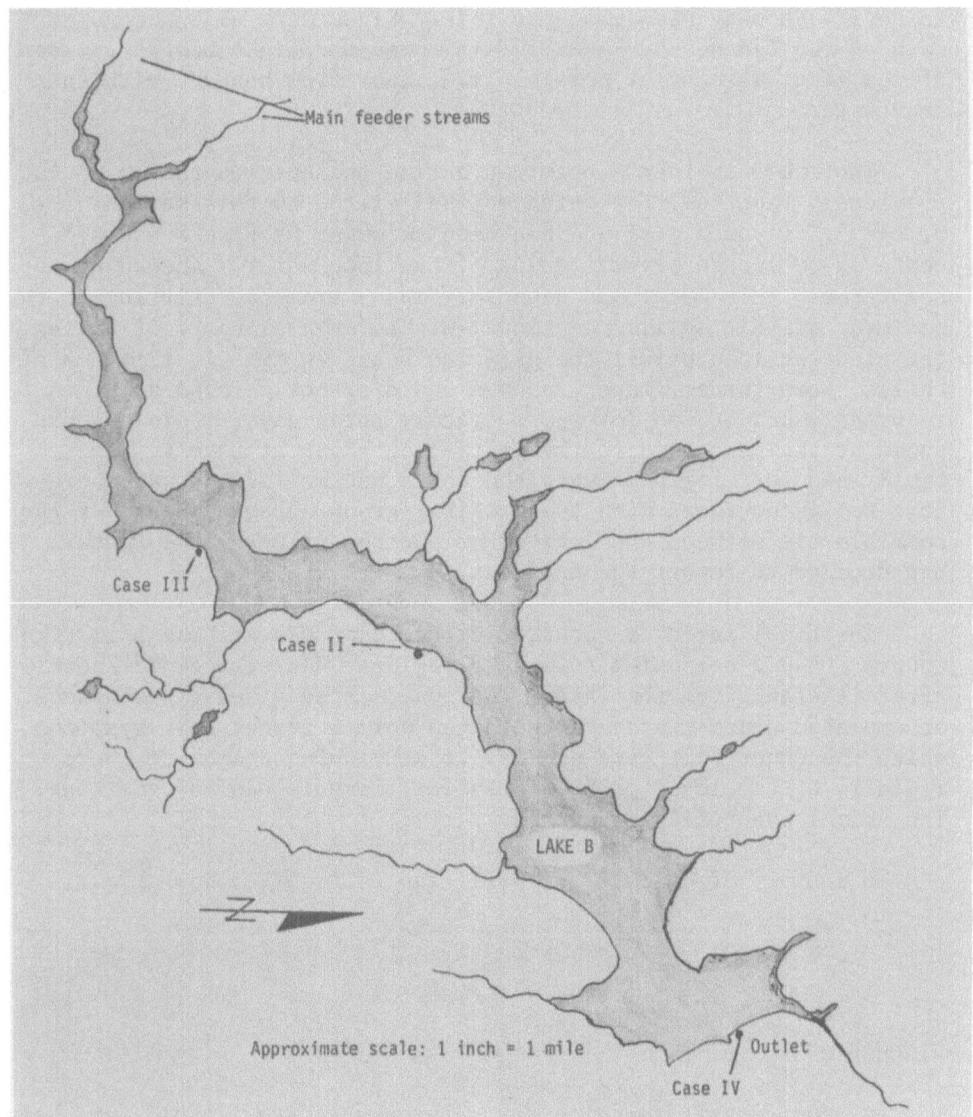

Fig. 2. General layout of Lake B and sites of Cases II, III, and IV.

that a heavy bloom of the blue-green algae *Anabaena* was occurring on
Lake B (Penn. Dep. Environ. Res., 1979). No bloom was then evident
at Lake A, the site of the first outbreak in Monroe County.

Although no additional problems had been reported from Lake A,
the bloom on Lake B had become visibly heavier. A decision was made
to close Lake B's beaches. As a heavy influx of people was expected

for the upcoming Labor Day weekend, multi-media coverage was used
to notify as many people as possible. Signs were also posted at
each of the beaches by Pennsylvania Power and Light Company, owner
of the lake, warning of possible reactions from contact with the
lake water.

Two minor incidents occurred during this period that were not
previously reported, but which may be related to this series of
problems. A young girl who had been swimming at Lake A had acci-
centally swallowed several gulps of the lake water. According to
her father, this water had been very muddy from her stirring up the
sediment of this shallow section. Within several hours of having
ingested the lake water, the girl had begun to exhibit symptoms of
chills, sore throat, fever, nausea and diarrhea. These symptoms
reportedly had lasted for approximately three days. This had hap-
pened at the same time as the initial outbreak at Lake A. The
family had left for home when the child became ill and was therefore not interviewed with the original group. Having heard of the
possible tie between the lake water and the illness, the father
had decided to report the incident.

The final incident occurred after Labor Day at Lake B in Pike
County. A dog belonging to a lake resident was strickened shortly
after it drank from the lake. The owner noted a heavy green scum
accumulated along the shore where the animal drank. The symptoms,
which the owner felt were similar to strychnine poisoning, were
rigidity of the legs, labored breating, convulsions and vomiting.
The animal eventually recovered.

A summary of Cases I, II, III, and IV is given in Table 2.

DISCUSSION

 Circumstances preceding the outbreak at Lake A could have been
conducive to the action of several different etiologic agents. In
the warm shallow waters of this lake, high fecal coliform counts
were expected. As noted, this was not true. The rise in the fecal
coliform level was found only for a short period at beach number 2.
This was suspected to be the result of the heavy rain, having
flushed septic tank effluent into the lake very near beach number
2. That this material had not reached this beach during the dry
weeks preceding this outbreak was evidenced by the low fecal coliform counts before the rain. These tests, taken in conjunction
with the negative tests for *Salmonella* and *Shigella* and the short
incubation period, are felt to rule out a bacterial agent in this
outbreak (Benenson, 1970).

Table 2. Comparison of Symptoms in Cases I, II, III, and IV.

Symptoms	Lake A	Lake B		
	Case I	Case II	Case III	Case IV
Headache	x			
Stomach cramps	x			
Nausea	x		x	
Vomiting	x		x	
Diarrhea	x		x	
Fever	x		x	
Rash	x			
Sore or red throat	x		x	
Earache		x	x	x
Sneezing		x		x
Runny nose		x		x
Eye irritation		x	x	x
Swollen lips				x

Evidence that introduction of water and organic material from the beaver dam had an effect on the occurrence of these symptoms was inconclusive.

An outbreak of giardiasis that occurred in McKean County, Pennsylvania was traced to fecal contamination of the public water supply by infected beaver (C. D. C. Report, 1980). As symptoms and incubation period shown at Lake A varied significantly from the symptomatology of giardiasis (Benenson, 1970; C. D. C. Report, 1980), this was eliminated as a possible agent.

A water-borne outbreak of gastrointestinal illness involving recreational use of a lake had occurred in Macomb County, Michigan, in July, 1979. This episode involving over 230 cases was suspected to have been caused by a viral agent (C. D. C. Report, 1979). Though these symptoms were similar, the incubation period was much longer (6 hours to 8 days vs. during contact to 12 hours) (C. D. C. Report, 1979). While there were families with more than one member ill in both outbreaks, all of those individuals at Lake A had

contact with the lake water no longer than 12 hours prior to onset
of symptoms. The possibility of these having been secondary infec-
tions was therefore felt very low. Without verified evidence of
secondary infection and considering the short incubation period of
the symptoms at Lake A, common viral agents (Benenson, 1970) were
ruled out.

After having considered the possibilities of various etiologic
agents for this case, including common bacterial, protozoal and
viral types, the following conclusions were reached:

1. The single common factor for the group was recent contact with
   the lake water.

2. Incubation times for the symptoms were shorter than those known
   for the common bacterial, protozoal or viral agents.

3. The agent involved in this outbreak appeared to have been
   deactivated by the recent weather change, i.e., temperature
   drop, wind and rain.

4. The rash experienced by the one adult woman involved did not
   appear to fit the general pattern of symptoms.

5. The rapid onset of symptoms might be explained by the action
   of a chemical agent.

The second outbreak, which involved only five people, appeared
to pose a different problem. Three family members had allergy-like
symptoms. The earaches experienced by two individuals could have
been the result of bacterial infection, swelling or inflammation
due to an allergenic response or direct contact with a chemical
irritant (Brunner and Suddarth, 1974). As in the first outbreak,
the very short incubation period tends to rule out any common bacterial or viral infection.

Working on the supposition that the green color of the lake
(a suspected algae bloom) might have been involved with the allergy-
like reactions of the father, son and daughter, contact was made
with Dr. Charles Reif, Professor of Biology at Wilkes College in
Wilkes-Barre, Pennsylvania. To Dr. Reif, allergic reactions to
algae blooms were not new. He had experienced symptoms, very similar to those of the three family members at Lake B, after swimming
in water containing a bloom of the blue-green algae *Microcystis*.
In addition, he and an associate had studied a problem involving
an allergic reaction to *Anabaena* in a Pennsylvania lake in 1953
(C. Reif, personal communication). Noted, were the findings that
allergic responses as rashes and hayfever-like symptoms, were possible from contact with water containing blooms of *Anabaena*
Microcystis (Cohen and Reif, 1953). Also reported was the production of a poisonous substance (hydroxylamine) from the decay of
algal proteins. This material, when isolated from either the water

or bottom sediment of the lake where the *Anabaena* bloom had
occurred, produced a primary irritant response in greater than 50%
of the individuals tested (Cohen and Reif, 1953).

Over the next two weeks, contact was made with various agencies
and groups in an attempt to gather information on algae-related disease. Although individual research groups were helpful in providing
copies of research papers on this subject, only one of four official
health agencies contacted had knowledge of illness caused by freshwater algae (S. G. Cohen, personal communication).

The symptoms of the third outbreak, reported from Lake B on
August 28 at the height of the algae bloom, were very similar to
those of the first outbreak from Lake A. Questioning of the owner
of the cottages had shown the lake water as the only known common
contact factor. As in the two previous outbreaks, a short incubation period was evident.

Literature references revealed studies where symptoms of
headache, nausea and gastrointestinal upset had occurred after
swimming in lakes blooming with *Anabaena* or *Microcystis*. One of
these individuals had reportedly swallowed an estimated half-pint
of lake water containing both *Anabaena* and *Microcystis*. His symptoms, which started three hours later, included stomach cramps,
nausea, vomiting, painful diarrhea, fever, headache, pains in
limbs and joints, and weakness (Dillenberg and Dehnel, 1960).

In light of this information, consideration must be given to
the possibility that Cases I and III were the result of toxins produced by one or more types of blue-green algae. All symptoms seen
in these two cases can be produced by contact with or ingestion of
water containing these agents (Cohen and Reif, 1953; Dillenberg and
Dehnel, 1960). The rapid action of these toxic substances would
also explain the short incubation periods found. Also clarified
would be the failure of standard test procedures to detect other
possible etiologic agents.

Case IV is an example of a problem faced by all health agencies
involved in epidemiological investigation. It was not promptly
reported. The symptoms experienced by the individuals of this group
appear to be a combination of allergic and toxic responses to the
Anabaena bloom. While symptoms such as eye irritation and earache
may be associated with allergic response, they may also be the
result of contact with primary irritants such as the algal toxins
(Cohen and Reif, 1953).

The discussion of Case III noted the symptoms exhibited when
water containing toxic blue-green algae had been ingested (Dillenberg and Dehnel, 1960). The young girl, mentioned as one of the

final two incidents, had accidentally ingested several swallows of
water from Lake A. Her symptoms, though not as severe or extensive
were similar to those reported in the literature (Dillenberg and
Dehnel, 1960). Though this supports the possibility that the outbreak at Lake A was the result of contact with toxic blue-green
algae, it does raise the question of the mode of entry of the toxin
into the bodies of the other children at this site and the adults
and children of Case III at Lake B.

It has been personally noted that many children tend to "drink"
varying amounts of the water in which they swim. This could account
for the mode of entry of the toxin in the outbreak at Lake A and
explain why only children experienced gastrointestinal symptoms at
this site. As fewer adults or teens would actually "drink" the
water, except by accident, gastrointestinal symptoms for the older
groups should be much lower. This conclusion is supported by what
is known of the symptoms of the different age groups in each of
these four case reports.

The final incident of this series was the only known report
of animal involvement. Reports are mentioned in the literature of
these same symptoms preceding animal death after ingestion of water
containing toxic *Anabaena* blooms (Dillenberg and Dehnel, 1960;
Carmichael and Gorham, 1978). In this case, the owner quickly
forced the animal to vomit, thus removing the toxin and algae from
the stomach.

PUBLIC HEALTH OBJECTIVES AND RECOMMENDATIONS

From the position of a public health official, simply recog-
nizing a "new" agent of human illness is not sufficient. The primary purpose of a public health agency must be to prevent disease.
Thus, when we have to investigate disease outbreaks, we have, in a
sense, failed.

To better utilize this knowledge in disease prevention, we
must accomplish the following objectives:

1. We must define the limits where toxic blooms can occur, i.e.,
   lakes, swimming pools or public water supply reservoirs, and
   when these occurrences are likely to happen.

2. We must develop a simple means of identifying a toxic algae
   bloom before it reaches the critical level of symptom production.

3. We must correlate the concentration of available toxins to the
   actual production of the various symptoms.

4. We must determine how long these materials remain toxic under
   various conditions in both the water and sediment.

5. We must explore the possibility of an alternate portal of entry
    of the toxin being absorption through the skin and its action
    on the skin and mucous membrane as a primary irritant.

6. If we are to eventually control this problem, we must define
    more distinctly why the toxic blooms occur and eliminate, as
    much as possible, the causes.

7. We must maintain an open line of communication between research-
    ers and public health officials. Thus, as new problems or solu-
    tions are found, the necessary persons can be promptly informed
    and implications determined.

8. Legislators and public health officials must assure that this
    new information is promptly and properly used to protect the
    health of those we serve.

ACKNOWLEDGMENTS

The author would like to thank all those who participated and
provided technical support and direction during the course of these
investigations.

REFERENCES

Benenson, A. S., ed. 1970. Control of Communicable Disease in
        Man. 11th Ed. The American Public Health Association,
        New York, New York. 316 pp.
Brunner, L. S., and D. S. Suddarth. 1974. The Lippincott Manual
        of Nursing Practice. J. B. Lippincott Company, Philadelphia,
        Pennsylvania. Page 705.
Carmichael, W. W., and P. R. Gorham. 1978. Anatoxins from clones
        of *Anabaena flos-aquae* isolated from lakes of western
        Canada. Mitt. Int. Ver. Limnol. *21*:285-295.
Cohen, S. G., and C. B. Reif. 1953. Cutaneous sensitization to
        blue-green algae. J. Allergy *24*:452-457.
Dillenberg, H. O., and M. K. Dehnel. 1960. Toxic waterbloom in
        Saskatchewan, 1959. Can. Med. Assoc. J. *83*:1151.
Pennsylvania Department of Environmental Resources. 1979. Unpub-
        lished Report.
U. S. Department of Health, Education, and Welfare, Center for
        Disease Control. 1979. Gastroenteritis associated with
        lake swimming - Michigan. MMWR 28:35:413.
U. S. Department of Health, Education, and Welfare, Center for
        Disease Control. 1980. Waterborne giardiasis - California,
        Colorado, Oregon, Pennsylvania. MMWR 29:11:122.

SOME ASPECTS CONCERNING REMOTE AFTER-EFFECTS OF

BLUE-GREEN ALGAE TOXIN IMPACT ON WARM-BLOODED ANIMALS[1]

Yu. A. Kirpenko, L. A. Sirenko,
and N. I. Kirpenko

Institute of Hydrobiology
Ukrainian Academy of Sciences
Str. Vladimirskaya, 44
P. O. Box 252003
Kiev, U. S. S. R.

ABSTRACT

Natural populations of the blue-green alga *Microcystis aeruginosa* were collected during the summer months. Toxic cells and toxin extracts were tested on laboratory rats to determine possible long term effects of the toxin.

From the investigations, it was ascertained that the toxin in a dose of 5×10^{-4} mg/kg and cell biomass in a dose of 10 mg/kg injected into pregnant rats over 19 days caused embryolethal, teratogenic and gonadotoxic effects. Algae biomass in the foregoing dose also possessed mutagenic properties provoking anomalies of chromosome and chromatid apparatus.

INTRODUCTION

Recent research on toxins of blue-green algae has emphasized mechanisms of action for rapidly acting neurotoxic alkaloid or

---

[1]This paper was reviewed and edited by Dr. Wayne Carmichael, Department of Biological Sciences and Dr. Jane Scott, Department of Anatomy, Wright State University. Portions of the text were edited for clarification and corrections by the authors were made in February 1981. The authors were not present at the conference proceedings.

hepatoxic peptide toxins on warm-blooded animals. These studies
(Carmichael and Gorham, 1977, 1978; Carmichael et al., 1977;
Skulberg, 1979; and Kirpenko et al., 1979) have ascertained that
these toxic algae metabolites manifest themselves as virulent
substances with a wide range of toxicological effects. However,
there is no data concerning remote after-effects of blue-green
algae toxic metabolites on warm-blooded organisms. It is known
from other work that a number of biologically active substances
can cause pathological deviations including gonadotoxic, embryotoxic, teratogenic and mutagenic effects.

It is possible that this state of affairs exists for the
toxins of blue-green algae, so far as they may exert similar
influences resulting from peculiarities of toxic effect (Kirpenko
et al., 1977). The aim of this paper is the investigation of
possible remote after-effects of blue-green algae toxin on warm-
blooded organisms.

MATERIALS AND METHODS

Material for the tests included biomass as well as algae
toxin extracted from it. Natural populations of the blue-greens
were gathered in a reservoir during summer water "bloom" periods
(Kirpenko et al., 1977). Samples were obtained from the Kremenchug
and Dneprodzerzhinsk reservoirs of the Dnieper.

Embryogenesis and Teratogenesis

Blue-green algae toxic effect on embryogenesis was evaluated
by embryolethality and teratogenicity. White inbred rats were
used in the work. Rats were chosen because of a short pregnancy
period (20 to 22 days), acute toxin sensitivity, omnivorous ability
and hemochorial type of placenta, which, to some extent, makes it
possible to extrapolate these data to other warm-blooded organisms.

During the experiment one group of pregnant rats was injected
with 5×10^{-4} and 5×10^{-7} mg/kg doses of toxin, while the other
group was injected with the biomass of *Microcystis aeruginosa* Kütz.
emend. Elenkin in a dose of 10 mg/kg body weight. Beginning with
the first day of gestation, injections were made daily until day
19 of gestation. Stage of gestation was monitored by use of vaginal
smears. On the 20th day experimental and control animals were
sacrificed by means of cervical dislocation, the uterus containing
feti removed, the feti extracted and the ovaries prepared for
experimental purposes. Three-hundred-twenty-five feti were studied.
Feti were examined externally with a dissecting microscope, weighed,
measured using the cranio-caudal index and fixed in Buaen's solu-

tion and in 96% ethanol. The composition of the Buaen's solution
is: 15 ml of a saturated aqueous picric acid solution, 5 ml of
formalin, 1 ml of glacial acetic acid (CH_3COOH). Feti were fixed
for 2 hours to 3 days and stored in 70 to 80% ethanol (C_2H_5OH).
Then the value of embryolethal effect, numbers of fetal mortality
for preimplantation and postimplantation periods and the relative
value of preimplantation and postimplantation mortality were calculated. Absolute value of preimplantation mortality was calculated
by means of subtracting the amount of implantation spots from the
amount of corpora lutea. The amount of implantation spots was calculated according to cicatrices in the uterus, left after embryo
implantation. Relative value of preimplantation mortality in percent was calculated by dividing the absolute preimplantation mortality value by the number of corpora lutea.

Absolute value of postimplantation mortality of feti was
calculated visually on the dissected uterus (living and dead feti
are easily distinguishable). Relative value of postimplantation
mortality was calculated by dividing the absolute postimplantation
mortality value by the number of implantation spots.

Absolute value of total embryo mortality was calculated by
means of adding absolute values of preimplantation and postimplantation mortality, relative value was determined by dividing this
value by the number of corpora lutea.

Teratogenic effect of blue-green algae toxin was determined
by internal examination of the embryo, by measuring weight and
cranio-caudal index and by macroscopic examination of internal
organs. Skeletal system anomalies were examined on clarified feti
(Dyban et al., 1970).

Gonadotoxic Effects

Gonadotoxic effect of blue-green algae toxin was investigated
over three months on 120 inbred male and female white rats. Toxin
extract and algae biomass were injected per mouth in the same doses
as listed in the previous section.

The estrus cycle in female rats was studied first using a
vaginal smear method (Yelizarova et al., 1968). This was followed
by examination of functional and morphological changes in ovaries.
Following sacrifice of the animals their ovaries were removed and
fixed in Cahnker solution for examination. Cahnker's fixative
includes: 100 ml of Muller's liquid (2.5 g $K_2Cr_2O_7$, 1 g Na_2SO_4,
100 ml distilled H_2O); 5 g of corrosive sublimate (Hg^{2+}); 5 ml of
glacial acetic acid (CH_3COOH), added ex tempore. Fixation duration
depends on the thickness of tissue slice. It should not exceed
5 mm, then fixation lasts for 1 to 24 hours.

Microscopic study in males was made of the substrate from genital appendages and testes. Histopreparations from testes were prepared for the evaluation of spermatogenesis. Functional and morphological changes in ovaries and testes were observed using 10 μ paraffin sections stained with hematoxylin-eosin.

Mutagenic Effects

The mutagenic effect of blue-green algae toxin was investigated on 60 white rats using bone marrow cell preparations. The toxin was injected per mouth at 5×10^{-4} and 5×10^{-7} mg/kg (toxin extract) and 10 mg/kg (cell biomass) daily over a six month period. Cytogenetic studies were made using the somatic cell (bone marrow) preparations from these orally dosed rats. Chromosome preparations were made using the metaphase method. The preparations were stained with hematoxylin-eosin. It was ascertained that chromosome analysis in the marrow cells makes it possible to judge mutation frequency of somatic cells. Aberrations of chromosome and chromatid types as well as their gaps were taken into account (Bochkov, 1971; Methodical Recommendations, U. S. S. R., 1974). Aneuploidy and polyploidy levels were not taken into account because the first anomaly might be the result of cell treatment, while the importance of the second for the marrow can not be determined at present.

RESULTS

Investigation of embryolethal effects for toxin extract and blue-green algae cells revealed that in the control group there were 132 corpora lutea in ovaries, 120 implantation sites, 8.3% preimplantation mortality, 4.1% postimplantation mortality, 12.8% total embryo mortality and 115 living feti without external abnormalities. From this it was possible to conclude that in the control group the indicators under study were within admissible limits.

Making a comparison between control and test results one may arrive at a conclusion that the number of corpora lutea in all groups was approximately the same, and there was only a slight difference in the amount of implantation sites and preimplantation mortality. However, postimplantation fetus mortality as well as total embryo mortality among test animals increased. There was also a tendency for fetal weight to decrease (Table 1).

Analysis of Table 1 leads to the conclusion that blue-green algae toxin provokes an increase in total embryo mortality mainly after implantation. This indicates an absence of adaptive mechanisms to toxins in pregnant rats as well as to toxin penetration through the placental barrier.

Table 1. Embryo Development of White Rat Feti in the Control and After Injection of Two Different Doses of Algae Toxin

Statistical Criteria	Number of Animals in Test Group	Number of Corpora Lutea	Number of Implantation Spots	Preimplantation		Postimplantation		Number of Living Feti	Total Embryo Mortality In %	Embryo Weight In mg
				Absolute	In %	Absolute	In %			
Toxin in a Dose of 5 x 10^{-4} mg/kg										
M	14	10.2	9.5	10	6.9	38	28.5	6.7	33.5	2600
± m		1.89	3.33					3.5		131
P		> 0.05	> 0.05					> 0.05		< 0.05
Toxin in a Dose of 5 x 10^{-7} mg/kg										
M	14	10.3	9.8	6	4.1	2.3	16.6	8.2	20.2	2700
± m		1.47	2.76					2.96		751
P		> 0.05	> 0.05					> 0.05		< 0.05
Control										
M	14	9.4	8.5	11	8.3	5	4.1	8.2	12.8	2860
± m		2.04	1.45					2.42		451

Key: M = Number; ± m = Standard Deviation; P = Confidence Level.

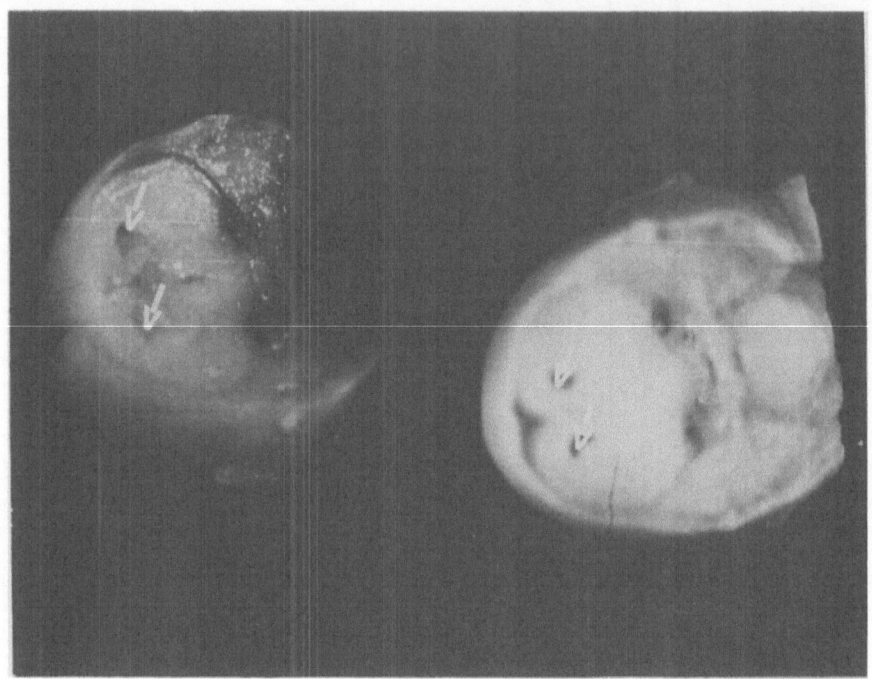

Fig. 1.   Cerebral ventricular enlargement in rat embryos after
          injection of females with algae toxin in a dose of
          5×10^{-4} mg/kg over a 19 day period. Lateral verticles
          (arrow) showing abnormal expansion.

     Teratogenicity was studied as an injuring effect of toxin and
algae biomass on animal embryos. Following injection of pregnant
rats with algae biomass (10 mg/kg) and toxin extract (5×10^{-4}
mg/kg), over a 19 day period, cuts through cerebral ventricles of
the brain revealed an increased size of the cavities (Figure 1).

     In this instance not acutely expressed hydrocephaly was
revealed, because other cuts revealed the normal state of cerebral
ventricles.

     In addition, injection of toxin and algae biomass in the same
doses, revealed that the weight and dimensions of some feti were
smaller than in the control (Figure 2). Eighteen feti from 325
were malformed. These feti were viable, but judging by external
appearance and dimensions they were abnormal.

     Injecting pregnant rats with toxin and algae biomass in the
same doses resulted in hemostasis in the organs plus a large number
of hemorrhages in internal organs of embryos (Figure 3).

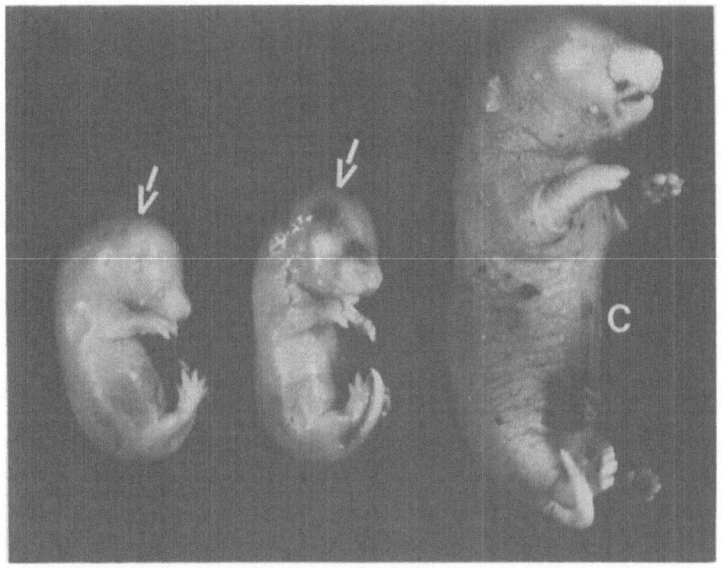

Fig. 2.   Diminution of fetal dimensions and weight on injecting
          white female rats with algae toxin extract (arrows) in
          a dose of 5 x 10^{-4} mg/kg over a 19 day period. Fetus
          on the right is the control.

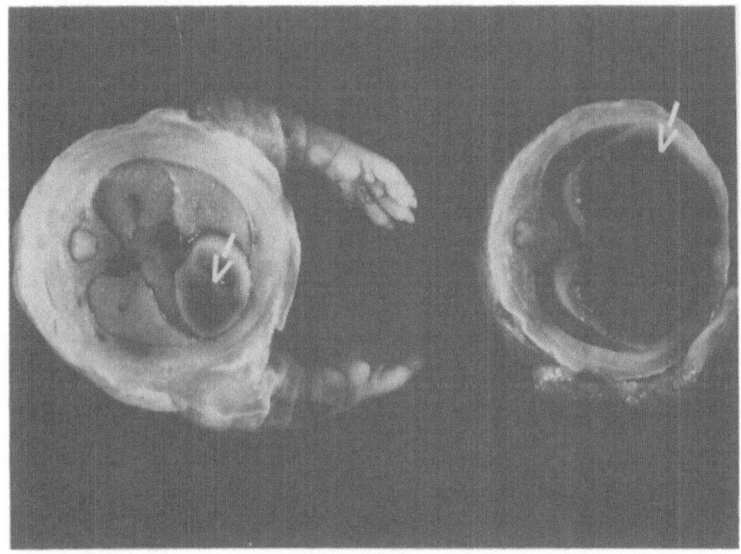

Fig. 3.   Hemostasis and a number of hemorrhages (arrow) in internal
          organs of embryos from pregnant rats injected with toxin
          at a dose of 5 x 10^{-4} mg/kg over a 19 day period.

Table 2. Ooogenesis in White Female Rats After Different Doses of Blue-green Algae Toxin Extract

Groups of Animals and Toxin Doses in mg/kg	Injection Period in Days	Statistical Criteria	Number of Follicles		Number of Follicles — Mature		Number of Corpora Lutea
			Primordial	Maturing	With Oocyte	Without Oocyte	
Control	–	M	9.45	8.76	8.63	1.63	8.91
		± m	1.18	0.96	1.36	0.36	1.32
5 x 10⁻⁴	90	M	13.73	4.01	3.68	4.82	5.33
		± m	1.21	1.08	0.96	1.33	1.48
		P	< 0.05	< 0.05	< 0.05	< 0.05	< 0.05
5 x 10⁻⁷	90	M	9.28	8.63	8.11	1.47	8.63
		± m	1.26	1.47	1.56	0.29	1.39
		P	> 0.5	> 0.5	> 0.5	> 0.5	> 0.5

Key: M = Number; ± m = Standard Deviation; P = Confidence Level.

No skeletal abnormalities were noted in control or experimental feti which were clarified and stained with alazarin dyes.

When comparing the weight and cranio-caudal index of control feti to those of feti exposed to the toxin extract and algae biomass injections at all doses, no statistically reliable changes were detected (Table 1).

The frequency of individual estrus disturbances was not higher in animals receiving the toxin at 5×10^{-7} mg/kg than in the control. However, the toxin in a dose of 5×10^{-4} mg/kg and algae biomass in a dose of 10 mg/kg frequently provoked estrus cycle changes in the estrus stage. In the majority of cases the estrus stage was absent and the diestrus stage predominated. Autopsy of animals at the end of the experiment corroborated these results: atrophic phenomena in uterus and genital appendages were revealed in the majority of cases. After 1.5 months of daily injections, some test animals were killed in order to investigate functional and morphological changes in ovaries. It was ascertained that after injection of the toxin in a dose of 5×10^{-4} mg/kg and algae biomass in a dose of 10 mg/kg, growth and maturation of oocytes were disturbed, the number of primoridal follicles increased while the number of mature ones decreased (Table 2). The toxin in a dose of 5×10^{-4} mg/kg provoked degeneration of oocytes in Graafian vesicles, follicle dimensions decreased and there was an increase in the number of involuted corpora lutea.

In male rats under the influence of the toxin at a dose of 5×10^{-4} mg/kg and algae biomass in a dose of 10 mg/kg sperm motility decreased (sperm motility was determined by observing the duration of movement for sperm while suspended in a warm physiological solution), percentage of dead sperm increased, quantity of spermatogonia and spermatocytes decreased while the number of tubules with shelled out epithelium increased (Table 3). Observations of histologic preparations of testes revealed a great deal of tubule deformation and epithelium shelled out from basal membranes. Sertoli's cells of triple-edged and oval forms frequently occurred and a considerable amount of secretion was revealed in tubular gaps indicating a reduction of phagocytosis in Sertoli cells. Also noted was an increase in the number of degenerating spermatogonia. It is necessary to underline the fact that the toxin extract first affected spermatogonia and spermatocytes, resulting in fission disturbance of spermatogonia and growth disturbance of spermatocytes.

After investigating mutagenic effects of the blue-green algae toxin in a dose of 5×10^{-4} mg/kg, over a six month injection period, four anomalous metaphases with chromatid deletions were noted out of a total of 500 marrow cell metaphases. Using a dose

Table 3.   Spermatozoid Motility and Spermatogenesis under the
Influence of Two Doses of Blue-green Algae Toxin

Indicators Under Study	Statistical Criteria	Control	Toxin in mg/kg	
			5×10^{-4}	5×10^{-7}
Spermatozoid Motility in Points				
In 1 minute after killing	M	3.86	1.32	2.66
	± m	0.18	0.09	0.81
	P		< 0.05	> 0.5
In 30 minutes after killing	M	3.56	1.16	2.91
	± m	0.72	0.06	0.72
	P		< 0.05	> 0.5
Spermatozoids Quantity in 1 mm^3	M	557.21	549.32	562.7
	± m	98.2	83.8	86.8
	P		> 0.5	> 0.5
Living	M	169.7	93.8	172.6
	± m	27.9	19.9	27.7
	P		< 0.05	> 0.5
Dead	M	389.9	427.3	382.7
	± m	63.7	59.6	48.7
	P		< 0.05	> 0.5
Spermatogonia Quantity	M	61.81	44.62	63.86
	± m	6.33	9.81	7.33
	P		< 0.05	> 0.5
Spermatocytes Quantity of 1-2 Order	M	56.84	39.41	55.68
	± m	8.73	4.92	8.36
	P		< 0.05	> 0.5
Spermatids Quantity	M	54.29	42.88	53.78
	± m	6.81	7.36	8.76
	P		< 0.05	> 0.5
Spermatozoids Quantity in Testes	M	46.92	44.82	48.66
	± m	8.21	6.88	5.33
	P		> 0.5	> 0.5
Quantity of Tubules with Shelled Out Epithelium	M	6.68	18.32	7.32
	± m	3.39	2.36	1.86
	P		< 0.05	> 0.5

Key: M = Number; ± m = Standard Deviation; P = Confidence Level.

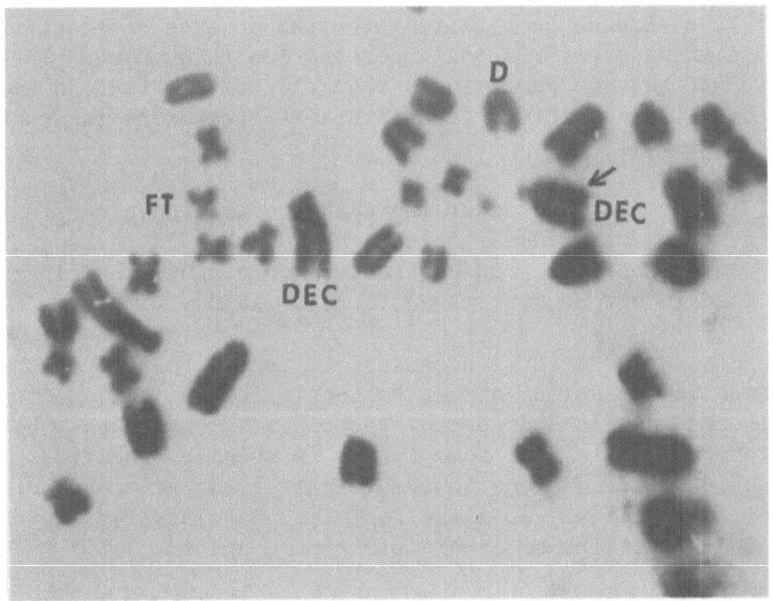

Fig. 4. Characteristic picture of chromosome and chromatid
         aberrations of marrow cells in white rats on injecting
         (per mouth) algae biomass in a dose of 10 mg/kg daily
         over a six month period (chromatid aberrations include:
         solitary fragments, deletions; chromosome aberrations
         include: free twin fragments, decentrical chromosomes).
         D = Deletions; FT = Free Twin Fragments; DEC = Decentrical
         Chromosomes.

of 5 x 10^{-7} mg/kg only one anomaly for 500 cell metaphases was noted.
The anomaly was deletion with a fragment.

    Injecting algae biomass in a dose of 10 mg/kg revealed 31
chromosomal aberrations including: chromatid and chromosome changes
(presence of solitary chromatid fragments, free twin fragments of
chromosomes and decentrical chromosomes). On an average, one metaphase equals 0.07 ± 0.012 aberrations (Figure 4). In the control
experiments 0.02 ± 0.008 chromosome aberrations, in the form of
deletions and gaps, were detected in one metaphase.

DISCUSSION

    Due to the performed experiments on white rats, the obtained
results related to remote after-effects of algae toxin impact
(viz., *Microcystis aeruginosa*) both in the form of purified extract

and biomass injection. It was ascertained that algae toxin provoked
an increase in total embryo mortality mainly after implantation.
Possibly the phenomenon resulted from the absence of adaptive mechanisms in animals or from toxin penetration through the placental
barrier. It can not be excluded that the toxin effect on embryos
is based on the possible cyclic nature of the toxins in its content
(Chebotar, 1974).

It is necessary to underline that the teratogenic effect of
the algae toxin and biomass is connected with mild teratogenic
properties of substances under study [according to the classification used in our country, Dyban et al., 1970].

Because of this it is necessary to take into account the
difference in the period of embryogenesis for various species of
warm-blooded animals and species responses to teratogens, because
the teratogenic effect of some toxins is species specific. The
teratogenic effect was due to the relatively large amount of the
toxin or toxic cells. So algae toxin may be classified as a mild
teratogen in relation to other teratogenic agents.

The toxin and algae mass in the foregoing doses depressed the
development of male and female gonads. The toxin disturbed growth
and oocyte maturation, increased primordial follicles, degener-
atively changed oocytes, increased the number of immobile sperm,
disturbed spermatogonia division, and caused degenerative changes
in the tubules.

Long-term injection of toxin extract in the foregoing doses
did not significantly increase the number of cells with chromosome
apparatus anomalies. The percentage of cells with anomalous
chromosomes did not exceed 0.8% which was within the limits of
spontaneous chromosome aberrations in marrow cells of rats
(Methodical Recommendations, U. S. S. R., 1974). When algae biomass was injected for a long time, the percentage of cells with
anomalous chromosomes was 10.3%. Intensification of algae biomass
mutagenic qualities in comparison with the toxin itself, was apparently due to the presence of substances other than the toxin. In
this study the synergistic effect of the combined substances could
not be excluded.

In conclusion it is necessary to underline that algae toxin
and biomass injected for an extended period of time, resulted in
remote after-effects which were expressed in the first generation
offspring.

REFERENCES

Bochkov, N. P. 1971. Chromosomes of Man and Irradiation.
    Atomizdat, M. 161 pp.
Carmichael, W. W., and P. R. Gorham. 1977. Factors influencing
    the toxicity and animal susceptibility of *Anabaena flos-
    aquae*. J. Phycol. *13*(2):97-101.
Carmichael, W. W., and P. R. Gorham. 1978. Anatoxins from clones
    of *Anabaena flos-aquae* isolated from lakes of western
    Canada. Mitt. Int. Ver. Limnol. *21*:285-295.
Carmichael, W. W., P. R. Gorham, and D. F. Biggs. 1977. Two
    laboratory case studies on the oral toxicity to calves of
    the freshwater cyanophyte (blue-green algae) *Anabaena flos-
    aquae* NRC-44-1. Can. Vet. J. *18*(3):71-75.
Chebotar, N. A. 1974. Embryotoxic and teratogenic effect of
    proguanyl, chloroproguanyl and cycloguanyl on albino rats.
    Bull. Exp. Biol. Med. *6*:56-57.
Dyban, A. P. et al. 1970. Principal methodical approaches to
    testing teratogenic activity of chemical substances. Arch.
    Anat. Histol. Embryol. *10*:89-100.
Kirpenko, Y. A. et al. 1977. Blue-green Algae Toxins and the
    Animal Organism. Nauova Dumka, Kiev. 250 pp.
Kirpenko, Y. A. et al. 1979. Comparative evaluation of toxic effect
    of blue-green algae biologically active agents at cellular and
    organismic levels. Hydrobiol. J. *15*(6):91-94.
Methodical Recommendations of IMG AMS USSR. 1974. Method of
    accounting chromosome aberrations as biological indicator
    of environmental factors' effect on man. M. 36 pp.
Skulberg, O. M. 1979. Giftvizkninger av blagronnalger ferste
    tilfelle av *Microcystis* - forgifthing registrert i Norge.
    Norsk Institutt for Vannforsking, Temarapport 4. 41 pp.
Yelizarova, O. N. 1968. Some methods of investigating generative
    function in sanitary-toxicological experiment. Pages 49-59
    in Methods of Sanitary-Toxicological Experiment. Medicine,
    Moscow.

TEMPORAL ASSOCIATION BETWEEN AN ALGAL BLOOM AND

MUTAGENICITY IN A WATER RESERVOIR

Michael D. Collins, C. S. Gowans[1], Frank Garro[2],
David Estervig[2], and Tracey Swanson[2]

Environmental Health Surveillance Center
University of Missouri
Columbia, Missouri

ABSTRACT

As a result of a human epidemiological investigation which
found an association between a certain reservoir water and a high
birth defect rate, a study of the reservoir water was undertaken.
The reservoir water was found to be mutagenic in the Ames/*Salmonella
typhimurium* assay over three consecutive sampling dates in 1978.
The period of mutagenicity was temporally matched with reservoir
algal bloom. In 1979, the water was not determined to be mutagenic
and there was no algal bloom. The mutagenic response was found
only in the *Salmonella* strain TA1537 and only in the presence of
the S-9 liver microsomal fraction. The predominant alga in the
algal bloom was found to be *Oscillatoria subbrevis* Schmidle.

The mutagenicity of the water was temporally associated with
the initiation of a mutant rat colony involving Charles River CD
(Sprague-Dawley) rats. Several of the rats given mutagenic reservoir water as a drinking source and mated consanguineously gave
birth to pups that later developed a characteristic neurological
abnormality (however, the pups were exposed to the mutagenic water
both during their gestational period and during their neonatal
period). Brother-sister matings of the abnormal rats has permitted

---

[1]Address: Biology Department, University of Missouri, Columbia,
 Missouri.

[2]Address: Dalton Research Center, University of Missouri,
 Columbia, Missouri.

the passage of the neurological abnormality for six generations. Only the F_0 and F_1 generations were exposed to the mutagenic drinking water. The pedigree for the neurological abnormality indicates that the trait is inherited as a single autosomal recessive gene.

Water taken from the reservoir during the mutagenic period was found to be ichthyotoxic to *Gambusia affinis*.

INTRODUCTION

An epidemiological investigation found an association between a high human birth defect rate and the use of two different water supply schemes for a medium sized community in Missouri (Collins, 1979). As a consequence of the implications of this finding, a study of several water quality parameters in the community's reservoir was undertaken. The findings expressed in this report represent portions of the aforementioned study which relate to the concept of an association between a reservoir algal bloom and mutagenicity of the reservoir water.

TEXT

Ames/*Salmonella* Testing

Water sampling procedure. Water samples were taken at the treatment plant from a pipe that transported reservoir water into the plant (a distance of about 60 meters). Samples were taken in 20 liter Nalgene containers and transported back to the laboratory (a period of several hours during which the samples were maintained at ambient temperature) where they were maintained at a temperature of approximately 0 to 4°C. The unconcentrated water samples were tested in the mutagenicity assay within 7 days after collection.

Prior to testing, the reservoir water sample was filter sterilized with a disposable 0.2 micron membrane filter. The control water, which was utilized to ascertain spontaneous reversion rates, consisted of Columbia, Missouri tap water that had been deionized and autoclaved.

Mutagen assay methodology. The unconcentrated reservoir water was tested for mutagenicity in the Ames/*Salmonella* assay using the histidine-dependent mutant strains of *S. typhimurium*. The strains used in testing the water samples consisted of the five strains recommended for general screening by Ames et al. (1975), TA1535, TA1537, TA1538, TA98, and TA100. The strains were obtained from Dr. Bruce Ames of the University of California at Berkeley. The strains were tested for the histidine requirement, deep rough (rfa) character, presence of the ampicillin resistent R factor and uvrB deletion as well as their response to five diagnostic mutagens (methyl methanesulfonate, 9-aminoacridine, 4-nitroquinoline-N-oxide, N-methyl-N'-nitro-N-nitrosoguanidine and 2-aminofluorine) according to the procedure outlined by Ames et al. (1975). The testing meth- odology followed the protocol of Pelon et al. (1977) with the fol- lowing modifications:

1. Scrapings from frozen bacterial stocks or from "master plates" were used to inoculate 40 ml volumes of water containing 320 mg of nutrient broth, 200 mg of yeast extract and 200 mg of sodium chloride. The inoculated media were incubated at 37°C for 12 to 15 hours. These cultures were used directly in the assay without the centrifugation and resuspension steps described by Pelon et al. (1977). This procedure consistently resulted in concentrations of 1 to 2 x 10^9 viable organisms per milliliter.

2. The liver microsome-activated testing method of Pelon et al. (1977) utilizes uninduced guinea pig livers for the production of the S-9 fraction, whereas the liver microsome fraction in the present study was obtained after the induction of rat liver enzymes (Collins, 1979).

3. The S-9 mixture in the Pelon et al. (1977) methodology has a concentration of 0.45 ml of liver S-9 fraction per milliliter of S-9 mixture. The present study followed the recommendation of Ames et al. (1975) for general screening by using 0.1 ml of S-9 fraction per milliliter of S-9 mixture.

Methodological differences between Ames assays performed in 1978 and those performed in 1979. Because of shifts in the lab- oratory methodology for performing the Ames/*Salmonella* assays, the assays of 1979 differed in two respects from the assays of 1978. The first difference involved the use of "master plates" (cultures of the tester strains on agar plates which can be stored at 4°C) which were used as a source of inoculum for the overnight cultures in the assays of 1979. Alternatively, the assays of 1978 were per- formed by inoculating overnight cultures from liquid cultures as described by Ames et al. (1975).

The second difference between the assays of 1978 and the assays of 1979 involved the method of liver enzyme induction for the production of the S-9 fraction. The S-9 fraction utilized in the assays performed in 1978 was prepared after inducing the rat liver enzymes with a combination of phenobarbitol and 3-methyl- cholanthrene. However, the S-9 fraction utilized in the 1979 assays was isolated from rats induced with Aroclor 1254. The rela- tive abilities of these three chemicals to induce the rat liver enzymes which activate two polycyclic hydrocarbons (benz(a)pyrene

and 3-methylcholanthrene) and an aromatic amine (2-acetylamino-
fluorine) are compared by Ames et al. (1975).

Results. The 1978 sampling schedule consisted of 15 sampling
dates spanning from 3/28/78 to 1/10/79 (Table 1). The unconcentrated water sample from each of these dates was assayed for mutagenicity using the Ames/*Salmonella* test with a modified Pelon et al.
(1977) methodology. Of the 15 water samples tested, three consecutive sampling dates had histidine reversion rates more than double
the spontaneous reversion rates (Table 2). The three samples which

Table 1. Reservoir Water Sampling Dates for the Mutagenicity Assay
          in 1978

Sampling Dates
3/28/78
4/27/78
5/12/78
5/29/78
6/19/78
7/07/78
7/21/78
8/04/78
8/26/78[a]
9/09/78[a]
9/13/78[a]
9/28/78
11/02/78
11/30/78
1/10/79

[a]These three consecutive sampling dates represent the only dates
on which the reservoir water was determined to be mutagenic.
Also, the algal bloom was visually determined to occur during
this same time span.

Table 2. Mutagenicity of the Reservoir Water in *Salmonella typhimurium* TA1537 with the Addition of S-9 Activation

Date of Collection	Spontaneous Reversion Rate	Reservoir Water Reversion Rate
8/26/78	19.0 (10)[a]	66.2 (10)
9/09/78	22.1 (10)	50.3 (10)
9/13/78	25.2 (5)	55.2 (5)

[a]Parenthetical values represent the number of plates used to calculate the reversion rate.

exhibited mutagenicity were mutagenic only in *S. typhimurium* strain TA1537 (a sensitive indicator of frameshift mutagens) and only when the S-9 mix was utilized via the preincubation technique. Thus, the unconcentrated water contained a promutagenic substance that was converted to a *S. typhimurium* mutagen via incubation at 37°C in the presence of the rat liver microsomal fraction.

Temporally associated with the period of mutagenicity in the reservoir water was a visually determined algal bloom. A rough estimate of the algal constituents of the bloom are listed in Table 3. The primary alga of the bloom was found to be *Oscillatoria subbrevis* Schmidle. The identification of the *O. subbrevis* Schmidle was made according to Desikachary (1959) who provides the following description of the alga: "Trichomes single, 5-6 microns broad, nearly straight, not attenuated at the apices; cells 1-2 microns long, not granulated at the crosswalls; end-cell rounded, calyptra absent."

Because of the predominance of the *O. subbrevis* in the algal bloom that was temporally associated with the mutagenic reservoir water, a unialgal culture of the *O. subbrevis* was grown in the laboratory and tested for mutagenicity. The alga was grown in Allen's modification (Allen, 1968) of the media of Hughes et al. (1958) supplemented with thiamine, vitamin B_{12} and soil extract. The algal cells were then homogenized to release any mutagenic substance within the cells. The homogenate was filter sterilized with a 0.2 micron disposable membrane filter and subsequently tested in the Ames/*Salmonella* assay. The homogenized filter-sterilized unialgal cultures of *O. subbrevis* Schmidle were not found to be mutagenic in the Ames assay (only four cultures were tested).

Table 3.   Estimated Composition of the 1978 Algal Bloom

Percentage of Total Algae	Algal Type
80%	*Oscillatoria subbrevis* Schmidle
19%	Wide variety of diatoms
1%	A variety of genera:
	Cosmarium sp.[a]
	Haematococcus sp.[a]
	Pediastrum sp.[a]
	Oocystis sp.[a]
	Chlorella sp.[b]
	Chlamydomonas sp.[b]
	Scenedesmus spp. (at least 3 spp.)[b]
	Microspora sp.[b]
	Ankistrodesmus sp.[b]
	Golenkinia sp.[b]
	Radiospheria sp.[b]

[a]These were the most abundant.

[b]These were identified from enrichment cultures.

The unconcentrated reservoir water was again sampled and tested for mutagenicity in the Ames assay in 1979. Twelve sampling dates between 7/13/79 and 10/18/79 were assayed for mutagenicity (Table 4). Unlike the period in 1978, there were no samples that were found to be mutagenic in 1979. Also, there were no algal blooms that could be visually perceived during the late summer bloom period in 1979. Although the lack of a mutagenic period for the reservoir water and no algal bloom in 1979 were disappointing in terms of attempting to identify the promutagenic agent, it is felt that the concomitant absence of the two events strengthens the association between them.

Multigeneration Rat Reproduction Pilot Study

The rationale for performing a reproduction study was to determine if the reservoir water was capable of inducing noticeable anomalies in a mammalian species. This particular pilot study was

Table 4. Reservoir Water Sampling Dates for the Mutagenicity Assay
          in 1979[a]

Sampling Dates

7/13/79

7/20/79

7/27/79

8/06/79

8/21/79

8/28/79

9/06/79

9/13/79

9/21/79

9/28/79

10/04/79

10/18/79

---

[a]None of the reservoir water samples taken in 1979 exhibited any
mutagenicity in the Ames/*Salmonella* assay. Also, no algal bloom
occurred during the four months of sampling/observation.

meant only to be an exploratory project to determine if a full-
scale rat reproduction study of the effects of the reservoir water
was warranted. However, since the most significant finding in the
rat pilot study was exhibited shortly after the rats were exposed
to the water which was determined to be mutagenic in the Ames
assay, and since the water from the reservoir has not exhibited
the Ames test mutagenicity since September of 1978, it has been
impossible to repeat the pilot study results in a full-scale rat
reproduction study.

The rats were obtained from Charles River Breeding Laboratories
(Wilmington, Massachusetts) and were maintained and bred at the
Dalton Research Center on the campus of the University of Missouri-
Columbia. The rats were of the Sprague-Dawley CD (COBS) strain.

The methodology for testing the reproductive effects of the
reservoir water consisted of maintaining several generations of
rats on reservoir water from the beginning of the pilot study in

April of 1978 until three generations of rats had been exposed
throughout their lifespan to the water (Collins, 1978). This
reproductive approach did not take into account the variability of
water quality over time because the human epidemiological data did
not indicate a seasonality associated with the human birth defects.
The water was collected on either bimonthly or monthly sampling
dates (according to the Ames testing schedule) in 20 liter Nalgene
bottles and was maintained at room temperature until it was used
to fill the rats' water bottles. The water was taken from the
reservoir prior to entering the water treatment plant, so the water
was untreated. This water was the same water that was found to be
mutagenic in the Ames test on three consecutive sampling dates,
namely 8/26/78, 9/9/78 and 9/13/78. The water was unconcentrated
because of the variety of chemical alterations which occur as a
result of the concentration process, as well as introducing problems in terms of the testing methodology (e.g., a lyophilized
sample could not be tested directly on the rats because of the
problem of salt toxicity).

    The rat breeding scheme consisted of both outcross and con-
sanguineous matings. Due to space limitations, the control animals
for the inbred portion of the study were not temporally matched
with the experimental group. Thus, the controls for the inbred
matings were bred after the experimental consanguineous matings
were completed. The outcrossed controls were run simultaneously
with the outcrossed experimental group.

    Although there were a number of abnormal birth outcomes asso-
ciated with the experimental rats (i.e., a runted monster, a rat
without a tail, a diabetic, a rat with a cataract, decreased litter
size, decreased fertility, resorptions and stillbirths), all of
these outcomes occurred either once or a very limited number of
times. The one abnormal outcome which occurred in a number of
cases was an ataxia or spasticity of the hindlimbs. The abnormality appeared in rats anywhere from about 7 days after birth to
several months after birth. The degree of severity of the abnormality varied over a wide range in different individual rats. The
spasticity in the limbs produced an uncoordinated gait, and the
ataxia in advanced cases eventually spread to the forelimbs. One
of the earliest symptoms of the abnormality was a lack of a right-
ing reflex. Pin pricking in advanced cases indicated an absence
of sensory perception. The abnormality which appears to be neurological in origin, but may also have a circulatory component,
eventually led to a loss of the digits and formation of a stump
on the hindlimbs (in advanced cases the forelimbs would also lose
their digits after hindlimb stumps had already developed). It was
originally hypothesized that the loss of the digits was due to
mechanical abrasion, however, a recent litter developed hindlimb
stumps almost simultaneously with the appearance of the ataxia

(at approximately 8 days of age). In this case, the stump formation was almost certainly not due to mechanical abrasion.

The histopathology of the rats with the neurological abnormality was performed by Dr. Ian Dunkin (Montreal General Hospital
Research Institute, Montreal, Quebec, Canada) (unpublished data,
1980). He found that the dorsal nerve roots and ganglia were
smaller in size for the abnormal rats than for controls. To a
lesser extent, the peripheral nerves were also found to be smaller
than the corresponding nerves found in control animals. In the
spinal cord, the dorsal columns appeared atrophic. Dr. Dunkin
hypothesized that the smallness of these structures might be con-
genital in origin. He states, "It is possible that there are too
few neurons in the dorsal root ganglia and hence fewer sensory
fibers. It appears unlikely that it is due to degeneration of
sensory fibers as there is little evidence of this on microscopic
examination."

Dr. Dunkin has stated that the neurologically abnormal rats
appear to have a very similar, if not identical, mutation to a rat
colony described by Jacobs et al. (1980). The mutation in that
case was spontaneous in origin.

An interesting time-sequence association was noted between
the period of mutagenicity in the reservoir water and the birth of
the rats with the neurological abnormality. Rats were maintained
on reservoir water as their only drinking source (ad libitum) and
were fed Purina lab chow (ad libitum) for the duration of the experiment. The rats were kept in polycarbonate cages containing
aspen litter for nesting. Experimental rats receiving reservoir
water were bred from April of 1978 until September of 1978 without
a single case of the neurological abnormality (this period included
4 groups of litters of which 2 groups included inbreeding). There
were two groups of litters that received the reservoir water promutagenic to *Salmonella* during portions of or the entirety of their
gestational periods. Each of these groups contained several rats
suffering from the abnormality. The first group of rats showing
symptoms of the neurological abnormality were the offspring of rats
that were mated about 9/7/78 and were born around 10/1/78. This
group of rats was probably the first group of rats to be exposed
to the promutagen for the early portion of their gestational period.
The incidence of the abnormality in this group of litters was five
affected animals from 179 births. The second group of litters was
the result of mating the same group of rats that gave birth to the
first set of animals with the abnormality. This second pregnancy,
immediately following birth of the first group via post-parturition
conception, led to a second round of births around 11/2/78. Aside
from the exposure of the mated rats during the previous pregnancy,
the parents received the promutagenic water only during the initial
portion of the second gestational period, not throughout the ges-

tational period. This group of litters exhibited an incidence of
the abnormality of five affected rats from 118 total offspring.
Subsequent litters that were exposed to nonmutagenic reservoir water
did not exhibit any cases of the neurological abnormality (except
that inbred crosses of the rats with the neurological abnormality
were maintained on nonmutagenic water and continued to produce
affected rats).

It has now been established that the neurological abnormality
is genetically transmitted. The abnormality has been passed for a
total of six generations in the absence of reservoir water and the
pedigree indicates that the abnormality is transmitted via a single
autosomal recessive gene. Primary cultures from the abnormal rats
of bone marrow, liver and lung cells all exhibited a normal karyo-
type.

Although a genetically recessive trait may be uncovered via
consanguineous brother-sister matings of a non-inbred strain such
as Charles River Sprague-Dawley CD (COBS) rats, it is not believed
that the neurological abnormality resulted solely from the inbreeding of the rats. The rats that were inbred prior to the occurrence
of the promutagenic water expressed no cases of the neurological
abnormality. The control rats that were inbred following the experimental groups' exposure to the reservoir water did not yield
any cases of the nuerological abnormality. Also, Dr. Carl Hansen
at NIH has developed an inbred strain from an outbred Sprague-
Dawley colony and he did not find any genetically determined abnormal ties in the strain, although two spontaneous mutants were discovered from the outbred stocks (Hansen, personal communication,
1979). Thus, the neurologically abnormal mutant in the Charles
River CD (COBS) rats is most likely the result of either a spontaneous mutation in the colony or the exposure to the promutagenic
water.

A fact that corroborates the hypothesis that the neurological
abnormality is the result of the reservoir water, is that one of
the five original rats to contract the neurological abnormality
was the offspring of an outcrossed mating of rats exposed to the
promutagenic water. In this case, inbreeding is ruled out as a
causative factor and the probability of the same gene spontaneously
mutating in both of the parents is very small.

Ichthyotoxicity of the Mutagenic Reservoir Water to *Gambusia affinis*

On the last of the three sampling dates where the water was
found to be mutagenic (namely 9/13/78), samples of the water were
taken for the purpose of attempting to perform a dominant lethal
assay using *Gambusia affinis*. Due to technical difficulties, the

Table 5. Ichthyotoxicity of the Reservoir Water Collected during
the Mutagenic Period[a] to *Gambusia affinis*

Water Type	Number of Fish Exposed	Number of Fish Surviving for Ten Days	Percent Mortality
Reservoir Water Tank 1	69	38	44.9%[b]
Reservoir Water Tank 2	69	48	30.4%[b]
Control Water	69	69	0.0%

[a]The reservoir water used in both Tank 1 and Tank 2 was collected on 9/13/78.

[b]By combining the Tank 1 and Tank 2 results, and calculating the chi-square value; $p < 0.0001$.

dominant lethal assay was never performed, however, the lethal effect of the mutagenic reservoir water on the *G. affinis* was recorded.

Samples of the reservoir water were taken on 9/13/78. The samples were transported back to the laboratory in Nalgene containers. On 9/14/78, two 20 gallon aquaria were filled with reservoir water and a third 20 gallon aquarium was filled with dechlorinated Columbia, Missouri tap water (control). Each tank was stocked with 69 male *Gambusia affinis*. After 10 days exposure, the surviving fish were counted. It was found that a significant percentage of the fish in the tanks containing the promutagenic reservoir water had died, whereas none of the control *Gambusia* had died (Table 5). No pathology was performed on the *Gambusia* to determine the cause of death. The implication of this observation is that the promutagenic reservoir water may also be ichthytoxic. Any conclusions from this observance will need to await further characterization of both the promutagen and the ichthyotoxin in the reservoir water.

DISCUSSION

The Ames/*Salmonella* assay has indicated that unconcentrated water from a pretreatment reservoir was mutagenic in *S. typhimurium* TA1537 only when the sample was activated with a rat liver microsomal fraction in three consecutive samples taken over a 19 day period in 1978. In two other reports where unconcentrated waters were found to be mutagenic in the Ames test (Pelon et al., 1977; Neeman et al., 1980), the level of mutagenicity did not represent a doubling of the spontaneous reversion rate, however all three of the promutagenic water samples in the present study were able to elicit a doubling of the background (spontaneous) reversion rate. The significance of this finding is that the mutagenicity found in the unconcentrated reservoir water represents a relatively high level of mutagenic activity.

The period when the water samples were found to contain promutagens was temporally associated with an algal bloom in the reservoir. The primary alga of the bloom was the cyanophyte *Oscillatoria subbrevis* Schmidle. Preliminary studies of unialgal laboratory cultures of this alga were not found to be mutagenic. However, the testing of a large number of cultures grown under a wide variety of environmental conditions will be necessary if the algal mutagens are produced under very specific conditions in the same manner as algal toxins (Collins, 1978). In the case of algal toxins, the production of a specific toxic substance varies with the environmental factors (light, temperature, inorganic and organic nutrition, etc.) as well as with genetic factors (*Microcystis aeruginosa* NRC-1 is genetically heterogeneous for toxin production) (Gorham, 1964). Thus, even if the correct culture conditions are found, some of the algae may not produce toxins (or mutagens) because of genetic differences within a particular strain.

The promutagenic unconcentrated water sample which was temporally associated with the cyanophyte bloom hints at the possibility that a factor associated with the bloom is responsible for the mutagenicity. A recent study found that five natural products from a marine alga were mutagenic in the Ames test (Leary et al., 1979), and one of the compounds was found to have 200 times the mutagenic activity of the well known mutagen ethyl methanesulfonate (EMS). Thus, the relatively high level of mutagenicity found in the unconcentrated reservoir water may be theoretically explained in terms of an algal natural product mutagen.

It could be argued that the difference in the mutagenicity of the reservoir water between 1978 (when mutagenic activity was present) and 1979 (when mutagenic activity was absent) is due to one or both of the methodological differences in performing the Ames assay in each of these years. The difference between inoculating overnight cultures with stored liquid cultures or "master plates" should not affect the determination of mutagenicity since control plates are poured in each of these techniques to provide the spontaneous reversion rate. However, it is possible that rat liver induction with phenobarbitol and 3-methylcholanthrene may induce activating enzymes that are not induced (or minimally so) by Aroclor 1254.

The multigeneration rat reproduction pilot study uncovered a mutant rat which is either the result of exposure to the mutagenic reservoir water or a spontaneous mutation. The temporal association between the occurrence of the mutation and the exposure to the mutagenic water, provides credibility for the hypothesis that the mutagenic water caused a mutation in a mammalian species that was expressed in a pathological abnormality.

A single sample of the mutagenic water proved to be ichthyotoxic in *Gambusia affinis*. The association between the ichthyotoxicity and the mutagenicity of the water is in need of further elucidation.

ACKNOWLEDGMENTS

The authors would like to acknowledge the technical assistance and support of Dr. Carl Marienfeld, Dr. Ian Dunkin, Dr. Greg Bartling, Dr. William Lower, Dr. Carl Hansen, the Dalton Research Center and the Cancer Research Center (Ellis Fischel State Cancer Hospital). The support for a portion of this work was derived from EPA Contract No. CA-8-2497-J and EPA Grant No. R805297.

REFERENCES

Allen, M. M. 1968. Simple conditions for growth of unicellular blue-green algae on plates. J. Phycol. *4*:1-4.

Ames, B. N., J. McCann, and E. Yamasaki. 1975. Methods for detecting carcinogens and mutagens with the *Salmonella/* mammalian-microsome mutagenicity test. Mut. Res. *31*:347-364.

Collins, M. 1978. Algal toxins. Microbiol. Rev. *42*(4):725-746.

Collins, M. 1979. Possible link between water supply and high birth defect rate. Proceedings of the American Water Works Association 1978 Annual Conference: Part II (Atlantic City, N. J.). AWWA, Denver, Colorado. Paper 24-5.

Collins, T. F. X. 1978. Multigeneration reproduction studies. Pages 191-214 *in* J. G. Wilson and F. C. Fraser, eds., Handbook of Teratology (Volume 4): Research Procedures and Data Analysis. Plenum Press, New York.

Desikacháry, T. V. 1959. Cyanophyta. Indian Council of Agricultural Research, New Delhi. 686 pp.

Gorham, P. R. 1964. Toxic algae. Pages 307-336 *in* D. F. Jackson, ed., Algae and Man. Plenum Press, New York.

Hughes, E. D., P. R. Gorham, and A. Zehnder. 1958. Toxicity of a unialgal culture of *Microcystis aeruginosa*. Can. J. Microbiol. *4*:225-236.

Jacobs, J. M., F. Scaravilli, L. W. Duchen, and J. Mertin. 1980. Hereditary sensory neuropathy in the rat: a new neurological mutant 'mutilated foot'. J. Anat. (Abstr.).

Leary, J. V., R. Kfir, J. J. Sims, and D. W. Fulbright. 1979. The mutagenicity of natural products from marine algae. Mut. Res. *68*:301-305.

Neeman, I., R. Kroll, A. Mahier, and R. J. Rubin. 1980. Ames' mutagenic activity in recycled water from an Israeli water reclamation project. Bull. Environ. Contam. Toxicol. *24*: 168-175.

Pelon, W., B. F. Whitman, and T. W. Beasley. 1977. Reversion of histidine-dependent mutant strains of *Salmonella typhimurium* by Mississippi River water samples. Env. Sci. Tech. *11*(6): 619-623.

CYANOBACTERIA AND ENDOTOXINS IN DRINKING WATER SUPPLIES

Jan L. Sykora and Georg Keleti

Graduate School of Public Health
University of Pittsburgh
Pittsburgh, Pennsylvania   15261

ABSTRACT

Field and laboratory studies were performed to evaluate the quantitative distribution and sources of endotoxins (lipopolysaccharides = LPS) in drinking water supplies. The *Limulus* amoebocyte lysate (LAL) test was used to measure total, bound and free endotoxins in six drinking water systems.

The major source of endotoxin, as measured by LAL, are uncovered finished water reservoirs which usually display high phytoplankton counts. LAL gelation in drinking water can be attributed to a specific reaction caused by Cyanobacteria (blue-green algae) and heterotrophic bacteria and to a nonspecific reaction with algae. The LAL test was also used to detect endotoxin and endotoxin-like substances in several species of cultured Cyanobacteria i.e., *Anabaena flos-aquae* (two strains), *A. cylindrica, Oscillatoria brevis* and *O. tenuis. Anabaena* sp. have caused LAL gelation at substantially lower cell concentrations than the *Oscillatoria* sp.

Lysozyme and alkali treatment of two *A. flos-aquae* strains (UTEX 1444 and NRC-44-1) followed by LAL testing indicated presence of "true" endotoxin in both. However, LPS was isolated from only one of them (UTEX 1444).

Statistical analyses of field data and laboratory experiments indicate that one of the significant sources of endotoxin in drinking water are Cyanobacteria. In order to maintain desirable drinking water quality through control of algae, Cyanobacteria and endotoxins, all open finished water reservoirs should be covered.

INTRODUCTION

Several epidemiological investigations have found high concentrations of blue-green algae in drinking water supplies concurrent with gastroenteritis of unknown etiology. Tisdale (1931) and
Veldee (1931) mentioned that unusually heavy blooms of Cyanobacteria
(blue-green algae) in water supplies might have been responsible
for an epidemic of water-borne gastroenteritis in the drainage basin
of the Ohio and Potomac Rivers during the drought year of 1930-31.
Dean and Jones (1972) suggested that algal toxins were a causative
agent in a repeated and prolonged gastroenteritis episode in the
Philippines. The occurrence and reoccurrence of diarrhea in
Udaipur, India, has been attributed to the presence of *Microcystis
aeruginosa* in drinking water (Gupta and Dashora, 1977).

A water-borne outbreak of gastroenteritis affected approximately 5,000 persons in Sewickley, Pennsylvania, during August
1975. Investigation of the Sewickley water system revealed an
accumulation of *Schizothrix calcicola* (Cyanobacteria) in open,
finished water reservoirs (Lippy and Erb, 1976; Sykora et al.,
1980). The water in one reservoir was contaminated by especially
high cellular counts of this species (400,000 cells/ml), and by
high endotoxin (lipopolysaccharide, LPS) concentration (2,500 ng/ml).

Most of the available information on algae and drinking water
contamination emphasize the possibility of the presence of toxic
substances unrelated to endotoxins. In this context, the simulta-
neous occurrence of *S. calcicola* and high concentrations of endotoxin during the Sewickley epidemic is worth investigating.

Endotoxin is a lipopolysaccharide (LPS) present in the outer
membrane of the cell walls of certain prokaryotic organisms,
notably Gram-negative bacteria. Therefore, its presence may
reflect the occurrence of Gram-negative bacteria in water, and
quantification of LPS may provide an estimate of bacterial densities. Several investigators who followed this approach reported
association between LPS, as measured by *Limulus* amoebocyte lysate
assay (LAL), and one or more forms of water-borne heterotrophic
bacteria. Watson et al. (1977) measured free (released) and
bound (in cells) LPS by LAL procedures in sea water samples, and
observed a good correlation between LPS and bacteria. The first
effort to relate LPS to specific types of bacteria was performed
by Evans et al. (1978) who demonstrated a statistically significant
correlation between bound and total (bound plus free) endotoxins
and several groups of bacteria in samples collected from clean
streams in Montana. However, Jorgensen et al. (1979) were unable
to establish a strong correlation between bacteria and endotoxins
in renovated wastewater. Di Luzio and Friedmann (1973), and
Jorgensen et al. (1976) surveyed the presence of LPS in drinking

water. Both investigations linked the occurrence of LPS in drinking water to the presence of Gram-negative bacteria.

The effect of ingested LPS and its occurrence in drinking
water has not been intensively studied and remains a subject of
controversy. However, numerous unexplained cases of endotoxic
reactions that have occurred in hospitals affecting mostly debilitated, immunosuppressed patients, are suspected to have originated
from the consumption of endotoxins in the milk or water shortly
before the affliction developed (Di Luzio and Friedmann, 1973).

It has been also suggested that a possible relationship may
exist between endotoxin levels in milk and the incidence of infant
Sudden Death Syndrome (Crib Death) (Di Luzio and Friedmann, 1973).
This study indicated that a substantial increase of endotoxin
levels occurred after prolonged exposure of the milk to room temperature. It was further noted that Crib Death is rare in breast-
fed infants. Enhanced gastrointestinal permeability in infants
suggests that absorption of exogenously derived endotoxin could
initiate illness under certain conditions. Interestingly, the
highest value of 800 ng/ml of LPS was detected by Di Luzio and
Friedmann (1973) in the tap water of Mexico City. This high endotoxin level does not seem to affect the indigenous population but
may cause "traveler's diarrhea" of newcomers (tourists) to the
area. Similarly, endotoxins may be responsible for diarrhea in
weaned breast-fed infants who can be characterized as "newcomers"
to a new environment.

The presence of endotoxin in drinking water and the effect of
injected LPS were observed by Hindman et al. (1975) while studying
an epidemic of pyrogenic reactions among kidney dialysis patients.
This epidemic coincided with the maximum concentration of algae in
the raw water source. Recently, several cases of endotoxin reactions (fever, chills, and hypotension) were observed in patients
undergoing cardiac catheterization (MMWP, 1979). The investigation suggested that the endotoxins were induced by contaminated
distilled water used for cleaning the catheters. The difficulty
with LPS is that most of the standard sterilizing procedures used
in hospitals effectively destroy organisms but leave endotoxin
residuals, because they are relatively heat stable.

Di Luzio and Friedmann (1973) and Jorgensen et al. (1976)
demonstrated the feasibility of the *Limulus* amoebocyte lysate (LAL)
procedure to quantify endotoxin levels in drinking water. Elevated
LPS levels were found in drinking water originating from surface
sources while lower concentrations were recorded from systems
treating groundwater. In this study, no attempt was made to test
the reliability of LAL for drinking water examination nor were the
sources of LPS discussed and explored. However, recent research
revealed that any relationship between LPS concentrations

(measured by LAL) and Gram-negative heterotrophic bacterial densities in drinking water are confounded by the sensitivity of LAL to
LPS derived from Cyanobacteria, to peptidoglycans from Gram-positive
bacteria and to additional organic compounds which may be produced
by algae or other organisms (Keleti et al., 1979; Sykora et al.,
1980).

This contribution examines distribution of algae and LPS in
drinking water, and explores the sources of LAL gelation in conjunction with algae and Cyanobacteria.

MATERIALS AND METHODS

Field Study

Five public water supply systems in Allegheny County, Pennsylvania, were studied during the summer of 1978. These facilities
were chosen on the basis of their source of raw water and the
presence of open reservoirs within their distribution system. The
samples from each water system were collected and analyzed weekly
at treatment plant effluent (TPE), open reservoir, and distribution
system(s). The water systems studied were:

Facility	Water Source	No. of Reservoirs Studied
1. Dixmont State Hospital	Ohio River	1
2. City of McKeesport	Youghiogheny R.	1
3. City of Pittsburgh	Allegheny River	1
4. Borough of Sewickley	Groundwater and infiltrated Ohio River	2
5. Borough of Springdale	Groundwater	1

Temperature, pH, total hardness, free and total chlorine
analyses were performed according to Standard Methods for the
Examination of Water and Wastewater, 14th Edition (1975). Biological measurements included standard plate count and Gram-negative
bacteria determinations. The bacterial plates were incubated for
48 hours and 8 days at 20° and 35°C. The delayed SPC incubation
was performed to detect slow growing bacteria. All the colony
types were tested using the standard gram staining. In addition,
the samples were analyzed for the occurrence of fecal and total
coliforms using the membrane filter technique described in Standard Methods (1975). Phytoplankton was concentrated in lugol pre-
served samples by settling and counted in a Palmer-Maloney nano-
plankton cell.

All the samples were also analyzed for the presence and concentrations of endotoxin utilizing the *Limulus* amoebocyte lysate
(LAL) method. Within one week of collection, analyses were conducted using freeze-dried *Limulus* lysate (Pyrotest [R], Difco Laboratories) capable of detecting as little as 2.5 ng/ml of endotoxin.
The LAL firm clot procedure for determination of total endotoxins
described by Sykora et al. (1980) was utilized.

A general review of the results derived from this field study
was published by Sykora et al. (1980). However, more recent statistical analyses of the data has revealed additional facts on the
effect of algae in drinking water on endotoxin concentrations.

Algal Cultures

Several species of Cyanobacteria and *Chlorella vulgaris* in
laboratory cultures were tested for possible presence of LPS using
the LAL gelation procedure. The list of species, their origin and
growth media are summarized in Table 1. All species were mass cultured in 9 liter, clear pyrex glass bottles, kept at 24°C, aerated
at 100 to 250 ml per minute of compressed air, and illuminated by
"white" fluorescent light on a 16-hour cycle. Each bottle was
aerated prior to inoculation with 200 ml of pure starter culture.
For LAL testing freeze-dried Pyrotest [R] (Difco Laboratories) and
Pyrotell [R] (Associates of Cape Cod) were used.

The number and biomass of bacteria occurring in these non-
axenic algae were determined by plating serial dilutions of algal
cultures on Trypticase Soy Agar in duplicates. After incubation
on the plates for 72 hours at 20°C and 35°C, the average number of
bacteria per ml was established by counting the colonies. A series
of dilutions in physiological saline containing *Salmonella typhimurium* was plated on Trypticase Soy Agar medium and the number of
bacteria present in the culture was calculated using this organism
as a standard. A known aliquot of the same culture was lyophilized
and the dry weight of bacteria was established. Based on this
standard, the average count of bacteria in our laboratory cultures
constituted only about 0.03% of the total biomass.

The alkali and lysozyme treatment was used to differentiate
between specific and nonspecific reaction of LAL test. Alkali
hydrolysis destroys LPS and lysozyme breaks down peptidoglycans
into inactive compounds. Samples of *C. vulgaris* and Cyanobacteria
were treated with three times crystallized egg white lysozyme (100
µg/mg at pH 6.3, in 0.15 N saline kept at 37°C for 18 hours) according to Wildfeuer et al. (1975).

Alkali treatment, as described by Suzuki et al. (1977),
required suspension in pyrogen free 0.1 N sodium hydroxide solution

Table 1.   Cyanobacterial and Algal Cultures

Species and Strain	Origin	Medium
Anabaena cylindrica UTEX B1611	University of Texas at Austin collection (Dr. Starr)	ASM-tricine
A. flos-aquae UTEX 1444	University of Texas at Austin collection (Dr. Starr)	ASM-tricine
A. flos-aquae NRC-44-1	Wright State University (Dr. Carmichael)	ASM-tricine
Chlorella vulgaris	University of Pittsburgh (drinking water)	Allen's
Oscillatoria brevis UTEX B1567	University of Texas at Austin collection (Dr. Starr)	ASM-tricine
O. tenuis UTEX 1506	University of Texas at Austin collection (Dr. Starr)	ASM-tricine
Schizothrix calcicola (brown stain)	University of Pittsburgh (Mosquito Creek Reservoir, Ohio)	Allen's

which was kept in a sealed tube maintained for four hours in a
water bath at 70°C.   After neutralization with pyrogen free hydrochloric acid, the samples were subjected to the LAL gelation test.

Statistical Analysis

Simple and multiple linear regression analyses were performed
comparing LPS concentrations, heterotrophic bacteria and total
phytoplankton.   All data including the LPS concentrations, were
log transformed to compute Pearson's correlation coefficients and
significance levels.   The 0.05 level of significance was designated
as the criterion for rejection.

RESULTS AND DISCUSSION

Elevated standard plate count bacteria (SPC), Gram-negative
bacteria (GNB) and phytoplankton counts were a common occurrence

in finished water reservoirs (Table 2) and their distribution systems, but were less abundant in finished water obtained directly at the plants. Especially, the cellular densities of phytoplankton organisms recorded from treatment plant effluents were low and originated from growth on filter walls and other facilities inside of treatment plants, and penetration of algae through the filtration system (McKeesport). Thus the major source of algae and Cyanobacteria in the distribution systems were open, finished water reservoirs. In the five water systems, *Chlorella vulgaris* (Chlorophyta) was the most common alga, whereas *Schizothrix calcicola* was the most dominant cyanobacterium. It is important to note that rarely was *S. calcicola* the dominant species in any reservoir as it was during the Sewickley outbreak in 1975. The one exception is the Pittsburgh Reservoir in 1978 when *S. calcicola* was the leading species and its population reached 5,000 cells/ml. In Sewickley during the outbreak the concentration was almost 100 times that number. It is not yet clear what occurred at Sewickley in 1975 that permitted *S. calcicola* to suppress *C. vulgaris*, the normally dominant phytoplankton organism in open, finished water reservoirs.

Total and fecal coliform bacteria were absent in treatment plant effluent samples and rare in distribution systems. The highest total coliform counts were recorded in Sewickley reservoirs after these storage facilities were provided with plastic, floating covers. Our results suggest that the rainwater collecting in the access hatches of the covers caused the contaminationg of Sewickley Reservoirs Numbers 3 and 4 by coliform bacteria.

Endotoxin values varied from nondetectable levels in some finished water samples (plant effluents) to as high as 630 ng/ml in Sewickley Reservoir No. 4. With the exception of McKeesport Reservoir (chlorinated) and Sewickley Reservoir No. 3 (low temperature), endotoxin levels recorded in the open reservoirs were higher than in plant effluents and distribution systems. The geometric means of the endotoxin concentrations detected in reservoirs were up to 100 times higher than those of LPS detected in finished waters. The greatest endotoxin concentrations occurred in small reservoirs; Sewickley No. 4 and Springdale (Sykora et al., 1980).

The endotoxin concentrations were higher in the samples from the distribution systems than in the plant effluents but slightly lower than the values obtained from the open reservoirs. The increase in endotoxin levels could be from bacterial contamination in the distribution systems, decomposition of algae and bacteria originating in open reservoirs, and/or decreasing chlorine levels in the mains.

Simple linear regression analysis of data from finished waters, reservoirs, and distribution systems indicated significant relationship between all SPC groups and endotoxin. The total algae counts

Table 2. 95% Confidence Intervals and Geometric Means of Phytoplankton Counts SPC, Bacteria and LPS

Sampling Location	SPC Bacteria Colonies/ml			Total Phytoplankton cells/ml			Cyanobacteria cells/ml			LPS ng/ml		
	N	C.I.	x̄g	N	C.I.	x̄g	N	C.I.	x̄g	N	C.I.	x̄g
Springdale Finished Waters	14	2.0 - 6.4	3.6	17	15. - 130.	44.	17	0.86- 2.8	1.6	17	0.61- 1.3	0.9
Springdale Reservoir	14	40. - 1000.	200.	15	35000. -150000.	73000.	15	2.9 - 34.	9.9	15	4.5 - 22.	9.9
Springdale Distribution	13	7.3 - 92.	26.	17	190. - 2500.	690.	17	0.84- 2.6	1.5	17	0.87- 2.0	1.3
McKeesport Finished Waters	12	4.6 - 31.	12.	17	26. - 500.	110.	17	1.3 - 10.	3.5	17	0.81- 3.6	1.7
McKeesport Reservoir	14	24. - 410.	100.	17	190. - 1700.	560.	17	28. - 290.	90.	17	0.72- 4.3	1.8
McKeesport Distribution	10	5.5 - 82.	21.	16	33. - 240.	88.	16	0.89- 4.4	2.0	16	0.98- 5.0	2.2
Pittsburgh Finished Waters	12	4.7 - 57.	16.	11	7.3- 100.	27.	12	0.69- 2.7	1.4	12	0.32- 1.0	0.5
Pittsburgh Reservoir	5	3.2 - 3400.	110.	12	1100. - 9100.	3200.	12	120. -5000.	770.	12	2.1 - 9.8	4.6
Pittsburgh Distribution	7	2.3 - 460.	33.	9	120. - 2100.	490.	9	7.0 - 140.	31.	9	0.31- 2.0	0.7
Dixmont Finished Waters	6	0.49- 14.	2.7	6	6. - 4400.	160.	6	0.55- 3.9	1.5	6	0.43- 1.5	0.8
Dixmont Reservoir	10	97. - 4400.	650.	15	920. - 1300.	3400.	15	0.95- 1.2	1.1	15	2.1 - 10.	4.6
Dixmont Distribution	11	110. - 2600.	530.	15	610. - 9000.	2300.	15	0.76- 16.	3.5	15	1.8 - 5.4	3.1
Sewickley Finished Waters	13	1.3 - 9.1	3.4	15	7.1- 140.	31.	15	0.76- 3.5	1.6	15	0.37- 1.5	0.7
Sewickley Reservoir No. 3 (open)	4	20. -21000.	650.	6	25. - 3900.	310.	6	1.8 - 120.	14.	6	0.52- 3.1	1.3
Sewickley Reservoir No. 3 (covered)	7	7.6 - 1400.	100.	7	1.6- 25.	6.3	8	0.75- 2.0	1.2	8	0.31- 1.3	0.6
Sewickley Reservoir No. 4 (open)	2	160. - 3400.	2300.	4	4300. -290000.	36000.	8	130. -6400.	900.	8	4.9 -3200.	130.
Sewickley Reservoir No. 4 (covered)	6	13. - 4900.	260.	8	3.5- 130.	21.	11	0.83- 1.6	1.2	11	2.7 - 16.	6.5
Sewickley Low Pressure System	10	1.5 - 39.	7.7	14	28. - 150.	65.	14	1.3 - 14.	4.3	14	0.55- 2.9	1.3
Sewickley High Pressure System	12	5.3 - 150.	28.	15	25. - 690.	130.	15	1.1 - 12.	3.7	15	1.1 - 6.7	2.7

Key: N = Number of Samples; C.I. = Confidence Interval; x̄g = Geometric Mean.

Table 3.   Pearson's Correlation Coefficients and Significance Levels
of Linear Regression Analyses of Bacteria, Phytoplankton
and Endotoxin

Y	X	R	S
Log LPS	Log SPC (20°C -- 2 days)	0.31761	0.00054
Log LPS	Log SPC (35°C -- 2 days)	0.35703	0.00001
Log LPS	Log SPC (20°C -- 8 days)	0.35910	0.00001
Log LPS	Log SPC (35°C -- 8 days)	0.42028	0.00001
Log LPS	Log GNB (20°C -- 2 days)	0.34528	0.00025
Log LPS	Log GNB (35°C -- 2 days)	0.41322	0.00001
Log LPS	Log GNB (20°C -- 8 days)	0.36312	0.00001
Log LPS	Log GNB (35°C -- 8 days)	0.37286	0.00001
Log LPS	Log Phytoplankton	0.47933	0.00001

Key: SPC = Standard Plate Counts; GNB = Gram-Negative Bacterial
Counts; Y = Dependent Variable; X = Independent Variable;
R = Correlation Coefficient; S = Significance Level.

as correlated to endotoxin were also found to be highly significant
($P < 0.00001$, Table 3).   This is to be expected as other environmental and chemical factors in the mains and treatment plants act
together effectively reducing bacteria, algae, and endotoxin.   The
observed relationship may, therefore, reflect the combined action
of many factors without revealing the origins or sources of contamination.

Sunlight provides energy for organisms in uncovered reservoirs,
the chlorine dissipates quickly and the excessive pressure common
in pumps and distribution systems is absent.   Therefore, the environment in the reservoirs eliminates many parameters present in
distribution systems and study of the former may give direct information about the origins of LAL gelation.

Simple linear regression analysis of the reservoir samples
showed a highly significant relationship between algae and endotoxin and between SPC (35°C, 8 days) bacteria and lipopolysaccharides.   The significance of the other parameters (including SPC
bacteria incubated for 2 days) as related to endotoxin was not as
high as the relationship found for algae (Table 4).

Table 4. Summary of Simple Linear Regression Analysis of Phyto-
          plankton Organisms and Bacteria Versus LPS in Reservoirs

Parameter	B	C	R	S	N
Phytoplankton	0.3207	-0.4945	0.59630	0.00001	58
Gram-negative 20°C - 8 days	0.2997	-0.2878	0.32795	0.00435	63
Gram-negative 35°C - 8 days	0.2938	-0.2315	0.30985	0.00636	64
SPC 20°C - 2 days	0.1712	0.2044	0.21337	0.03807	70
SPC 35°C - 2 days	0.1439	0.2220	0.21089	0.06930	73
SPC 20°C - 8 days	0.3348	-0.4556	0.36724	0.00103	68
SPC 35°C - 8 days	0.3990	-0.5690	0.45235	0.00005	68

Key: R = Correlation Coefficient; S = Significance Level; B =
      Slope; C = Intercept; N = Number of Samples.

Further analyses of the reservoir data has been undertaken
using a multiple regression between the most important biological
and chemical constituents; LPS, phytoplankton, SPC bacteria (incubated at 35°C for 2 and 8 days), free and total chlorine.

The significance between biological contaminants and LPS in
individual reservoirs was inconclusive because of the small sample
size. However, when multiple regression analysis was applied to
the pooled data from all rexervoirs, the phytoplankton became
again the most significant variable influencing LPS concentrations.
The second most important constituent affecting endotoxin concentrations were SPC bacteria incubated at 35°C for 8 days. At the
selected level of significance ($S = 0.05$), the remaining variables
do not contribute significantly to the observed variances in LPS
(Table 5).

The laboratory experiments followed the field study and were
designed to determine the LAL reaction with selected species of
Cyanobacteria and to estimate their LPS content.

Anabaena flos-aquae (NRC-44-1) culture caused a positive LAL
gelation at the highest dilution (180 cells/ml) whereas at least

Table 5. Multiple Regression Analysis Reservoir Samples

Parameter	B	T	S	ΔR^2	R^2
Phytoplankton	0.2729	4.59	0.0001	0.35669	0.35669
SPC 35°C – 2 days	-0.1016	0.93	0.3524	0.00734	0.36403
SPC 35°C – 8 days	0.3916	2.20	0.0278	0.05535	0.41938
Free Chlorine	1.5492	1.26	0.2076	0.01783	0.43721
Total Chlorine	-1.4010	1.17	0.2420	0.01495	0.45216

Multiple R = 0.6724, R^2 = 0.4522, C = 1.8282

Key: B = Slope; T = T Test on B; S = Significance; R = Correlation Coefficient; C = Intercept; ΔR^2 = Contribution to R^2; R^2 = Explains Variance.

358 cells/ml of *A. flos-aquae* (UTEX 1444) were needed for positive *Limulus* lysate reaction. *A. cylindrica* (UTEX B1611) culture caused gelation at 10^4 cells/ml and 550 cells/ml of green alga *C. vulgaris* were needed for positive gelation test. Both species of *Oscillatoria* demonstrated a limited ability to cause LAL gelation (Table 6).

Sykora et al. (1980) concluded that *C. vulgaris* is an important causative agent of non-specific LAL gelation in drinking water samples. Table 7 shows that the LAL test gives positive reaction with endotoxins, thrombin compounds, polynucleotides, peptidoglycans (formerly mucopolysaccharides occurring in Gram-negative and Gram-positive bacteria), and dextran derivatives. This table also indicates that several of the chemical substances causing positive gelation reaction could be effectively removed or confirmed by sample pretreatment.

Lysozyme and alakli treatments performed with two strains of *A. flos-aquae* (UTEX 1444 and NRC-44-1) indicated presence of true LPS, which was documented by removal of gelation capacity after NaOH treatment. On the other hand, there was no change in LAL gelation after lysozyme was applied to both *Anabaena* cultures (Table 8, 9, and 10). This result is unexpected as true LPS were

Table 6. LAL Gelation of Cyanobacterial and Algal Cultures

Species	Batch	Cells/ml	LPS in undiluted sample ng/ml	Highest positive dilution	Sensitivity of LAL ng/ml	No. of cells at highest pos. dilution cells/ml	Total endotoxin/cell pg
Anabaena flos-aquae NRC-44-1 (toxic)	1	3.1×10^6	500	2×10^{-3}	.25	1.5×10^3	.16
Anabaena flos-aquae NRC-44-1 (toxic)	2	1.8×10^5	125	10^{-3}	.125	1.8×10^2	.69
Anabaena flos-aquae UTEX 1444	1	3.6×10^5	125	10^{-3}	.125	3.58×10^2	.35
Anabaena cylindrica UTEX B1611	1	2×10^6	50	2×10^{-2}	.25	1×10^4	.02
Oscillatoria brevis UTEX B1567	1	2.4×10^6	50	2×10^{-2}	.25	1.2×10^4	.02
Oscillatoria brevis UTEX B1567	2	2.4×10^6	12.5	10^{-2}	.125	2.4×10^4	.005
Oscillatoria tenuis UTEX 1506	1	1.9×10^6	50	10^{-2}	.25	9.8×10^3	.02
Oscillatoria tenuis UTEX 1506	2	1.7×10^6	12.5	10^{-2}	.125	1.7×10^4	.007
Chlorella vulgaris	1	1.1×10^6	500	2×10^{-3}	.25	5.5×10^2	.46

Table 7.   Substances Suspected of Causing LAL Gelation[a]

Positive Reaction	Reaction Eliminated By	Negative Reaction
Lipopolysaccharide		Dextran
Glycolipid	Mild alkaline hydrolysis	Chitosan
		Lentinan
Lipid A (solubilized by BSA[b] and tri-ethylamine)		Starch (soluble)
		Sucrose fatty acid esters
		Whole Blood
		Plasma
Thrombin		Calcium
		Streptolysin
Thromboplasmin		Streptokinase
		Streptodormase
Poly (A):Poly (U)	Ribonuclease treatment	Cl. tetani toxin
Poly (I):Poly (C)		Hemoglobin
		Serotonin
Peptidoglycan	Lysozyme treatment	Histamine
		Epinephrine
		Norepinephrine
Dextran phosphate		Bradykinin
Palmitoyldextran phosphate		Str. pyogenes
		Str. pneumoniae
		Staph. aureus
Palmitoyldextran		

[a]Table compiled from Suzuki et al. (1977), Rotta (1975), Wildfeuer et al. (1975), and Elin and Wolff (1973).

[b]Bovine Serum Albumin.

not isolated from one strain (*A. flos aquae* NRC-44-1) but were found in the other (UTEX 1444) (Keleti et al., 1980).

Wang and Hill (1977) could not isolate LPS from another strain of *A. flos-aquae* (A-37), and therefore, the presence of true endo-toxins in *A. flos-aquae* seem to be variable. Healey (1973) indi-cated that UTEX 1444 strain and *A. variabilis* may be morphologically identical. Weckesser et al. (1974) isolated LPS from *A. variabilis* which may support Healey's conclusions concerning the identity of the UTEX 1444 strain and *A. variabilis*. However, the differences between chemical composition of macromolecular substances and mor-phology of individual species of the genus *Anabaena* are substantial and more research in this area is needed before any final conclusion can be formulated.

Table 8. LAL Gelation of Lysozyme and NaOH Treated *Anabaena flos-aquae* (UTEX 1444 Strain)

Cells/ml	No Pretreatment	Treated
Lysozyme Treatment		
3.6×10^5	++	++
3.6×10^4	++	++
3.6×10^3	++	++
3.6×10^2	+	+
36	−	−
Alkali Treatment (NaOH)		
6.5×10^5	++	+
6.5×10^4	++	−
6.5×10^3	−	−

Key: ++ = Firm Clot; + = Clot Breaks; − = No Reaction.

In conclusion, the results of the field and laboratory study show that many of the drinking water endotoxin analyses reflected presence of eukaryotic algae and Cyanobacteria more than the bacterial contamination. Until now, the LAL technique was considered by many to to be quite specific indicating only the presence of Gram-negative bacteria. Evans et al. (1978) and Watson et al. (1977) found highly significant relationships between bacteria and LPS. However, both papers deal with environments where the phytoplankton population is sparse or practically nonexistent (small streams and ocean waters). This does not hold for a eutrophied

Table 9. *Limulus* Lysate Gelation of *Anabaena flos-aquae* (Strain NRC-44-1) (Lysozyme Treatment)

Cells/ml	No Treatment	Treated
1.8×10^5	++	++
1.8×10^4	++	++
1.8×10^3	++	++
1.8×10^2	+	+
18	−	−

Key: ++ = Firm Clot; + = Clot Breaks; − = No Reaction.

Table 10. *Limulus* Lysate Gelation of *Anabaena flos-aquae* (Strain NRC-44-1) (NaOH Treatment)

	Free Endotoxin		Total Endotoxin	
Cells/ml	Untreated Sample	Treated Sample	Untreated Sample	Treated Sample
6.8×10^5	++	++	++	++
6.8×10^4	++	±	++	++
3.4×10^4	++	−	++	++
1.7×10^4	++	−	++	++
8.5×10^3	±	−	++	−
6.8×10^3	−	−	++	−
3.4×10^3	−	−	++	−
1.7×10^3	−	−	−	−

Key: ++ = Firm Clot; + = Clot Breaks; − = No Reaction

freshwater environment, a frequent source of drinking water where the algae and Cyanobacteria dominate. Development of a modified LAL technique employing lysozyme and mild alkaline hydrolysis treatment is recommended before LAL technique is used for quantification of bacteria in drinking water.

In addition, the field study produced results which suggest deterioration of water quality in open drinking water reservoirs. The laboratory experiments indicate that one of the significant sources of endotoxin in drinking water are blue-green algae. In order to maintain the desired water quality, and control algae and Cyanobacteria, all open, finished water reservoirs should be covered.

ACKNOWLEDGMENTS

The work was supported by the Environmental Protection Agency Research Grant No. 805368.

REFERENCES

Dean, A. G., and T. C. Jones. 1972. Seasonal gastroenteritis and
    malabsorption at an American military base in the Philip-
    pines. Am. J. Epidemiol. 95:111-127.
Di Luzio, N. R., and T. J. Friedmann. 1973. Bacterial endotoxins
    in the environment. Nature (London) 244:49-51.
Elin, J. R., and S. W. Wolff. 1973. Nonspecificity of the Limulus
    amoebocyte lysate test: positive reaction with polynucleo-
    tides and proteins. J. Infect. Dis. 128:349-352.
Evans, T. M., J. E. Schillinger, and D. G. Stuart. 1978. Rapid
    determination of bacteriological water quality by using
    Limulus lysate. Appl Environ. Microbiol. 35:376-382.
Gupta, R. S., and M. S. Dashora. 1977. Algal pollutants and
    potable water. Pages 431-459 in R. B. Pajasek, ed., Drink-
    ing Water Quality Enhancement through Source Protection.
    Ann Arbor Press, Ann Arbor.
Healey, F. P. 1973. Characteristics of phosphorus deficiency in
    Anabaena. J. Phycol. 9:383-394.
Hindman, S. M., M. S. Favero, and A. Peterson. 1975. Pyrogenic
    reactions during haemodialysis caused by extramural endo-
    toxins. Lancet 18:732-737.
Jorgensen, J. H., J. C. Lee, G. A. Alexander, and H. W. Wolf.
    1979. Comparison of Limulus assay, standard plate count
    and total coliform count for microbiological assessment of
    renovated wastewater. Appl. Environ. Microbiol. 37:928-931.
Jorgensen, J. H., J. C. Lee, and H. R. Pahren. 1976. Rapid detec-
    tion of bacterial endotoxins in drinking water and renovated
    wastewater. Appl. Environ. Microbiol. 32:347-351.
Keleti, G., J. L. Sykora, E. C. Lippy, and M. A. Shapiro. 1979.
    Composition and biological properties of lipopolysaccharides
    isolated from Schizothrix calcicola (Ag.) Gomont (Cyano-
    bacteria). Appl. Environ. Microbiol. 38(3):471-477.
Keleti, G., J. L. Sykora, L. A. Maiolie, D. L. Doerfler, and I. M.
    Campbell. 1980. Isolation and characterization of endo-
    toxin from Cyanobacteria (blue-green algae). (In press)
    in W. W. Carmichael, ed., The Water Environment: Algal
    Toxins and Health. Plenum Press, New York.
Lippy, E. C., and J. Erb. 1976. Gastrointestinal illness at
    Sewickley, Pa. J. Am. Water Works Assoc. 68:606-610.
MMWR Morbidity and Mortality Weekly Reports. 1979. Endotoxic
    reactions associated with the reuse of cardiac catheters -
    Massachusetts. 23:25-27.
Rotta, J. 1975. Endotoxin-like properties of the peptidoglycan.
    Z. Immun. Forsch. 149:230-244.
Standard Methods for the Examination of Water and Wastewater.
    1975. 14th Edition. American Public Health Association,
    American Water Works Association, and Water Pollution Control
    Federation

Suzuki, M., T. Mikanu, I. Matsumoto, and S. Suzuki. 1977. Gelation
    of *Limulus* lysate by synthetic dextran derivatives. Micro-
    biol. Immunol. *21*:419-425.

Sykora, J. L., G. Keleti, R. Roche, D. R. Volk, G. P. Kay, R. A.
    Burgess, M. A. Shapiro, and E. C. Lippy. 1980. Endotoxins,
    algae and *Limulus* amoebocyte lysate test in drinking water.
    Water Research *14*:829-839.

Tisdale, E. S. 1931. Epidemic of intestinal disorders in Charles-
    ton, W. Va., occurring simultaneously with unprecedented
    water supply conditions. Am. J. Public Health *21*:198.

Veldee, M. V. 1931. Epidemiological study of suspected water-
    borne gastroenteritis. Am. J. Public Health *21*:1227-1235.

Wang, A. W., and A. Hill. 1977. Chemical analysis of the phenol
    water-extractable material from *Anabaena flos-aquae*. J.
    Bacteriol. *130*:558-570.

Watson, S. W., T. L. Novitsky, H. L. Quinby, and F. W. Valois.
    1977. Determination of bacterial number and biomass in the
    marine environment. Appl. Environ. Microbiol. *33*:940-946.

Weckesser, J., A. Katz, G. Drews, H. Mayer, and I. Fromme. 1974.
    Lipopolysaccharide containing L-acofriose in the filamentous
    blue-green alga *Anabaena variabilis*. J. Bacteriol. *120*:672-
    678.

Wildfeuer, A., B. Heymer, D. Spilker, K. H. Schleifer, E. Vanek,
    and O. Haferkamp. 1975. Use of *Limulus* assay to compare
    the biological activity of peptidoglycan and endotoxin.
    Z. Immun. Forsch. *149*:258-264.

POTENTIAL FOR GROUNDWATER CONTAMINATION

BY ALGAL ENDOTOXINS

Charles P. Gerba and Sager M. Goyal

Department of Virology and Epidemiology
Baylor College of Medicine
National Center for Groundwater Research
Houston, Texas 77030

ABSTRACT

Groundwater is an important resource which must be protected
from contamination. Currently, 20% of the nation's water supply
is taken from groundwater. By the year 2000, this is expected to
increase to 33% of the total water used in the United States. From
1946 to 1977 there were 264 outbreaks and 62,273 cases of illness
related to contaminated groundwater. In half of these outbreaks
no responsible agent was identified. In recent studies we have
shown that bacterial endotoxins can reach groundwater in high concentrations, especially after simulated rainfall events. Previously
absorbed endotoxins and other phycotoxins could potentially exhibit
similar behavior, resulting in periodic high concentrations of these
toxins in groundwater supplies which may be difficult to detect.
Potential sources of algal toxins include sewage oxidation ponds,
groundwater recharge operations, wastewater irrigation, surface
reservoirs, etc.

TEXT

Groundwater is an important resource which must be protected
from contamination. Already over 19% of the nation's water supply,
and upwards of 46% in some areas, is taken from groundwater sources.
The use of groundwater is increasing, and by the year 2000 this
proportion is expected to increase to 33% of the total water used
in the United States of America (Freeze and Cherry, 1979). As the
demand for clean water increases and supply decreases, it is evident
that artificial recharge of the present sources will become more

prevalent.   The re-charge of groundwater sources with wastewater
is an attractive alternative source of groundwater and an economically favorable method of wastewater disposal.   However, care must
be taken to ensure the safety of this practice by preventing contamination with disease-causing bacteria and viruses passed with
human excreta.   The microbial contamination of groundwater is a
serious problem that can result in large outbreaks of waterborne
disease.   For example, from 1946 to 1977 there were 264 outbreaks
and 62,273 cases of illness related to contaminated (untreated or
inadequately treated) groundwater (Craun, 1979).   This represents
48 and 58%, respectively, of all waterborne outbreaks and cases of
illness for that period.   Thus, 20% of the supply accounts for 50%
of the illness.   Overflow from septic tanks and cesspools was
responsible for 42% of outbreaks, and 71% of illness was caused by
using untreated groundwater in non-municipal systems.   Furthermore,
in 356 (65%) of the 550 documented outbreaks of waterborne disease
from 1946 through 1977 no specific agent could be identified (Craun,
1979).   The percentage of outbreaks categorized as acute gastrointestinal illness is slightly higher in groundwater systems (Craun,
1979).   It is important to note that a significantly higher number
of outbreaks due to groundwater occur during the summer months
(Craun, 1979).

    In the majority of waterborne disease outbreaks no protozoan
or bacterial agent is identified as the cause.   Most of these outbreaks, usually characterized as gastroenteritis, are thought to
have a viral etiology, although attempts are seldom made to isolate
a virus agent because of the difficulties involved.   Thus, the possibility of algal or bacterial toxins as responsible agents cannot
be ruled out.   There are numerous reports of a possible association
between human gastrointestinal disorders and ingestion of algae
(Schwimmer and Schwimmer, 1964; Keleti et al., 1979).   In a recent
case in Sewickley, Pennsylvania, 62% of the population contracted
a gastrointestinal illness, characterized by diarrhea and abdominal
cramps, from the drinking water.   The outbreak of the disease
occurred during the appearance of a heavy growth of filamentous
blue-green algae *Schizothrix calcicola* in one of the reservoirs.
Keleti et al. (1979) later found that lipopolysaccharide extracted
from *S. calcicola* induced Limulus lysate gelation, Schwartzman
reaction, and was pyrogenic in rabbits.   Algae has also been implicated as a source of respiratory and dermatologic disorders
(Schwimmer and Schwimmer, 1964).   As observed by Gorham and
Carmichael (1979) poisoning of livestock and other animals attributed to toxic blooms of freshwater cyanophytes occurs with considerable frequency in western Canada and the mid-western United
States.

    Blue-green algae produce a number of toxins (Gorham and
Carmichael, 1979; Moore, 1977) which have a potential for finding

their way into groundwater, but this report will largely concern the potential for groundwater contamination by endotoxins. Algal endotoxins have been implicated in at least one outbreak of endotoxemia in patients undergoing hemodialysis (Hindman et al., 1975) and possibly with the gastroenteritis outbreak in Sewickley, Pennsylvania (Keleti et al., 1979; Lippy and Erb, 1976).

It has been observed that cities which obtain their drinking water from wells generally have lower levels of endotoxin than those cities which obtain their water from surface sources (Di Luzio and Friedmann, 1973; Jorgensen et al., 1976). The concentration of endotoxin in surface water was found to vary from 1 to 400 μg/ml (Di Luzio and Friedmann, 1973), but for most drinking water the concentration varied from 1.25 to 12.5 ng/ml. The lower concentration of endotoxin in groundwater was hypothesized to be due to the sieving action of soil.

Our laboratory has been involved for several years in studies concerning the fate of microbial pathogens during the land application of wastewater (Gerba et al., 1975). Land application of wastewater is popularly regarded as a valuable means for advanced sewage treatment, for water recycling through crop irrigation, and for groundwater recharge.

Because of the presence of large numbers of gram-negative bacteria, sewage effluents contain high levels of endotoxin. We set out initially to determine the concentration of endotoxin in groundwater underlying sites where wastewater land application was being practiced. In many cases wastewater is held in holding ponds before land application, allowing the opportunity for algae growth. This is especially true when wastewater stabilization ponds are used before land application. The resulting algae growth can contribute algal endotoxins and possibly phytotoxins to the water before land application. Two common systems for land treatment involve the use of rapid infiltration basins and/or slow-rate systems. In rapid infiltration systems most of the applied wastewater percolates through the soil and the treated effluent eventually reaches the groundwater. The wastewater is applied to rapidly permeable soils, such as sands and loamy sands, by spreading basins or by sprinkling. Vegetation is not usually used as a soil cover in these systems and algal mats tend to develop at the soil surface. The basins used in such systems are usually intermittently flooded on various flooding and drying cycles to reduce clogging of the soil surface and concurrent reduction in infiltration rates. During the drying cycle, the algal mat is allowed to cake and decay. The annual application rate for rapid infiltration varies from 20 to 560 feet of wastewater per year.

Problems with algae clogging the soil surface of infiltration basins (EPA, 1977) and penetration of algae into the groundwater

(Folkman and Wachs, 1970) have been reported. Experience in Israel
has shown that percolation through dune sand does not remove all
the algae from stabilization pond effluents. Experiments on filtration of stabilization pond effluents have shown that pond liquid
percolating through a 3.5 m sand layer is still greenish in color
due to the presence of high concentrations of algae (Folkman and
Wachs, 1970). These results gave rise to the fear that significant
concentrations of algae could be carried by the percolating water
to depths at which they might affect the quality of the water
recharged in the aquifer.

     Lower soil permeability characterizes slow-rate land applica-
tion systems, which are usually characterized as irrigation systems
for crop production. In these systems application rates vary from
2 to 20 feet per year. Again, if waste stabilization ponds are
used to treat the wastewater or if holding ponds open to the sunlight are used before land application, high concentrations of
algae are usually present.

     We have examined a number of wells beneath both slow and rapid
infiltration sites for the presence of endotoxins with the use of
Limulus amebocyte lysate (LAL Assay, 1979) (Table 1). The range
of endotoxin in the groundwater beneath these sites varied from 0.3
to 480 ng/ml. The sites at Phoenix, Arizona, and Ft. Devens, Massachusetts, are both rapid infiltration sites characterized by
sandy soils. At Phoenix, the wastewater receives activated sludge
treatment before discharge into infiltration basins. At Ft. Devens,
the raw sewage is held in Imhoff tanks before being released to
infiltration basins. The site at Lubbock, Texas, is a slow-rate
application system where secondarily treated wastewater is being
applied to cotton crops and pasture lands. The treated wastewater
is held in several large lagoons before land application. As a
result, large growths of algae develop in the wastewater which
could account for the high concentration of endotoxin in the sewage
at this site.

     Generally, endotoxin concentrations were higher in wells
underlying wastewater application sites, but it appears that a 90
to > 99% reduction of endotoxin occurs during the percolation of
wastewater through the soil. Large reductions in gram-negative
bacteria, and undoubtedly, algae occur during infiltration by a
combined process of physical filtration and adsorption (Gerba et
al., 1975; Folkman and Wachs, 1970). The well depths at the Phoenix
site varied from 60 to 100 feet, while those at the Lubbock site
were screened at depths from 90 to 120 feet. The higher levels of
endotoxins in the wells at the Lubbock site than the Arizona site
could reflect differences in the nature of the soil and/or endotoxin.

Table 1. Endotoxin Concentrations Beneath Wastewater Land Application Sites

Site	Sample	Endotoxin (ng/ml)	Depth (m)
Ft. Devens, Massachusetts	Tapwater	0.3	–
	Well No. 4	0.6	3
	Well No. 5	0.3	3
	Well No. 13	30	3
	Sewage	300	–
Lubbock, Texas	Well No. 10	120	27–37
	Well No. 11	480	27–37
	Well No. 27	240	27–37
	Well No. 22	120	27–37
	Well No. 2	480	27–37
	Sewage	6000	–
Phoenix, Arizona	North Well	3	30.5
	Center Well	3	18.3
	Sewage	600	–

Many factors are known to control the migration of bacteria and viruses through the soil (Gerba et al., 1975). These include pH, salt concentration, flow rate, soluble organics (especially humic substances), nature of the soil, nature of the microorganism, etc. It has been shown that many of these same factors also influence the removal of algae through sand (Folkman and Wachs, 1970).

Bacterial and algal removal by soil is due to both filtration and adsorption, whereas virus removal is dependent entirely on adsorption (Gerba et al., 1975; Folkman and Wachs, 1970). Understanding the physical-chemical properties of soil as related to microbial adsorption has been seen as an aid in understanding the potential for groundwater contamination. Virus adsorption to soil surfaces is believed to be largely governed by electrostatic double-layer interactions and van der Waal's forces (Gerba et al., 1975). Bacteria and algae may also possess adhesive surfaces and specialized attachment systems (Corpe, 1980). Thus, the surface charge on both the organism and the soil and factors controlling the net charge are believed to be important in determining the efficiency of microbial adsorption.

Three factors have a large influence on both microorganism survival and their vertical migration through soil. These are climate, the nature of the soil, and the nature of the microorganism. Rainfall has been shown to have a major influence on the subsurface migration of virus, bacteria, and even algae (Gerba et al., 1975; Folkman and Wachs, 1970; Lance et al., 1976; Hagedorn et al., 1978). Rainfall acts to lower the ionic concentration in the soil, resulting in the apparent desorption of microorganisms and their subsequent further travel through the subsurface. The organisms appear to move in a burst or wave as the ionic conductivity of the soil water is lowered (Duboise et al., 1976). The maximal level of virus elution occurs immediately before the minimal conductivity is reached (Duboise et al., 1976). The concentration of viruses or bacteria in the subsurface during these events is much greater than during the normal application of wastewater. Such events have been observed in both laboratory column studies (Lance et al., 1976; Duboise et al., 1976) and under field conditions (Hagedorn et al., 1978; Wellings et al., 1974). A marked increase in the total organic carbon content occurs during rainfall events in an almost identical pattern to the virus (Duboise et al., 1976). A marked increase in turbidity and coloration of the leachate also coincides approximately with the release of virus. The importance of these events is that abnormally high concentrations of pathogenic microorganisms or possibly toxic substances could occur in the groundwater for limited periods of time, increasing the potential for a public health hazard which would not be detected if sampling of the water had not coincided with the time of the greatest hazard.

Microorganisms released from near the soil surface during a rainfall event will eventually be readsorbed at greater depths (Lance et al., 1976) but the possibility exists that they could again desorb in response to another rainfall event (Duboise et al., 1976) and move through the soil in a chromatographic fashion. In the case of viruses the degree of elution which occurs is also dependent on both the type and specific strain of virus (Landry et al., 1979). This difference could be due to differences in the configuration of proteins in the viral capsid, resulting in variations in the net charge on the virus. Thus, the effects of rainfall on the migration of organisms and their products might be expected to vary greatly.

Finally, the nature of the soil also influences the migration of organisms and their products. Soils with a high clay content, large soil surface area, and high cation capacity have all been shown to favor organism removal (Gerba et al., 1975). The amount of organic matter and soil pH also affect the degree of removal.

We have recently conducted a series of laboratory experiments to determine the degree of endotoxin removal from secondarily

Table 2.   Endotoxin Concentration at Various Depths of 250-cm Soil
           Column Flooded with Sewage[a]

Column Depth (cm)	Endotoxin Concentration (ng/ml)		
	Expt. 1	Expt. 2	Expt. 3
0	60	300	300
2	60	300	300
5	60	300	300
10	60	300	300
20	ND[b]	300	300
40	6	300	300
80	6	300	ND
160	6	ND	ND
240	3	ND	ND
250	3	30	30

[a]Reproduced courtesty of the American Society for Microbiology
(Appl. Environ. Microbiol. *39*:544-547, 1980).

[b]ND = Not done

treated sewage (activated sludge) and the effect of rainfall on
the mobilization of previously adsorbed endotoxin (Goyal et al.,
1980).   Soil columns 100 or 250 cm long were packed with loamy
sandy soil from a wastewater rapid infiltration site near Phoenix,
Arizona.   Identically designed laboratory columns have been shown
to be good models for field studies involving the removal of various pollutants (Lance et al., 1976; Lance and Gerba, 1977).   During
these studies, it was found that 90 to 99% of the endotoxin was
removed after travel of the sewage through 100 to 250 cm of the
loamy sand soil (Tables 2 and 3).   When distilled water was allowed
to infiltrate into the soil to simulate rainfall, the endotoxin was
mobilized and moved in a concentrated band through the soil column
(Figure 1).   It thus seems that endotoxin can move through soil in
a manner similar to that observed for bacteria and viruses.

     In addition to land application systems, other sources can
contribute to algae and algal toxin contamination of groundwater

Table 3. Endotoxin Concentration at Various Depths in a 100-cm
          Soil Column[a]

Column Depth (cm)	Endotoxin Concentration (ng/ml)	
	Expt. 1	Expt. 2
0	3000	3000
10	3000	2400
15	2400	2400
25	2400	2400
35	1200	1200
45	600	1200
55	600	600
75	300	600
95	30	300

[a]Reproduced courtesy of the American Society for Microbiology
(Appl. Environ. Microbiol. *39*:544-547, 1980).

sources. Contamination of groundwater can occur from infiltration
near rivers, lakes, ponds, etc. Sewage stabilization ponds usually
contain high concentrations of algae. Leaks often develop from
such ponds which allow for contamination of groundwater. As an
example, a recent large outbreak of gastroenteritis was found to
be associated with a wastewater leak from a municipal wastewater
oxidation lagoon (C. D. C., 1978). The terrain surrounding the
lagoon was hilly and composed of porous limestone with numerous
subterranean streams, permitting rapid movement of the groundwater. A number of wells in the surrounding area were found to be
contaminated by the leak from the lagoon. The fact that stool
cultures were negative when tested for bacterial enteric pathogens
suggests a possible viral or other etiological agent.

    If endotoxin does reach the groundwater, is there any need
for concern? Landy and Braun (1964) have shown that adsorption of
minute quantities of endotoxin from the gastrointestinal tract of
a normal human may not pose a health threat. Wolf and Jorgensen
(1979) found no pyrogenic activity when rabbits were inoculated
with drinking water containing 1.2 to 25 ng/ml. However, when
they were inoculated with effluents from advanced wastewater

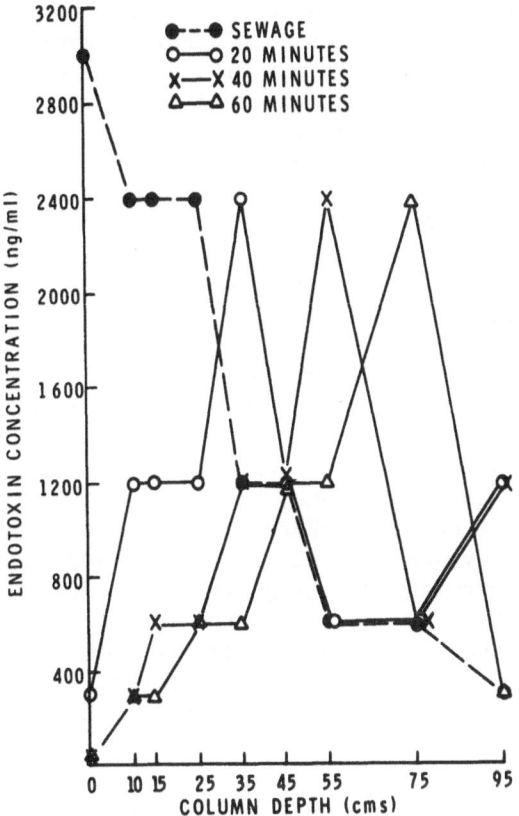

Fig. 1. Movement of endotoxin through the soil column following
         simulated rainfall. Note the movement of endotoxin in
         the form of a concentrated band as a function of time.
         Reproduced courtesy of the American Society for Micro-
         biology (Appl. Environ. Microbiol. *39*:544-547, 1980).

treatment plants containing 6 to 250 ng/ml free endotoxin, a pyrogenic response was seen. These authors further reported that
ingestion of a wastewater effluent containing 3000 ng/ml endotoxin
resulted in an elevated temperature in 5 of 8 rabbits. The
increases were, however, too small to satisfy the requirements of
the United States Pharmacopeia test for pyrogenicity. On the other
hand, if absorption increases as a result of increased permeability
of the gastrointestinal tract or if there is impairment of the
normal detoxification mechanism for endotoxin, detrimental effects
could result (Di Luzio and Friedmann, 1973). Increased sensitivity
to endotoxin following blockage of the reticuloendothelial system
or altered adrenal status has also been reported (Di Luzio and
Friedmann, 1973). Also, humans can be 100-fold more sensitive to
the effects of endotoxin than rabbits (Greisman et al., 1964).

In land application systems the effect of endotoxin on animals grazing on sewage-laden pastures should not be overlooked. It has been postulated that endotoxins may play a role in the pathogenesis of certain diet-induced diseases in ruminants, e.g., lactic acidosis, bloat and sudden death syndrome (Nagaraja et al., 1978). In pigs, absorption of *Escherichia coli* endotoxin from the small intestine has been demonstrated (Nagaraja et al., 1978). That endotoxin can be absorbed by damaged and inflamed epithelium has been shown repeatedly. Once absorbed, endotoxin may produce endotoxic or ana-phylactic shock either alone or in combination with specific anti-bodies or complement.

Another important consideration is the inhalation of endotoxin aerosols during spray irrigation of wastewater directly or after land treatment. It has been postulated that inhalation exposure to endotoxin from Gram-negative bacilli is an important factor in occupational diseases like asthma, bagassosis, byssinosis, mill fever, and detergent enzyme workers' asthma (Hudson et al., 1977). In a recent study, Rylander et al. (1976) observed attacks of high fever and a pustulant discharge from the eyes in 50% of sewage workers. About 20% were reported to have efflorescenses of skin. These symptoms were attributed to the presence in sewage sludge of a variety of antigenic materials of microbiological origin, particularly endotoxins. No such symptoms were seen in matched controls.

The data presented in this review suggest the potential for contamination of groundwater with algal endotoxin. However, the importance of these observations needs to be ascertained.

ACKNOWLEDGMENTS

This work was supported by the Environmental Protection Agency through research project R-805,292 and the National Center for Groundwater Research at Rice University, Houston, Texas.

REFERENCES

Associates of Cape Cod, Inc. 1979. *Limulus* amoebocyte lysate
        (Pyrotell) for the detection and quantitation of endotoxins.
        Woods Hole, Mass. 8 pp.
Center for Disease Control. 1978. Morbidity and Mortality Weekly
        Report, Vol. 27, No. 183.
Corpe, W. A. 1980. Microbial surface components involved in
        adsorption of microorganisms onto surfaces. Pages 105-144
        in G. Bitton and K. C. Marshall, eds., Adsorption of Micro-
        organisms to Surfaces. Wiley-Interscience, New York.

Craun, G. F. 1979. Waterborne disease -- a status report empha-
    sizing outbreaks in ground-water systems. Ground Water *17*:
    183-191.
Di Luzio, N. R., and T. J. Friedmann. 1973. Bacterial endotoxins
    in the environment. Nature *224*:49-51.
Duboise, S. M., B. E. Moore, and B. P. Sagik. 1976. Poliovirus
    survival and movement in a sandy forest soil. Appl.
    Environ. Microbiol. *31*:536-542.
Environmental Protection Agency. 1977. Office of Technology
    Transfer, Cincinnati, Ohio.
Folkman, Y., and A. M. Wachs. 1970. Filtration of *Chlorella*
    through dune-sand. J. Sanit. Eng. Div. Am. Soc. Civ. Eng.
    96:675-689.
Freeze, R. A., and J. A. Cherry. 1979. Groundwater. Prentice-
    Hall, Englewood Cliffs, New Jersey. 604 pp.
Gerba, C. P., C. Wallis, and J. L. Melnick. 1975. Fate of waste-
    water bacteria and viruses in soil. J. Irrig. Drain. Div.,
    Am. Soc. Civ. Eng. *181*:157-174.
Gorham, P. R., and W. W. Carmichael. 1979. Phycotoxins from blue-
    green algae. Pure Appl. Chem. *52*:165-174.
Goyal, S. M., C. P. Gerba, and J. C. Lance. 1980. Movement of
    endotoxin through soil columns. Appl. Environ. Microbiol.
    39:544-547.
Greisman, S. E., H. N. Wagner, Jr., M. Iio, R. B. Hornick, F. A.
    Carozza, Jr., and T. E. Woodward. 1964. Mechanisms of
    endotoxin tolerance in man. Pages 567-574 *in* M. Landy and
    W. Braun, eds., Bacterial Endotoxins. Quinn and Boden Co.,
    Inc., Rahway, New Jersey.
Hagedorn, C., D. T. Hansen, and G. H. Simonson. 1978. Survival
    and movement of fecal indicator bacteria in soil under
    conditions of saturated flow. J. Environ. Qual. *7*:55-59.
Hindman, S. H., M. S. Favero, L. A. Carson, N. J. Petersen, L. B.
    Schonberger, and J. T. Solano. 1975. Pyrogenic reactions
    during haemodialysis caused by extramural endotoxin.
    Lancet *2*:732-734.
Hudson, A. R., K. H. Kilburn, G. M. Halprin, and W. N. McKenzie.
    1977. Granulocyte recruitment to airways exposed to endo-
    toxin aerosols. Am. Rev. Resp. Dis. *115*:89-95.
Jorgensen, J. H., J. C. Lee, and H. R. Pahren. 1976. Rapid
    detection of bacterial endotoxins in drinking water and
    renovated wastewater. Appl. Environ. Microbiol. *32*:347-351.
Keleti, G., J. L. Sykora, E. C. Lippy, and M. A. Shapiro. 1979.
    Algae and occurrence of LPS (lipopolysaccharide) in drinking
    water. Page 231 in Abs. Ann. Mtg. Am. Soc. Microbiol.
Lance, J. C., and C. P. Gerba. 1977. Nitrogen, phosphate and
    virus removal from sewage water during land filtration.
    Prog. Water Technol. *9*:157-166.
Lance, J. C., C. P. Gerba, and J. L. Melnick. 1976. Virus move-
    ment in soil columns flooded with secondary sewage effluent.
    Appl. Environ. Microbiol. *32*:520-526.

Landry, E. F., J. M. Vaughn, M. Z. Thomas, and C. A. Bechwith. 1979. Adsorption of enteroviruses to soil cores and their subsequent elution by artificial rainfall. Appl. Environ. Microbiol. *38*:680-687.

Landy, M., and W. Braun, eds. 1964. Bacterial Endotoxins. Quinn and Boden Co., Inc., Rahway, New Jersey. 431 pp.

Lippy, E. C., and J. Erb. 1976. Gastrointestinal illness at Sewickley, Pa. J. Am. Water Works Assoc. *68*:606-610.

Moore, R. E. 1977. Toxins from blue-green algae. Bioscience *27*(12):797-802.

Nagaraja, T. G., E. E. Bartley, L. R. Fina, H. D. Anthony, S. M. Dennis, and R. M. Bechtle. 1978. Quantitation of endotoxin in cell-free rumen fluid of cattle. J. Anim. Sci. *46*:1759-1767.

Rylander, R., K. Anderson, L. Belin, G. Berglund, R. Bergstrom, L. A. Hanson, M. Lundholm, and I. Mattsby. 1976. Sewage workers' syndrome. Lancet *2*:478-479.

Schwimmer, D., and M. Schwimmer. 1964. Algae and medicine. Pages 368-412 *in* D. F. Jackson, ed., Algae and Man. Plenum Press, New York.

Wellings, F. M., A. L. Lewis, and C. W. Mountain. 1974. Virus survival following wastewater spray irrigation of sandy soils. Pages 253-260 *in* J. F. Malina, Jr. and B. P. Sagik, eds., Virus Survival in Water and Wastewater Systems. University of Texas Press, Austin.

Wolf, H. W., and J. H. Jorgensen. 1979. Pyrogenic activity of carbon-filtered waters. EPA-600/1-79-009, U. S. Environmental Protection Agency, Cincinnati, Ohio.

TOXICITY STUDIES WITH BLUE-GREEN ALGAE

FROM NORWEGIAN INLAND WATERS

Øyvin Østensvik, Olaf M. Skulberg,[1]
and Nils E. Søli[2]

Department of Food Hygiene
The Veterinary College of Norway
Oslo, Norway

ABSTRACT

Extracts from the 1978 and 1979 water bloom from Lake Frøylandsvatn, both of which were dominated by *Microcystis aeruginosa* Kütz., and a laboratory clone culture of *Microcystis aeruginosa* NIVA CYA 57 isolated from Lake Frøylandsvatn 1978 were toxic to mice. Freeze-dried algal material from the 1978 water bloom and the laboratory culture both have MLD_{100} of about 40 mg/kg. Extracts from the 1979 water bloom were less toxic, with MLD_{100} of about 200 mg/kg. Extracts from the laboratory cultures of a red colored clone of *Oscillatoria agardhii* Gom Var., NIVA CYA 18 isolated from Lake Gjersjøen 1971, was toxic to mice. The extract had a MLD_{100} of about 400 mg/kg. Water bloom from Gjersjøen dominated by the red colored *Oscillatoria agardhii* var., sampled April 1980, killed mice in a similar manner, and these algal extracts had a MLD_{100} of 200 mg/kg. Extracts from water blooms dominated by *Microcystis aeruginosa* Kütz. were administered either intravenously or intraperitoneally to rats. The most striking effects were decreased blood pressure, increased liver weight and increased plasma levels of the liver enzyme aspartateaminotransferase (ASAT).

---

[1]Address: Norwegian Institute for Water Research, Oslo, Norway.

[2]Address: Department of Pharmacology and Toxicology, The Veterinary College of Norway, Oslo, Norway.

INTRODUCTION

Human activities are causing extensive and striking fertility changes in Norwegian inland waters which were originally oligotrophic. The distribution of the human population and the waste disposal together with the modern agricultural practices are causing a rapid eutrophication (Baalsrud, 1975). Among the several undesirable consequences are dramatic increases in algal biomass as well as changes in the dominating algae species (Skulberg, 1964, 1968). Consequently the basis of the normal food chain in lakes and rivers is upset. This unbalance leads to obnoxious effects (e.g. unsatisfactory tastes and odors) on water used for drinking water. The rotting of algal masses is followed by a depletion of oxygen in stagnant waters. Water blooms of blue-green algae with toxic properties have caused death in cattle (Skulberg, 1979b).

Some reports on mortalities of fish, birds and mammals in Norway during the 1970's concluded that toxins from blue-green algae could have been the source of poisoning. However, follow up investigations were not carried out to confirm the diagnosis.

In August 1978, four heifers were found dead. These cattle drank from a lake which had a massive population of *Microcystis aeruginosa* (Skulberg, 1979b). Post-mortem examinations showed small hemorrhages in many organs. The livers were enlarged and dark, with hemorrhages and massive necrosis of the liver parenchyma. Other cattle in the same herd showed symptoms of edema in the ears and perianal region. Two to three weeks later some of these showed photosensitization.

This paper deals with toxicological and pathological studies with extracts from freeze-dried algal material. Assays for toxicity were performed on water blooms with blue-green algae, as well as with laboratory clone cultures of blue-green algae isolated in Norway. The strains isolated are kept in the culture collection at the Norwegian Institute for Water Research (Skulberg, 1979a).

RESULTS

Toxicity of the Tested Extracts

From freeze-dried algal material, extracts were made in sterile 0.9% NaCl, equivalent to 50 mg algal material per milliliter.

The results of the toxicity tests using the extracts from blue-green algae isolated in Norway are given in Table. 1. Water blooms from two localities, Lake Frøylandsvatn and Lake Gjersjøen,

Table 1. Acute Toxicity of Extracts from Blue-green Algae

Type of Extract	MLD_{100} (Approximate)[a]
Water Blooms	
Lake Frøylandsvatn, August 5, 1978.	40 mg/kg
Lake Frøylandsvatn, August 14, 1979.	200 mg/kg
Lake Gjersjøen, April 19, 1980.	200 mg/kg
Cultures	
Microcystis aeruginosa Kutz. NIVA CYA 57 isolated from Lake Frøylandsvatn, August 5, 1978.	40 mg/kg
Oscillatoria agardhii Gom. NIVA CYA 18 isolated from Lake Gjersjøen, 1971.	400 mg/kg

[a]The toxicity of the extracts were determined by intraperitoneal administration to mice, and expressed as minimal lethal dose (MLD_{100}) in mg/kg.

have shown toxic effects in mice. Samples from the water blooms in Lake Frøylandsvatn were taken at the time that deaths occurred in the cattle which used this lake as drinking water. During 1978 and 1979 water blooms showed toxic effects on mice after intraperitoneal injection of extracts. The MLD_{100} ranged from 40 to 200 mg/kg.

The water blooms in Lake Frøylandsvatn during 1978 and 1979 were dominated by *Microcystis aeruginosa* Kütz. A clone of *Microcystis aeruginosa* NIVA CYA 57, isolated from the 1978 water bloom caused toxic effects in mice.

Lake Gjersjøen has an annual water bloom dominated by a red colored variety of the alga *Oscillatoria agardhii* var. Gom. (Skulberg, 1978). During winter this alga maintained a prominent overwintering population under the ice. The sample from April 19, 1980, was taken at the time of the ice breakup. Surface water contained high concentrations of *Oscillatoria agardhii* var. This alga was virtually present as a monoculture. Extracts from algal

material killed mice, and the MLD_{100} was about 200 mg/kg. A clone
of the red colored *Oscillatoria agardhii* var. NIVA CYA 18, isolated
from Lake Gjersjøen 1971, also showed toxic effects, with MLD_{100}
about 400 mg/kg.

Water blooms with *Oscillatoria agardhii* in Lake Årungen and
Anabaena flos-aquae in Lake Bakkavatn, did not show toxic effects
in mice following intraperitoneal administration.

Acute Toxicity in Mice

Water blooms and algal cultures from Lake Frøylandsvatn. The
acute toxicity tests showed that extracts from 1978 and 1979 water
blooms both of which were dominated by *Microcystis aeruginosa* Kütz.,
and a laboratory clone culture of *Microcystis aeruginosa* NIVA CYA 57
isolated from Lake Frøylandsvatn August 5, 1978, were toxic to mice.
Freeze-dried algal material from the 1978 water bloom and the laboratory culture both had a MLD_{100} of about 40 mg/kg. Extracts from
the 1979 water bloom were less toxic, with a MLD_{100} of about 200
mg/kg.

Following the intraperitoneal administration of a lethal dose,
there was a latent period of about 30 minutes. The main symptoms
which initially developed were uncoordination and paralysis of the
hind limbs, lethargy and pallor of the ears and tail. The lethargy
alternated with short periods of tremor and mild convulsions. Subsequently the animals became weak and respiration became labored
and infrequent. Deaths occurred in about 60 minutes after administration.

Post-mortem examinations revealed a pale carcass, enlarged
dark red liver, the surface of which was mottled. The kidneys and
lungs were pale. Histological examinations were carried out with
liver, lungs, myocardium and kidneys. Histological changes were
most pronounced in the livers. The livers were engorged with
blood, and there were wide-spread hemorrhages. The liver parenchyma showed degenerations and dissociation of the liver cords.
Some hepatocytes were necrotic. The blood vessels contained necrotic, eosinophilic masses. The kidneys showed early nephrotic
changes with slight eosinophilic masses within the lumina of the
tubules. The myocardium showed slight degenerative changes with
indistinct cross-striation and granular sarcoplasm. The lungs
revealed focal hemorrhages, and in some areas the tissue was
atelectatic.

Water blooms and algal cultures from Lake Gjersjøen. Extracts
from the laboratory culture of a red colored clone of *Oscillatoria
agardhii* Gom. var., NIVA CYA 18, isolated from Lake Gjersjøen 1971
was toxic to mice. The extracts had a MLD_{100} of about 400 mg/kg.

The clone was used for repeated culturing in the laboratory, and new extracts were prepared. The latter revealed a similar toxicity, and had a MLD_{100} of about 400 mg/kg. Recently, we have found that a water bloom dominated by the red colored *Oscillatoria agardhii* var., sampled April 19, 1980 killed mice in a similar manner, and had a MLD_{100} of 200 mg/kg.

The symptoms which developed in mice after the injection of a lethal dose i.p. mimicked the symptoms produced by extracts of *Microcystis*. There was a latent period of about 30 minutes after which the animals showed an uncoordination and paralysis of the hind limbs. Lethargy and pallor of the ears and tail were present. Periods of tremor and mild convulsions seemed to occur more frequently with extracts from *Oscillatoria* compared with extracts from *Microcystis*.

Later the mice weakened and respiration became labored and infrequent. Deaths occurred in 60 to 120 minutes.

Post-mortem examinations have not been carried out in detail. The carcasses were pale and the livers were swollen with a dark red mottled surface. The hepatic tissue was soft and partly liquefied, and it was difficult to remove the livers for weighing them without causing a rupture. The liver weights accounted for about 9% of the body weight. No macroscopically visible changes were observed in other organs.

Acute Toxicity in Rats

Determinations of MLD_{100}. Intraperitoneal injections of extracts of algal material from the water bloom in Lake Frøylandsvatn, August 5, 1978, were toxic to rats. The extracts had a MLD_{100} of about 40 mg/kg. The symptoms that developed after the administration of a lethal dose were similar to those observed in mice. However, the latent periods were somewhat longer, about 45 minutes, and deaths occurred within 3 to 4 hours.

Post-mortem examinations showed the same changes as in mice. The livers were swollen with dark red mottled surface. Histolological examinations were not carried out.

Effect on blood pressure. The effect of the extract from the water bloom dominated by *Microcystis aeruginosa* Kütz. on blood pressure, was studied after intravenous administration to rats (200 to 250 g). The results are given in Figures 1 and 3. One milliliter of the extract was slowly injected intravenously for about four minutes. During the first 1 1/2 minutes the blood pressure lowered markedly. The blood pressure began increasing before

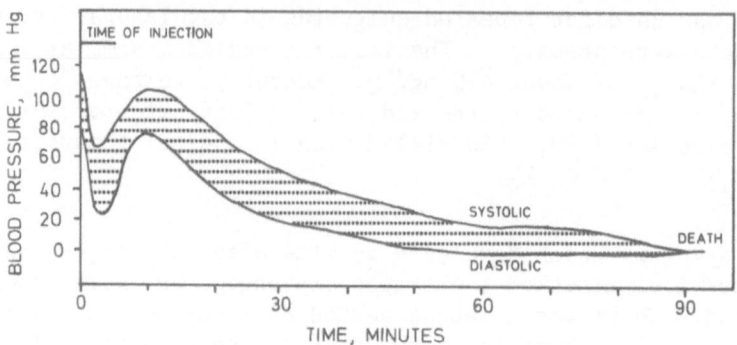

Fig. 1. Systolic and diastolic blood pressure decrease in the
         rat in response to an intravenous injection of 1 ml
         extract from the waterbloom dominated by *Microcystis
         aeruginosa* Kütz. equivalent to 50 mg freeze-dried algal
         material (approximately 200 mg/kg body weight).

the injection was finished and increased again to normal levels
about 10 minutes after the injection had started. During the next
40 minutes the blood pressure decreased slowly to a low level.
This low blood pressure remained almost constant, until the rats
died about 90 minutes after the injection.

    Liver weights and liver enzymes in plasma. In one experiment
18 rats were given i.p. 50 mg/100 g body weight of the extract from
the waterbloom dominated by *Microcystis aeruginosa* Kütz. Two rats
were killed at 15, 30, 45, 60, 90, 120, 180, and 240 minutes after
the administration. In cases where the rats showed symptoms indicating an impending death, the rats were killed and the time after
the administration was registered. The animals were anesthetized
with ether and decapitated. Blood was collected in glass tubes,
the tubes were centrifuged and the plasma concentrations of
aspartateaminotransferase (ASAT), alaninaminotransferase (ALAT)
and bilirubin were determined. The livers were rinsed in 0.9%
NaCl and blotted dry on filter paper. Relative liver weights were
expressed as percent of the total body weight.

    About 30 minutes after administration the livers showed in-
creasing weights (Figure 2 and Figure 3). The mean relative liver
weight of eight control rats was 3.7%. Rats killed immediately
and 15 minutes after the administration showed no increase in liver
weights but between 30 and 90 minutes post-injection the relative
liver weights increased from 3.7 to more than 6% of body weight.
At that time the plasma ASAT began increasing (Figure 2). Beyond
90 minutes post-injection no further gains in liver weights were
observed but the continually increasing amounts of plasma ASAT

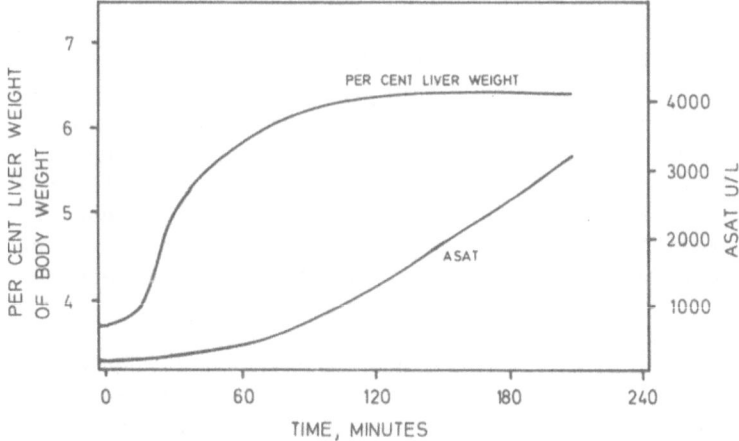

Fig. 2. The changes in relative liver weights and plasma ASAT
(aspartateaminotransferase) in rats following intraperitoneal injections of 1 ml extract/100 g body weight
equivalent to 50 mg freeze-dried algal material/100 g.
Extract from the waterbloom dominated by *Microcystis
aeruginosa* Kütz. was administered.

indicated increasing liver damage. Neither ALAT nor bilirubin in
the plasma showed significant increases in concentration.

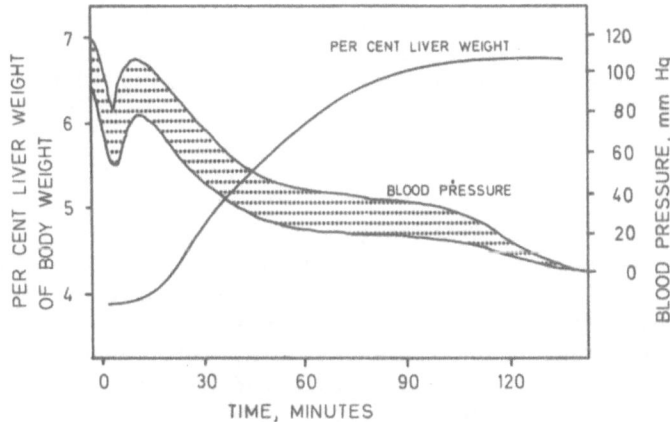

Fig. 3. Changes in blood pressure following an intravenous injection of extract from the waterbloom dominated by *Microcystis aeruginosa* Kütz. together with changes in relative
liver weights following intraperitoneal injections.

Effect on isolated rat liver. A simple technique of isolated
rat liver perfusion was used to test the direct effect of toxin on
the liver. Three ml extract from the waterbloom dominated by
Microcystis aeruginosa Kütz. (equivalent to 150 mg dried algal
material) were injected into the perfusate (1/3 rat blood + 2/3
albumin containing perfusion buffer). Preliminary studies show
that the perfusate flow decreased to half during the first 15 min.
The liver weight doubled during an hour of perfusion due to accumulation of perfusate. The bile flow was relatively normal during
the first minutes then decreased to about zero after about 30
minutes.

Effect on isolated rat heart. Extracts of freeze-dried algal
material in Kreb's solution were tested on isolated rat hearts
perfused with Kreb's solution. Extracts from the waterbloom dominated by *Microcystis aeruginosa* Kütz. (equivalent to 20 mg algal
material, 0.4 ml) was injected during 2 minutes. Significant negative chronotropic effects were observed after about 1 minute of
injection. In addition the extracts seemed to show moderate negative inotropic effects. These effects on the heart lasted for
about 2 minutes after the end of injection.

Effect on isolated phrenic nerve diaphragm. The extract from
the waterbloom dominated by *Microcystis aeruginosa* Kütz. did not
result in effects on the isolated phrenic nerve diaphragm preparation of the rat.

Testing for Mutagenicity

The extract from the waterbloom in Lake Frøylandsvatn, August
5, 1978, showed no mutagenic effects in Ames *Salmonella*/mammalian
hepatic microsomes mutagenicity test (Ames et al., 1975).

DISCUSSION

Extracts of freeze-dried algal material originating from
waterblooms and cultures of blue-green algae isolated in Norway
showed toxic effects on mice and rats. The samples of algal material from the waterbloom in Lake Frøylandsvatn, August 5, 1978,
were taken at a time when cattle that used this lake as drinking
water, died. The waterbloom was dominated by the blue-green alga
Microcystis aeruginosa Kütz., and this alga was so dominant that
it nearly resembled a monoculture.

The symptoms which developed in mice receiving a lethal dose
were almost identical with other reports on toxic effects of *Micro-*

cystis aeruginosa (Gorham, 1960; Konst et al., 1965; Heaney, 1971). The latent period and death after about 60 minutes post-injection indicate the presence of microcystin, or FDF, in the extracts. It should be noted that both the 1978 and the 1979 waterbloom, as well as the laboratory culture of *Microcystis aeruginosa* isolated from the 1978 waterbloom, showed identical toxic effects on mice. The post-mortem examinations of mice killed with extract from the 1978 waterbloom revealed the same changes as described in the literature (Ashworth and Mason, 1946; Konst et al., 1965; Heaney, 1971). The main hepatic changes included varying degrees of degeneration and necrosis in the parenchyma. The liver was engorged with blood and revealed hemorrhages. These findings suggest that the toxic principle present in our extract have hepatotoxic properties. Gorham and Carmichael (1979) suggest that the toxic principle present in extracts of *Microcystis* can act either as a hepatotoxic agent or as a neurotoxic agent.

We did some preliminary experiments with rats in order to throw light upon the possible mechanism of toxicity of extract from the waterbloom dominated by *Microcystis aeruginosa* Kütz. The most striking effects were on blood pressure, liver weights and liver enzymes in the plasma. The first transient fall in blood pressure following intravenous injection of extracts could have been caused by a transient direct effect on the heart. The second and permanent fall in blood pressure seems to have been caused by shunting of blood to the liver.

The plasma levels of the liver-enzyme ASAT which began to increase about 60 minutes post-injection indicate a liver cell damage. The effect on the liver could have been explained in two ways. There could be a direct effect of the toxic agent to the liver or it could be caused by effects on the circulatory system. The results from the isolated liver test indicate that the toxic agent acts as a direct hepatotoxic agent, since the hemostatic regulatory mechanisms were eliminated in this preparation.

Our extracts did not contain toxin that effected the neuro-muscular system.

The red colored *Oscillatoria agardii* Gom. var., NIVA CYA 18, isolated from Lake Gjersjøen and the extracts of algal material from the waterbloom April 19, 1980, dominated by the red colored *Oscillatoria*, produced nearly identical symptoms on mice that are described for extracts of *Microcystis aeruginosa*. This indicates that the toxic principle produced by *Oscillatoria* in this case may be closely related to the toxic principle produced by *Microcystis*. Further studies will be carried out to confirm the results presented here.

REFERENCES

Ames, B. N., J. McCann, and E. Yamasaki. 1975. Methods for
    detecting carcinogens and mutagens with *Salmonella*/mammalian
    - microsome mutagenicity test. Mutat. Res. *31*:347-364.
Ashworth, C. T., and M. F. Mason. 1946. Observations on the patho-
    logical changes produced by a toxic substance present in
    blue-green algae (*Microcystis aeruginosa*). Am. J. Pathol.
    22:369-383.
Baalsrud, K. 1975. Oversikt over eutrofieringsproblemer (Survey
    of eutrophication problems). Scandinavian Symposium of
    Water Research. Nordforsk, Helsingfors. *10*:5-9.
Gorham, P. R. 1960. Toxic waterblooms of blue-green algae. Can.
    Vet. J. *1*:235-245.
Gorham, P. R., and W. W. Carmichael. 1979. Phycotoxins from blue-
    green algae. Pure Appl. Chem. *52*(1):165-174.
Heaney, S. I. 1971. The toxicity of *Microcystis aeruginosa* Kütz.
    from some English reservoirs. Water Treat. Exam. *20*:235-244.
Konst, H., P. D. McKercher, P. R. Gorham, A. Robertson, and J.
    Howell. 1965. Symptoms and pathology produced by toxic
    Microcystis aeruginosa NRC-1 in laboratory and domestic
    animals. Can. J. Comp. Med. Vet. Sci. *29*:221-228.
Skulberg, O. M. 1964. Algal problems related to the eutrophica-
    tion of European water supplies. Pages 262-299 *in* D. F.
    Jackson, ed., Algae and Man. Plenum Press, New York.
Skulberg, O. M. 1968. Studies on eutrophication of some Norwegian
    inland waters. Mitt. Int. Ver. Limnol. *14*:187-200.
Skulberg, O. M. 1978. Some observations on red-coloured species
    of *Oscillatoria* (Cyanphyceae) in nutrient-enriched lakes of
    southern Norway. Societas Internationalis Limnologiae, XX
    Congress, Copenhagen August 7-17, 1977. Verh. Int. Ver.
    Limnol. *20*:776-787.
Skulberg, O. M. 1979a. NIVA's kultursamling av alger (Culture
    collection of algae at Norwegian Institute for Water
    Research [NIVA]). Norsk Institutt for Vannforsknings Årbok
    1978, Olso. 67-71.
Skulberg, O. M. 1979b. Toxic effects of blue-green algae, first
    case of *Microcystis* - poisoning reported from Norway.
    Temarapport No. 4, Norwegian Institute for Water Research
    (NIVA), Oslo.

ISOLATION, CHARACTERIZATION AND PATHOLOGY OF THE TOXIN FROM

THE BLUE-GREEN ALGA *MICROCYSTIS AERUGINOSA*

M. T. Runnegar and I. R. Falconer

Department of Biochemistry and Nutrition
University of New England
Armidale, New South Wales, 2351
Australia

ABSTRACT

A toxic peptide was isolated from a bloom of *Microcystis aeruginosa* by alkaline extraction of lyophilized algae. It was purified 200-fold by a procedure involving ammonium sulphate fractionation, solvent extraction, acid precipitation, ion exchange chromatography with DEAE-Sephadex, gel filtration with Sephadex G-25 and high voltage electrophoresis at pH 6.5. The peptide had no free amino group and on hydrolysis gave equimolar amounts of L-methionine, L-tyrosine, D-alanine, D-glutamic acid, erythro B-methyl aspartic acid and methylamine. The LD_{50} of the purified toxin was 0.056 mg/kg of mice. Inoculation of mice with toxin (> LD_{100}) caused death within one to three hours. Histologically, changes in liver were observed at 15 minutes after injection; by 30 to 60 minutes there was massive pooling of blood in the liver. Scanning and transmission electron microscope studies showed that normal sinusoids had almost disappeared, the overall structure of the tissue was difficult to recognize, and numerous necrotic hepatocytes were also observed. These changes were accompanied by a large increase in glutamate oxaloacetate transaminase and lactate dehydrogenase activities in the serum. The peptide toxin was shown to have no effect on protein, RNA or DNA synthesis as measured by the incorporation of radioactive substrates by liver slices; it also had no effect on oxygen consumption by liver slices. The results from electron microscopy combined with the mainly negative biochemical findings point to an effect of the toxin on liver cell membranes.

INTRODUCTION

It is now just over 100 years since publication in Nature of
Francis' report of the occurrence of a toxic bloom consisting of
blue-green algae in Lake Alexandrina, South Australia. His was the
first paper to describe the scientific investigation of such an
occurrence (Francis, 1878). His description cannot be faulted:

> *"The bloom being very light it floats on the water.*
> *Thus floating it is wafted to the lee shores forming*
> *a thick scum like green oil paint: some two to six*
> *inches thick, and as thick and pasty as porridge, it*
> *is swallowed by cattle when drinking. This acts*
> *poisonously and rapidly causes death."*

Toxic algal blooms are still occurring and at the University
of New England in New South Wales, Australia, we have been involved
since 1973 in an investigation of various properties of toxic
blooms consisting of the blue-green alga *Microcystis aeruginosa*.
The occurrence of such blooms is widespread in Australia. Between
March and May 1980, we investigated almost a dozen occurrences of
Microcystis blooms within 50 kilometers of the University of New
England. Although these blooms occurred within a short period with
little climatic change, and in a relatively homogenous geographical
area, we found that there was a large variation between these different samples in toxicity to mice. The approximate LD_{100} varied
from about 10 mg to over 200 mg of dry weight of bloom killing the
equivalent of 1 kg of mice when injected intraperitoneally. Only
one bloom consisting of *Microcystis* was found to be non-toxic.
Blooms that were most toxic when tested with mice (i.e. had the
lowest LD_{100}) were also responsible for the most extensive stock
losses.

The bloom of *Microcystis aeruginosa* that we used for isolation,
purification, and characterization of the toxin occurred in 1973 on
Malpas Dam (Figure 1). This reservoir is situated approximately
35 km northeast of Armidale on the Northern Tablelands of New South
Wales, Australia. It is used as the main water supply for the town
of Armidale and has a storage capacity of 13,000 megaliters. The
catchment area for the dam is hilly country of basaltic origin and
is used for pastoral activities; aerial topdressing with super-
phosphate is carried out routinely by most landowners in the area.
Since 1972 unpleasant odors and tastes in the town water of Armidale (which is purified conventionally in a water treatment plant)
have been traced back to blooms of *Microcystis aeruginosa* in the
reservoir. One such bloom that occurred in October 1973 was found
to be toxic, when extracts obtained by the sonic disruption of the
algal cells were injected into mice. The mice died between one to
two hours after injection. We report here on our subsequent work

Fig. 1. Malpas Dam, northern New South Wales, Australia.

in purifying and characterizing the toxin, and in determining the pathological effects of the toxin isolated from that bloom of *Microcystis aeruginosa*.

TEXT

Purification of the Toxin from *Microcystis aeruginosa*

In 1973 the most recently published method for the purification of *Microcystis aeruginosa* toxin was that of Murthy and Capindale (1970). The method consisted of the extraction of 100 g of freeze-dried algal cells by a bicarbonate/carbonate solution. This step was followed by butanol extraction, dialysis and DEAE Sephadex chromatography in 0.1 M to 0.2 M NH_4HCO_3. This method was stated to give about 100 mg purified toxin per 100 g of cells and a quantitative yield of 100% of all toxicity present in the algal cell extract, although the purification procedure including the chromatography step had to be repeated ten times. The purified toxin was found to be a peptide with a LD_{100} of 0.1 mg/kg of mice.

We attempted to reproduce Murthy and Capindale's procedure and we found that we could not recover toxicity. We therefore

Table 1. Purification of the Toxin from a Bloom of *Microcystis*
          aeruginosa

Purification Step	Total Toxicity (kg of mice)	Recovery %
$Na_2CO_3/NaHCO_3$ Extract	970	100
0-40% $(NH_4)_2SO_4$	550	56
Butanol Extract	475	49
Sephadex G25 Chromatography	358	37
Precipitation at pH 3.0	280	29
DEAE-Sephadex Chromatography 0.1 - 0.3 M NH_4HCO_3	213	22
Electrophoresis at pH 6.5	148	15

1. 25 g of freeze-dried algal cells were extracted at 5°C for 1 hr
   in a buffer of 375 ml 0.1 M Na_2CO_3 and 250 ml 0.05 M $NaHCO_3$.
2. After clearing the suspension by centrifugation (35000 x g for
   40 min) solid $(NH_4)_2SO_4$ was added to give 40% saturation, the
   suspension was again centrifuged (35000 x g for 40 min) and the
   precipitate suspended in 100 ml 0.05 M $NaHCO_3$ and dialysed
   against 4 liters of 0.05 M $NaHCO_3$ overnight at 5°C (in Visking
   cellulose dialysis tubing, size 32/32).
3. The toxin was retained insice the dialysis case; any precipitate
   left was removed by centrifugation (35000 x g for 30 min) before
   freeze-drying and extracting with 150 ml butanol water (9:1) at
   5°C for 2 hours.
4. The butanol extract obtained by filtering the suspension was
   dried by rotary evaporation at 40°C.
5. The brown gummy residue was dissolved and made up to 10 ml with
   01. M NH_4HCO_3 and applied to a Sephadex G25 column (1.3 x 60 cm).
6. The toxic fractions eluted by 0.1 M NH_4HCO_3 were pooled and
   acidified to pH 3.0 by 1 N HCl.
7. The precipitated toxin was separated by centrifugation (35000 x
   g for 30 min) dissolved in 60 ml of 0.1 M NH_4HCO_3 and applied to
   a DEAE-Sephadex A25 column in 0.1 M NH_4HCO_3.
8. The toxin was eluted with 0.3 M NH_4HCO_3 at pH 8.0.
9. After pooling and dialysis against water the toxic fractions
   were freeze-dried to concentrate them.
10. Electrophoresis followed on Whatman No. 3 paper (loading 0.4
    mg/cm) at pH 6.5 in 1% pyridine/acetic acid.
11. Toxin was eluted by 1% w/v aqueous ammonia.
12. Bloom, partially and fully purified toxin were stored indefi-
    nitely at −15°C without loss of toxicity.
13. Toxicity was assayed by intraperitoneal injection into male
    white laboratory mice. The minimum dose sufficient to kill
    each member of a group of mice was found (approximate LD_{100}).
    This was done at each step to monitor purification and recovery
    of toxicity.

modified the procedure to suit our material (Elleman et al., 1978).
The procedure we developed and the results we obtained are shown in
Table 1.   Initially 15 to 30 mg of freeze-dried cells caused death
to approximately 1 kg of mice.   The final purified toxin was a
white powder with an approximate LD_{100} of 70 µg/kg, while the LD_{50}
was 56 µg/kg of mice.

Characterization of the Toxin from *Microcystis aeruginosa*

On hydrolysis the toxin, purified as described in the previous
section, was found to contain five amino acids and methylamine in
approximately equimolar amounts (Table 2).

Table 2.   Hydrolyses and Amino Acid Analyses of Toxin from *Microcystis aeruginosa**

Amino Acids	3 M p-toluene sulphonic[a] acid or 5.7 M HCl[b]	Performic acid + 5.7 M HCl[c]	HI[d]	4.2 M NaOH[e]
β-Methyl aspartic acid	1.0	0.9	0.9	1.0
Glutamic acid	1.0	1.0	1.0	0.8
Alanine	1.1	1.1	1.1	1.2
Methionine	0.6	0.7	0.8	0.7
Tyrosine	0.9	0.8	0.9	1.0
Methylamine	0.9	0.9	1.0	0.7

*Amino acids were identified directly by a Beckman 120C amino acid
  electrophoresis (Gray, 1967) and thin layer chromatography (Woods
  and Wang, 1967).

Toxin was hydrolyzed:

[a] In 3 M p-toluene sulphonic acid and 0.2% (w/v) 3-(2- amino ethyl)
    – indole at 115°C for 25 hours to detect any tryptophan.

[b] In 5.7 M HCl with β-mercaptoethanol at 108°C for 4 hours.

[c] Under the same conditions, after oxidation by performic acid,
    methionine is converted to the sulphone.

[d] In constant boiling HI at 130°C for 24 hours, methionine is con-
    verted to homocysteine lactone.

[e] In 4.2 M NaOH at 108°C for 24 hours.

L-amino acid oxidase oxidized methionine and tyrosine (Lichtenberg and Wellner, 1968); alanine was oxidized by D-amino acid oxidase; glutamic acid was not oxidized by L-glutamate decarboxylase,
indicating that in the toxin glutamic acid and alanine are present
as the D-isomers while methionine and tyrosine are present as
L-isomers.

On acid hydrolysis the amino acid later identified as β-methyl
aspartic acid was found to be present as two interchangeable forms
differing in electrophoretic mobility and eluting on amino acid
analysis in the same place as aspartic acid. Periodate and performic acid oxidations had no effect on the amino acid indicating
the absence of proximal hydroxyl and amino groups, and of unsaturated [C-C] bonds. Mass spectroscopy of the dansyl methyl ester
gave a molecular ion at 408. The two forms of the unknown amino
acid and synthetic DL β-methyl aspartic acid were found to be
identical on chromatography and electrophoresis. Partial acid
hydrolysis of the toxin gave only the component of lower electrophoretic mobility and greater solubility indicating that β-methyl
aspartic acid present in the toxin has the erythro configuration.

The toxin was found to be stable on boiling in a neutral solution, but boiling in 1 N HCl or 1 N NaOH resulted in loss of toxicity. As indicated from the purification procedure the toxin was
insoluble in acid. It was adsorbed by activated charcoal. At pH
6.5 it was negatively charged with an electrophoretic mobility 0.6
that of aspartic acid. It had no primary amide group since no
ammonia was released on hydrolysis by hydroiodic acid. Alanine,
tyrosine and methionine have no free carboxyl groups while the
α-carboxyls of glutamic and β-methyl aspartic acid were shown by
selective tritiation to be free (Holcomb et al., 1968). The toxin
did not react with ninhydrin, nor were any amino groups dansylated
indicating the absence of any free $-NH_2$. The phenolic -OH of tyrosine did dansylate.

The amino acid composition of our toxin, purified from a bloom
of *Microcystis aeruginosa*, differed from other toxin preparations
isolated from the same organism (Table 3), although aspartic,
glutamic acids and alanine were common.

At this stage we do not have enough information to know the
structure of the toxin from *Microcystis aeruginosa*. From the amino
acid analysis a minimum molecular weight of 654 was derived; there
were two free α-carboxyl groups and no free amino groups; electrophoretic mobility at pH 6.5 was consistent with the presence of
two negative charges if the molecular weight was indeed 654 (Offord,
1966). The toxin could then have a blocked N-terminal group since
no free amino groups were present, and one carboxyl group could be
substituted by methylamine in amide linkage. Nevertheless, it

Table 3. Composition of Toxin Preparations from *Microcystis*
aeruginosa

Composition						
L-Asp 1	L-Glu 2	D-Ser 1	L-Val 1	L-Orn 1	L-Ala 2	L-Leu 2[a]
Asp*	Glu	Ser	Val	Pro	Ala	Leu[b]
Gly	Tyr	Thr	Phe	Arg	Ile	NH$_3$
Asp 3	Glu 5	Ser 3	Val 1	Orn 2	Ala 2	Leu 1[c]
Gly 6	Pro 1	Thr 1	Lys 1	Ile 1		

Peptide[d]

β-CH$_3$Asp 1 D-Glu 1 L-Met 1 L-Tyr 1 D-Ala 1 CH$_3$NH$_2$ 1[e]

*Amino acids underlined were more abundant on amino acid analysis.

[a]Bishop et al., 1959.

[b]Murthy and Capindale, 1970.

[c]Rabin, 1976.

[d]Toerien et al., 1976.

[e]Elleman et al., 1978.

would still be possible for the toxin to have a cyclic structure
or to be a dimer, since the arrangement of amino acids would not
necessarily be constrained by conventional bonding as in proteins.
Sephadex G25 chromatography indicated a small peptide, but the
technique did not allow us to distinguish between a genuinely small
molecule (monomer) of unusual shape, which could elute from the gel
as if it had a larger molecular weight, and a larger molecule
(dimer) retarded because of adsorption. All the published results
on molecular weight are consistent with our findings that the toxin
is a small peptide (Table 4).

Pathological and Biochemical Effects of the Toxin from *Microcystis* *aeruginosa*

With the aim of determining the physiological sites and types
of damage by toxic *Microcystis aeruginosa* blooms on animals

Table 4.   Molecular Weights of Toxin Preparations from *Microcystis*
           aeruginosa

Reference	Preparation and Molecular Weight (M.W.)
Bishop et al. (1959)	Dialysis and A.A. Analysis: M.W. 1300–2600
Murthy and Capindale (1970)	A.A. Analysis?: M.W. 1200?
Rabin Ph.D. Thesis London (1976)	A.A. Analysis, S.D.S. Gel Electrophoresis, Sephadex G25: M.W. 2400 approx.
Toerien et al. (1976)	Sephadex G25: M.W. 1000 approx.
Elleman et al. (1978)	A.A. Analysis: Minimum M.W. 654, Sephadex G25: M.W. < 1400 (KD=0.55)
Amann and Eloff (1980)	Sephadex G25: M.W. $\leq$ 1300

consuming the algae, we have investigated the pathological effects
on mice of algal bloom extracts and of the purified peptide toxin
that we have isolated.

Mice injected intraperitoneally with samples of algal extracts
or with toxin solution in excess of the minimum lethal dose (MLD)
died within 1 to 2 hours. Death was due to hemorrhagic shock
caused by the pooling of blood in the liver which became very
enlarged and dark red in color. This was accompanied by pallor in
the rest of the body, which was especially obvious in the tail,
ears and eyes. The enlargement and pooling of blood in the liver
was progressive and was already significant $(0.01 > p > 0.001)$ 15
minutes after toxin injection (Table 5).

Histological changes were only observed in the liver. At
death there was massive liver hemorrhage mainly in the mid and
peripheral zones of the lobules. The progressive pooling of
blood in the liver after toxin injection was studied by scanning
electron microscopy. Figure 2 shows the increase in the number
of red blood cells in the liver, even though the tissue was perfused before processing. We can also see the progressive disruption of the normal hepatic architecture due to the disintegration
of the sinusoidal network, and possibly of cell to cell adhesion
between hepatocytes.

Transmission electron microscopy showed the disappearance of
the membranes between hepatocytes and sinusoids, so that erythrocytes can be seen in direct contact with the cytoplasm. This

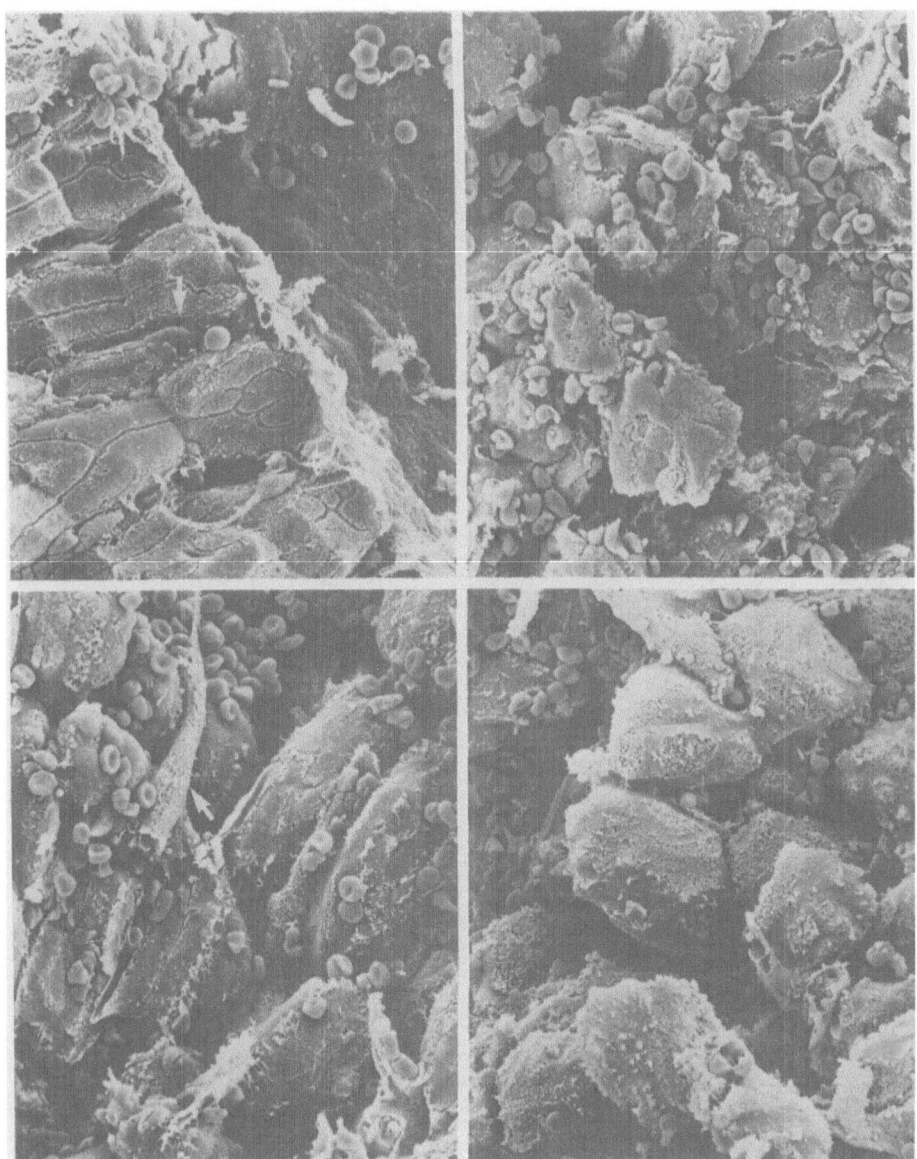

Fig. 2.   Scanning electron microscopy of mouse liver following
          injection of toxic extract (5 x MLD) from *Microcystis
          aeruginosa* bloom showing the progressive disruption of
          the normal hepatic architecture, the disintegration of
          the sinusoids (arrows) and engorgement with erythrocytes;
          x 1000, (Top Left) 5 minutes, appears normal; (Top Right
          and Bottom Left) 30 minutes; (Bottom Right) 60 minutes
          after treatment.

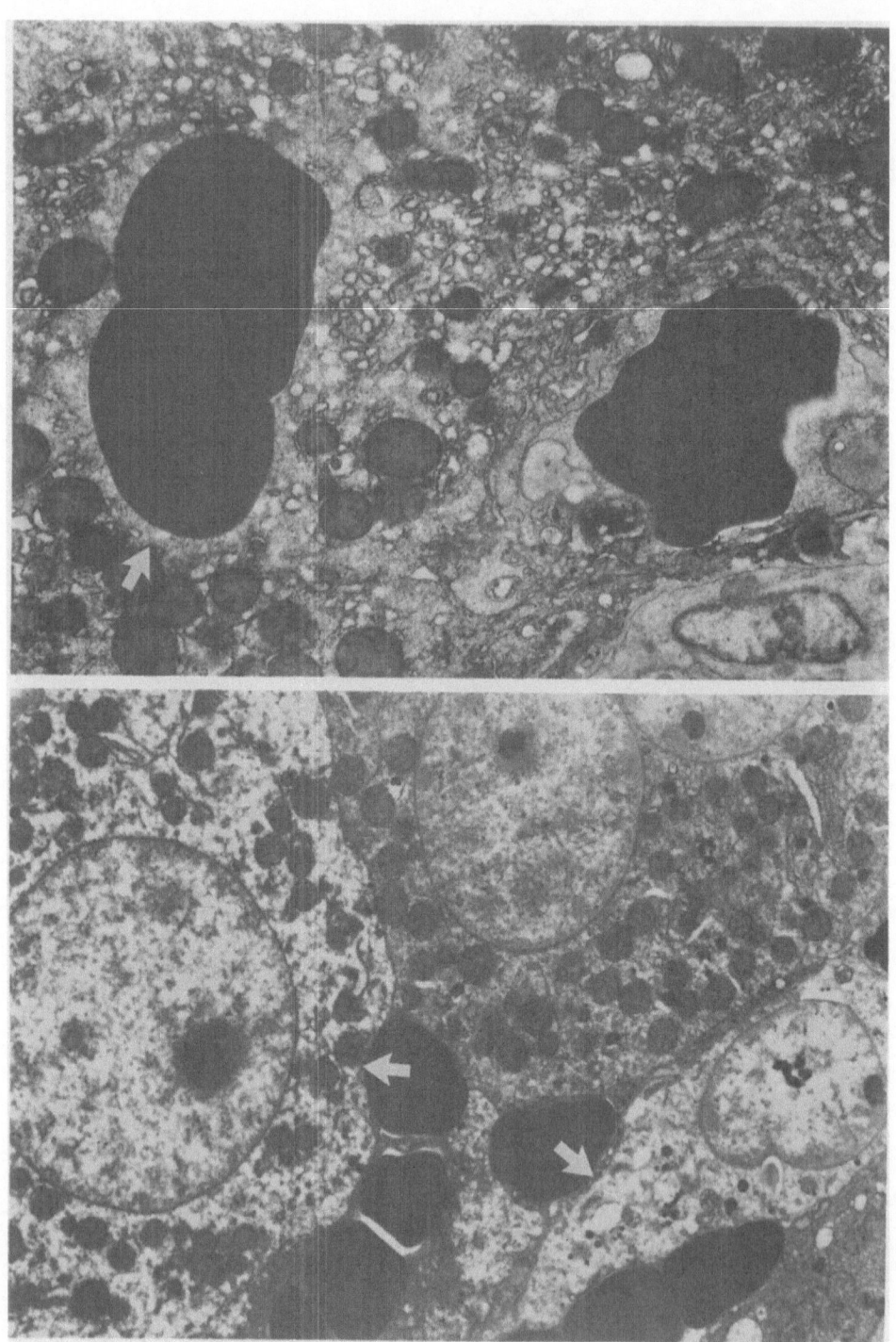

Table 5. Increase of Weight and Blood Content of Liver with Time
          of Mice Injected with Toxin from a Bloom of *Microcystis*
          aeruginosa

Mice[a] Killed At	No. of Mice	Weight of Liver as g per 100 g Body Weight	Blood Content of Liver as ml per 100 g Tissue
Control	6	4.27 ± 0.11	6.55 ± 0.92
15 minutes	7	5.63 ± 0.26	22.13 ± 4.28
30 minutes	7	6.23 ± 0.18	43.47 ± 3.76
45 minutes	7	6.99 ± 0.32	53.40 ± 2.73

[a]Mice injected intraperitoneally with approximately 5 x MLD.

change was first seen 15 minutes after toxin injection in a very
small proportion of cells (Figure 3A). We also found evidence of
progressive necrotic changes to hepatocytes confirming our previous findings by light microscopy (Elleman et al., 1978). The
plates show hepatocytes which are clearly necrotic but still bound
by membranes (Figure 3B) next to other hepatocytes whose intracellular structure still appears little damaged. The structures
of other hepatocytes have disappeared completely and red blood
cells can be seen floating in disintegrated cytoplasm (Figure 4).
These changes become progressively more prevalent with increasing
survival time after injection of individual mice.

The necrotic changes to hepatocytes observed microscopically
were reflected in large increases in the activities in serum of
hepatic enzymes (Table 6).

Changes in GOT and LD activities were already apparent 15
minutes after toxin injection and by 30 minutes they were highly

Fig. 3. Transmission electron microscopy of mouse liver following
         injection of toxic extract (5 x MLD) from *Microcystis*
         aeruginosa bloom. (Top) 15 min after treatment showing
         an erythrocyte directly in contact with the hepatocyte
         cytoplasm (arrow); x 12500. (Bottom) 60 min after treat-
         ment showing membrane bound necrotic hepatocytes (arrows);
         x 5000.

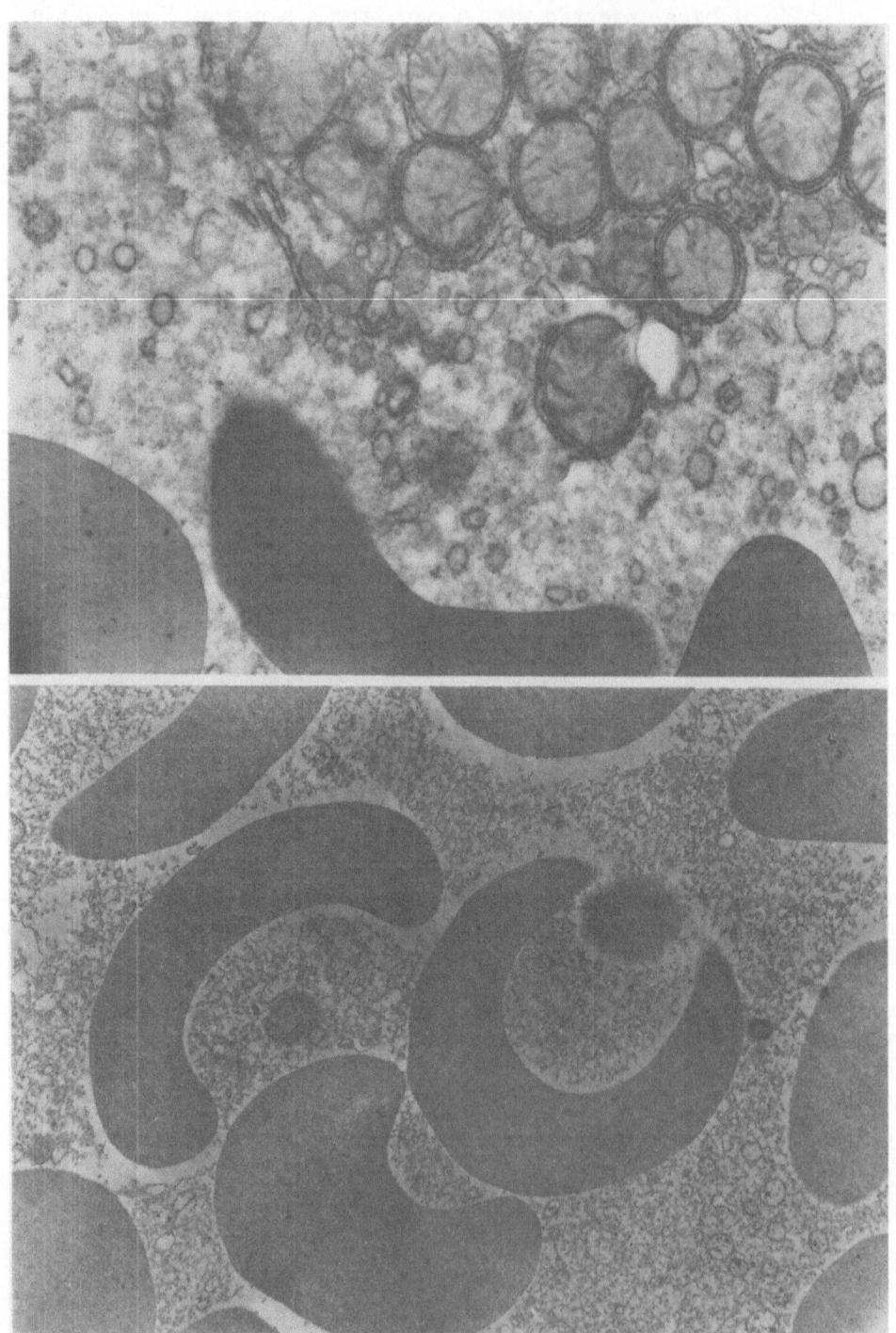

Table 6. Glutamate-oxaloacetate Transaminase (GOT) and Lactate
Dehydrogenase (LD) Activity of Serum in Mice Injected
With Toxin from a Bloom of *Microcystis aeruginosa*

Mice Killed At Intervals After Injection[a]	No. of Mice	GOT U/ml	LD U/ml
Control	6	0.13 ± 0.04	2.70 ± 0.29
15 minutes	7	3.66 ± 2.00	13.33 ± 5.97
30 minutes	7	7.37 ± 1.85	33.36 ± 7.96
45 minutes	7	9.70 ± 0.90	37.95 ± 3.33

[a]Mice injected intraperitoneally with approximately 5 x MLD.

significant ($0.01 > p > 0.001$) having increased over 50-fold and 10-fold respectively.

When we looked at the effect of repeated sublethal doses of toxin injected into mice, we found again that the liver appeared to be the only organ affected. Repeated dosing of mice with 0.75 and 0.50 of the approximate MLD of toxin led to loss of weight, and at the highest dose to the death of most mice within one week of daily injections (Table 7). Histologically the severity of the liver changes observed was proportional to the dose of toxin given. The livers of mice given 0.25 of MLD of toxin showed very few changes, although mitotic figures were observed in an increased proportion of cells. At 0.5 MLD of toxin the liver showed progressive hepatocyte necrosis which was accompanied by fibrosis and mononuclear cellular infiltration. Similar changes were observed in the livers of mice given 0.75 MLD of toxin which survived the first few days of the experiment. The livers of mice receiving 0.75 MLD of toxin daily, that died within the first few days of

Fig. 4. Transmission electron microscopy of mouse liver following injection of toxic extract from *Microcystis aeruginosa* bloom. (Top) 60 minutes after treatment (5 x MLD) showing a disintegrating hepatocyte with erythrocytes in the cytoplasm; x 20000. (Bottom) 150 minutes after treatment (approximate MLD) showing erythrocytes floating in disintegrated cytoplasm.

Table 7. Effect on Mice[a] of Repeated Doses of *Microcystis*
          aeruginosa Toxin

Dose	No. of Mice	Body Weight	Weight of Liver As g per 100 g Body Weight	Clinical Observations
Control	12	Gain	5.3	Normal
0.25 MLD	12	Gain	5.4	Apparently normal
0.5 MLD	14	Loss	6.9	Subcutaneous edema and jaundice
0.75 MLD	35	Loss	7-10	Subcutaneous edema and jaundice. Most dead within 1st week of injections

[a]Mice were injected (intraperitoneally) daily with toxin for a
maximal period of 6 weeks. Two mice were sacrificed each week
for histological examination.

treatment, showed massive hepatic hemorrhages which were similar
to those observed in the livers of mice that had received the peptide toxin in excess of the MLD.

The pathological changes produced by the peptide toxin from
Microcystis aeruginosa are consistent with the acute (Ashworth and
Mason, 1946; Konst et al., 1965), and chronic effects (Tustin et
al., 1973) of *Microcystis aeruginosa* extracts reported by other
workers. From our pathological findings we can conclude that the
toxin caused death by:

a) Disintegration of hepatocyte cell membranes

b) Loss of epithelial lining of sinusoids

c) Necrosis of hepatocytes

d) Subsequent accumulation of red blood cells in the liver.

We have started to investigate the mechanism of action of the
toxin at the cellular level. At present we can only report largely
negative findings (Table 8). Toxin incubated with mouse liver
slices had no statistically significant effect on the rate of
incorporation of leucine, uridine or thymidine into protein, RNA
and DNA respectively. It did not have a significant effect on the
oxygen consumption of liver slices. There was a small increase

Table 8.   Metabolic Effects of the Toxin from *Microcystis aeruginosa*

Liver Function Tested	Tissue Preparation	Method	Effect on Function
Protein synthesis	Slices	[³H] Leucine incorporation into protein	None
Protein synthesis	Microsomes	[³H] Leucine incorporation into protein	None
RNA synthesis	Slices	[6-³H] Uridine incorporation into RNA	None
DNA synthesis	Slices	[Methyl-³H] Thymidine incorporation into DNA	None
Respiration	Slices	Oxygen uptake	None
Respiration	Mitochondria	Oxygen uptake	20% increase

(20%) in oxygen consumption when isolated mitochondria were incubated with toxin.

The similarity in chemical structure between our toxic peptide from *Microcystis aeruginosa* and phalloidin, a bicyclic peptide hepatotoxin from the poisonous mushroom *Amanita phalloides* (Figure 5, Wieland and Wieland, 1970), together with their similarity in

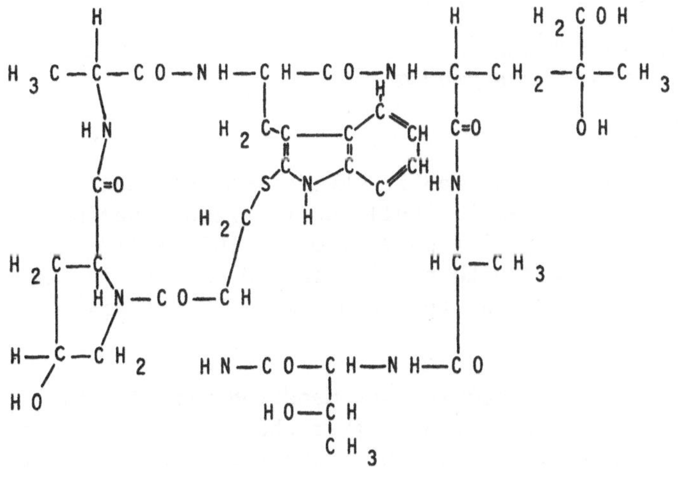

Fig. 5.   Phalloidin isolated from *Amanita phalloides*.

gross pathological effects led us to look at whether the similarity
extended to toxic effects at the cellular level.

Specifically phalloidin inhibits the depolymerization of F-actin
to G-actin by a variety of treatments (e.g. sonic vibration, potassium iodide) which normally loosen the structure of the actin filament causing its depolymerization. The F-actin to G-actin change
can be followed by measuring changes in viscosity of actin solutions. We compared the effect of our peptide toxin and of phalloidin
on depolymerization of F-actin by potassium iodide (Table 9).

We found that the hepatotoxin from *Microcystis aeruginosa* unlike
phalloidin did not protect F-actin from depolymerization by potassium
iodide. From this we must conclude that superficial similarities
between the structure and the physiological effects of phalloidin
and of the toxic peptide from *Microcystis aeruginosa* are fortuitous
and do not extend to the molecular level.

Table 9. Effect of Toxin from *Microcystis aeruginosa* and of
Phalloidin on the Depolymerization of F-actin by
Potassium Iodide[a]

Solution	Specific Viscotity[b]	
	−KI	+KI
Actin	0.66	0.09
Actin + Phalloidin	0.72	0.83
Actin + Toxin from *M. aeruginosa*	0.68	0.14

[a]Actin was prepared from rabbit skeletal muscle according to
Iyengar and Weber (1964). Fully polymerized F-actin (1.2 mg/ml
in 1×10^{-3} M Tris buffer pH 7.4, 0.1×10^{-3} M ATP, 1.0×10^{-3} M
$MgCl_2$) was incubated with phalloidin (70 µM) or with *Microcytis
aeruginosa* toxin (200 µM for an assumed M.W. of 654) overnight
at 5°C.

[b]The viscosity of the samples was measured with a capillary
viscometer at 20°C before and after the addition of solid KI to
a final concentration of 0.6 M. Specific viscosity is defined
as t/t_0-1, where t = flow time of samples and t_0 = flow time of
the protein free medium.

These negative biochemical findings taken together with the histological evidence indicate that the toxin's primary attack on the liver is likely to be at the membrane level, perhaps causing permeability changes which lead to cell death. We are actively investigating this possibility at the present time.

We have presented here an overview of some aspects of our work with toxic *Microcystis aeruginosa*, but in no way has our work been comprehensive. Many aspects of the toxicology of this alga have not been considered, as we have concentràted our investigations on the isolation, identification and some pathological effects on mice of the small peptide hepatotoxin associated with blooms and cultures of *Microcystis aeruginosa*. It is also obvious that the aspects of *Microcystis* toxicity that we have investigated will benefit from further work by ourselves and by other groups bringing a fresh perspective and approach to the problem. Many other facets of *Microcystis aeruginosa* toxicity are discussed at this conference, providing a comprehensive account of the present understanding of this most important field in algology.

ACKNOWLEDGMENTS

The work described in this paper was carried out in collaboration with Dr. T. C. Elleman of the CSIRO Division of Protein Chemistry, Parkville, Victoria, Australia, and Dr. A. R. B. Jackson of the New South Wales Department of Agriculture District Veterinary Laboratory, Armidale, New South Wales, Australia. Financial support by the Australian Research Grants Committee is gratefully acknowledged. We also wish to thank the U. S. Environmental Protection Agency for sponsoring, the Wright State University for hosting, and Dr. Wayne Carmichael for organizing this conference.

REFERENCES

Amann, M. J., and J. N. Eloff. 1980. A preliminary study of the toxins of different *Microcystis* strains. S. Afr. J. Sci. (In press).

Ashworth, C. T., and M. F. Mason. 1946. Observations of the pathological changes produced by a toxic substance present in blue-green algae (*Microcystis aeruginosa*). Am. J. Pathol. *22*:369–383.

Bishop, C. T., E. F. L. J. Anet, and P. R. Gorham. 1959. Isolation and identification of the fast-death factor in *Microcystis aeruginosa* NRC-1. Can. J. Biochem. Physiol. *37*:453–471.

Dancker, P., I. Löw, W. Hasselbach, and T. H. Wieland. 1975. Interaction of actin with phalloidin, polymerization and stabilization of F-actin. Biochim. Biophys. Acta *400*:407–414.

Elleman, T. C., I. R. Falconer, A. R. B. Jackson, and M. T. Runnegar.
      1978. Isolation, characterization and pathology of the toxin
      from a *Microcystis aeruginosa* (= *Anacystis cyanea*) bloom.
      Aust. J. Biol. Sci. *31*:209-218.

Francis, G. 1878. Poisonous Australian lake. Nature (London) *18*:
      11-12.

Gray, W. R. 1967. Dansyl chloride procedure. Pages 139-151 *in*
      C. H. W. Hirs, ed., Methods in Enzymology. Vol. 11. Aca-
      demic Press, New York.

Holcomb, G. N., S. A. James, and D. N. Ward. 1968. A critical
      evaluation of the selective tritiation method of determining
      C-terminal amino acids and its application to luteinizing
      hormone. Biochemistry *7*:1291-1296.

Iyengar, M. R., and H. H. Weber. 1964. The relative affinities
      of nucleotides to G-actin and their effects. Biochim.
      Biophys. Acta *86*:543-553.

Konst, H., P. D. McKercher, P. R. Gorham, A. Robertson, and J.
      Howell. 1965. Symptoms and pathology produced by toxic
      Microcystis aeruginosa NRC-1 in laboratory and domestic
      animals. Can. J. Comp. Med. Vet. Sci. *29*:221-228.

Lichtenberg, L. A., and D. Wellner. 1968. A sensitive fluori-
      metric assay for amino acid oxidase. Anal. Biochem. *26*:
      313-319.

Murthy, J. R., and J. B. Capindale. 1970. A new isolation and
      structure for the endotoxin from *Microcystis aeruginosa*
      NRC-1. Can. J. Biochem. *48*:508-510.

Offord, R. E. 1966. Electrophoretic mobilities of peptides on
      paper and their use in the determination of amide groups.
      Nature (London) *211*:591-593.

Rabin, P. 1976. Studies on the endotoxin from the blue-green
      alga *Microcystis aeruginosa* NRC-1. Ph.D. Thesis. Univer-
      sity of London King's College.

Toerien, D. F., W. E. Scott, and M. J. Pitout. 1976. *Microcystis*
      toxins: isolation, identification, implications. Water
      S. A. (Pretoria) *2*:160-162.

Tustin, R. C., S. J. van Rensburg, and J. N. Eloff. 1973. Hepatic
      damage in the primate following ingestion of toxic algae.
      Pages 383-385 *in* S. J. Saunders and J. Terblanche, eds.,
      Liver: Proceedings International Liver Congress. Pitman
      Medical, London.

Wieland, T., and O. Wieland. 1970. The toxic peptides of *Amanita*
      species. Pages 249-280 *in* A. Ciegler, S. Kadis, and S. Ajl,
      eds., Microbial Toxins. Vol. 8. Academic Press, New York.

Woods, K. R., and K. T. Wang. 1967. Separation of dansyl-amino
      acids by polyamide layer chromatography. Biochim. Biophys.
      Acta *133*:369-370.

TOXICOLOGICAL STUDIES ON *MICROCYSTIS*

J. N. Eloff and A. J. Van Der Westhuizen

University of the Orange Free State
Bloemfontein, Republic of South Africa

ABSTRACT

Experiments were carried out to determine whether varying toxicity of natural blooms of *Microcystis* is due to: a) *Microcystis*-bacteria interactions b) an environmental effect c) the presence of more than one strain or species of *Microcystis*. Partial characterization of bacteria associated with unialgal cultures and natural blooms indicated that no relationship exists between *Microcystis* and the occurrence of one or more types of bacteria. Examination of different axenic toxic cultures proved that toxicity resided in *Microcystis* only. To examine possibilities b and c, toxin extraction and mouse assay were examined and standardized. Environmental parameters had an influence on toxicity of toxic isolates, but even the highest possible dose of a non-toxic isolate injected (800 mg/kg) was innocuous whereas 20 mg/kg from a toxic isolate generally killed mice. Both toxic and non-toxic isolates lost colony habit soon, but only non-toxic isolates lost the ability to form gas vacuoles. Gas vacuole-less mutants of toxic *Microcystis* isolates were however, still toxic. A preliminary comparison of toxins extracted from several toxic isolates indicated that similar toxins were present. Many observations and experimental results supported the conclusion that toxic and non-toxic isolates of *Microcystis* belong to different taxons.

TEXT

Waterblooms of *Microcystis* vary in toxicity, can persist seemingly unchanged and yet lose toxicity (Deem and Thorpe, 1939).

Olson (1951) was the first to grow toxic *Microcystis* in the laboratory and found that toxicity varied to a large degree.

Various suggestions have been put forward to explain this
change in toxicity. Some of the possibilities will be discussed
here with reference to results found in our laboratory.

Microcystis-Bacterial Interaction Responsible for Toxicity

It is possible that toxin is produced as a result of a *Microcystis*-bacterial interaction or even by the bacteria associated
with *Microcystis*. Any change taking place in the bacterial population would thus be reflected in changes in toxicity but not
necessarily in the appearance of the waterbloom.

Thomson et al. (1957) found Gram-negative rods to be the most
common bacteria when a survey of 25 algal collections of *Microcystis* were made. A neurotoxin was produced by 22 of the 26 Gram-
negative isolates.

A large variety of bacteria were isolated from 17 laboratory
cultures and natural blooms of *Microcystis* (Eloff and Sadie, 1979).
The bacteria were grown on nutrient agar or casein-peptone-starch
agar, both of which are suitable for freshwater bacteria. Of the
49 isolates, 28 were Gram-negative (13 rods) and 21 were Gram-
positive (10 rods). The color of 23 colonies was light grey, of
10 light yellow, of 9 white, and of 7 yellow. Only eight of the
isolates were distinctly mobile. The morphology of the bacteria
also varied. A large number of the isolates were cocci with an
average size of ca. 0.5 μm or short rods with sizes of 1.2 to 1.4
x 0.5 to 0.6 μm. The rest of the bacteria were highly variable in
size and form very small to large cocci, short wide rods, diplo-
cocci, streptococci and diplococci/short rods.

To partially characterize the bacteria biochemically, twenty-
two microscale API 20E tests (Analytab Products Inc., New York)
were run. The results were: 2 isolates reacted positively in
only one test, 1 in two, 7 in three, 4 in four, 5 in five, 7 in
six, 5 in seven, 2 in eight, 3 in nine, 1 in ten, 1 in eleven, 1
in thirteen, 1 in fifteen, 1 in sixteen, 2 in seventeen, and 1 isolate reacted positively in eighteen tests.

An analysis of the number of bacterial isolates possessing
various enzyme activities, that may possibly be correlated with
compounds excreted by *Microcystis*, is given in Table 1.

Table 1. Analysis of Reactions of 49 Bacterial Isolates from 17
Microcystis Blooms or Cultures[a]

Reaction	% Positive Response
β-galactosidase	58
Arginine dehydrolase	72
Lysine decarboxylase	81
Ornithine decarboxylase	84
Citrate metabolism	61
H_2S production from thiosulphate	4
Urea hydrolysis	24
Tryptophan deaminase	6
Tryptophan to indole	4
Sodium pyruvate to acetoin	33
Gelatine liquefication by proteolytic enzymes	58
Glucose metabolism	28
Mannose metabolism	24
Inositol metabolism	7
Sorbitol metabolism	13
Rhamnose metabolism	13
Saccharose metabolism	24
Melibiose metabolism	16
Amylase metabolism	26
Arabinose metabolism	36
Cytochrome oxidase	11
Nitrate reduction	28

[a]Eloff and Sadie, 1979.

There were very few isolates from different *Microcystis* cultures that reacted similarly. It does not seem as if there is a definite association between *Microcystis* and any one or more bacterial species.

Many lines of evidence have been presented to show that the fast death factor (FDF) in bacterized *Microcystis* cultures is due to *Microcystis* and not to contaminating bacteria, e.g., a *Microcystis*-rich fraction obtained by differential centrifugation contained the FDF whereas the bacteria-rich fraction contained only the slow death factor (SDF). Five of the six recognized bacterial contaminants when isolated and cultured did not contain the FDF. Contamination of a non-toxic culture with bacteria from a toxic culture did not produce FDF. Isolates from three different areas all contained FDF although their bacterial composition was probably different (Gorham, 1960). This deduction was supported when we used the methods developed by Pretorius (1977) to purify three toxic isolates of *Microcystis*, i.e. UV-001 (NRC-1), UV-006 isolated from Hartbeespoort Dam and UV-010 isolated from Witbank Dam (see also paper by Eloff at this conference). Although plating on different agar media was employed we found, as did Stanier et al. (1971) that careful examination (phase optics, 100x objective) of old cultures, in which some cells had already started to lyse, gave best results in determining whether cultures were axenic. Young healthy cultures frequently had a small number of bacteria present which were only visible after examining a number of fields, but contamination of old cultures (8 to 12 weeks) was usually very evident. We found that this was not due to an antibiotic effect of young cultures but rather to the unavailability of carbon sources in the medium.

All of the axenic cultures were as toxic as the bacterized cultures and the symptoms were the same as described by Hughes et al. (1958). The time of death was not similar to that described by Gorham (1964), but this aspect will be discussed later. In the laboratory of M. Shilo at Jerusalem we could show that two axenic gas vacuolated cultures obtained from the Paris Culture Collection (PCC 7806 and PCC 7820) were also toxic.

Both Gorham (1964) and Elleman et al. (1978) found that their bacterized cultures caused slow deaths. The deduction that the slow death factor was of bacterial origin (Gorham, 1960) was also confirmed because no slow deaths were observed when sub-lethal concentrations of axenic *Microcystis* were injected into mice. If toxic *Microcystis* isolates produce a SDF, it must have a lower LD_{100} than the FDF. Very high doses of non-toxic isolates also did not lead to any slow deaths.

Bacteria may influence the toxicity of natural blooms if some bacteria are able to break down or inactivate the released toxin.

Some indications have been found that ornithine is a component of
microcystin (Hughes et al., 1958). The fact that 84% of the bacterial isolates associated with *Microcystis* had ornithine decarboxylase activity (Table 1) could be significant in this regard.

Presence of More than One Taxon of *Microcystis* in Blooms

Stephens (1949) investigated a toxic *Microcystis* bloom on the
Vaal Dam that killed thousands of stock (Steyn, 1945) and compared
it to apparently non-toxic blooms of *Microcystis*. She found enough
differences in the colony structure and the odor of decomposing
blooms to merit the description of a new species, M. *toxica*, but
Komárek (1958) considered M. *toxica* to be synonomous with M.
aeruginosa forma *aeruginosa*. In 1954, Ingram and Prescott suggested
that changes in toxicity may be due to the presence of different
algal subspecies.

Toerien et al. (1976) stated that toxin production in *Microcystis* was associated with M. *aeruginosa* forma *aeruginosa* and not
with M. *aeruginosa* forma *flos-aquae*. These authors showed that
both occurred simultaneously and that the colony proportion changed
from 100% of the one form to 100% of the other form within two
months (see also paper by Scott et al., this conference).

Before axenic cultures of *Microcystis* became available it was
impossible to determine whether *Microcystis* also has a slow death
factor. A suspension of broken cells of the non-toxic isolate M.
UV-007 was injected into white mice in doses of up to 800 mg/kg
(more or less the highest dose we could manage to get in the 0.5
ml in a consistency that could still be injected). Even doses as
high as this had no influence on the mice whatsoever, indicating
that the difference between toxic and non-toxic strains is a qualitative, not a quantitative one.

When broken cell suspensions of M. *incerta* and M. *wesenbergii*
(= M. *marginata*) (Kessel and Eloff, 1975) were tested, no signs of
toxicity were ever found. This means that only M. *aeruginosa* has
both toxic and non-toxic isolates.

If M. *aeruginosa* forma *aeruginosa* is indeed always toxic and
forma *flos-aquae* always non-toxic, it is of utmost importance to
be able to distinguish between them. Unfortunately, the colonial
habit which is used as the main diagnostic parameter varies to such
an extent that it is frequently difficult for the inexperienced
observer to distinguish between the two forms. Another diagnostic
parameter is cell size. Measurements made by Komárek (1958) indicated that M. *aeruginosa* forma *aeruginosa* had the larger cells
with > 90% of the cells of 58 natural populations having cell diameters of 4 to 6 μm (range 3 to 9.4 μm). He found that with M.

aeruginosa forma *flos-aquae* > 90% of the cells had diameters of 2.5 to 5 μm. On the other hand, other authors (Huber-Pestalozzi, 1929; Hortobágyi, 1943, 1947) considered forma *aeruginosa* to have the smaller and forma *flos-aquae* the larger cells. W. E. Scott observed that toxic blooms of *Microcystis* generally had cells with a diameter of > 5 μm (personal communication).

We found that environmental conditions influenced the size of *Microcystis* cells (see paper by Krüger and Eloff, this conference). When morphological changes in toxic and non-toxic *Microcystis* isolates grown at different light intensities were determined (Krüger et al., 1980) the size of *M.* UV-006 (toxic) varied from 1.8 to 6.4 μm whereas the *M.* UV-007 (non-toxic) had cell diameters of 3.4 to 8.2 μm.

When the cell diameters of different *M. aeruginosa* isolates in our culture collection were determined, in an effort to examine the possible correlation between cell size and toxicity, the results presented in Table 2 were obtained.

Table 2 shows that toxic cells of *Microcystis* were larger (4.43 μm, standard deviation (SD) = 1.02) than non-toxic cells (3.64 μm, SD = 0.64), but that this is not a useful diagnostic character.

In order to determine whether toxic and non-toxic isolates represent two species, studies such as those of Stanier et al. (1971) and Rippka et al. (1979) have been started on our isolates of *Microcystis*. It already seems as if many differences occur even among the toxic isolates.

One of the interesting aspects evident from Table 2 is the good correlation between toxicity and gas vacuole content (except for the UV-009, UV-020, UV-021, UV-022 and UV-024 isolates). The latter four have been in culture for only a few months and may still lose the ability to form gas vacuoles.

Stanier et al. (1971) assigned the sole representative of *Microcystis* they had available to group II-C on the wrong supposition that bright refractile granules in this organism were gas vacuoles. Rippka et al. (1979) assigned this organism to *Synechocystis* because it did not produce gas vacuoles in pure culture "although it cannot be stated categorically that PCC 7005 did not at one time produce gas vacuoles". In our hands the clonal isolate *M.* UV-004 has lost the ability to produce gas vacuoles since its isolation in 1973.

This isolate is similar to UV-007 (= PCC 7005) in many ways and we believe that PCC 7005 is still the same gas-vacuolated *M. aeruginosa* 1036 isolated by Gerloff from Lake Mendota, Wisconsin

Table 2. Average Cell Diameter[a], Gas Vacuole Content, and Toxicity[b] of Clonal *Microcystis* Isolates[c]

Culture Number	Gas Vacuole	Cell Diameter	Toxic
UV-001	✓	3.36 (0.29)	✓
UV-006	✓	4.81 (0.52)	✓
UV-009	–	4.14 (0.58)	✓
UV-010	✓	6.02 (0.62)	✓
UV-019	✓	5.75 (0.65)	✓
UV-023	✓	3.46 (0.25)	✓
UV-025	✓	3.32 (0.26)	✓
UV-026	✓	3.79 (0.44)	✓
UV-027	✓	4.22 (0.56)	✓
UV-028	✓	5.83 (0.71)	✓
UV-029	✓	4.04 (0.56)	✓
UV-004	–	3.37 (0.29)	–
UV-007	–	3.68 (0.43)	–
UV-008	–	3.26 (0.30)	–
UV-011	–	3.49 (0.30)	–
UV-018	–	3.37 (0.22)	–
UV-020	✓	5.70 (0.52)	–
UV-021	✓	3.89 (0.52)	–
UV-022	✓	3.56 (0.35)	–
UV-024	✓	2.94 (0.30)	–

[a] The parenthetical figure is the standard deviation.

[b] At ca. 300 mg/kg.

[c] See paper by Eloff, Autecological Studies on *Microcystis*, this conference for more information on isolates.

in 1946. The % of gas vacuole-less cells in isolates UV-020, UV-021, UV-022, and UV-024, compared to that of UV-023 and UV-029 isolated at the same time, lost the ability to form gas vacuoles under labo- ratory conditions more readily. This questions the validity of assigning all non-gas-vacuolated isolates to *Synechocystis* and typ- ifies the objections raised by Golubic (1979). The presence in a bloom of toxic and non-toxic taxons of *Microcystis* is clearly a very probable explanation of varying toxicity in natural blooms over long periods. Toerien et al. (1976) showed that between December 1973 and February 1974, *M. aeruginosa* forma *aeruginosa* was completely supplanted by forma *flos-aquae* and the chlorophyll a content changed only from ca. 66 to 62 µg per liter. It is not easy for an inexpe- rienced observer to distinguish between the two types because tran- sitional forms occur. This type of succession may very well have led to deductions that the toxicity of the bloom is highly variable if only one type were toxic.

Possibility of Plasmids Regulating Toxicity

The possibility has been raised that plasmids may be involved in gas vacuole formation (Walsby, 1977) as in *Halobacterium* (Simon, 1978), or in toxin production (Lau and Doolittle, 1979) in Cyano- bacteria. This possibility became more likely once it was realized that there is a good correlation between possession of gas vacuoles and toxicity, at least in our cultures.

Gas-vacuolated and non-gas-vacuolated mutants of four toxic isolates of *Microcystis* (*M*. UV-001, *M*. UV-006, *M*. UV-010, *M*. UV-017) were isolated in Jerusalem in cooperation with Bob Simon and Moshe Shilo. Some problems were encountered in lysing the *Microcystis* isolates under mild conditions before analysing for plasmids.

When toxicity of the mutants was determined, it was found that there was no correlation between production of toxin and presence of gas vacuoles (Table 3). The non-gas-vacuolated mutants were generally as toxic as the gas-vacuolated mutants. Differences were encountered in the mutation rate between the different toxic iso- lates and also in the rate of reversion to gas-vacuolated cells.

Gorham (1964) showed that in two clonal isolates, one was toxic and the other non-toxic. This would mean that toxicity in *Micro- cystis* is genetically unstable. Unfortunately, the original cul- ture may not have been unialgal (Stanier et al., 1971; Gorham and Carmichael, 1979) so that possibly the non-toxic clone obtained from the original NRC-1 was originally also non-toxic and not a mutant.

To investigate this aspect, 28 clones of toxic isolates were prepared, cultured and tested for toxicity. Although there were

Table 3. Approximate Toxicity and Percent Gas Vacuoles (% GV) in
          Cultures (Microscopically Determined)[a]

Isolate	LD_{100} mg/kg	% GV
UV-010/1	< 18	0
UV-010/2	14-27	0
UV-010/3	26-28	0
UV-010/4	27-56	99
UV-010/5	18-19	73
UV-010/6	40-65	13
UV-010/7	33-37	62
UV-010/8	60-83	13

[a]Four months earlier all the isolates had 0% gas vacuoles.

differences in the level of toxicity, all the isolates were toxic.
Furthermore, the NRC-1 culture isolated from Little Rideau Lake,
Ontario nearly 26 years ago and received from P. R. Gorham as a
gift in 1973, is still toxic with an LD_{100} of ca. 28 mg/kg (compared to the highest toxicity of 40 mg/kg obtained for the original
isolate, Hughes et al., 1958). The retention of toxicity over at
least 150 reinoculations indicates a high genetic stability as far
as toxin production is concerned.

     This conclusion was confirmed when toxin was extracted from
several toxic *Microcystis* isolates and separated by gel chromatography. Elution characteristics showed that similar toxins, or at
least toxins with a similar molecular mass (ca. 1300), are present
in the different isolates (Amann and Eloff, 1980). Toxins with a
similar molecular mass were also present in the 25 year old isolate
and in freshly isolated *Microcystis* (UV-019). Amann and Eloff
(1980) also found more than one toxin in *M.* UV-006 which is interesting in the light of a type-c toxin found for *Microcystis* (Gorham
and Carmichael, 1979).

Release of Endotoxin in Nature or in Laboratory Tests

     Louw (1950) and Steyn (1945) indicated that intact cells of
Microcystis were not toxic and Grant and Hughes (1953) stated that

natural blooms had to reach a certain critical stage of decomposition before they were toxic to animals drinking the water.

It is not easy to break the cells of *Microcystis*, especially if cell contents are to be investigated, and mild conditions should be used. Wheeler et al. (1942) reported that freezing solubilized the toxin and caused a ten-fold increase in toxicity of *M. aeruginosa* obtained from blooms.

Hughes et al. (1958) found that freezing, incubating (15 hours at 36°C in dark) and ultrasonification yielded the same LD_{100} which was more than four times lower than for non-treated cells. Microscopic investigation of the effect of ultrasonification (Branson B12-microprobe, 3 minutes at setting 5, without glass beads), French press treatment (133 KPa) or freezing (liquid N_2) and thawing (room temperature) in our hands yielded 70%, 49% and 13% breaking of fresh cells of *M*. UV-006, respectively. Although rupture of the plasmalemma would release the endotoxin without necessarily breaking the cell wall our results indicate that if some change in the resistance of the cell wall took place, artifacts on the toxicity of natural blooms may be experienced. On the other hand, we found that freeze-drying of *Microcystis* apparently changes the permeability of the plasmalemma to such an extent that no difference in LD_{50} was observed when freeze-dried cells were ultrasonified or directly extracted with water. This may also explain why we have never been able to grow freeze-dried material of *Microcystis*.

Artifacts Due to Problems with Toxicity Bioassay

Many workers doing toxicological research do not realize the pitfall inherent in using a bioassay, especially if their data is to be compared to those of other authors (see Paget, 1970 for a good general review).

Mice have been generally used to test for toxicity in *Microcystis*. It is important that inbred mice should be used as large differences may occur in genetically heterogeneous strains. The LD_{50} for thiourea ranged from 5 mg/kg to 1830 mg/kg for two strains of rats (Dieke and Richter, 1945). The sex of the test organism may play an important role. The age of the animals also may influence toxicity. Rats younger than two months proved to be 200 to 400 times as resistant as adult rats towards thiourea (Dieke and Richter, 1945). The nutrition of the animals should be standardized. Withdrawal of food from mice 4 to 6 hours before dosage decreased the LD_{50} for methohexidone from 354 to 162 mg/kg and 20 hours of starvation decreased it to 66 mg/kg (Quinton and Reinert, 1968).

The conditions of the animals with regard to microbial infec-
tions and social factors, e.g. crowding, influences toxicity in
some cases (Chance, 1946).

The laboratory environment is important. Increase in temperature from 18°C to 25°C increased the LD_{50} values of chlorpromazine
from 20 mg/kg to 310 mg/kg (Berti and Cima, 1955). Similarly,
parameters such as humidity, light, season and insecticides may
have some effect in specialized cases.

It is therefore possible that artifacts in the bioassay may
have led to the conclusion that the toxicity of natural blooms are
highly variable especially if cell suspensions, not treated to
break cells effectively, are injected into a mounse at only one
dosage level.

Finding range of toxicity. We have spent some effort in
devising a system that gives reproducible results. The LD_{50} determination (Bliss, 1935) is most probably the most accurate measure
of toxicity but it requires many animals and the approximate LD_{50}
has to be known. If the toxic reaction is very fast, the Up-and-
Down method of Dixon and Mood (1948) which depends on sequential
treatment gives very good results and requires fewer animals. The
method of moving averages (Thompson, 1947) is very useful for determining the LD_{50}, especially if the tables of Weil (1952) are
used for the calculation.

For most purposes it is sufficient to have the range of the
toxicity to within 0.1 mg per mouse. We have never had a toxic
culture in which 6 mg dry mass was not sufficient to kill mice.
The following procedure was developed to get an approximation of
the LD_{100}. This procedure is a combination of the methods of
Deichmann and Leblanc (1943) and Maynard in Smith et al. (1960).

Three mice were injected with the equivalent of 2.0, 4.0, and
6.0 mg of dry *Microcystis*. If only the 4.0 and 6.0 mg treatments
died, then the equivalent of 2.5, 3.0, and 3.5 mg was injected.
If e.g. all of these mice died, then four mice were injected with
the equivalent of 2.1, 2.2, 2.3, and 2.4 mg. It was possible to
get an estimate of the LD_{100} within a long working day and using
10 mice (Table 4). The test mice must have a reasonably uniform
mass (ca. 5 to 10%) otherwise e.g. the 2.1 and 2.3 mg treatment
may die while the 2.2 mg treatment may survive. This would
require the injection of two more mice.

Determination of LD_{50}. We investigated various parameters
of the bioassay to determine LD_{50} because we were interested in
determining the influence of environmental conditions on toxicity.

Table 4. Protocol for Determining Range of Toxicity[a]

First Injection (mg; i.p.)	Second Injection (mg; i.p.)	Third Injection (mg; i.p.)
		2.1
		2.2
	→	2.3
		2.4
2.0	2.5	2.5
		2.6
		2.7
	→	2.8
		2.9
4.0	3.0	3.0
		3.1
		3.2
	→	3.3
		3.4
6.0	3.5	3.5
		3.6
		3.7
	→	3.8
		3.9

[a]The arrow indicates lowest level at which mice died. If mice also died at 2.0 mg level, then second injection would have been 0.5, 1.0, and 1.5 mg etc. By three sequential injections using ten mice, 60 doses may be covered.

The mice used were BALB-cAn and their body mass generally ranged from 19 to 24 g. Mice were injected intraperitoneally after diluting extracts with double distilled water. Syringes and needles were washed in double distilled water. Mice that did not die from *Microcystis* toxin always remained alive until they were sacrificed and one even bore a litter.

    To determine the LD_{50}, a preliminary "titration" was made to
get an indication of the approximate LD_{100}. Five concentrations
were chosen around the approximate LD_{100} and five groups of ten
mice were injected with the five concentrations. The dead mice
were noted, weighed and the dose calculated in mg/kg. Due to variation in weight, mice that were injected with more toxin frequently
had a lower dose (mg/kg) than lighter mice that were injected with
less toxin.

    Subsequently the results were arranged from the lowest to the
highest dose and this was divided in five groups of ten mice. The
percentage of dead mice in each group was plotted against the average dose received by the ten mice in each group. The result was
the typical sigmoid curve found with many toxins (Figure 1). The
dose that would have yielded 50% deaths was read from the curve.
In cases where a curve could not be plotted easily, a regression
line was calculated for values between 1% and 99% deaths.

    In the region of 13 mg/kg, which corresponds to a difference
of 0.286 mg for a 22 g mouse, the dose response was relatively
steep. This means that if a test with only a few mice is carried
out, the LD_{50} can only be estimated with a limited accuracy ($\pm$ 6.5
mg/kg).

    Volume and concentration injected. We investigated whether
it was necessary to inject a constant volume (0.5 ml) into the
mice or if different volumes (0.1 to 0.5 ml) could be injected
containing the same amount of toxin. The resulting LD_{50} values
of 10.5 vs 11.0 for the two methods showed that the volume used
for injecting the toxin made no difference.

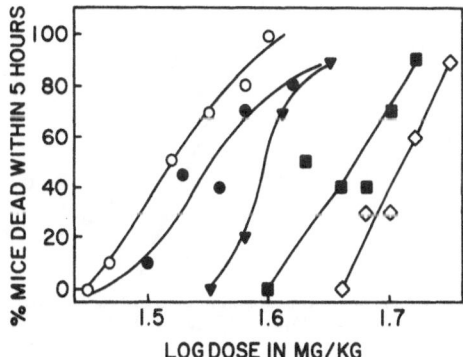

Fig. 1. Determination of LD_{50} on cultures of *M.* UV-006 grown
         at different pH values from left to right: 7.0 (o),
         10.5 (●), 9.5 (▼), 7.5 (■), 8.0 (◇).

Influence of sex of mice. A comparison was made between male
and female mice by injecting a constant volume (0.5 ml) of *M*. UV-
006 freeze-dried cells ultrasonified in water. The results showed
that female mice were more sensitive to microcystin than male mice
with LD_{50} values of 10.5 vs 13.9 mg/kg, respectively.

Dosage/death time relation. In the above experiment the influence of the dose on survival time was examined. Cases have been
found where survival time and toxin concentration correlate well,
as for example the influence of NaCl on young adult albino rats
(Boyd and Shanas, 1963). From the regression lines (not significant at $P = 0.05$ but significant at $P = 0.10$ level) it appeared
that the higher the toxin concentration, the sooner the mice died
(Figure 2). However, survival time is of little use in evaluating
toxin concentration. The survival time in our experiments (70 to
210 minutes for female and 90 to 290 minutes for male mice) was
longer than that found elsewhere, e.g., 30 to 60 minutes (Gorham,
1964), 60 to 90 minutes (Wheeler et al., 1942), and 30 to 180 minutes (Hughes et al., 1958) although > 90% of the female mice died

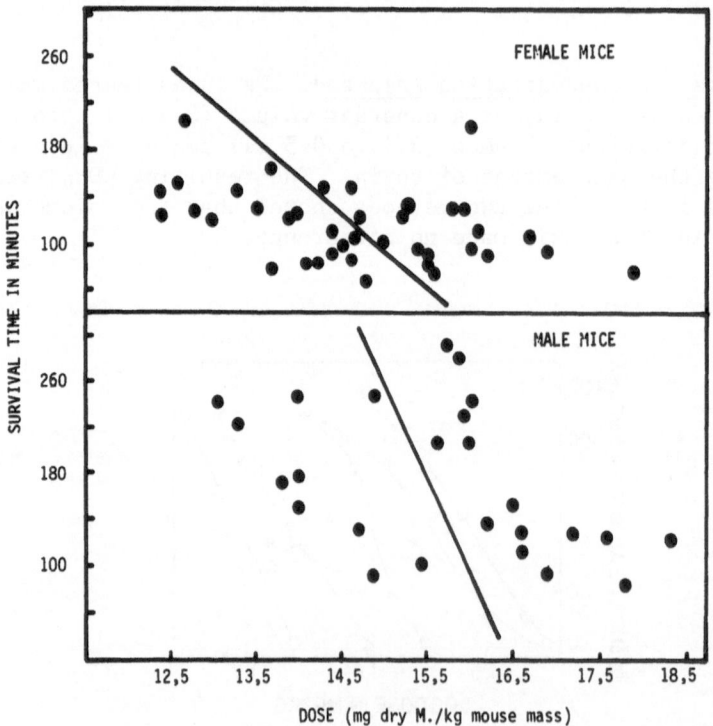

Fig. 2.   The survival time of male and female mice injected with
          different doses of *M*. UV-006.

within 70 to 150 minutes. Subsequently we decided to include all mice dying within 5 hours in our assay.

Extraction procedure. When the influence of ultrasonification and ultracentrifugation on the toxicity of M. UV-006 freeze-dried cells was determined, it was found that ultrasonification did not increase the quantity of toxin extracted. All the toxin was present in the supernatant when the suspension was centrifuged at 74000 g for 15 minutes. In the ultrasonicated cells, some of the toxin was either inactivated during the extraction or precipitated during ultracentrifugation. The procedure of extracting freeze-dried cells and removing cell debris by ultracentrifugation has the advantage of removing not only cell components, that may be toxic to the mice or that may inactivate the toxin, but also any bacteria possibly contaminating the *Microcystis* culture.

Immunity development. Some animals apparently develop resistance towards *Microcystis* poisoning which suggests that microcystin may be antigenic and opens up the possibility of immunoanalysis. This seems doubtful in view of the small size of the toxic peptide in extracts of *Microcystis*. To test this possibility preliminarily, sixty female mice were injected with a centrifuged water extract of axenic M. UV-006 freeze-dried cells in a dose of ca. LD_{10}. The surviving mice and equal groups of mice (that were not injected with the toxin) were injected with the equivalents of 0.6, 0.5, 0.4, and 0.3 mg of the same freeze-dried cells 14 days later. The results (LD_{50} for mice previously injected was 17.1 mg/kg vs LD_{50} of 17.4 mg/kg for mice that were not previously injected) showed no change in the sensitivity of mice towards microcystin.

The results demonstrate that the method of determining LD_{50} was highly reproducible. Mice, which have been injected with bacteria-free extracts, may be used more than once if they survived previous injections, although this should only be done with great circumspection.

The Influence of Environmental Parameters on Toxicity

Microcystis aeruginosa forma *aeruginosa* blooms are not always toxic (W. E. Scott, personal communication) therefore changes in the type of *Microcystis* present is probably not the only explanation for varying toxicity. If environmental parameters influence toxicity they may cause short term changes in toxicity observed in natural blooms.

Parameters that influence toxin production, extractability of toxin or ease of decomposition of toxin would all have an effect on the toxicity of cultures.

Using large quantities of toxic cells, Hughes et al. (1958)
determined the influence of culture age on the toxicity of bacterized cultures. They found variable results, in one experiment
toxicity decreased from day 4 to day 14 (LD_{100} 40 mg/kg to 480 mg/kg)
and in another experiment toxicity stayed more or less constant and
then increased with LD_{100} of 80 to 160 mg/kg.

Harris and Gorham (1956) found that toxicity increased until
the lag phase was reached, then decreased. The aeration rate and
temperature had a large effect on toxicity with cultures yielding
5 to 10 times more toxin at 25°C than at 20° or 30°C. Light intensity also influenced toxicity, especially at higher temperatures.
The highest toxicity obtained in their experiments (590 mouse units
per liter of a 0.9 g per liter cell density) corresponds to an
LD_{100} of ca. 60 mg/kg.

Because we are interested in the biosynthesis of the toxin
one of the aspects we have recently started to investigate is the
influence of growth parameters on toxicity.

Culture age. Because of the conflicting results reported by
other workers, we repeated an examination of the effect of culture
age on toxicity by growing axenic *Microcystis* UV-010 in the 60 liter
mass culture system (Krüger and Eloff, 1978 and paper, these Conference Proceedings) at 26°C, pH 9.0 with a light intensity of 41
$\mu E./m^2/s$ up to day 2 and then 105 $\mu E./m^2/s$. Toxicity was determined by injecting 2 male mice with each of 4 to 7 different concentrations of each successive sample until the lowest lethal concentration (ca. = LD_{100}) could be determined. The results (Figure
3) corroborate the work of Harris and Gorham (1956). The toxicity
increased up to day 4, the exponential growth phase of the culture,
and then decreased when the cells entered the stationary phase.

In this system, pH, CO_2 concentration, temperature and light
intensity were kept constant. Although the light intensity per
cell decreased due to increased turbidity, the decrease in toxicity
is most probably an effect of culture age.

Cells in the stationary or death phase of growth were less
toxic than exponentially growing cells and this could be related
to the varying toxicity measured in natural blooms.

pH of the culture. Axenic cells of *M.* UV-006 were grown at
27° to 29°C at an average light intensity of 100 $\mu E./m^2/s$ in the
mass culture system (Krüger and Eloff, 1978, and paper, these Conference Proceedings) at the following pH values: 6.5, 7.0, 7.5,
8.0, 8.5, 9.0, 9.5, 10.0, and 10.5. The pH was monitored by a gel-
filled electrode which had no influence on the growth of *Microcystis*
(see paper by Eloff, these Conference Proceedings). Whenever pH

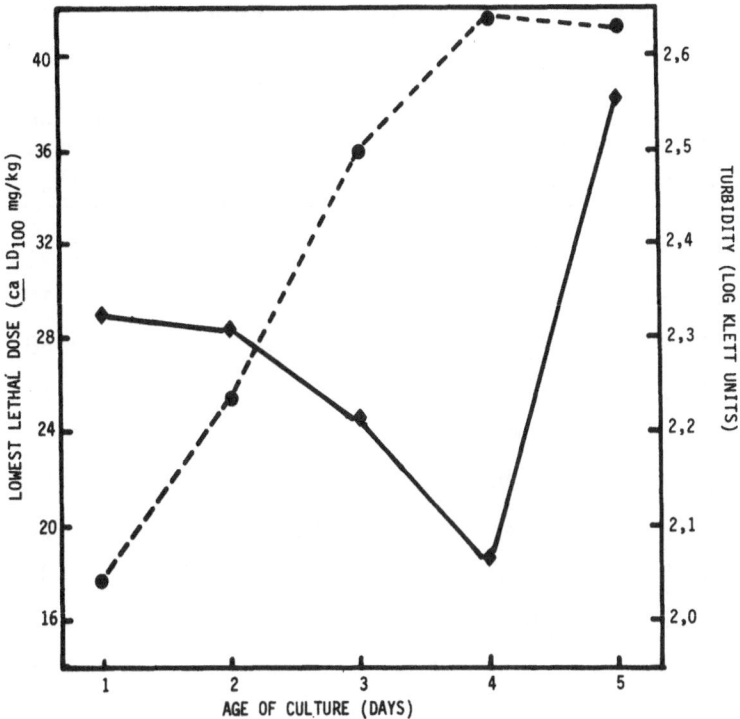

Fig. 3. The influence of culture age on growth and toxicity of
         M. UV-010.

rose above the required value, it was controlled by automatic
addition of CO_2. *M.* UV-006 would not grow at pH 11 or 6.0 under
these conditions. The experiment was started with high pH values
and in late logarithmic growth phase at a turbidity value of 200,
ca. 95% of the culture was cropped. The pH control setting was
changed, new medium added and 5% of the previous culture used as
inoculum until all the required cells at the different pH values
were obtained. At the pH 9.0 and 10.5 treatments, the cultures
went into the stationary growth phase overnight and were immediately cropped. Considering the effects of culture age (see §
titled Culture age) this could mean that the toxicity values
determined for pH 9.0 and 10.5 could be too low. The cells were
collected by centrifugation and stored at -20°C after freeze-
drying.

No bacteria were found when cells were observed (100 x phase-
contrast optics) at the end of the experiment. The LD_{50} was
determined (see § titled Determination of LD_{50}) using ca. 60 mice
in each treatment. The results are presented in Table 5 and
Figure 4.

Table 5.   Influence of pH on Toxicity and Growth of *Microcystis* UV-006.

pH	LD$_{50}$ mg/kg	Regression Coefficient (r^2)	Specific Growth Rate (μ)	Doubling Time Days (t_d)
10.5	36.20	0.91	0.30	2.31
10.0	38.73	0.95	0.33	2.10
9.5	39.99	0.92	0.39	1.80
9.0	47.01	0.52	0.60	1.16
8.5	42.00	0.91	0.30	2.29
8.0	51.47	0.98	0.31	2.22
7.5	46.49	0.82	0.34	2.04
7.0	33.65	0.98	0.29	2.35
6.5	32.57	0.90	0.20	3.39

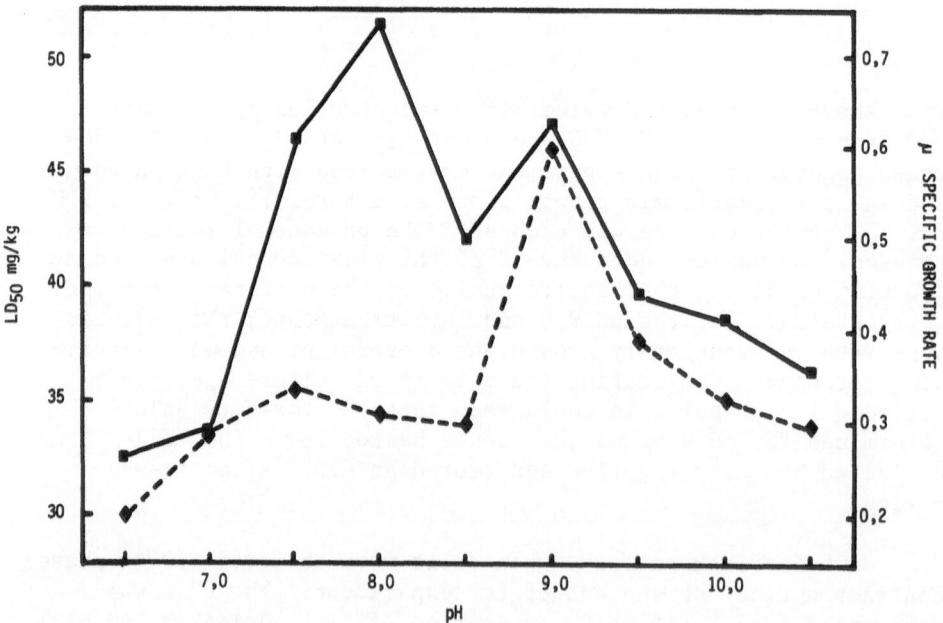

Fig. 4.   The influence of pH on specific growth rate and toxicity of *M.* UV-006.

The results showed that pH influences the toxicity of *M.* UV-006. Cells were most toxic at the higher and lower pH values. There was a close correlation between the reciprocal of toxicity (LD_{50}) and the growth rate, i.e., the slower the cells grew the more toxic they were.

Changes in pH may influence the toxicity of *M.* UV-006, and this is another factor that may influence the toxicity of natural blooms although toxicity did not change much at the pH values above 9.5 where *Microcystis* usually grows.

CONCLUSION

The changes in toxicity of natural blooms over a long period could be ascribed to succession of non-toxic taxons of *Microcystis*. Bacteria do not seem to play an important role in the toxicity of *Microcystis*, but may be able to decompose the toxin.

The short time changes in toxicity of natural blooms of *Microcystis* may be due either to errors in bioassay or to the influence of environmental parameters such as light intensity, temperature, age of cells and pH.

ACKNOWLEDGMENTS

The financial assistance of the National Institute for Water Research, Council for Scientific and Industrial Research and of the University of the Orange Free State is gratefully acknowledged.

REFERENCES

Amann, M. J., and J. N. Eloff. 1980. A preliminary study on the toxins of different *Microcystis* strains. S. Afr. J. Sci. (In press).

Berti, T., and L. Cima. 1955. Einfluss der temperatur auf die pharmakologische wirkung des chlorpromazins. Arzneimittel-Forsch. *5*:73.

Bliss, C. I. 1935. The calculation of the dosage-mortality curve. Ann. Appl. Biol. *22*:134-167.

Boyd, E. M., and M. M. Shanas. 1963. The acute oral toxicity of sodium chloride. Arch. Int. Pharmacodyn. Ther. *144*:86.

Chance, M. R. A. 1946. Aggregation as a factor influencing the toxicity of sympathomimetic amines in mice. J. Pharmacol. Exp. Ther. *87*:214.

Deem, A. W., and F. Thorpe. 1939. Toxic algae in Colorado. J. Am. Vet. Med. Assoc. *95*:542-544.

Deichmann, W. B., and J. J. Leblanc. 1943. Determination of the
    approximate lethal dose with about six animals. J. Ind.
    Hyg. Toxicol. 25:415-417.

Dieke, S. H., and C. P. Richter. 1945. Acute toxicity of thiourea
    to rats in relation to age, diet, strain and species vari-
    ation. J. Pharmac. Exp. Ther. 83:195.

Dixon, W. J., and A. M. Mood. 1948. A method for obtaining and
    analyzing sensitivity data. J. Am. Stat. Assoc. 43:109-126.

Elleman, T. C., I. R. Falconer, A. R. B. Jackson, and M. T.
    Runnegar. 1978. Isolation, characterization, and pathology
    of the toxin from a Microcystis aeruginosa (= Anacystis
    cyanea) bloom. Aust. J. Biol. Sci. 31:209-218.

Eloff, J. N., and D. N. Sadie. 1979. Unpublished results.

Golubic, S. 1979. Cyanobacteria (blue-green algae) under the
    bacteriological code? An ecological objection. Taxon 28(4):
    387-389.

Gorham, P. R. 1960. Toxic waterblooms of blue-green algae. Can.
    Vet. J. 1:235-245.

Gorham, P. R. 1964. Toxic algae. Pages 307-336 in D. F. Jackson,
    ed., Algae and Man. Plenum Press, New York. 434 pp.

Gorham, P. R., and W. W. Carmichael. 1979. Phycotoxins from blue-
    green algae. Pure Appl. Chem. 52(1):165-174.

Grant, G. A., and E. O. Hughes. 1953. Development of toxicity in
    blue-green algae. Can. J. Public Health 44:334-339.

Harris, R. E., and P. R. Gorham. 1956. (Unpublished) Cited in
    P. R. Gorham, 1964. Toxic algae. Pages 307-336 in D. F.
    Jackson, ed., Algae and Man. Plenum Press, New York.

Hortobágyi, T. 1943. Beiträge zur Kenntnis der im Boglarer
    Seston, Psammon und Lasion lebenden algen des Balaton-Sees.
    Arb. Ung. Biol. Forschungsinstitut 15:75-127.

Hortobágyi, T. 1947. New observations on some algae from Lake
    Balaton. Bot. Közlemenyek 44:39-54.

Huber-Pestalozzi, G. 1929. Algologische mitteilungen. VI. Algen
    aus dem Lago di Muzzano. Arch. Hydrob. 20:413-426.

Hughes, E. O., Gorham, P. R., and A. Zehnder. 1958. Toxicity of
    a unialgal culture of Microcystis aeruginosa. Can. J.
    Microbiol. 4:225-236.

Ingram, W. M., and G. W. Prescott. 1954. Toxic fresh-water algae.
    Am. Mid. Nat. 52:75-87.

Kessel, M., and J. N. Eloff. 1975. The ultrastructure and devel-
    opment of the colonial sheath of Microcystis marginata.
    Arch. Microbiol. 106:209-214.

Komárek, J. 1958. Die taxonomische revision der planktischen
    blaualgen der Tschechoslowakei. In J. Komárek and M. Ettl,
    eds., Algologische Studien. Tschechoslowakische Akademie
    der Wissenschaften, Prague.

Krüger, G. H. J., and J. N. Eloff. 1978. Mass culture of Micro-
    cystis under sterile conditions. J. Limnol. Soc. S. Afr.
    4(2):119-124.

Krüger, G. H. J., J. N. Eloff, and J. A. Pretorius. 1980. Mor-
    phological changes in toxic and non-toxic *Microcystis* iso-
    lates at different irradiance levels. (In press).
Lau, R. H., and W. F. Doolittle. 1979. Covalently closed circular
    DNAs in closely related unicellular Cyanobacteria. J. Bac-
    teriol. *137*:648-652.
Louw, P. G. J. 1950. The active constituent of the poisonous
    algae, *Microcystis toxica* Stephens. S. Afr. Ind. Chemist
    4:62-66.
Olson, T. A. 1951. Toxic plankton. Pages 86-95 *in* Proceedings
    of Inservice Training Course in Water Works Problems,
    February 15-16. University of Michigan, School of Public
    Health, Ann Arbor, Michigan.
Paget, G. E., ed. 1970. Methods in Toxicology. Blackwell
    Scientific Publications, Oxford. 390 pp.
Pretorius, J. A. 1977. Groei van *Microcystis aeruginosa* op Agar-
    medium. M.Sc. Thesis. University of the Orange Free State,
    Bloemfontein, Republic of South Africa.
Quinton, R. M., and H. Reinert. 1968. Cited in J. K. Morrison,
    R. M. Quinton, and H. Reinert, 1968. The purpose and value
    of LD_{50} determinations. Pages 1-17 *in* E. Boyland and R.
    Goulding, eds., Modern Trends in Toxicology, Vol. 1. Butter-
    worths, London.
Rippka, R., J. Deruelles, J. B. Waterbury, M. Herdman, and R. Y.
    Stanier. 1979. Generic assignments, strain histories and
    properties of pure cultures of Cyanobacteria. J. Gen.
    Microbiol. *111*:1-61.
Simon, R. D. 1978. *Halobacterium* strain 5 contains a plasmid
    which is correlated with the presence of gas vacuoles.
    Nature (Lond.) *273*:314-317.
Smith, F. A., W. L. Downs, H. C. Hodge, and E. A. Maynard. 1960.
    Screening of fluorine-containing compounds for acute toxicity.
    Toxicol. Appl. Pharmacol. *2*:54-58.
Stanier, R. Y., R. Kunisawa, M. Mandel, and G. Cohen-Bazire. 1971.
    Purification and properties of unicellular blue-green algae
    (order Chroococcales). Bacteriol. Rev. *35*:171-205.
Stephens, E. L. 1949. *Microcystis toxica* sp. nov.: a poisonous
    alga from the Transvaal and Orange Free State. Trans. R.
    Soc. S. Afr. *32*(1):105-112.
Steyn, D. G. 1945. Poisoning of animals by algae (scum and water-
    bloom) in dams and pans. Union of South Africa, Department
    of Agriculture and Forestry, Government Printer, Pretoria.
    9 pp.
Thompson, W. R. 1947. Use of moving averages and interpolation
    to estimate median-effective dose. I. Fundamental formulas,
    estimation of error, and relation to other methods. Bac-
    teriol. Rev. *11*:115-145.

Thomson, W. K., A. C. Laing, and G. A. Grant. 1957. Toxic algae.
    IV. Isolation of toxic bacterial contaminants. Can. Defence
    Research Board, Defence Research Kingston Laboratory. Report
    No. 51, Project No. D52-20-20-18. Ottawa. 9 pp.
Toerien, D. F., W. E. Scott, and M. J. Pitout. 1976. *Microcystis*
    toxins: isolation, identification, implications. Water S.
    A. (Pretoria) 2:160-162.
Walsby, A. E. 1977. Absence of gas vesicle protein in a mutant of
    Anabaena flos-aquae. Arch. Microbiol. 114:167-170.
Weil, C. S. 1952. Tables for convenient calculation of median
    effective dose (LD_{50} or ED_{50}) and instructions in their use.
    Biometrics 8:249-263.
Wheeler, R. E., J. B. Lackey, and S. A. Schott. 1942. Contribution
    on the toxicity of algae. Public Health Rep. 57:1695-1701.

EFFECTS OF A HEPATIC TOXIN FROM THE CYANOPHYTE

MICROCYSTIS AERUGINOSA

Thomas L. Foxall and John J. Sasner, Jr.

Zoology Department
Spaulding Life Science Building
University of New Hampshire
Durham, New Hampshire 03824

ABSTRACT

Microcystis aeruginosa is commonly involved in freshwater
blooms and one of its toxins (microcystin) causes liver damage in
birds and mammals. This study determined the site of action of
microcystin and characterized the hepatic damage at the ultrastructural level. Histological changes in centrilobular regions of
liver tissue were noted after i.p. administration in mice. Hepatic
sinusoids expanded, parenchymal cords disintegrated, and cells
lysed resulting in extensive tissue damage and subsequent death
within one hour. Ultrastructural studies showed that sinusoidal
epithelium and hepatocyte plasma membranes ruptured with the release of cellular components that pooled with blood. Mitochondria
appeared swollen but there were no obvious distortions of organelles. Extensive vesiculation of membrane fragments was observed.
Hepatic damage caused hemorrhaging into the liver where blood and
cell debris accumulated and produced a significant increase in
liver weight.

Primary cultures of hepatocyte microexplants from pre- and
postnatal mice and rats were exposed to microcystin but in vitro
effects were not observed. In vivo experiments demonstrated that
young animals were not sensitive to the toxin but developed sensitivity as they matured.

Microcystin was shown to be very specific in its site of
action since it had no antibiotic activity against green algae,
yeast, or bacteria and was non-toxic to certain zooplankton,
crayfish, amphibians and teleosts. Electrical or mechanical

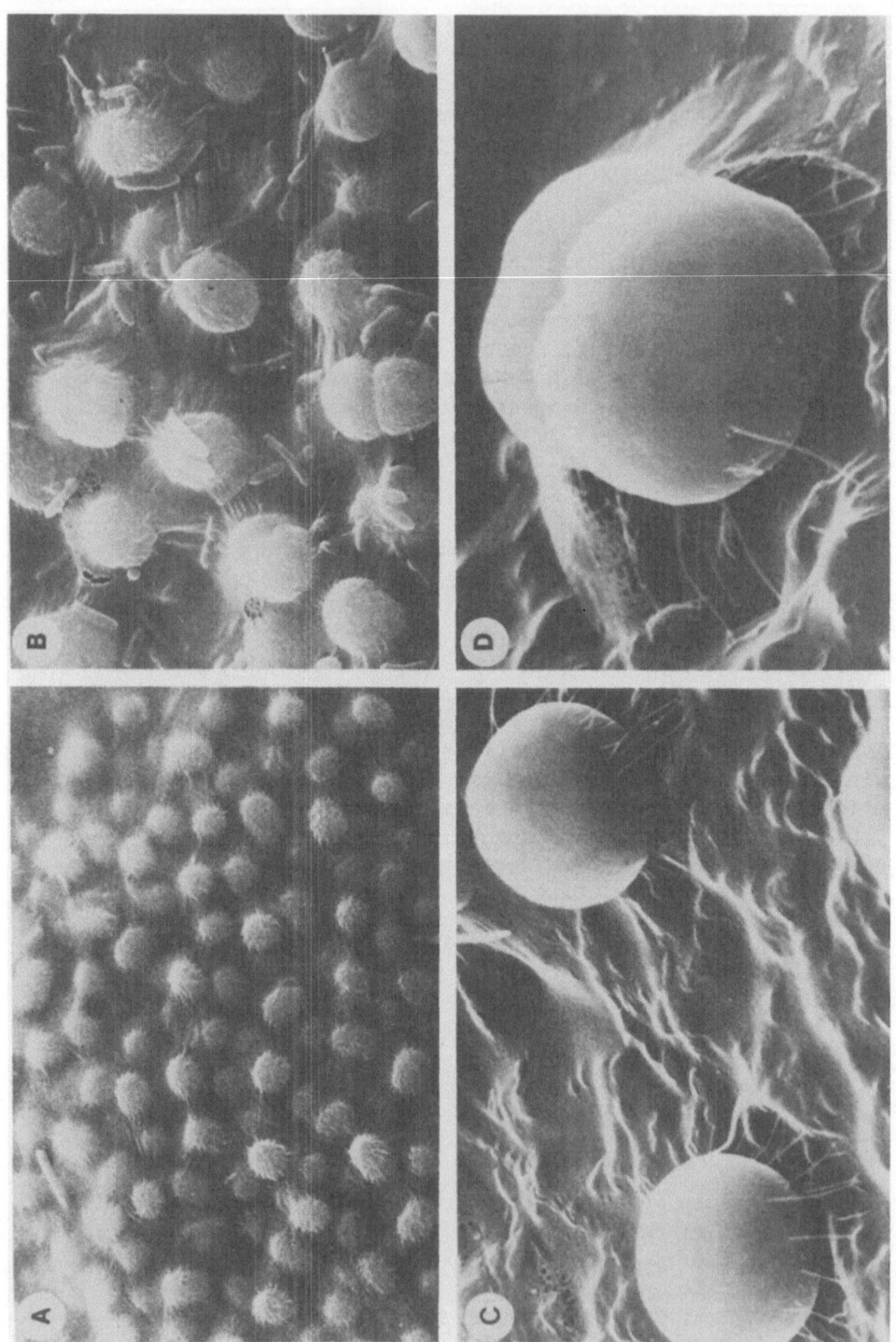

activity in isolated nerve, nerve-muscle and cardiac preparations
also was not effected. The toxins employed in this study came from
cultured cells and naturally occurring blooms and gave identical
results in all experiments.

INTRODUCTION

When freshwater ponds, lakes and reservoirs become eutrophic,
whether naturally or through man's intervention, phytoplankton
blooms of various algae and/or Cyanobacteria (blue-green algae)
are likely to occur. Freshwater phytoplankton blooms may be essentially monospecific or consist of several species and some of these
may be the source of potent toxins. These blooms are worldwide in
occurrence (Gorham, 1964).

Cyanophytes that occur in toxic form and that appear to be the
major organisms responsible for toxic blooms are *Aphanizomenon
flos-aquae*, *Anabaena flos-aquae* and *Microcystis aeruginosa* (Echlin,
1966; Gorham, 1964; Moore, 1977). All three of these species occur
when toxic Cyanobacteria are involved in freshwater blooms.

Unlike *Aphanizomenon* and *Anabaena* that produce neurotoxins
(Carmichael, 1975; Sasner, 1973), *Microcystis aeruginosa* (Figure 1)
produces a toxin called Fast Death Factor (FDF) (Hughes et al.,
1955; Bishop et al., 1959) and later called microcystin (Konst et
al., 1965). The further elucidation of the site of action of this
toxin is the focus of this study.

Blooms of *M. aeruginosa* are not always toxic. Field samples
may vary in potency with season and location. Laboratory monospecific cultures vary in toxicity with several physical, chemical
and biological factors (Simpson et al., 1958; Gorham, 1962, 1964).

There are at least three different toxins associated with
Microcystis: Fast Death Factor (FDF) that produces symptoms and
death in mice within one hour, Slow Death Factor (SDF) that produces death in mice within 48 hours, and a diarrheagenic factor.

Fig. 1. (A) A colony of toxic *(Microcystis aeruginosa* cells in a
gelatinous mass (1200X), (B) Toxic *M. aeruginosa* cells
showing the presence of bacteria living within the gelatinous mass (3500X), (C) A higher magnification of two
single cells anchored to the gelatinous mass with filaments but not covered by it in this preparation (4500X),
(D) A pair of toxic cells from which the sheath material
is missing (7200X).

FDF is an endotoxin produced by *M. aeruginosa* while SDF and the diarrheagenic factor are most likely products of the bacteria associated with *Microcystis* (Aziz, 1974; Gorham, 1962, 1964; Hughes et al., 1955; Thompson et al., 1957). The cyanobacterial species of interest in this study is the source of FDF, but may have other harmful materials associated with it, i.e., SDF and diarrheagenic factor.

Schwimmer and Schwimmer (1964) reviewed cases of poisoning in wild and domestic mammals and birds that consumed water containing dense concentrations of cyanophytes. Domestic animals most effected were cattle, sheep and birds (Stephens, 1949). Laboratory studies revealed that sheep, cattle (including calves), guinea pigs, rabbits, cats, rats, mice, and chickens were sensitive to the effects of the toxin (Gorham, 1962; Mason and Wheeler, 1942; Konst et al., 1965).

A narrow range of values for minimum lethal dose, LD_{100} and LD_{50} appears in the literature. Gorham (1964) established that the survival time for a minimum lethal dose of FDF in 20 gram mice was 30 to 60 minutes. For bioassay purposes in comparative studies one mouse unit (MU) was defined as the amount of toxin (FDF) required to kill 20 gram mice in 60 minutes after intraperitoneal injection.

The first attempt at isolation and chemical characterization of FDF was done by Louw (1950) using a strain of *Microcystis* from a South African lake. The cells were lysed and extracted and the resulting toxin gave a positive reaction for alkaloids and a molecular weight of 220 daltons was determined for the hydrochloride salt. Another isolation of toxin from the NRC-1 strain of *Microcystis* was accomplished from laboratory cultures by Bishop, et al. (1959). This work yielded an electrophoretically pure toxic peptide (later called microcystin, Konst et al., 1965) with an LD_{50} of 0.46 ± 0.013 mg/kg body weight in mice. All the toxicity in crude extracts was accounted for by this peptide. Additional work indicated that this was a cyclic peptide and contained D-serine. Murthy and Capindale (1970) used the same NRC-1 strain to produce an ammonium salt of the toxin that was chromatographically and electrophoretically homogeneous. In addition to the seven amino acids reported by Bishop et al. (1959), they found another seven amino acids. This toxic peptide had an LD_{50} of 0.1 mg/kg body weight (i.p.) in mice. The accumulated information thus far suggests that microcystin is a homeomeric peptide with an estimated molecular weight between 1300 and 2600 daltons. In a more recent study (Kirpenko et al., 1975) a molecular weight of 19400 daltons was ascribed to a biologically active material extracted from mixed algal samples containing *M. aeruginosa*.

Studies on the physiological effects and pathology of toxic *Microcystis* extracts were done at least 25 years ago. Mason and Wheeler (1942) reported that the administration of toxin (i.p.) to mice produced a 20 minute latent period, followed by pallor, lowered blood pressure and body temperature, hyperglycemia and tachycardia, and death by respiratory failure. During the latter stages of involvement the hematocrit, hemoglobin concentration, RBC count and total serum protein were all lowered. There was evidence for some involvement of the heart in both mice and amphibians, but the toxin(s) did not alter oxygen consumption values of liver, kidney, diaphragm or brain tissues.

Ashworth and Mason (1945) challenged adult rats with a "maximum" sublethal dose of *Microcystis* toxin and measured gross organ changes over a period of 30 days. They noted an enlarged, tense or turgid, "redder than normal" liver within 30 minutes after toxin administration. Liver size increased along with the blood volume surrounding parenchymal cells after 3 to 6 hours, at which time the tissue was soft and friable. No gross changes were noted in other organs. The liver continued to degenerate and other organs showed slight pathological changes. After 5 days the liver appeared more normal, as did other organs, and within 30 days no gross changes were noted in any organs. The liver showed neither nodular regeneration, fibrosis, nor cirrhosis. These same authors also reported on the histological changes versus time after toxin administration (i.p.) in rats. Liver cells became swollen, hepatocyte cytoplasm was more granular and in some areas was hydropic, clear halos were noted in the perinuclear cytoplasm, fat droplets appeared in the center of lobules and the sinusoids were distended with RBC's. There was marked necrosis in the centrilobular areas where there was dissociation of liver cords with remnants of liver cells lying free in the blood in confluent sinusoids. Isolated hepatocytes were swollen and rounded and the cytoplasm was hyalinized and eosinophilic. Work by Konst et al. (1965) showed no pronounced changes in lungs, intestine, or kidneys and no hematological changes in mice. They also noted a close correlation between the symptoms from naturally occurring *M. aeruginosa* and laboratory cultures (NRC-1). Histological work showed that cells in the central zone of each lobule were necrotic or replaced by pools of blood.

Heaney (1971) also demonstrated hepatic damage in white mice. The livers became dark red, had congestion in the centrilobular areas, swollen cells, and localized subcapsular hemorrhaging.

Prescott (1949) sites Steyn (1945) who reported that the toxin from *Microcystis* species produced pathological conditions such as constipation and decreased milk production in cattle and sheep. Orlovskii and Kirpenko (1976) reported that the biological

oxidation of carbohydrates in warm-blooded animals was altered by
toxins(s) from a dense growth of blue-green algae.

The past work has left ambiguous and conflicting information
about microcystin (FDF) in the literature, some of which does not
agree with present knowledge. These anomalies will be discussed
below, in light of the results of this study. Most of the former
research was done with materials from cultures, field samples or
extracts that may have contained mixtures of microcystin, SDF, and
diarrheagenic factor. Although histological changes resulting from
acute and chronic doses of toxin(s) were described, the cellular,
subcellular and biochemical changes that are the ultimate cause of
tissue damage have not been elucidated.

MATERIALS AND METHODS

Samples of *Microcystis aeruginosa* used in this study were
gathered from several sources. A starter culture was obtained
from Dr. Paul Gorham, Edmonton, Alberta, Canada via Dr. John H.
Gentile, National Marine Water Quality Laboratory, West Kingston,
Rhode Island and cultured in our laboratory between 1970 and 1972.
Field samples of toxic cells were harvested from Kezar Lake, North
Sutton, New Hampshire and used for comparative studies versus the
NRC-1 strain. Samples of non-toxic strains of *M. aeruginosa* were
also collected at Kezar Lake, New Hampshire and were obtained from
Dr. Alan L. Baker, Botany Department, University of New Hampshire,
Durham, New Hampshire. Cells were concentrated by centrifugation
and lysed by freezing and thawing the concentrate 3 times. The
lysed cell mixture was then centrifuged to yield a clear blue supernatant that was used as the crude extract. Toxicity was checked by
i.p. injection of 1.0 ml into mice (strain B6D2F1/J, Jackson Laboratory, Bar Harbor, Maine).

Partial purification of the FDF was accomplished using Diaflo
Ultrafiltration Membranes (Amicon Corp., Lexington, Massachusetts)
and Immersible Molecular Separators (Millipore Corp., Bedford,
Massachusetts). Toxic samples were filtered through the molecular
sieves using a magnetic stirrer for constant mixing and nitrogen
gas at 30 psi as a driving force. The following sieve sizes were
used: 100,000, 50,000, 30,000, 20,000, 1,000 and 500 daltons. A
10,000 dalton immersible separator was used on a vacuum line.

A gross anatomical and histological survey of mouse organs
was conducted in an attempt to locate the lesions caused by both
toxic crude extracts and partially purified extracts of *Microcystis*.
All mice were maintained under controlled conditions of temperature,
daylength and diet. All animals were injected and sacrificed at
approximately the same time of day. Non-toxic extracts were used

as controls and were compared to tissues from untreated mice. Tissue samples were excised, fixed in 10% phosphate buffered formalin (pH 7.0), and stained with hematoxylin and eosin.

Gross liver changes were observed and measured. Two groups of 15 mice (20 g ± 2 g) were injected, i.p., with 1.0 ml of aqueous extract from 1.0 mg of lyophilized toxic or non-toxic *Microcystis*. All mice either died or were sacrificed by cervical separation at 60 minutes from the time of injection. The weights and volumes of the whole livers from each animal were recorded to the nearest 0.01 g and 0.1 ml, respectively. The percent of liver:body weight was calculated for each animal and routine statistical analysis of the data was performed for each group. Significance was checked at the 95 and 99 percent confidence levels.

To assess the development of structural damage in liver tissue, a toxin-time study was conducted. A group of mice (20 g ± 2 g) were challenged, i.p., with 1 MU of toxin and were sacrificed by cervical separation after 10, 20, 40, and 60 minutes. At least 4 mice were used for each time interval and liver tissues were prepared for histological and ultrastructural study. An equal number of control animals were injected with an equivalent amount of non-toxic *M. aeruginosa* extract and control liver samples were processed along with the experimental preparations described above. Thus, there were normal untreated mouse liver samples, preparations from mice given non-toxic extracts, and preparations from mice given a lethal dose of toxin for varying time periods.

Immediately after sacrificing the mice, tissue samples were removed for standard histological and ultrastructural transmission electron microscope (T.E.M.) analysis. The liver tissues for T.E.M. were cut into 1 mm^3 pieces in a primary fixation of 3% cacodylate (1.0 M) buffered glutaraldehyde (pH 7.2 to 7.4). The glutaraldehyde contained approximately 0.05 ml of 1% $CaCl_2$ for each 10 ml of fixative. Primary fixation was for 2 hours at room temperature. The tissues were then washed with cold cacodylate-sucrose solution (0.1 M disodium cacodylate buffer, pH 7.0, 400 milliosmols adjusted with sucrose) and kept overnight at 4°C in the wash. The wash was replaced with a post-fixative of 1% osmium tetroxide (OsO_4) in 0.1 M Millonig's phosphate buffer (pH 7.2) for 1 hour. The tissues were dehydrated in ethanol. Propylene oxide was added and then replaced with a 50-50 mixture of complete embedding resin (Epon 812, Polysciences, In.) and propylene oxide. This mixture was replaced with complete resin in which the tissues were left overnight. After embedding, the pieces of liver were thin sectioned on a Reichert OM-2 thermal advance ultramicrotome; the sections were mounted on copper grids and stained with saturated uranyl acetate, washed and dried, and stained again with Reynold's lead citrate. Grids were scanned and sections photographed on both

the Philips 200 and the JEOL 100S transmission electron microscopes
at Electron Microscope Facility at University of New Hampshire.

Primary hepatocyte cultures of monolayers and microexplants
(5 to 10 cells) were prepared using established procedures. Hepatocytes were seeded into T flasks and Leighton tubes and maintained
in MAB 87/3 medium (GIBCO, Inc.) with the addition of 10% fetal
calf serum (heat inactivated). The hepatocytes were exposed to
toxic and non-toxic extracts in vitro for up to 90 minutes, fixed
and stained.

Experiments using rats and mice of various ages were conducted
to determine if neonatal, juvenile and mature mammals had the same
or different sensitivities to microcystin. All animals were challenged with i.p. injections of toxic crude extracts. Death times
were noted, weights were recorded and liver tissue was removed from
each animal for histological examination.

Experiments were conducted to determine if previous exposure
to sublethal doses of toxin might affect animal (mouse) sensitivity
to microcystin. All mice in this study were of approximately 20 g
(± 2 g) and were of the same sex. Mice were given i.p. injections
of sublethal dilutions of toxic crude or partially purified extracts.
The mice were then challenged with a lethal dose of toxic extract
and symptoms and/or death times were noted. Tissues were again
removed for histological examination.

Electrophysiological and mechanical measurements on nerve
and muscle tissues can be a valuable means of determining the site
and mode of action of aquatic biotoxins. The following standard
preparations were challenged with the toxin from *Microcystis* by
direct bathing or injection: frog and mouse heart, frog sartorius
muscles, frog sciatic nerves and mouse ileum.

Fig. 2.   Untreated control mouse liver cells. (A) Part of a single
          hepatocyte showing the normal components of such cells.
          Note the pools of glycogen (GLY), the numerous mitochondria
          (M), and areas containing large amounts of lamellar rough
          endoplasmic reticulum (RER) normally attached ribosomes.
          Lipid droplets (L) are present and one bile canaliculus
          (BC) is in view. The nucleus (N) shows normal morphology
          and arrangement of euchromatin, heterochromatin and nucle-
          olus (6000X). (B) A section showing a sinusoid (SIN) with
          its endothelial lining on top of hepatocytic microvilli
          (7500X). (C) A higher magnification showing the surface
          detail of one hepatocyte (H) with its surface microvilli
          (MV). Endothelium (ENDO) forms the lining of the sinu-

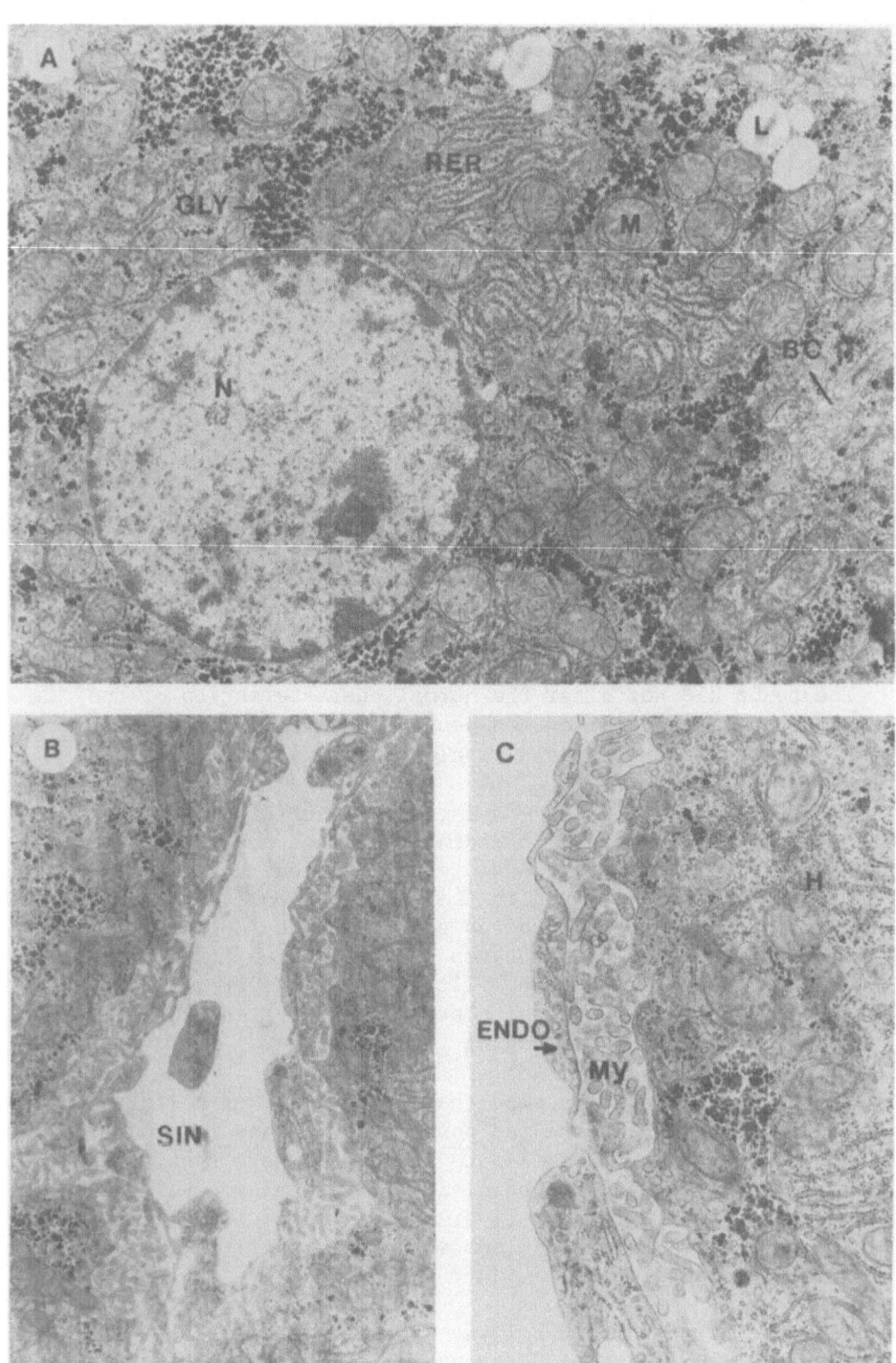

Since microcystin's proposed cyclic peptide structure may be
similar to certain peptide antibiotics, tests were designed to
measure effects on the growth of microorganisms. The test organisms
included Gram-negative and Gram-positive bacterial species, a
fungus, and a green alga.

Several aquatic animals, that might come in contact with toxic
blooms of *M. aeruginosa* in nature, were challenged with toxin via
immersion and/or intraperitoneal injection. Tadpoles and adult
frogs, crayfish, fish, and cladocerans were used in these assay
experiments. All animals were observed for changes in mobility,
response to tactile stimulation and general behavior, for several
days.

RESULTS

The cyanobacterium, *M. aeruginosa*, produces a unique biotoxin
that is, apparently, very selective and specific with regard to
the animal taxa and organs(s) that it affects. The results of the
assay tests performed in this study showed that selected bacteria,
green algae *(Chlorella)*, yeast *(Saccharomyces cerevisiae)*, arthro-
pods, amphibians, and teleosts were not affected by *Microcystis*
toxin in the concentrations used.

The toxin did not alter the normal characteristics of the
electrical and mechanical events associated with excitation phenomena in vertebrate nerve and muscle preparations.

After i.p. injection of *Microcystis* toxin into mice, the
symptoms of poisoning prior to death, the gross anatomical changes
and the histological liver changes were identical to the descrip-
tions in the literature cited in the introduction. Hepatocyte
degeneration and necrosis occurred and progressed outward from
centrilobular regions. Hepatocytes became swollen and their cytoplasm became cloudy and eosinophilic. Sinusoids were distended,
hepatocytes reptured, there was a breakdown in parenchymal cord
structure and massive hemorrhaging into the liver.

The in vitro studies showed that isolated liver cells and
micro-explants from prenatal and postnatal mice and rats were not
affected by the toxin. This led to the experiments that tested
mice of different ages for sensitivity. Neonatal and very young
mice did not show symptoms or liver lesions after i.p. injections
of toxic extracts. Injected mice did not die until they were
approximately 20 days old.

The development of mouse liver damage is shown at the ultrastructural level in Figures 3 through 7. Figure 2 is of untreated
mouse liver and is shown as a control for comparison.

At 10 minutes after i.p. injection all hepatocytes appeared as in the control except for some slight mitochondrial swelling and the presence of vesiculated membrane fragments in the sinusoids (Figure 3).

At 20 minutes the hepatocytic organelles and inclusions still appeared as in the control except for slight mitochondrial swelling. Bile canaliculi and microvilli in the space of Disse appeared distorted and endothelial tissue was either fragmented (many membrane fragments and myelinoid-like bodies were observed) or it was swollen (Figure 4).

At 40 minutes, sinusoids that were still intact were swollen, many of them were distended and contained fragmented endothelium. Many RBC's were present, spaces of Disse, bile canaliculi and microvilli were either absent in centrilobular areas or greatly distorted. Cell disruption had taken place in centrilobular areas and there was much pooling of RBC's and the cell organelles and inclusions. The released cell structures were not grossly changed except for mitochondrial swelling, some distorted nuclei and fragmentation of RER (Figure 5.).

At 60 minutes, after the mice had died, the cell degeneration and necrosis was extensive. In the centrilobular areas the parenchymal cord structure was missing as cells rounded up, pulled apart and lysed. There was massive hemorrhaging as evidenced by the presence of many RBC's that had pooled with cell contents. Lysed hepatocytes showed discontinuities in the plasma membrane or no membrane at all (Figure 6 and Figure 7).

DISCUSSION

Reports on microcystin (FDF) are often conflicting and difficult to interpret. These anomalies may be due to several factors: a) a lack of standardized materials for testing and/or specific assay methods for the qualitative and quantitative determination of FDF, b) the unavailability of more sophisticated techniques at the time the research was done, c) the use of extracts from mixed algal-bacterial samples, and d) different toxins being produced by different strains. The present study looked at some of these areas of conflict and more specifically describes the site and mode of action of microcystin.

One area of uncertainty concerned the species effected by FDF. This study shows that cladocerans, amphibians, crustacea and teleosts are not effected by either immersion or injection of FDF. Mammals and birds were previously reported sensitive to the toxin and this work confirms the sensitivity of mammals.

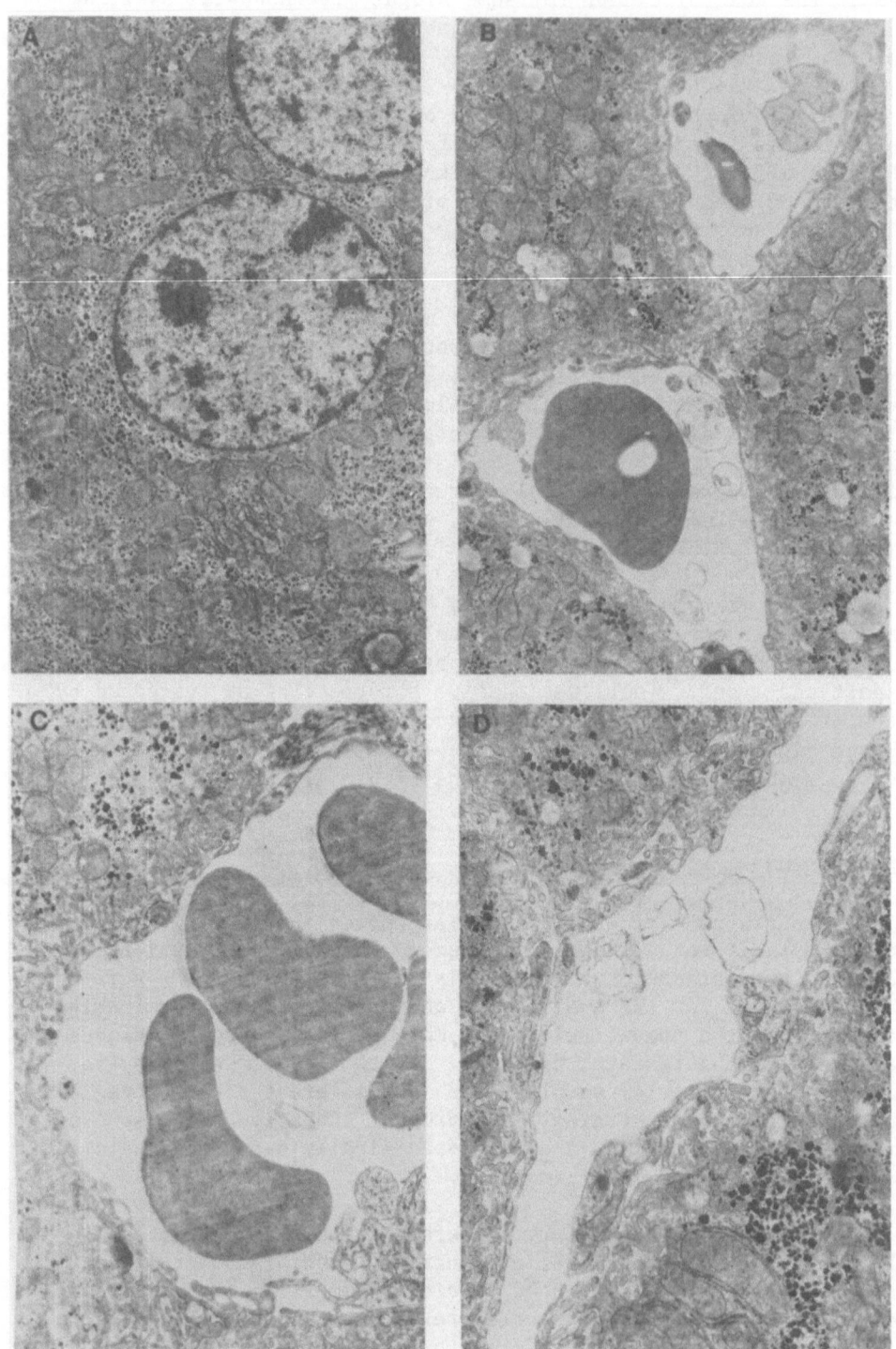

Whether or not FDF effects excitable membranes had not been determined conclusively by previous work. It has been shown by the
present research that neither crude nor partially purified toxic
Microcystis extracts have any effect on smooth, cardiac or skeletal muscle, or on isolated nerve preparations.

Young hepatocytes in vitro are not effected and neither are
young mice or rats. Since sensitivity to the toxin appears to
develop with age, it may be that specific receptors are needed
that are not present or available to bind the toxin until the
liver has developed more completely.

Although past studies of the histopathology resulting from
Microcystis poisonings have indicated that organs other than the
liver are effected (Gorham, 1962; Heaney, 1971), no lesions were
ever observed in other organs or tissues in this study. FDF
appears to act specifically on liver tissue. Changes in other
organs, especially in chronically poisoned animals, may be due
to the extensive liver damage that occurs. It would seem likely
that such massive liver destruction would have many secondary
and/or indirect effects on other tissues.

FDF destroys hepatocytes and hepatic endothelial cells. The
only cellular/subcellular alterations noted prior to cell rupture
are a slight mitochondrial swelling and cell swelling. Organelles
and inclusions in mammalian hepatocytes are not altered. This
observation, plus the extensive fragmentation and vesiculation of
membranes, the presence of myelin figures and plasma membrane
discontinuities indicates that the site of action may be the cell
membrane. FDF appears to be a very specific cytolytic toxin.
Further investigation is needed to discover the mechanism by which
these membranes are lysed.

Fig. 3. Mouse liver tissue exposed to toxic NRC-1 extract for
         10 minutes. (A) No noticeable changes in the nuclei,
         RER, mitochondria, glycogen content, or lysosomes are
         observed (4500X). (B) Sinusoids contain numerous
         membrane vesicles. Endothelium is intact as are
         microvilli and hepatocytes. No distortions of these
         structures are visible (4500X). (C) A higher magni-
         fication of a sinusoid containing several RBC's
         (7500X). (D) A more magnified micrograph of a sinu-
         soid showing the membranous vesicles that are common
         even at this time (10000X).

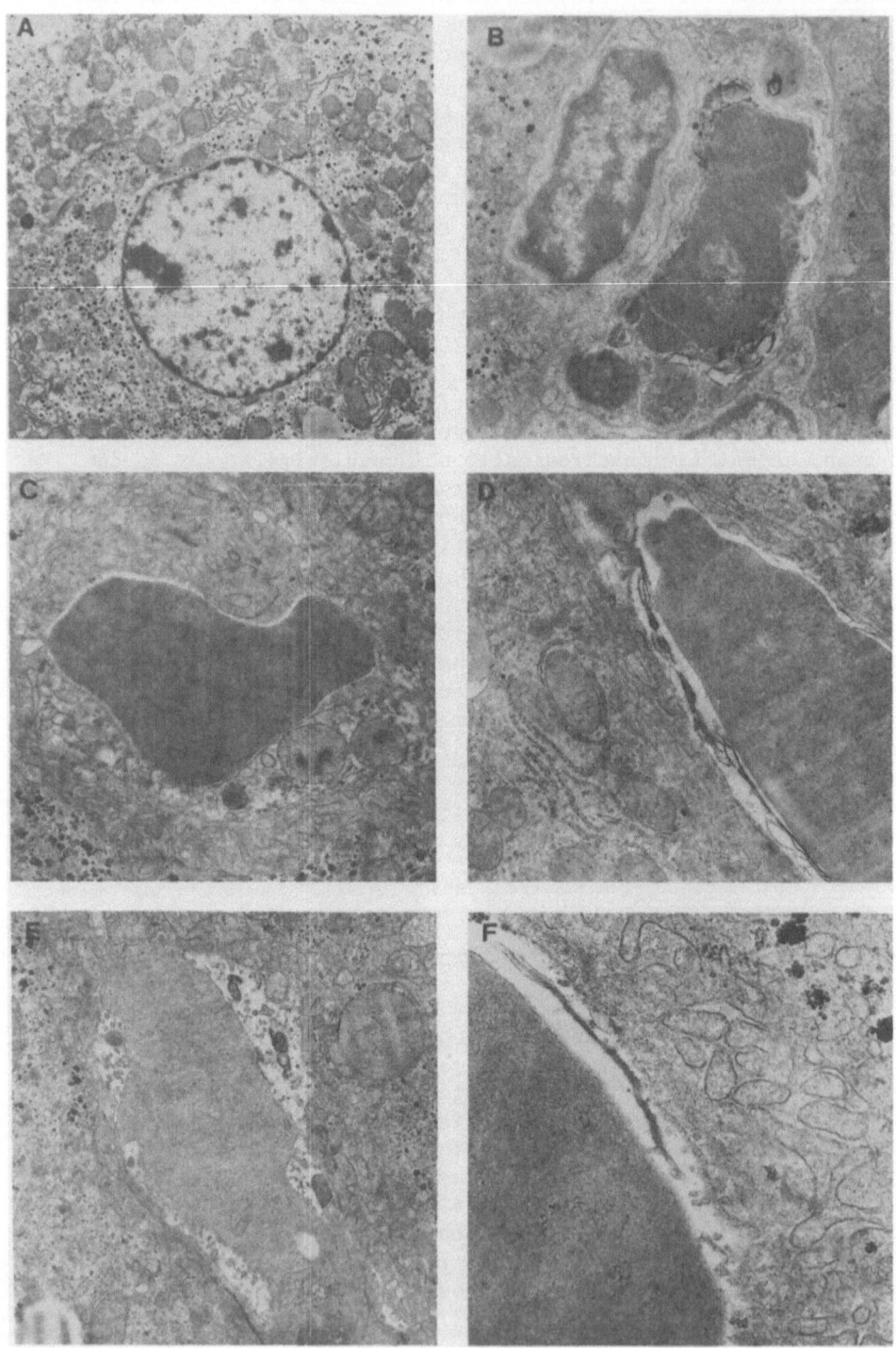

Fig. 4. Mouse liver exposed to toxic NRC-1 extract for <u>20 minutes</u>.
         (A) No changes in hepatocyte organelles or inclusions
         are observed (4500X). (B) Shows a RBC in a sinusoid
         that has swollen, distorted and fragmented endothelium
         (6000X). (C) A RBC in a sinusoid where the changes in
         epithelium are seen more clearly. The endothelium is
         greatly swollen, the cells appear washed out, darkly
         staining material can be seen adjacent to the RBC and
         the microvilli in the space of Disse appear to be com-
         pressed. The plasma membrane of the endothelial cell(s)
         appears discontinuous (9000X). (D) Note that the cyto-
         plasm and its contents do not appear distorted in the
         hepatocyte, the RBC in the sinusoid is intact but the
         endothelium has fragmented into discontinuous, darkly
         staining pieces of membrane. Microvilli are distorted
         and there is darkly staining material around them
         (7500X). (E) The endothelium appears to have completely
         fragmented, hepatocyte plasma membranes are more diffi-
         cult to distinguish and microvilli are greatly distorted
         (4500X). (F) A high magnification showing a RBC (lower
         left) and a portion of one hepatocyte (upper right).
         Microvilli are swollen and surrounded by amorphous
         material. Note the long fragment of membrane between
         the RBC and the microvilli (25000X).

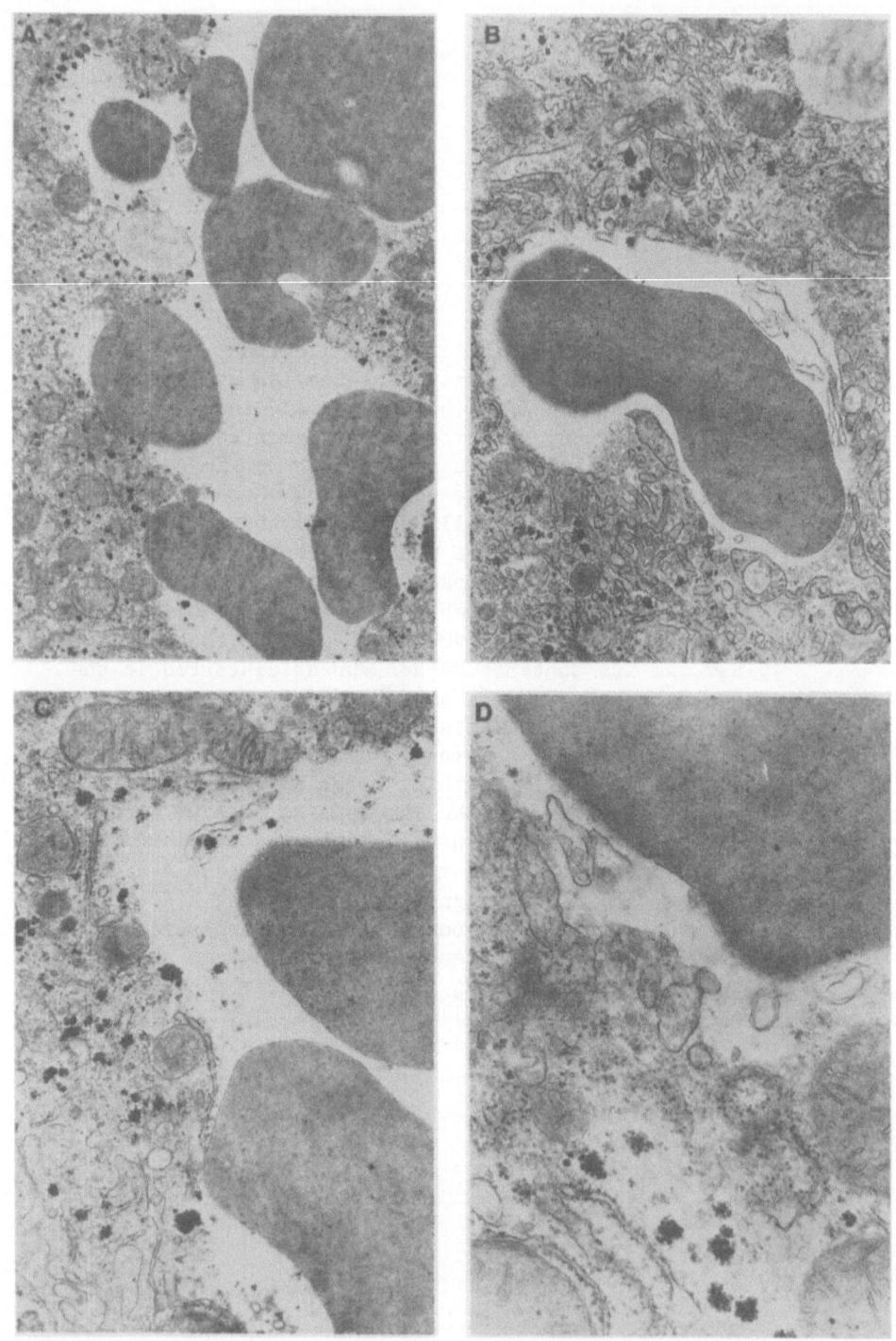

Fig. 5.   Mouse liver exposed to toxic NRC-1 extract for <u>40 minutes</u>. (A) Sinusoidal area where endothelial and hepatocytic membranes have lysed and there is a pooling of blood cells and cytoplasmic components. This micrograph illustrates the massive hemorrhaging that takes place in poisoned liver tissue (6000X). (B) A single RBC and membrane fragments that appear to have vesiculated. There is loss of cytoplasm around contorted microvilli (10000X). (C) A higher magnification of the damage but note the intact mitochondria, fragments of RER, glycogen rosettes, etc. that have spilled from the cells (15,000X). (D) This is a good illustration of the nature of toxin damage. The RBC and cell components are intact but distorted while the plasma membranes are broken apart and vesiculated. Note mitochondria, endoplasmic reticulum with attached ribosomes (RER), glycogen particles, etc. (25000X).

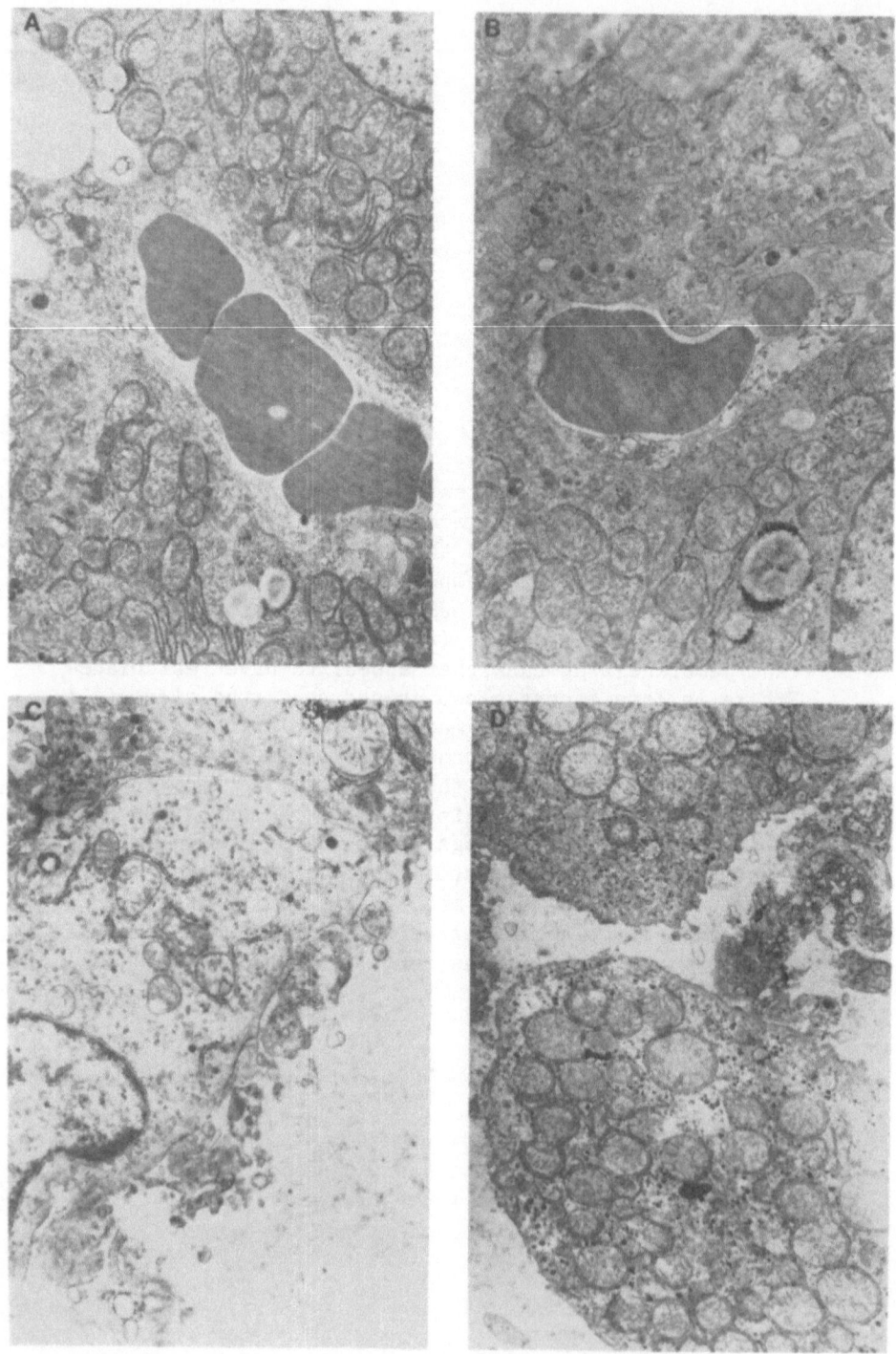

Fig. 6. Mouse liver exposed to toxic NRC-1 extracts for <u>60 minutes</u>.
         (A) Upper right and lower left show parts of hepato-
         cytes. Note intact organelles. In the center is a
         sinusoid containing RBC-s and lined with distended
         endothelium that contains rather sparse cytoplasm.
         The spaces of Disse are compressed and few micro-
         villi are present; those few are very distorted
         (4500X). (B) Similar to "A" and illustrating the
         degeneration of endothelium (5900X). (C) Portions
         of the surfaces of two hepatocytes showing surface
         destruction, absence of microvilli, vesiculated mem-
         branes and what appear to be discontinuous areas in
         the plasma membrane (8900X). (D) Distorted hepato-
         cytes pulled away from each other (5900X).

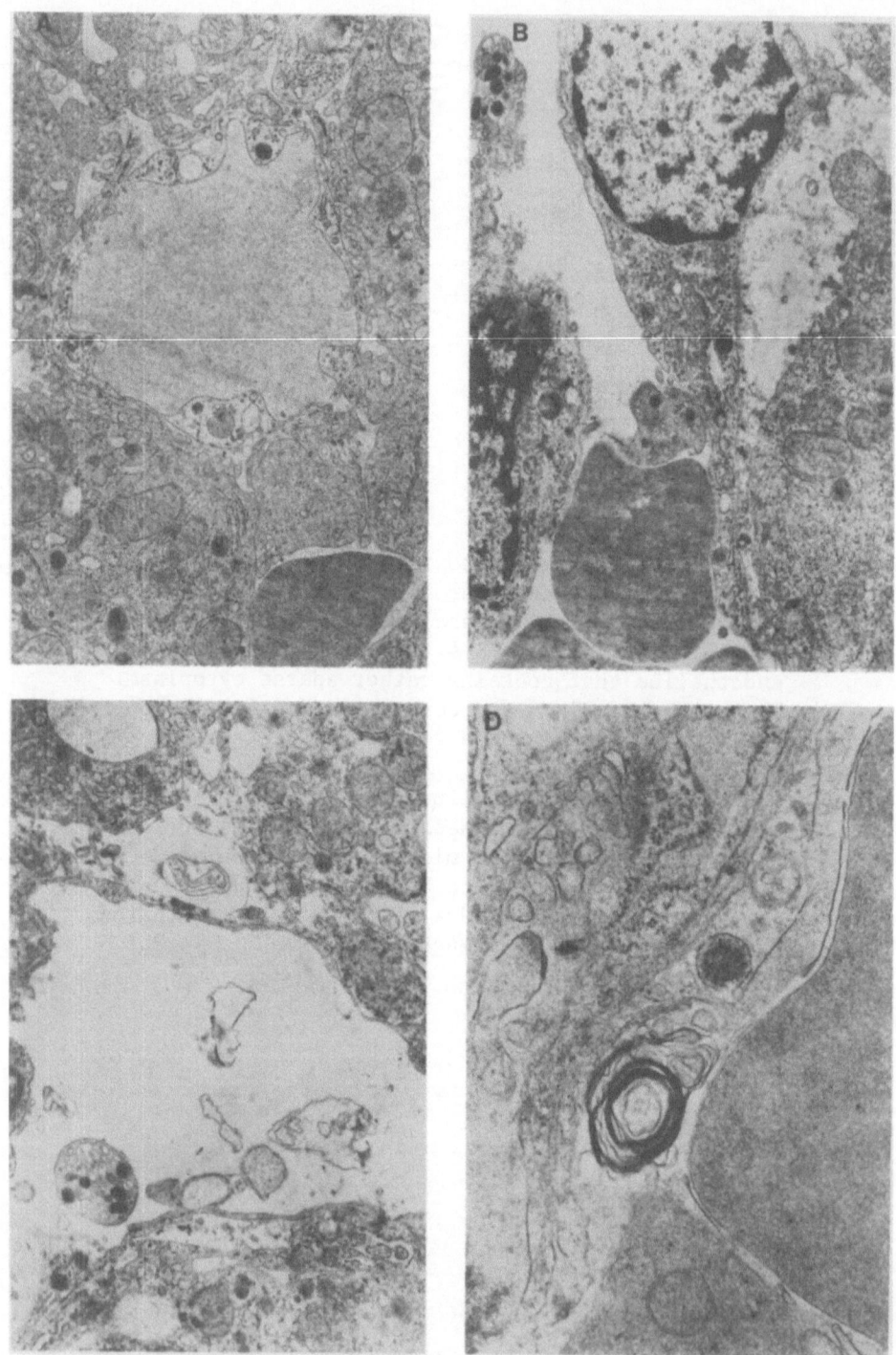

Fig. 7. Mouse liver acutely poisoned (60 minutes) by toxic
extracts of *M. aeruginosa* from Kezar Lake, New Hampshire. (A) A sinusoid containing a granular, amor-
phous material and lined by a swollen, "ghost-like"
endothelial cell in which a few organelles can still
be distinguished. Hepatocytes surrounding the sinusoid appear swollen and are becoming detached, microvilli are fewer and compressed into a distorted space
of Disse (5000X). (B) The lower right shows a swollen
hepatocyte containing several mitochondria, some RER,
and a lysosome but lacking normal cytoplasmic appearance. The surface of this cell is absent of microvilli and in the space between the cell and the
adjacent Kupfer cell (large nucleus in the upper
middle) there appears to be much cell debris. Several
RBC's are visible in the lower portion of this micrograph (5000X). (C) Remnants of endothelium line this
sinusoid containing several membranous vesicles. Surrounding hepatocytes have greatly distorted surfaces
lacking microvilli. A lipid droplet is seen in the
upper left and several swollen mitochondria are
observed in association with RER (6000X). (D) A
high magnification of a RBC (right) and a portion
of a hepatocyte (upper left) with the remnants of
microvilli on its surface. Note the large myelin
figure (dark circular object) composed of lamellar
rings of membrane fragments. Also, note the long,
interrupted membrane next to the RBC. The endothelial lining is gone and normal cellular organization in the space of Disse is absent. The plasma
membrane does not appear to be completely continuous
(20000X).

ACKNOWLEDGMENTS

This research was supported by a grant from the Office of Water Research and Technology, United States Department of the Interior as authorized under the Water Research and Development Act of 1978, Public Law 95-467, through the Water Resources Research Center of the University of New Hampshire.

REFERENCES

Ashworth, C. T., and M. F. Mason. 1946. Observations on the pathological changes produced by a toxic substance present in blue-green algae (*Microcystis aeruginosa*). Am. J. Pathol. *22*(2):369-383.

Bishop, C. T., E. F. L. J. Anet, and P. R. Gorham. 1959. Isolation and identification of the fast-death factor in *Microcystis aeruginosa* NRC-1. Can. J. Biochem. Physiol. *37*:453-471.

Echlin, P. 1966. The blue-green algae. Sci. Am. *214*(16):75-81.

Gorham, P. R. 1962. Laboratory studies on the toxins produced by waterblooms of blue-green algae. Am. J. Public Health *52*(12):2100-2105.

Gorham, P. R. 1964. Toxic algae. Pages 307-336 *in* D. F. Jackson, ed., Algae and Man. Plenum Press, New York. 434 pp.

Grant, G. A., and E. O. Hughes. 1953. Development of toxicity in blue-green algae. Can. J. Public Health *44*:334-339.

Heaney, S. I. 1971. The toxicity of *Microcystis aeruginosa* Kütz. from some English reservoirs. Water Treat. Exam. *20*(4):235-244.

Hughes, E. O., P. R. Gorham, and A. Zehnder. 1955. Toxicity of *Microcystis aeruginosa* in pure culture. Phycol. News Bull. *8*(5):5.

Hughes, E. O., P. R. Gorham, and A. Zehnder. 1958. Toxicity of a unialgal culture of *Microcystis aeruginosa*. Can. J. Microbiol. *4*:225-236.

Kirpenko, Y. A., I. I. Peruvozchenko, L. A. Sirenko, and L. F. Lukina. 1975. Isolation of toxin from blue-green algae and some of its physico-chemical properties. Akademia Nauk URSR, Kiev. Dopovidi. Serri B. Geologiia, GeoFizika, Kluimiia ta Brologiia, Kiev. No. 4. pp. 359-361.

Konst, H., P. D. McKercher, P. R. Gorham, A. Robertson, and J. Howell. 1965. Symptoms and pathology produced by toxic *Microcystis aeruginosa* NRC-1 in laboratory and domestic animals. Can. J. Comp. Med. Vet. Sci. *29*:221-228.

Louw, P. G. J. 1950. The active constituent of the poisonous algae, *Microcystis toxica* Stephens. S. Afr. Ind. Chemist *4*:62-66.

Maloney, T. E., and R. A. Carnes. 1966. Toxicity of a *Microcystis* waterbloom from an Ohio pond. Ohio J. Sci. *66*(5):514-517.

Malyarevskaya, A. Y., T. I. Birger, O. M. Arsau, and V. C. Solomatina. 1972. Metabolic relationships between blue-green algae and fish. Gidrobiol. Zh. *8*:47-55.

Mason, M. F., and R. E. Wheeler. 1942. Observations upon the toxicity of blue-green algae. Fed. Proc. *1*:124.

May, V. 1970. A toxic alga in New South Wales and its distribution. Contrib. N. S. W. Natl. Herb. *4*(3):84-86.

McBarron, E. J., and V. May. 1966. Poisoning of sheep in New South Wales by the blue-green alga *Anacystis cyanea* (Küetz.) Dr. and Dail. Aust. Vet. J. *42*:449-453.

McLachlan, J., and P. R. Gorham. 1961. Growth of *Microcystis aeruginosa* Kütz. in a precipitate-free medium buffered with Tris. Can. J. Microbiol. *7*:869-882.

McLachlan, J., and P. R. Gorham. 1962. Effects of pH and nitrogen sources on growth of *Microcystis aeruginosa* Kütz. Can. J. Microbiol. *8*(1):1-11.

Moore, R. E. 1977. Toxins from blue-green algae. Bioscience *27*(12):797-802.

Murthy, J. R., and J. B. Capindale. 1970. A new isolation and structure for the endotoxin from *Microcystis aeruginosa* NRC-1. Can. J. Biochem. *48*:508-510.

Orlovskii, V. M., and Y. A. Kirpenko. 1976. Biologically active metabolites of blue-green algae and their effect on experimental animals. Gig. Sanit. *3*:13-17.

Prescott, G. W. 1948. Objectionable algae with reference to the killing of fish and other animals. Hydrobiologia *1*:1-13.

Simpson, B., and P. R. Gorham. 1958. Source of the fast-death factor produced by unialgal *Microcystis aeruginosa* NRC-1. Abstr. Phycol. Soc. Am. News Bull. *11*:59-60.

Solomatina, V. C., and S. F. Matchinskaya. 1972. Alteration of the amino acid content of the peptide induced by blue-green algae. Hydrobiologia J. *8*:46-49.

Stangenberg, M. 1968. Toxic effects of *Microcystis aeruginosa* Kg. extracts on *Daphnia longispina* O. F. Müller and *Eucypris virens* Jurine. Hydrobiologia *32*:81-88.

Stephens, E. L. 1949. *Microcystis toxica* sp. nov.: a poisonous alga from the Transvaal and Orange Free State. Trans. R. Soc. S. Afr. *32*(1):105-112.

STUDIES ON APHANTOXIN FROM *APHANIZOMENON FLOS-AQUAE*

IN NEW HAMPSHIRE

John J. Sasner, Jr., Miyoshi Ikawa[1],
Thomas L. Foxall, and Winsor H. Watson

Department of Zoology
Spaulding Life Science Building
University of New Hampshire
Durham, New Hampshire 03824

ABSTRACT

Toxic Cyanobacteria (blue-green algae) bloom in eutrophic, freshwater lakes and ponds in New England and have caused environmental, health, legal and recreational problems over the past 15 years. Although several species have been implicated with animal kills and water fouling, a common offender was *Aphanizomenon flos-aquae*. Representative strains of *A. flos-aquae* bloom in New Hampshire intermittently, in both toxic (aphantoxins) and non-toxic forms. Research has focused on methods of: a) toxin accumulation from natural blooms and laboratory cultivation, b) toxin assay, using the mouse bioassay and a modified fluorometric technique developed for paralytic shellfish poisons, c) toxin characterization and purification, using solvent separation and molecular weight filters, and d) testing active extracts on nerve and muscle preparations to determine the specific sites and modes of action of aphantoxins.

Aphantoxin samples were passed through molecular weight filters (10,000 and 500 daltons), lyophilized and weighed, prior to physiological testing. Microgram quantities of toxin reversibly blocked compound action potentials in amphibian nerves as well as mechanical activity in skeletal muscle. No effect was measured on the transmembrane resting potential or on spontaneous miniature

[1]Address: Department of Biochemistry, Spaulding Life Science
Building, University of New Hampshire, Durham, New Hampshire 03824

end-plate potentials (mepps). Tests on lateral and medial giant
axons from crayfish gave similar results. The Na$^+$ dependence of
the crayfish preparation was verified. The aphantoxins (4 µg/ml)
reversibly blocked intracellular recordings of action potentials
with no alteration of the resting potential. Amphibian and crustacean cardiac activity was blocked in diastolic arrest, while
bivalve hearts were unaffected at increased dose levels. Aphantoxins may block excitability by affecting ion conductance pathways as do toxins from several marine dinoflagellates and may be
useful in basic studies on membrane systems.

INTRODUCTION

     Freshwater blooms of toxic Cyanobacteria (blue-green algae)
are common in many countries of the world. Animal, including human,
involvement with toxicity problems have been reported for at least
12 countries, 4 Canadian provinces, and 10 of the United States
(Schwimmer and Schwimmer, 1964 and 1968; Moore, 1977; Collins,
1978). Although many microorganisms have been implicated with
water fouling and animal kills, a frequent offender is *Aphanizomenon flos-aquae*. Representatives of this species occur in New
England intermittently, in both toxic and non-toxic forms. Environmental problems may arise when these Cyanobacteria are involved
in phytoplankton blooms. A critical time during bloom conditions
occurs when dense cell masses decompose naturally or with the aid
of algicides (e.g., copper sulphate) commonly used to enhance water
quality. The decomposition products plus toxic cellular materials
released into the water when the cells lyse may cause death or
illness to mammals, birds, and fishes, and may reduce water quality
for animal (including human) consumption and recreational purposes
(Collins, 1978; Palmer, 1964; Schwimmer and Schwimmer, 1964).

     Blooms of toxic Cyanobacteria occur in several freshwater
lakes, ponds and reservoirs in New Hampshire, some of which are
used for water supplies and/or recreational purposes. The environmental effects of these noxious blooms have caused concern from
state water quality control agencies (New Hampshire Water Supply
and Pollution Control Commission Staff Reports, Nos. 59, 62, 63,
64, 70; 1973-1975). In addition, at least one legal case in New
Hampshire focused attention on sewage treatment effluent and toxic
cyanophyte blooms in Kezar L., North Sutton, New Hampshire (W. A.
Sundall et al., versus Town of New London, 1977). The environmental parameters associated with toxic blooms in this lake have
been described by Haynes (1971) and in the reports enumerated
above.

     Biotoxins attract interest and attention from researchers
because of their specificity, potency, and potential utility as

physiological and/or pharmacological "tools" (O'Brien, 1969). This is particularly true when the toxin's effects are reversible, i.e., when a system can be blocked and then restored to normal activity. Aphantoxins meet most, if not all, of these criteria.

In 1968, Sawyer et al., demonstrated the presence of a very fast death factor (VFDF) from *A. flos-aquae* cells collected during blooms in two New Hampshire lakes. The toxin was dialyzable, heat and acid stable, alkali labile, and was soluble in water and ethanol, but non-soluble in less polar solvents. The same year (1968) Jackim and Gentile, using laboratory cultures, reported the partial purification and properties of aphantoxin. Three toxic fractions were obtained using acid extraction, preparative paper chromatography and silica gel column chromatography. The most potent of these gave three Weber reagent-positive spots, one of which corresponded in R_f value, reactions with Weber, ninhydrin and Jaffe reagents, and infrared spectrum with saxitoxin (STX). The latter (STX) is the paralytic shellfish poison (PSP) produced by marine dinoflagellates of the genus *Gonyaulax* and found in bivalves that act like "biological storage depots" for toxin accumulation in nature. The active material contained 1.5 to 2.0 µg per mouse unit (MU) which is equal to 500 to 667 MU/mg.

In 1973, Alam et al., reported the partial purification and properties of aphantoxin obtained from natural blooms in Kezar Lake, North Sutton, New Hampshire. Extraction and purification was done using acid, alcohol, and chloroform extractions and preparative high voltage electrophoresis. A toxic ninhydrin and Weber reagent positive zone was eluted and chromatographed on IRC-50 resin. An active fraction was eluted with acetic acid and purified further using preparative thin-layer chromatography (TLC). This resulted in a chromatographically homogeneous material with a potency of 745 MU/mg. Positive reactions of this material with Weber, diacetyl-oc-naphthol, and Benedict-Behre reagents indicated that aphantoxin may be a substituted guanidine derivative. However, TLC in various solvent systems, color reactions given with various spray reagents, electrophoretic comparisons, and infrared spectra indicated that aphantoxin was not identical with saxitoxin (STX). Recent work by Alam et al. (1978), demonstrated that aphantoxin was a complex mixture of toxins containing saxitoxin and other related, but still unknown, substances. This mixture has not been completely characterized. In addition, Shoptaugh (1978) and Carter (1980) have shown that these toxic materials from *Aphanizomenon* may lend themselves to qualitative and quantitative analysis using alkaline-H_2O_2 oxidation and fluorometry; a method that is currently being tested with the aim of replacing the mouse bioassay.

Research at the University of New Hampshire has focused on the chemical properties, assay, and stability of the biotoxins from

Aphanizomenon flos-aquae. In addition, the physiological effects
of the toxins were studied particularly as they effect neuromuscular
systems (including cardiac) in vertebrate and invertebrate animals.
Support for the research was provided by the New Hampshire Water
Resources Research Center, G. Byers, Director.

MATERIALS AND METHODS

     State agencies monitor algal bloom conditions regularly through-
out New Hampshire and Vermont and their reports are available to us.
We have routinely collaborated with the New Hampshire Water Pollution
Control Commission on potential toxicity problems and on assessment
of toxic cyanophytes prior to algicide treatment that could produce
animal kills. In recent summers essentially unialgal blooms of
Cyanobacteria occurred, intermittently, in Kezar Lake, Winnisquam
Lake, Skatutakee Lake, Marsh Pond, Enfield Reservoir, Exeter Reservoir, several farm ponds and other freshwater environments. Cell
concentrations generally exceeded 5×10^4 per milliliter, and during
dense bloom conditions were $> 10^6$ per ml. An effective method for
obtaining bulk quantities of material occurring during bloom conditions in remote locations employed DeLaval Separators, at lake-
side, to spin and concentrate *Aphanizomenon* cells from large volumes
of water. The crude materials were then stored in the frozen state,
either wet or lyophilized. Samples retained potency under these
conditions for more than seven years. Unialgal, but not bacteria-
free, cultures were initiated from toxic clones of *Aphanizomenon*
using serial dilution methods with solid (agar), then liquid media.
The cultures were expanded to 20 liter carboys and grown in the
synthetic, modified ASM-1 medium of Carmichael and Gorham (1974),
under controlled conditions of temperature and illumination. Back-
up cultures were maintained in incubators separate from the culture
room to ensure against accidental equipment failure.

     We have recently found that Amicon and Millipore molecular
weight filters were useful for "cleaning up" samples of aphantoxin
(Shoptaugh, 1978). This method separated the aphantoxins from
high and intermediate molecular weight contaminants. We have successfully passed small amounts of the aphantoxin through the 500
dalton filter, prior to lyophilization, in preparation for chemical
and physiological studies.

     Mice (B6D2F1/J) were obtained from the Jackson Laboratory,
Bar Harbor, Maine. Standard methods employed for biotoxins from
marine microorganisms were used for the bioassay of the active
materials (Halstead, 1965; Prakash et al., 1971). The mouse unit
(MU) for aphantoxin was the same as that used to evaluate amounts
of paralytic shellfish poison (PSP) in marine bivalves, i.e., the
amount of material that killed mice (18 to 22 g) in 15 minutes =
1 MU. When only small amounts of purified aphantoxin were availa-

able, assay was accomplished using electrophysiological methods by measuring action potentials and tension development in muscle preparations. The mouse bioassay was also used in preliminary tests to determine if freshwater bivalves (*Elliptio camplanatus*) act as "biological storage depots" for aphantoxin accumulation, i.e., like marine bivalves exposed to *Gonyaulax* toxins. In addition, tests were run on the sensitivity of *Daphnia magna* to aphantoxins from laboratory cultures.

Shoptaugh (1978) found that aphantoxins form fluorescent derivatives when treated with H_2O_2 (like saxitoxin and other PSP derivatives) and a promising fluorescence assay for PSP and aphantoxin was developed (see Ikawa et al., 1980, these proceedings).

The toxic extracts from the field and laboratory were tested on standard nerve and nerve-muscle preparations, from mammalian, amphibian, and crustacean species. Control preparations were challenged with similarly treated material from non-toxic extracts. Compound action potentials were recorded externally from amphibian sciatic nerves and crayfish medial and lateral giant axons (sheathed and desheathed) using Ag-AgCl electrodes, Grass or Tektronix stimulator, and Tektronix preamp and dual beam oscilloscope. Transmembrane resting potentials and unicellular action potentials employed 3M KCl-filled glass microcapillaries, Grass P-16 Amplifier and Tektronix CRO. In the muscle tests dual multi-electrode assemblies were used for direct stimulation with recordings displayed on a dual beam oscilloscope or write-out recorder. Isotonic/isometric measurements on skeletal and cardiac muscles employed Grass FT-03C Mechano-electrical transducers. Previous work in this laboratory (Sasner, 1973) demonstrated the utility of the amphibian sartorious nerve-muscle preparation as an assay tool particularly because of its sensitivity to several aquatic biotoxins. This preparation provides a consistently reproducible system widely utilized in muscle physiology. The sartorious muscle is composed of long, parallel fibers, is thin enough to allow simple gas exchange when excised, and performs well at low temperatures where transient physico-chemical phenomena associated with excitability are slowed.

Kezar Lake is located in the rural community of North Sutton, New Hampshire. It is appropriate to include information about this particular lake because: a) it has provided starter cultures and toxic materials to several research groups participating in this conference, and b) it is one of few lakes where physical, chemical and biological parameters have been monitored for extended periods of time. This information was compiled by Terrence P. Frost, Ronald E. Towne and the late Harry J. Turner of the New Hampshire Water Supply and Pollution Control Commission. Their Staff Reports Nos. 59, 64 and 79 (N.H.W.S.P.C.C.) provide a comprehensive history of the recent stages of the eutrophication process in Kezar Lake as well as attempts at coping with annual blooms of toxic Cyano-

bacteria. The lake has an area of approximately 180 acres with a
maximum depth of 8.0 meters and an average depth of 3.7 meters.
Over the years Kezar Lake was a popular recreational site and in
1934 Wadleigh State Park was established on the southeastern shore.
During the early 1930's, the New London sewage treatment plant
began discharging secondary waste-water into Lion Brook a few miles
upstream from the lake. The nitrogen and phosphate concentrations
greatly increased in the lake and this additional nutrient load
accelerated the natural eutrophication processes. Copper sulphate
treatment of microorganism blooms were successfully administered
during the early 1960's to combat *Anabaena* blooms. Similar attempts
at controlling *Aphanizomenon* blooms in the mid-1960's were not as
effective and in a couple of instances produced massive fish mortalities. One particular treatment resulted in tons of dead fish
(mostly perch). During the late 1960's the recreational utility
of Kezar Lake diminished, property values decreased and the New
Hampshire State Tax Commission reduced property appraisals by 30
percent. Attempts at mixing and destratification of the lake with
large air compressors was moderately successful for several years
and the recreational utility of Wadleigh State Park increased.
Nutrient stripping and the addition of advanced waste-water treatment at the New London sewage plant reduced the discharge levels
of phosphorus. However, toxic blooms still occur in Kezar Lake.
The last few summers bloom periods of *Aphanizomenon* were shortened
and replaced by *Microcystis aeruginosa* (= *Anacystis cyanea*). Thus
the history of cyanophyte blooms in this lake includes *Anabaena*,
Aphanizomenon and *Microcystis*; the organisms of major concern in
these conference proceedings.

RESULTS AND DISCUSSION

     The effects exhibited by whole organisms, either injected with
or bathed in toxic samples of *Aphanizomenon flos-aquae*, were qualitatively similar to those reported for several marine poisons. The
characteristic symptoms in mice challenged with aphantoxins include
spastic twitching, irregular ventilation, gaping mouth, coordination
loss, violent tremors and subsequent death by respiratory failure.
Mouse bioassay of bivalve tissue (*Elliptio camplanatus*), either
collected during a bloom or fed *A. flos-aquae* from lab cultures,
produced these same symptoms. Twarog and Yamaguchi (1975) showed
that *Elliptio* were more sensitive to saxitoxin and tetrodotoxin
than several marine bivalves and our preliminary studies with
aphantoxin may support these findings. Laboratory fed *Elliptio*
were used because of the scarcity of these animals in Kezar Lake,
where they were formerly abundant. Examination of bivalve gut
contents showed a large percentage of broken *Aphanizomenon* cells
indicating at least partial digestion of the algal material. After
a two day exposure to lab cultures, the bivalves were themselves

Table 1. Summary of Toxin Characteristics from *Aphanizomenon flos-aquae*

Name(s) Form in Nature and Lab Culture	Aphantoxin; very fast death factor (VFDF); endotoxin. Single trichromes, 25 to 70 cells long.
Method of Collection	DeLaval Separators at lakeside concentrate *Aphanizomenon* from large volumes of water.
Chemical	H_2O and ETOH soluble, $CHCl_3$ insoluble; acid and heat stable, alkali labile, low mol. weight < 500, guanidine derivative, may contain saxitoxin (STX) + 3 unknown substances; forms fluorescent derivatives upon alkaline-H_2O_2 oxidation.
General Effects	Mammals and Fish: spastic twitching, coordination loss, respiratory irregularity; freshwater bivalves may store aphantoxin like PSP in marine molluscs. Some planktonic crustaceans (*Daphnia*) paralyzed -- reversible.
Bioassay (i.p.)	1 MU = amount of toxin to kill 20 ± 2 g mice in 15 minutes. Similar to mouse bioassay for paralytic shellfish poison (PSP) from marine bivalves (*Gonyaulax*).
Dose Range (mammals)	Lyophilized cells = 10 mg/kg; purified = 745 MU/mg.
Excised Tissues	0.4 to 100 µg/ml, to elicit a muscle block.
Nerve-Muscle Action Potentials	Blocked in desheathed nerves and muscle -- reversible.
Resting Trans- Membrane Potential	No effect.
Muscle Mechanical Activity	Vert. skeletal-block to indirect then direct stimulation; Vert. and Crustacean heart block -- diastolic arrest; cardiac A.P. reduced, reversible; Mollusc heart -- no effect at increased doses (x 100).
Site of Action	Nerve and muscle membranes.
Mode of Action	Ion conductance pathways blocked.
References	Sawyer et al., 1968; Alam, 1972, 1973, 1978; Gentile, 1971; Thurberg, 1972; Sasner, 1973; Sasner & Ikawa, 1975; Shoptaugh, 1978; Carter, 1980.

affected by the toxin in a manner similar to soft-shell clams after
exposure to massive concentrations of the "red tide" dinoflagellate
Goynaulax tamarensis. The *Elliptio* exhibited flaccid paralysis of
the foot and mantle tissue; a condition that was reversed by replacing the *Aphanizomenon* culture with clean medium. Since freshwater
bivalves are consumed by humans in certain parts of California,
further research in this area may be of more than academic interest.
It was estimated that 50,000 to 100,000 pounds per year of the
freshwater bivalve (*Corbicula fluminea*) are available in the California marketplace without prior testing for consumer safety (Jerome
Jenkin, Pacific Shellfish Company, personal communication).

The cladoceran *Daphnia magna* was also affected when placed in
aerated cultures of *Aphanizomenon*. Within a few minutes, the characteristic movements of the second antennae were reduced, erratic
and subsequently blocked. This paralysis of the swimming appendages
caused the *Daphnia* to settle to the bottom of the container and
perish in approximately 24 hours. Animals removed to clean water
after exposure for 12 to 16 hours recovered within 24 hours. Zooplankton were usually absent in plankton samples collected from the
upper one meter, during natural blooms of *Aphanizomenon*.

Table 1 presents a summary of the chemical and physiological
characteristics of the toxins from *Aphanizomenon flos-aquae*. This
information was compiled using unextracted lyophilized materials
and partially purified samples, after passage through 500 dalton
molecular weight filters. The acute sensitivity of neuromuscular
systems and the potency of aphantoxins were demonstrated in both
vertebrate and invertebrate preparations. Figure 1 shows the isometric tension developed in amphibian sartorious muscle challenged
with 50 µg aphantoxin per milliliter. The mechanical response to
indirect (via nerve) stimulation decreased faster than to direct
(via muscle) stimulation, indicating a greater sensitivity of the
nerves or a greater diffusion barrier in the muscle tissue. The
effects of the toxin were readily reversible and showed a dependency on the exposure time in toxin. No change in the transmembrane
resting potential of the sartorious muscle fibers was recorded
during tension reduction and recovery. In addition, no effect was
measured on spontaneous miniature end-plate potentials (mepps).
Sheathed sciatic nerves were not affected by the toxin in large
doses, whereas partially desheathed nerves were blocked with 10 µg
per milliliter within a few minutes. The decrease in amplitude of
compound action potentials in toxin-treated sciatic nerve preparations could have resulted from either a graded blockage of many
units or a progressive but complete blockage of individual axons
(Figure 2). To gain insight into these possibilities, we chose
the medial and lateral giant axons from crayfish (*Cambarus* sp.)
and measured action potentials with both extracellular and intracellular electrodes. Sheathed and partially desheathed axons were
prepared and arranged in a chamber divided into three parts (A, B,

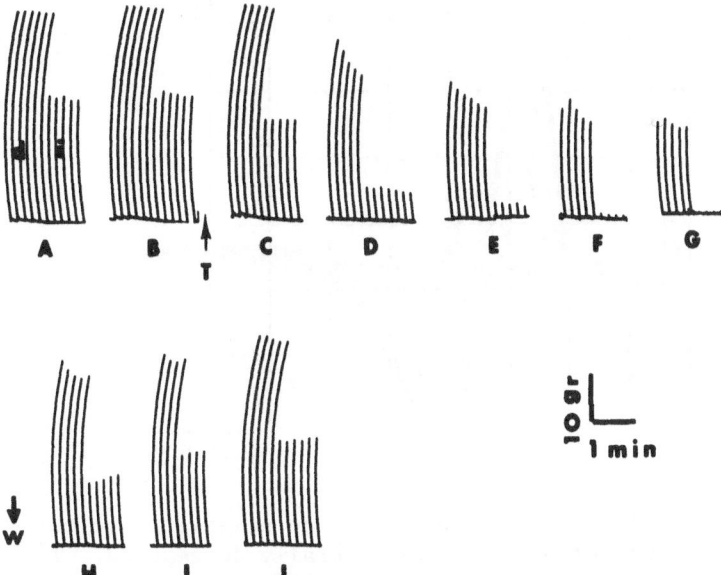

Fig. 1.   Effect of aphantoxin on isometric twitches from the
amphibian sartorius nerve-muscle preparation. Alternate
series of stimuli: direct stimulation (d) of 5 msec
duration, and indirect stimulation (i) of 50 μsec duration. A = control; B = control 20 minutes after A;
C, D, E, F, G = responses after 2, 5, 10, 15 and 20 min
in aphantoxin (50 μg/ml); H, I, J = recovery after 15,
30, and 60 minutes in amphibian Ringer. All stimuli
were given at 10 second intervals. Temperature 20°C.

and C), each separated by a paraffin and oil barrier. When the
axons were stimulated at A, toxin treated in B, and action potentials recorded at C, the waveform shifted to the right and then
abruptly disappeared. This blockage could be reversed 15 or more
times in a single preparation. The shift to the right in the
recorded action potential was interpreted as either a reduced conduction velocity or an increased latency time to reach firing
threshold. When the toxin (4 μg/ml) and recording intracellular
microelectrode were placed in the same chamber (B), the action
potential amplitude gradually decreased, had a slower sodium-
dependent rise time and was readily recoverable (Figure 3). The
same results were obtained with both sheathed and desheathed axons.
However, the former took approximately 5 times longer to block.
In all records from toxin treated axons, the rise time from base
line to peak amplitude increased 3 to 5-fold before complete block
of the action potential occurred. The slope of the falling or
recovery phase, however, remained essentially the same. In Figure
3, for example, the rise time increased from 0.5 msec to 1.8 msec,

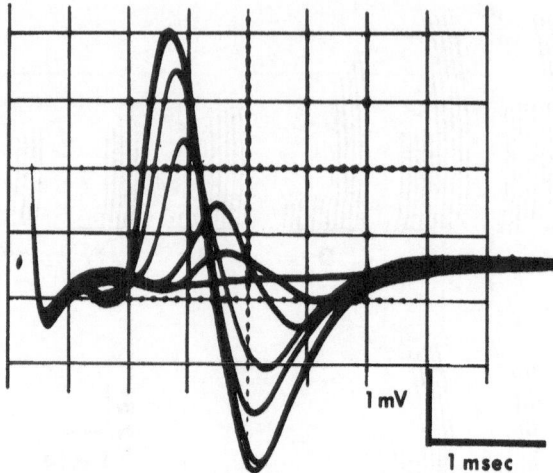

Fig. 2. Effect of aphantoxin on amphibian sciatic nerve compound
         action potentials. Nerve partially desheathed. Traces
         show progressive decline in amplitude and rise rate to
         complete block in 5 minutes. Wash in amphibian Ringer
         returned normal action potential in 15 minutes. Stimulus
         amplitude 3 V, duration 30 µsec. Grid scale 0.5 msec/di-
         vision horizontal, and 1 mV/division vertical. Temp. 20°C.

while the recovery times were approximately 1.3 msec in all traces.
The rising or depolarizing phase of the action potential is associated with a transient increase in sodium ion conductance. The
membrane permeability to sodium ions increases upon stimulation
and these cations flow inward, down concentration and electrical
gradients toward the sodium equilibrium potential, E_{Na^+}. This is
the familiar positive feedback loop called the Hodgkin cycle. The
resting impermeability of the membrane to sodium ions returns,
breaking the cycle, and potassium ions flow outward, down concentration and electrical gradients toward E_{K^+}, and restoration of
the transmembrane resting potential, E_{mem}. The data from crayfish giant axons suggests that aphantoxin alters the depolarizing
or sodium-dependent phase of the action potential and has little
or no effect on the potassium-dependent repolarization phase. The
sodium dependency was verified by challenging the axons with Na^+
free Ringer, in which case records similar to Figure 3 (with
aphantoxin) were obtained. Calcium-free Ringers only slightly
affected the action potential waveform and the addition of toxin
decreased the amplitude and rise rate, as in Figure 3. Aphantoxin
treatment did not alter the transmembrane resting potential (-80 mV)
or the membrane resistance.

     In comparative studies the myogenic hearts of amphibians
and molluscs and the neurogenic hearts of crustaceans may offer

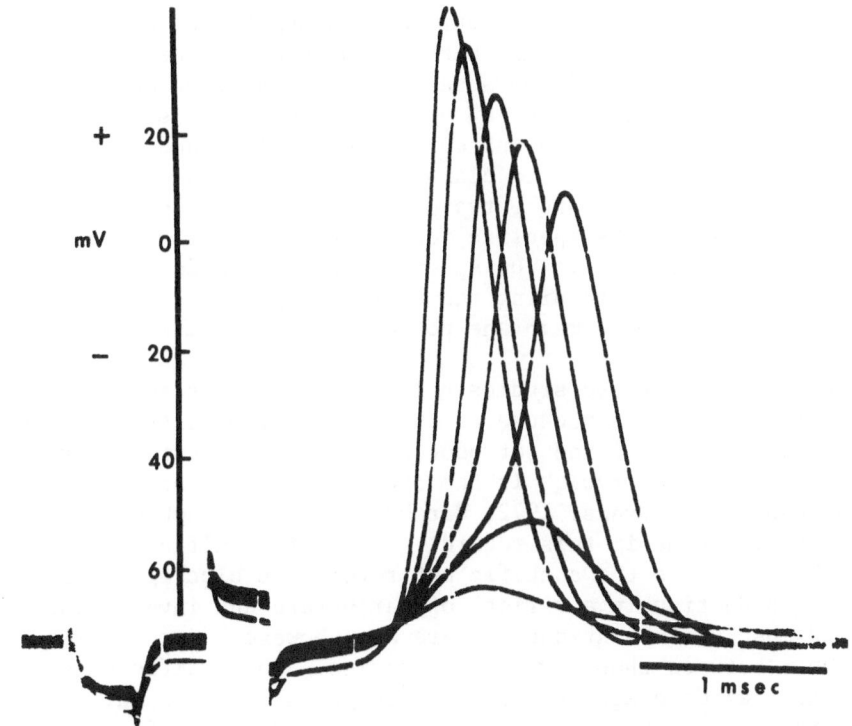

Fig. 3. Effect of aphantoxin on action potentials from crayfish
         medial giant axons. Desheathed preparation. Toxin (0.8
         µg/ml) and intracellular recording electrode in same
         chamber. Top trace control, progressive decline in
         amplitude and rise rate after 4, 10, 14, 17, 20 and 25
         minutes in toxin. Complete recovery within 15 minutes
         after wash in crayfish Ringer. Stimulus amplitude 2
         times threshold (2 V); duration 250 µsec pulse width.

useful preparations for the testing of biotoxins. Thurberg (1972)
showed that the marine poisons tetrodotoxin (TTX) and saxitoxin
(STX) and the freshwater aphantoxin (4 µg/ml) all produced reversible diastolic arrest in the hearts of the crab, *Cancer irroratus*.
None of these toxins, however, altered the normal mechanical or
electrical activity of bivalve hearts (*Mercenaria mercenaria*),
even at increased dose levels (100 X). The sodium dependence of
the bivalve heart was verified by substitution with sucrose and
lithium. This lack of sensitivity to sodium blocking toxins has
not been explained. Amphibian heart preparations were slowed,
then reversibly blocked in diastolic arrest by aphantoxin (Sawyer
et al., 1968).

    In previous work (Sasner and Ikawa, 1975) we have described
aphantoxin as a non-depolarizing, reversible, membrane blocking
agent that may alter ion conductance pathways associated with

excitation. This hypothesis is currently being tested using volt-
age-clamp methods on individual axons. The goal in these studies
is to measure the physical characteristics of the axon membrane
and transmembrane current flow on voltage clamped cells challenged
with purified aphantoxin (Ehrenstein, 1976; Adelman and French,
1976). The specific aim is to clamp the transmembrane voltage and
measure the inward (Na^+) and outward (K^+) currents to determine
whether the toxin blocks specific ion channels or all cation flow,
as it blocks excitability. The specific site and mode of action
should be revealed by measuring the current density, I_{mem}, (mA/cm^2)
as a function of clamped membrane potential, E_{mem}, (mV).

The results of these studies hold more than just academic
interest for biologists because of the potential utility of aphantoxin. Saxitoxin (STX) and tetrodotoxin (TTX) are currently used
in basic research as "tools" in the study of Na^+ dependent membrane
systems (Kao, 1966; Evans, 1972; Narahashi, 1975). Aphantoxin may
be equally important in this regard. The most significant role of
STX and TTX involves the specific but reversible blockage of action
potential conduction in a variety of vertebrate and invertebrate
nerve and muscle preparations. There are, however, more subtle
differences between these two marine toxins. These differences are
related to: a) dose-survival relationships in injected animals,
b) resistance of amphibian nerves (*Taricha*) and puffer fish nerves
(*Tetradon*) to TTX but not STX, c) recovery time of nerve-muscle
preparations after poisoning, and d) the differential effect on
evoked end-plate potentials , i.e., STX causes gradual decrease,
while TTX produces abrupt blockage. If our hypothesis is correct,
and aphantoxin specifically alters the ion conductance properties
of the membrane, then more extensive comparative work will be done
to determine if the cyanobacterial toxin is more STX-like or more
TTX-like. In addition, it would be important to include the variety of toxins from the "red tide" dinoflagellate, *Gonyaulax tamarensis*, since these materials are similar in their effects on membrane
systems (Evans, 1975; Narahashi, 1975; Shimizu, 1978).

ACKNOWLEDGMENTS

This research was supported through the New Hampshire Water
Resources Research Center of the University of New Hampshire by a
grant from the Office of Water Research and Technology, United
States Department of the Interior as authorized under the Water
Research and Development Act of 1978, Public Law 95-467.

REFERENCES

Adelman, W. J., Jr., and R. J. French. 1976. The squid giant axon. Oceanus *19*(2):6-16.

Alam, M. 1972. Algal Toxins. Ph.D. Thesis. University of New Hampshire, Durham, New Hampshire.

Alam, M., M. Ikawa, J. J. Sasner, Jr., and P. J. Sawyer. 1973. Purification of *Aphanizomenon flos-aquae* toxin and its chemical and physiological properties. Toxicon *11*:65-72.

Alam, M., Y. Shimizu, M. Ikawa, and J. J. Sasner, Jr. 1978. Reinvestigation of the toxins from the blue-green alga, *Aphanizomenon flos-aquae*, by a high performance chromatographic method. J. Environ. Sci. Health *A13*(7):493-499.

Carmichael, W. W., and P. R. Gorham. 1974. An improved method for obtaining axenic clones of planktonic blue-green algae. J. Phycol. *10*:238-240.

Carter, P. W. 1980. Comparison of Assay Methods for *Gonyaulax* and *Aphanizomenon* Toxins. M.Sc. Thesis. University of New Hampshire, Durham, New Hampshire.

Collins, M. 1978. Algal toxins. Microbiol. Rev. *42*(4):725-746.

Ehrenstein, G. 1976. Ion channels in nerve membranes. Physics Today (October) pp. 33-39.

Evans, M. H. 1972. Tetrodotoxin, saxitoxin, and related substances: their applications in neurobiology. Pages 83-166 *in* C. F. Pfeiffer and J. R. Smythies, eds., International Review of Neurobiology Vol. 15. Academic Press, New York.

Evans, M. H. 1975. Saxitoxin and related poisons: their actions on man and other mammals. Pages 337-345 *in* V. R. LoCicero, ed., Proceedings of the First International Conference on Toxic Dinoflagellate Blooms. Massachusetts Science Technology Foundation, Massachusetts.

Foxall, T. L. 1980. Studies on FDF from *Microcystis aeruginosa*. Ph.D. Thesis. University of New Hampshire, Durham, New Hampshire.

Gentile, J. H. 1971. Blue-green and green algal toxins. Pages 27-66 *in* S. Kadis, A. Ciegler and S. J. Ajl, eds., Microbial Toxins Vol. VII. Algal and Fungal Toxins. Academic Press, New York. 401 pp.

Gentile, J. H., and T. E. Maloney. 1969. Toxicity and environmental requirements of a strain of *Aphanizomenon flos-aquae* (L.) Ralfs. Can. J. Microbiol. *15*:165-173.

Gorham, P. R. 1962. Laboratory studies on the toxins produced by waterblooms of blue-green algae. Am. J. Public Health *52*(12):2100-2105.

Gorham, P. R. 1964. Toxic algae. Pages 307-336 *in* D. F. Jackson, ed., Algae and Man. Plenum Press, New York.

Halstead, B. W. 1964. Poisonous and Venomous Marine Animals of the World Vol. 1. U. S. Government Printing Office, Washington, D. C.

Haynes, R. 1971. Some Ecological Effects of Artificial Circulation on a Small, Eutrophic New Hampshire Lake. Ph.D. Thesis. University of New Hampshire, Durham, New Hampshire.

Jackim, E., and J. Gentile. 1968. Toxins of a blue-green alga: similarity to saxitoxin. Science *162*:915-916.

Kao, C. Y. 1972. Pharmacology of tetrodotoxin and saxitoxin. Fed. Proc. *31*:1117-1123.

Moore, R. E. 1977. Toxins from blue-green algae. Bioscience *27*(12):797-802.

Narahashi, T. 1972. Mechanism of action of tetrodotoxin and saxitoxin on excitable membranes. Fed. Proc. *31*:1124-1132.

Narahashi, T., M. S. Brodwick, and E. J. Schantz. 1975. Mechanism of action of a new toxin from *Gonyaulax tamarensis* on nerve membranes. Environ. Letters *9*(3):239-247.

O'Brien, R. D. 1969. Poisons as tools in studying the nervous system. Pages 1-59 *in* F. R. Blood, ed., Essays in Toxicology. Academic Press, New York.

Palmer, C. M. 1964. Algae in water supplies of the United States. Pages 239-261 *in* D. F. Jackson, ed., Algae and Man. Plenum Press, New York.

Prakash, A., J. C. Medcof, and A. D. Tennant. 1971. Paralytic shellfish poisoning in eastern Canada. Fish. Res. Bd. Can. Bull. *177*:1-87.

Sasner, J. J., Jr. 1973. Comparative studies on algal toxins. Pages 127-177 *in* D. F. Martin and G. M. Padilla, eds., Marine Pharmacology. Academic Press, New York.

Sasner, J. J., Jr., and M. Ikawa. 1975. Comparative studies on toxins from microorganisms. Pages 433-442 *in* V. R. LoCicero, ed., Proceedings of the First International Conference on Toxic Dinoflagellate Blooms. Massachusetts Science and Technology Foundation, Massachusetts.

Sawyer, P. J., J. H. Gentile, and J. J. Sasner, Jr. 1968. Demonstration of a toxin from *Aphanizomenon flos-aquae* (L.) Ralfs. Can. J. Microbiol. *14*:1199-1204.

Schwimmer, D., and M. Schwimmer. 1964. Algae and medicine. Pages 368-412 *in* D. F. Jackson, ed., Algae and Man. Plenum Press, New York.

Schwimmer, M., and D. Schwimmer. 1968. Medical aspects of phycology. Pages 279-358 *in* D. F. Jackson, ed., Algae, Man and the Environment. Syracuse University Press, Syracuse, New York. 369 pp.

Shimizu, Y. 1978. Dinoflagellate toxins. Pages 1-42 *in* P. J. Scheuer, ed., Marine Natural Products. Academic Press, New York.

Shoptaugh, N. H. 1978. Fluorometric Studies on the Toxins of *Gonyaulax tamarensis* and *Aphanizomenon flos-aquae*. Ph.D. Thesis. University of New Hampshire, Durham, New Hampshire.

Twarog, B. M., and H. Yamaguchi. 1975. Resistance to paralytic shellfish toxins in bivalve molluscs. Pages 381–393 *in* V. R. LoCicero, ed., Proceedings of the First International Conference on Toxic Dinoflagellate Blooms. Massachusetts Science and Technology Foundation, Massachusetts.

CHEMICAL STUDIES ON TOXINS FROM THE

BLUE-GREEN ALGA *APHANIZOMENON FLOS-AQUAE*

Maktoob Alam and Kenneth L. Euler

Department of Medicinal Chemistry and Pharmacognosy
College of Pharmacy
University of Houston
Houston, Texas 77004

ABSTRACT

Natural blooms of the blue-green alga *Aphanizomenon flos-aquae*
occur periodically in the lakes of the northern United States and
in certain provinces of Canada. These blooms have been reported to
cause poisoning in fish and cattle. Investigation of the toxins of
A. flos-aquae indicates that this alga contains three toxins in
addition to saxitoxin. The aphantoxins appear to have a chemical
structure similar to saxitoxin.

INTRODUCTION

Blooms of freshwater cyanophytic algae are world-wide in
occurrence (Gorham, 1964; Aziz, 1974; Vinberg, 1954; Branco, 1959).
Frequently, these blooms are associated with the mass poisoning of
fish and of cattle which drink lake water containing algal blooms.
The deleterious effects of algal blooms to fish could be attributed
to: a) poisonous compound(s) produced by the alga, or b) secondary conditions, such as imbalance of nutrients, oxygen deficiency,
hydrogen sulfide production from decomposing organic matters and
increased bacterial concentrations (Ballantine and Abbott, 1957).
Seven species of blue-green algae have been reported for their
toxic effects. These are *Aphanizomenon flos-aquae* (Sawyer et al.,
1968), *Microcystis aeruginosa, Anabaena flos-aquae* (Gorham, 1964),
Gloetrichia echinulata, Coelospharium kutzingianum (Gorham, 1964),
and *Synechococcus* sp. (Amann, 1977). Of these seven species the
first three have been responsible for the most dramatic and serious outbreaks of poisoning.

405

The cyanophyte *Microcystis aeruginosa* has been extensively
studied by P. R. Gorham's group. Bishop and Gorham (1959) reported
the isolation and chemical nature of the *Microcystis* toxin, microcystin. It was proposed to be a cyclic peptide consisting of seven
amino acids: aspartic acid (1 mole); glutamic acid (2 moles);
leucine (2 moles); ornithine (1 mole); alanine (2 moles); valine
(1 mole); and D-serine (1 mole), with a molecular weight of 1300 to
2600 daltons. Murthy and Capindale (1970) later reported that
microcystin from *Microcystis aeruginosa* NRC-1 had arginine, proline,
threonine, tyrosine, phenylalanine, isoleucine and glycine in addition to the original seven amino acids reported by Bishop and Gorham, (1959).

The toxin from another cyanophyte, *Anabaena flos-aquae* has
been studied extensively. Using X-ray crystallography Huber (1972)
elucidated the structure of the N-acetyl derivative (Figure 1, I)
of the *Anabaena flos-aquae* toxin [anatoxin-a, 2-acetyl-9-azabicyclo
(4.2.1) non-2-ene (Figure 1, II)].

Subsequently, Devlin et al. (1977) reported the isolation and
chemical structure of anatoxin-a (Figure 1, II) which was further
confirmed by the synthesis of nor-anatoxin-a and anatoxin-a (Campbell et al., 1977).

Our interest in the toxins of *Aphanizomenon flos-aquae* stemmed
from the observation of Professor Sawyer at the University of New

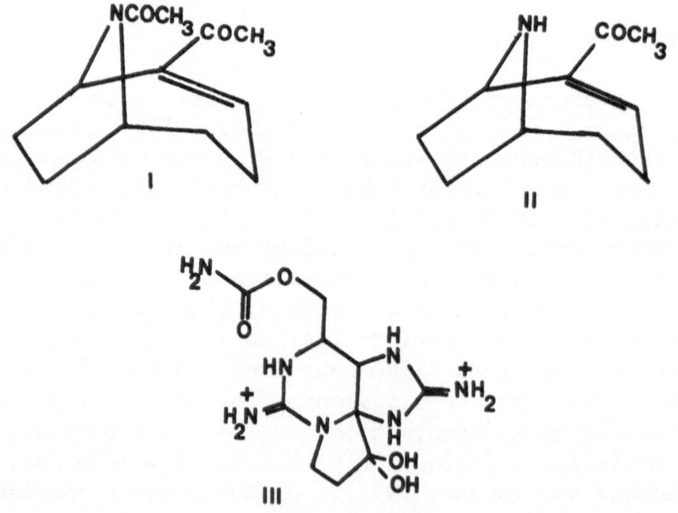

Fig. 1. Structure of: (I) N-acetyl anatoxin-a; (II) anatoxin-a
         hydrochloride; (III) saxitoxin.

Hampshire who reported (Sawyer et al., 1968) a serious outbreak of fish poisoning by waterblooms of cyanophytes in Kezar Lake, North Sutton, New Hampshire in 1964. The predominant algal species in Kezar Lake was identified as *A. flos-aquae*[1]. Similar algal blooms in Lakes Winnesguam and Skatulakee in New Hampshire also resulted in the mass poisoning of fish in 1966 (Sawyer et al., 1968). Later, Jackim and Gentile (1968) reported the isolation of a toxin, which was found to be similar (IR, color reactions, and physical activity), if not identical, to saxitoxin (Figure, III).

An earlier investigation into the toxic component(s) of *A. flos-aquae* had shown that they behaved differently on an ion exchange column than did saxitoxin. An attempt to purify these toxins by the method of Schantz et al. (1957), as reported for saxitoxin, met with little success. Alam et al. (1973) later reported the isolation of a toxin from *A. flos-aquae* and showed by TLC and color reactions that the toxin was not saxitoxin. Alam et al. (1978) reinvestigated the toxins of *A. flos-aquae* and proved that the blue-green alga contains, in fact, three toxins in addition to saxitoxin.

EXPERIMENTAL PROCEDURES AND RESULTS

Thick concentrates of *A. flos-aquae* were obtained by centrifugation of the lake water during the height of the blooms in Kezar Lake, North Sutton, New Hampshire during the summers of 1967-1970. The algal concentrates were taken to the laboratory at the end of the day's run and stored in a freezer until ready to be processed. The frozen concentrate (about 0.5 to 1 liter) was thawed and diluted with about 1 liter of distilled water. The suspension was acidified to pH 3.0 with HCl and heated to 80 to 90°C. The heated mixture was chilled in an ice bath and centrifuged at 10,000 x g for 15 minutes. The supernatant portion was separated and the residue was re-extracted an additional two times. The supernatant portions were combined and evaporated to dryness, and excess HCl was removed by placing the flask containing the residue in a vacuum desiccator over NaOH pellets. The residue was then extracted with ethanol; the ethanol extract upon evaporation gave a toxic residue which was partitioned between chloroform and water. The toxic aqueous layer was lyophilized to yield crude aphantoxin. At each step of purification the toxicity was determined by the standard mouse bioassay method (Schantz, et al., 1958).

---

[1]*A. flos-aquae*, in all cases in this text, refers to *Aphanizomenon flos-aquae*.

Table 1. Column Chromatography of *Aphanizomenon flos-aquae* Crude
          Toxin on Sephadex G-15

Fraction #	Solvent	Volume (ml)	Toxicity (MU)
1	Water	2.5	–
2	Water	2.5	–
3	Water	2.5	–
4	Water	2.5	–
5	Water	2.5	–
6	Water	2.5	–
7	Water	2.5	–
8	Water	2.5	65
9	Water	2.5	185
10	Water	2.5	300
11	Water	2.5	115
12	Water	2.5	23
13	0.05 N AcOH	2.5	–
14	0.05 N AcOH	2.5	–
15	0.05 N AcOH	2.5	–
16	0.05 N AcOH	2.5	–
17	0.05 N AcOH	2.5	–
18	0.05 N AcOH	2.5	1000
19	0.05 N AcOH	2.5	65
20	0.05 N AcOH	2.5	25
21	0.05 N AcOH	2.5	–
22	0.05 N AcOH	2.5	–
23	0.05 N AcOH	2.5	–
24	0.05 N AcOH	2.5	–
25	0.05 N AcOH	2.5	–

The crude aphantoxin (2100 MU) was dissolved in 1.5 ml of water
and the pH was adjusted to pH 5 with NaOH. The solution was then
layered on a Sephadex G-15 column (0.6 x 60 cm). The column was

Table 2. Column Chromatography of the Unabsorbed Toxic Fractions on Bio Rex 70 (H^+)

Fraction #	Solvent	Volume (ml)	Toxicity (MU)
1	Water	2.5	-
2	Water	2.5	-
3	Water	2.5	-
4	Water	2.5	-
5		2.5	-
6		2.5	-
7		2.5	-
8		2.5	-
9		2.5	-
10		2.5	-
11		2.5	20
12		2.5	70
13	Gradient Water – 1.0 N AcOH	2.5	29
14		2.5	-
15		2.5	17
16		2.5	55
17		2.5	12
18		2.5	-
19		2.5	-
20		2.5	-
21		2.5	-
22		2.5	-
23		2.5	-
24		2.5	-
25		2.5	-

eluted with water (30 ml) followed by 0.05 N AcOH (Table 1). About 40% of the toxin was eluted with water while the remaining 60% was bound to the gel and was eluted with 0.05 N AcOH. Fractions con-

Table 3. Column Chromatography of the Absorbed Toxic Fractions on
          Bio Rex 70 (H^+)

Fraction #	Solvent	Volume (ml)	Toxicity (MU)
1	Water	2.0	–
2	Water	2.0	–
3	Water	2.0	–
4	Water	2.0	–
5	Water	2.0	–
6	Water	2.0	–
7	Water	2.0	–
8	Water	2.0	–
9	Water	2.0	–
10	Water	2.0	–
11		2.0	–
12		2.0	–
13		2.0	–
14		2.0	–
15		2.0	–
16		2.0	–
17		2.0	–
18		2.0	–
19		2.0	–
20		2.0	–
21		2.0	–
22	Linear Gradient of 0.05 M – 1.0 M AcOH	2.0	–
23		2.0	–
24		2.0	–
25		2.0	–
26		2.0	–
27		2.0	–
28		2.0	–
29		2.0	–
30		2.0	–
31		2.0	–
32		2.0	12
33		2.0	17
34		2.0	39
35		2.0	47
36		2.0	20
37		2.0	21
38		2.0	43
39		2.0	17
40		2.0	–
41		2.0	–
42		2.0	–
43		2.0	–
44		2.0	–
45		2.0	–

taining unbound toxin(s) (688 MU) were combined and lyophilized, and the residue was rechromatographed on Bio Rex 70 (H^+). Chromatography was carried out at approximately 40 psi using a Fluid Metering RP-G Constant Flow pump. The column (0.6 x 60 cm) was eluted with water (10 ml) followed by a linear gradient of water to 1.0 N AcOH (Table 2). Two toxins (tentatively named aphantoxin 1 and aphantoxin 3) were separated by the Bio Rex 70 (H^+) chromatography. Fractions containing bound toxins (1090 MU) were combined and lyophylized. The residue was dissolved in water, then the pH was adjusted between 5.0 to 5.5 and rechromatographed on a Bio Rex 70 (H^+) column (0.6 x 60 cm). In a typical run about 200 MU of toxin mixture was placed on the column, which was eluted first with water and then followed by a gradient of 0.05 to 1.0 M acetic acid (Table 3). Two toxins were eluted with 0.6 N AcOH. One of the toxins was identified as saxitoxin by thin layer chromatography (TLC) on two different plates and solvent systems (Table 4), while the other was named aphantoxin 2. Thin layer chromatographic behavior of aphantoxins, saxitoxin, and ganyautoxin II in two solvent systems and on two different plates is shown in Table 4. The RF value of saxitoxin isolated from *A. flos-aquae* was found to be identical with that of an authentic sample of saxitoxin. To further confirm the presence of saxitoxin in *A. flos-aquae*, hydrolysis of the standard and isolated samples was performed on a TLC

Table 4. Thin Layer Chromatographic Behavior of Aphantoxins, Saxitoxin, and Gonyautoxin II

Toxin	R_f	
	S.system I[a]	S.system II[b]
Aphantoxin 1	0.80	0.45
Aphantoxin 2	0.69	0.47
Aphantoxin 3	0.47	0.43
Saxitoxin	0.63	0.45
Gonyautoxin II	0.81	0.62

[a]S.system I: Pyridine, ethyl acetate, acetic acid, water (75:25:15:30 v/v) used with Silica Gel 60 (E.M.) plate.

[b]S.system II: t-Butanol, acetic acid, water (2:1:1 v/v) used with Silica Gel GF (Analtech, Inc.) plate.

plate. The TLC plates were spotted with both samples sprayed with 1.0 percent hydrogen peroxide followed by 0.5 N NaOH (Oshima et al., 1977) to cause hydrolysis and developed (solvent system I, Table 4). The two samples gave products which were equal in number and Rf values. The ratio of saxitoxin and aphantoxin 1, 2, and 3 was 10:6:49:5 MU per 100 MU. The toxicities (one determination only) of aphantoxin 1, 2, and 3 were 4500 ± 1000 MU/mg, 2000 ± 1000 MU/mg, and 4000 ± 1000 MU/mg, respectively. The toxicity of saxitoxin isolated from *A. flos-aquae* was approximately 5000 MU/mg.

DISCUSSION

Our results indicated that in addition to saxitoxin there are at least three major toxins present in *A. flos-aquae*. In an earlier communication, Alam et al., 1973 reported that a toxin isolated from *A. flos-aquae* was not saxitoxin as determined by the Rf values of the toxin and saxitoxin in seven different solvent systems. We did not comment on the presence of saxitoxin in *A. flos-aquae*, as our earlier attempts were geared towards the isolation of a toxin or toxins other than saxitoxin. It was also noted that the toxin lost almost all of its toxicity during storage. One of the newly isolated toxins (aphantoxin 2) degrades in a relatively short time during storage in a freezer. In this respect aphantoxin 2 behaves similarly to the toxin isolated in 1973. All three aphantoxins gave fluorescent spots when the TLC plate was sprayed with 0.5% H_2O_2 indicating that at least a part of these molecules is similar to saxitoxin and the toxins isolated from *Gonyaulax tamarensis*. The presence of saxitoxin in *A. flos-aquae* has also been reported by Jackim and Gentile (1968). The partial similarity of aphantoxins to saxitoxin and gonyautoxin II is also supported by the fact that i.p. injections of aphantoxins into killifish (*Fundulus heteroclitus*) resulted in a characteristic response with darkening in anterior portion of the body. As the pigmentation spread posteriorly, loss of laterial equilibrium followed and death occurred within minutes. The darkening of the body was similar to the conditions that have been described for saxitoxin and tetrodotoxin (Dawson, 1972). It is noteworthy that Kirpenko et al. (1976) reported the isolation of eight toxic metabolites from Cyanobacteria (*M. aeruginosa, Anabaena flos-aquae,* and *Aphanizomenon flos-aquae*) which were proteinous in nature. Toxicity to rats was reported to be in the range of 0.74 to 0.78 mg/kg. One of the most important features of the toxins was the presence of a thiocyanate group. They also suggested that it is the protein part of the molecule which is involved in the toxic effect. The complete chemical identification of aphantoxins awaits the culturing of the blue-green alga in our laboratory. Thus far we have attempted to culture two strains of *A. flos-aquae*, but both have proved to be non-toxic in nature.

ACKNOWLEDGMENTS

Supported in part by a grant (E-745) from the Robert A. Welch Foundation, Houston, Texas.

REFERENCES

Alam, M., M. Ikawa, J. J. Sasner, Jr., and P. J. Sawyer. 1973. Purification of *Aphanizomenon flos-aquae* toxin and its chemical and physiological properties. Toxicon *11*:65-72.

Alam. M., Y. Shimizu, M. Ikawa, and J. J. Sasner, Jr. 1978. Reinvestigation of the toxins from the blue-green alga, *Aphanizomenon flos-aquae*, by a high performance chromatographic method. J. Environ. Sci. Health *A13*(7):493-499.

Amann, M. 1977. Untersuchungen über ein pteridiz als bestandteil des toxischen prinzips aus *Synechococcus*. D. Naturioiss. Dissertation. Eberhard-Karls Universitat, Tübingen. 64 pp.

Aziz, K. M. S. 1974. Diarrhea toxin obtained from a waterbloom-producing species, *Microcystis aeruginosa* Kütz. Science *183*:1206-1207.

Ballantine, D., and B. C. Abbott. 1957. The toxin from *Gymnodinium veniticum* Ballantine. J. Mar. Biol. Assoc. U. K. *36*:169-189.

Bishop, C. T., E. F. L. J. Anet, and P. R. Gorham. 1959. Isolation and identification of the fast-death factor in *Microcystis aeruginosa*. Can. J. Biochem. Physiol. *37*:453-471.

Branco, S. M. 1959. Algas toxicas controle das toxinas em aguas de abasticimento. Rev. Dep. Aguas Esgatos Sao Paulo (Brasil) *20*(33):21-30, *20*(34):29-42.

Campbell, H. F., O. E. Edwards, and R. Kolt. 1977. Synthesis of noranatoxin-a and anatoxin-a. Can. J. Chem. *55*:1372-1379.

Devlin, J. P., O. E. Edwards, P. R. Gorham, N. R. Hunter, R. K. Pike, and B. Stavric. 1977. Anatoxin-a, a toxic alkaloid from *Anabaena flos-aquae* NRC-44h. Can. J. Chem. *55*:1367-1371.

Down, R. J. 1972. *Fundulus heteroclitus* Melanophore system responses to dinoflagellate toxin (saxitoxin), tetrodotoxin, neurotropic drugs, and ions. Pages 317-346 *in* L. R. Worthen, ed., Food-Drug from the Sea Proceedings. Marine Technology Society, Washington, D. C.

Gorham, P. R. 1964. Toxic algae. Pages 307-336 *in* D. F. Jackson, ed., Algae and Man. Plenum Press, New York.

Huber, C. S. 1972. The crystal structure and absolute configuration of 2,9-diacetyl-9-azabicyclo (4,2,1) non-2, 3-ene. Acta Crystallogr. Sect. B *28*:2577-2582.

Jackim, E., and J. Gentile. 1968. Toxins of a blue-green alga: similarity to saxitoxin. Science *162*:915-916.

Kirpenko, Y. A., L. A. Sirenko, L. F. Lukina, N. I. Kirpenko, and
        E. N. Danilovskaya. 1976. Relation of the biological
        activity of toxic metabolites of blue-green alga to struc-
        tural features [in Russian]. Form. Kontrol Kach. Poverkhn.
        2:33-37; Chem. Abstr. *91*:71439q.
Murthy, J. R., and J. B. Capindale. 1970. A new isolation and
        structure for the endotoxin from *Microcystis aeruginosa*
        NRC-1. Can. J. Biochem. *48*:508-510.
Oshima, Y., L. J. Buckley, M. Alam, and Y. Shimizu. 1977. Hetero-
        geneity of paralytic shellfish poisons. Three new toxins
        from cultured *Gonyaulax tamarensis* cells, *Mya arenaria* and
        Saxidomus giganteus. Comp. Biochem. Physiol. *57c*:31-34.
Sawyer, P. J., J. H. Gentile, and J. J. Sasner, Jr. 1968. Demon-
        stration of a toxin from *Aphanizomenon flos-aquae* (L.)
        Ralfs. Can. J. Microbiol. *14*:1199-1204.
Schantz, E. J., E. F. McFarren, M. L. Schafer, and K. H. Lewis.
        1958. Purified shellfish poison for bioassay standardi-
        zation. J. Assoc. Off. Anal. Chem. *41*:160-168.
Schantz, E. J., J. D. Mold, D. W. Stanger, J. Shavel, F. J. Riel,
        J. R. Bowden, J. M. Lynch, R. S. Wyler, B. Reigel, and H.
        Sommer. 1957. Paralytic shellfish poison VI, a procedure
        for the isolation from toxic clam and mussel tissues. J.
        Am. Chem. Soc. *79*:5230-5235.
Vinberg, G. G. 1954. Toksicheskii fitoplankton. Uspecki Sour.
        Biologii *38*:216-226.

STUDIES ON THE FLUOROMETRIC DETERMINATION OF THE TOXINS OF

THE BLUE-GREEN ALGA *APHANIZOMENON FLOS-AQUAE*

Miyoshi Ikawa, Karin Wegener[1], Thomas L. Foxall[1], John
J. Sasner, Jr.[1], Philip W. Carter, and Nancy H. Shoptaugh

Department of Biochemistry
University of New Hampshire
Durham, New Hampshire 03824

ABSTRACT

The Bates and Rapoport fluoremetric method for the detection
os saxitoxin (STX) had been modified and used for the determination
of the *Gonyaulax tamarensis* toxins in New England shellfish, and
both a tube assay and a pre-oxidized column procedure have been
described (Shoptaugh, 1978). In the tube assay the shellfish
extracts are treated with alkaline hydrogen peroxide and the resulting fluorescence read. In the pre-oxidized column procedure the
extracts are treated with hydrogen peroxide, the oxidized extracts
placed on a cation exchange resin column, fractions eluted with a
system of buffers, and the fluorescence of the fractions read. In
view of the similarities between the physiological effects and the
chemical nature of the toxins of *Aphanizomenon flos-aquae* and the
toxins produced by the marine dinoflagellate *Gonyaulax tamarensis,*
the causative agent of paralytic shellfish poisoning in New England
shellfish, the applicability of the procedures worked out for the
Gonyaulax toxins was tried for the detection and determination of
the *Aphanizomenon* toxins.

The algae were extracted either with water or by boiling with
0.1 N HCl for five minutes. The tube method tended to give higher
values of *Aphanizomenon* toxins than the mouse assay, and high background pigment fluorescence was encountered. Filtration of the

---

[1]Address: Department of Zoology, University of New Hampshire,
Durham, New Hampshire 03824.

algal extracts through a 10,000 molecular weight membrane filter
cleanly removed interfering fluorescence, but the results now
tended to be low. Using the pre-oxidized column procedure the
results agreed more closely with the mouse assay. The elution
pattern of oxidized *Aphanizomenon* toxins were similar to the pattern
of oxidized STX. The elution behavior of the unoxidized toxins on
the cation exchange resin column was also studied. Oxidation by
hydrogen peroxide was carried out on the fractions obtained after
chromatography and the fluorescence read. The unoxidized *Aphanizomenon* toxins, like STX, were not eluted until HCl was used as the
eluting agent. Shellfish extracts gave earlier eluting peaks as
did the *Gonyaulax* toxins GTX_2 and GTX_3.

INTRODUCTION

     Recent investigations have led to the development of fluoro-
metric methods for the detection and determination of paralytic
shellfish poisons produced by species of the marine dinoflagellate
genus *Gonyaulax* (Bates et al., 1975, 1978; Buckley et al., 1976,
1978). Similarities between the physiological effects (Sawyer et
al., 1968) and the chemical nature (Jackim and Gentile, 1968; Alam
et al., 1978) of the toxins of the blue-green alga *Aphanizomenon
flos-aquae* with those of the *Gonyaulax* species have suggested the
possibility of the application of fluorometric methods for the
detection and determination of the *Aphanizomenon* toxins. This
paper describes our results in this direction.

MATERIALS AND METHODS

Materials

     Toxic *A. flos-aquae* cells collected in 1967 and in subsequent
years by continuous centrifugation of dense blooms in Kezar Lake,
North Sutton, New Hampshire, lyophilized, and stored in the freezer
were used. Non-toxic *A. flos-aquae* cells collected in 1969 from a
bloom in the Old Reservoir, Durham, New Hampshire, lyophilized, and
stored in the freezer were used as controls.

Mouse Bioassay

     Bioassays were carried out using mouse strain C57BL/6J (Jack-
son Laboratories, Bar Harbor, Maine) or strain CD-1 (Charles River
Breeding Laboratories, Wilmington, Massachusetts). As with saxitoxin (STX), one mouse unit (MU) was defined as the amount of toxin
required to kill a 20 g mouse in 15 minutes. Sommer's tables,

correlating MU of paralytic shellfish poison to death times and
mouse weights, were used (Horwitz, 1965). The mice were standardized with STX (5500 MU/mg), and the results were expressed as μg
STX equivalents per g of dry weight of alga.

Fluorescence Measurements

A Farrand MK-1 spectrofluorometer was used. The excitation
was set at 330 nm with a 5 nm slit width and the emission at 380 nm
with a 10 nm slit width.

Aqueous Extraction and Molecular Filtration Procedure

Lyophilized cells (100 mg) were added to 10 ml water and the
mixture sonicated in a 60 watt M.S.E. Ultrasonic Disintegrator for
two minutes, then centrifuged at 2000 rpm for ten minutes. The
dark blue-green supernate was then pulled by vacuum through a
Millipore 10,000 molecular weight exclusion ultrafiltration membrane,
giving a pigment-free filtrate. The filter was washed with water
to give a total filtrate volume of 20 ml. The pH of the filtrate
was 6.7.

Hot Acid Extraction Procedure

Lyophilized cells (100 mg) were added to 10 ml of 0.02 M HCl,
and the mixture heated in a boiling water bath for five minutes,
cooled, adjusted to pH 4.0 to 4.5 with NaOH, adjusted to a volume
of 20 ml with water, and centrifuged at 2000 rpm for 10 minutes.
The clear very pale yellow-green supernate was used without ultrafiltration.

Non-Column Tube Procedure

The procedure described by Shoptaugh (1978) for the assay of
paralytic shellfish poisons produced by *Gonyaulax tamarensis* is
as follows:

1. Aliquots of extracts adjusted to pH 4.0 to 4.5 (2 ml maximum)
    were diluted to 2 ml with water. Blanks containing the same
    amount of extracts were also prepared.

2. Two ml of 1 N NaOH were added to each tube.

3. A 0.6 ml amount of 1% H_2O_2 was added to each sample tube and
    the contents mixed with a vortex stirrer; 0.6 ml of water was
    added to each blank tube.

4. All tubes were incubated in the dark at room temperature for 40 minutes.

5. Glacial acetic acid (0.2 ml) was added to each tube.

6. The samples and blanks were read in the spectrofluorometer. From the fluorescence of each sample tube the fluorescence of the corresponding blank was subtracted, and the corrected fluorescence (μamp) was plotted vs. volume of extract. If the curve was not linear, smaller volumes of extracts were used.

7. Activity was calculated in terms of μg saxitoxin (STX) per ml of extract by dividing the μamp/ml slope of the curve by the μamp/μg STX slope of a standard curve obtained with STX.

Pre-Oxidation Column Procedure[1]

Dowex AG 50W-X4, 200 to 400 mesh (hydrogen form) (Bio-Rad Laboratories, Richmond, California) was converted to the sodium form by adding 1 N NaOH to the resin in water until the pH was higher than 10. The resin was washed with distilled water until the pH dropped to 8.0 to 8.5 and then converted back to the hydrogen form by adding 1 N HCl until the pH dropped below 4. The resin was again washed with distilled water until the pH rose above 5, then reconverted back to the sodium form by adding 1 N NaOH until the pH rose above 10. After a final rinsing with distilled water to a pH of 8.0 to 8.5, the resin was equilibrated with 0.01 M pH 5 sodium acetate buffer. The resin was packed in a disposable Pasteur pipet to form a 0.6 x 7.5 cm (2 ml) column and equilibrated with 50 ml of 0.01 M pH 5 sodium acetate buffer. The sample to be applied was prepared by mixing 0.5 ml of STX standard or 0.5 ml of alga extract with 0.5 ml of 1 N NaOH and 0.15 ml of 1% H_2O_2 (H_2O for the blank), incubating the mixture at room temperature for 40 minutes, acidifying with 0.05 ml glacial acetic acid, and bringing the volume to 2 ml with buffer. The 2 ml of solution was applied to the column and the column developed with 30 ml of 0.01 M pH 5 sodium acetate buffer, followed by 90 ml of 0.05 M pH 5 sodium acetate buffer, followed by 30 ml of 1 M HCl. Five ml fractions were collected using a peristaltic pump delivering 1.5 ml/min. Fluorescence of the fractions was measured at 380 nm (excitation wavelength 330 nm).

Post-Oxidation Column Procedure

The column in this procedure was the same as used in the pre-oxidation procedure. The column was equilibrated with 0.01 M sodium

---

[1]N. H. Shoptaugh, 1978.

Table 1. Comparison of Mouse Assay and Fluorometric Assays

Alga	Extraction Procedure	μg STX equiv./g dry wt.		
		Mouse	Non-Column Tube	Pre-Oxidation Column
A. *flos-aquae*	Hot 0.1 M HCl[a]	1170	2100	1500
A. *flos-aquae*	Hot 0.1 M HCl[a]	880	1400	1300
A. *flos-aquae*	Hot 0.1 M HCl[a]	700	1300	1000
A. *flos-aquae*	Hot 0.1 M HCl[a]	940	2100	1800
A. *flos-aquae*	Hot 0.1 M HCl[a]	< 40	2	0
A. *flos-aquae*	Hot 0.1 M HCl[a]	< 40		14
A. *flos-aquae*	Hot 0.1 M HCl[a]	693		799
C. *pyrenoidosa*	Hot 0.1 M HCl[a]	< 40		3
A. *flos-aquae*	Hot 0.1 M HCl[a]	937	613[b]	
A. *flos-aquae*	Water		607[b]	
A. *flos-aquae*	0.1 M pH 5 acetate buffer		219[b]	
A. *flos-aquae*	0.5 M pH 5 acetate buffer		75[b]	

[a]Same as hot acid extraction procedure in text except 0.1 M HCl instead of 0.02 M HCl.

[b]Extracts filtered through 10,000 mol. wt. filter. Represent same batch of cells.

acetate adjusted to pH 5 with HCl. Two ml of alga extract of pH 4
to 4.5 was applied to the column and developed as follows, 5 ml
fractions being collected. Tubes 1 through 7 were collected using
the 0.01 M pH 5 acetate-Cl⁻ equilibration buffer, tubes 8 through
51 with 0.1 M sodium acetate which had been adjusted to pH 5 with
HCl, and tubes 52 through 80 with 0.75 M HCl. Two 2 ml aliquots
of each fraction were pipeted out and 2 ml of 1 N NaOH added to
each aliquot. 0.15 ml of 1% H_2O_2 was added to each sample tube
and 0.15 ml water to each corresponding blank tube. The tubes
were allowed to stand at room temperature in the dark for 40 min.,
then 0.2 ml of glacial acetic acid was added to each tube and the
fluorescence read at 380 nm with excitation at 330 nm.

RESULTS AND DISCUSSION

Extraction Methods and the Non-Column Tube Procedure

     Aqueous extraction of A. *flos-aquae* cells led to very dark
blue-green extracts which had high background fluorescence. The
pigmentation and fluorescence could be effectively eliminated by
passing the extracts through a 10,000 molecular weight exclusion
ultrafiltration membrane but with an apparent loss in activity
(Table 1, last 4 rows). By using a hot acid extraction the pigmentation was virtually eliminated. Fluorometric assay of this
extract, however, gave results which were approximately double the
mouse assay results (Table 1). The reason is not known. Since
non-toxic A. *flos-aquae* extract showed almost no difference in
fluorescence between the sample and its blank, it does not seem
likely that fluorescence due to non-toxic components is responsible
for the high results.

     At any rate, hot acid extraction appears to result in a higher
recovery of activity than is realized by milder procedures followed
by ultrafiltration, but, as will be shown below, hot acid extraction chemically alters the toxins.

Pre-Oxidation Column Procedure

     The pre-oxidation column procedure is based on forming the
fluorescent derivative first, then doing a partial column purification on a cation exchange resin and reading the intensity of the
fluorescence peak. When STX was taken through this procedure, the
fluorescence elution pattern shown in Figure 1A was obtained. By
oxidizing varying amounts of STX, a linear relationship could be
obtained between the amount of STX and the peak height, which was
used in order to obtain the STX equivalents in the sample. As
shown in Figure 1B, pre-oxidized A. *flos-aquae* toxin gave an

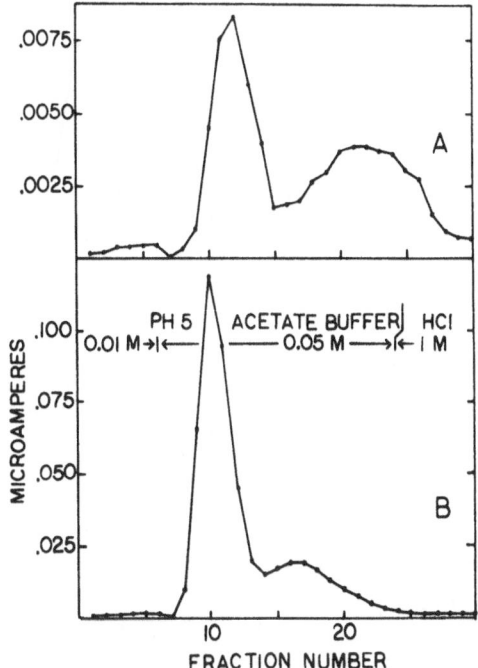

Fig. 1.   Pre-oxidation column procedure fluorescence elution pro-
          files.   (A) STX (0.4 μg);   (B) Hot 0.1 M HCl extract of
          A. flos-aquae (ca. 5 μg STX equivalents).

elution curve with a similar first principal peak as STX.   However,
the second peak was greatly reduced and was eluted earlier than the
second peak of pre-oxidized STX.   Non-toxic cell extracts showed
no fluorescence peaks.   By comparing peak heights of the *A. flos-
aquae* extracts with those of STX, the results shown in Table 1
(last column) were obtained.   Although the results were lower than
the values obtained by the non-column tube method, they still
appeared to be significantly higher than the mouse assay results.
Chlorella pyrenoidosa cells showed no significant amount of fluorescence.

     The pre-oxidation column procedure most likely can be adapted
to high performance liquid chromatography (HPLC) with a fluorescence detector for a more routine method for the detection, determination, and identification of *A. flos-aquae* toxins.   The high
resolving power of the HPLC method may also serve to show the
identity or the difference between the STX oxidation product and
the *A. flos-aquae* toxin oxidation product.

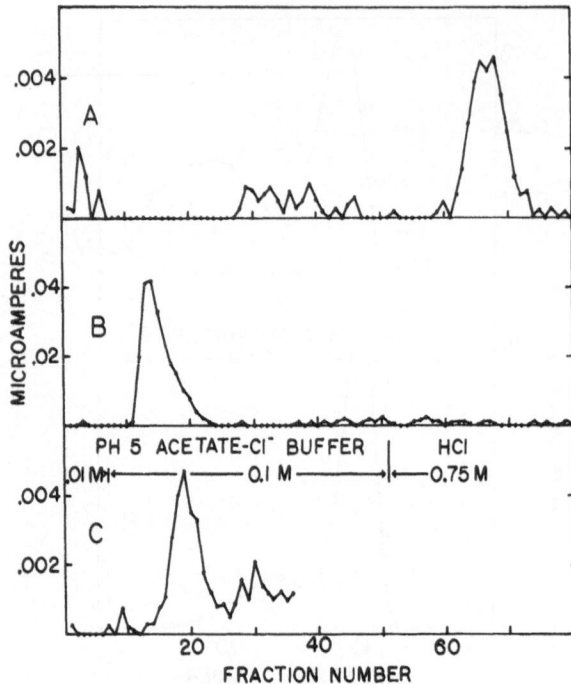

Fig. 2.   Post-oxidation column procedure fluorescence profiles.
           (A) STX (2 µg);   (B) GTX$_2$ (20 µg);   (C) GTX$_3$ (2 µg).

Post-Oxidation Column Procedure

        Because the pre-oxidation procedure destroys the original
toxins in the cells, the post-oxidation procedure was developed in
order to gain a greater insight into the nature of the natural *A.
flos-aquae* toxins.   Since the *A. flos-aquae* toxins appear to be
similar to the *Gonyaulax* toxins, STX, gonyautoxin II (GTX$_2$), and
gonyautoxin (GTX$_3$) were taken through this procedure.   The GTX$_2$
and GTX$_3$ were eluted early in the procedure with the GTX$_2$ slightly
preceding the GTX$_3$ (Figure 2B and 2C).   The STX was retained
tightly to the resin and was not eluted until considerable HCl had
been employed (Figure 2A).   When water-extracted and ultrafiltered
toxic *A. flos-aquae* extract was taken through the procedure, a
large initial peak, roughly where GTX$_2$ was eluted, was observed
and a smaller peak near the STX position (Figure 3A).   When a hot
0.02 M HCl extract was taken through the procedure, the initial
peak was much smaller and the later peak larger, indicating that
the hot acid treatment was converting the initial peak to the later
peak and that the initial peak was rather labile to hot acid.   This
was demonstrated when the ultrafiltered aqueous extract was heated
with 0.02 M HCl then taken through the procedure.   The result, as
shown in Figure 3B, demonstrates the hot acid lability of *A. flos-
aquae* toxin.   When GTX$_2$ was subjected to the same hot acid treat-

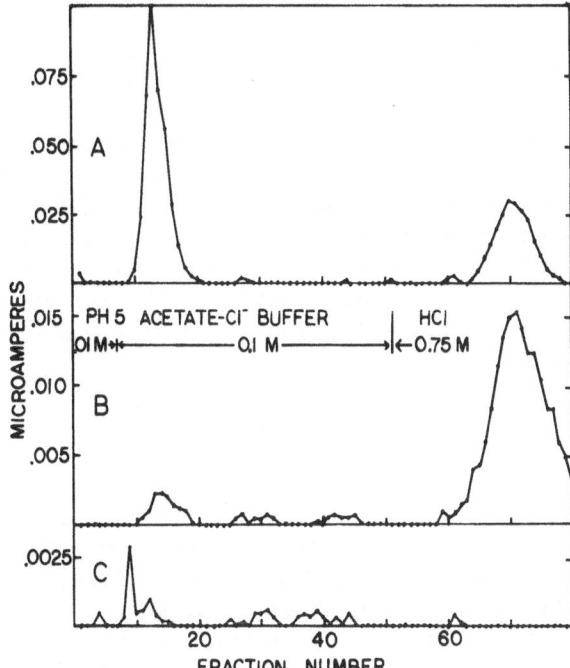

Fig. 3.   Post-oxidation column procedure fluorescence profiles.
(A) Aqueous and ultrafiltered extract of toxic *A. flos-aquae* cells (equivalent to 20 mg cells or 20 μg STX);
(B) Aqueous and ultrafiltered extract of toxic *A. flos-aquae* cells heated 5 minutes in 0.02 M HCl (equivalent to 10 mg cells or 10 μg STX);   (C) Hot 0.02 M HCl extract of non-toxic *A. flos-aquae* cells (equivalent to 10 mg cells).

ment, it appeared to be stable and no peak in the STX position was observed, indicating its acid stability and the non-identity of *A. flos-aquae* toxin to GTX_2 (unpublished results). Again non-toxic *A. flos-aquae* extract did not show any fluorescence peaks (Figure 3C).

The results of this study show that fluorescence might be of value both in the detection and determination of *Aphanizomenon* toxins and of value in their isolation and identification. Similarities in the behavior of the toxins and their H_2O_2 oxidation products again confirm the chemical similarities between some of the *Aphanizomenon flos-aquae* toxins and the *Gonyaulax* toxins. Some of the *Aphanizomenon* toxins, however, seem to differ from the *Gonyaulax* toxins in the greater lability of the former, and hot dilute acid treatment appears to convert all the toxins of the former to a single entity or to a more homogeneously behaving mixture. Even such mild conditions as concentrating aqueous solutions in a rotary evaporator caused a similar partial conversion of the toxins. This property may be of value in the eventual structure determination of the *Aphanizomenon* toxins.

The apparent tendency to overestimate these toxins in terms of STX equivalents may be due to a higher fluorescence yield per MU for these toxins than is realized for STX. Buckley et al. (1976, 1978) has shown the wide variation in fluorescence yields per MU for the various *Gonyaulax tamarensis* toxins and how they can affect the fluorometric determination. The resolution of this difficulty will have to await the isolation of the *Aphanizomenon* toxins in pure form.

SUMMARY

The *Aphanizomenon flos-aquae* toxins were converted to fluorescing derivatives by oxidation with hydrogen peroxide under alkaline conditions. Using a non-column tube procedure, the results in saxitoxin equivalents tended to be high, when compared with the mouse assay, when the cells were extracted by hot acid and low when cells were extracted by water and the extracts passed through ultrafiltration membranes. Using a column procedure where the sample was first oxidized, then chromatographed on strong cation exchange resin and the fractions monitored for fluorescence, the results also tended to be high for the hot acid extracts. When the oxidation reaction was performed on the fractions after column chromatography, two major fluorescent peaks were observed. The major peak was the first peak, but it was labile and was readily converted to the second peak.

ACKNOWLEDGMENTS

This work was supported by Hatch Project 205 of the New Hampshire Agricultural Experiment Station and by a grant from the United States Department of the Interior, Office of Water Research and Technology, as authorized under the Water Resources Act of 1964, Public Law 88-379, through the Water Resources Research Center of the University of New Hampshire.

REFERENCES

Alam, M., Y. Shimizu, M. Ikawa, and J. J. Sasner, Jr. 1978. Reinvestigation of the toxins from the blue-green alga, *Aphanizomenon flos-aquae*, by a high performance chromatographic method. J. Environ. Sci. Health *A13*(7):493-499.

Bates, H. A., R. Kostrikan, and H. Rapoport. 1978. A chemical assay for saxitoxin. Improvements and modifications. J. Agric. Food Chem. *26*:252-254.

Bates, H. A., and H. Rapoport. 1975. A chemical assay of saxitoxin, the paralytic shellfish poison. J. Agric. Food Chem. *23*:237-239.

Buckley, L. J., M. Ikawa, and J. J. Sasner, Jr. 1976. Isolation of *Gonyaulax tamarensis* toxins from soft shell clams (*Mya arenaria*) and a thin-layer chromatographic-fluorometric method for their detection. J. Agric. Food Chem. *24*:107-111.

Buckley, L. J., Y. Oshima, and Y. Shimizu. 1978. Construction of a paralytic shellfish toxin analyzer and its application. Analyst. Biochem. *85*:157-164.

Horwitz, W., ed. 1965. Official Methods of Analysis, 10th ed. Association of Official Agricultural Chemists, Washington, D. C. Page 282.

Jackim, E., and J. Gentile. 1968. Toxins of a blue-green alga: similarity to saxitoxin. Science *162*:915-916.

Sawyer, P. J., J. H. Gentile, and J. J. Sasner, Jr. 1968. Demonstration of a toxin from *Aphanizomenon flos-aquae* (L.) Ralfs. Can. J. Microbiol. *14*:1199-1204.

Shoptaugh, N. H. 1978. Fluorometric Studies on the Toxins of *Gonyaulax tamarensis* and *Aphanizomenon flos-aquae*. Ph.D. Thesis. University of New Hampshire, Durham, New Hampshire.

A FLUOROMETRIC TECHNIQUE FOR THE DETECTION AND

DETERMINATION OF PARALYTIC SHELLFISH POISONS

Nancy H. Shoptaugh[1], Miyoshi Ikawa,
Thomas L. Foxall[2], and John J. Sasner, Jr.[2]

Department of Biochemistry
University of New Hampshire
Durham, New Hampshire 03824

ABSTRACT

The occurrence of outbreaks of paralytic shellfish poisoning (PSP) along the New England Coase has generated research to develop a rapid, sensitive method for determining the toxin content in clams, *Mya arenaria*, and in mussels, *Mytilus edulis*. The oxidation of saxitoxin (STX) with alkaline hydrogen peroxide followed by acidification produced a solution fluorophor with an excitation of 332 nm and an emission of 381 nm. Measurement of the fluorescent intensity of the STX fluorophor is used to quantitate as little as 0.005 to 0.01 µg STX. Application of this solution-fluorometric procedure to samples of *M. arenaria* and *M. edulis* has enabled quantitation of toxin contents (as STX) that have corresponded with the mouse bioassay. In addition, levels of toxicity less than the lower limit of the bioassay, 40 µg STX per 100 g shellfish meat, are detectable by this tube assay.

INTRODUCTION

The unpredictable occurrence of toxic shellfish has dictated the necessity of constant monitoring of shellfish from harvesting

---

[1]Present address: Food Research Institute, University of Wisconsin, Madison, Wisconsin 53706.

[2]Address: Department of Zoology, University of New Hampshire, Durham, New Hampshire 03824.

areas to ensure their safety for human consumption.   The mouse bioassay for the detection of PSP was first used by Sommer and Meyer
(1937) and subsequently modified to the present standardized procedure for detecting and quantitating the poison (Schantz et al.,
1958; McFarren, 1959).   Although the mouse bioassay is the preferred
PSP quantitation method, the standard error of the mean is ± 20% and
the poison content of samples of low toxicity may be underestimated
by as much as 60% (McFarren, 1959).   The errors associated with the
mouse bioassay are multiplied in high toxicity extracts, by the
dilutions necessary to obtain a median death time of 5 to 7 minutes.
In addition, the mouse bioassay has been shown to be non-specific,
as other non-related marine neurotoxins have produced similar paralytic symptoms in mice (Mosher et al., 1964).

The inherent variability and lack of precision that accompanies
mouse bioassay monitoring has resulted in attempts to replace the
method with assays for STX that are both more sensitive and more
specific than the standardized method.   Perhaps the most promising
chemical assays, in terms of sensitivity and specificity, have been
those producing a fluorescent product (Shoptaugh, 1978).   Schantz
first reported the formation of a fluorescent compound from STX in
the presence of oxygen under alkaline conditions (Schantz, 1961).
A chemical assay for STX in West Coast shellfish has been developed,
using alkaline hydrogen peroxide oxidation, which is 100 times more
sensitive than the existing mouse bioassay (Bates et al., 1975,
1978).   The weak absorption of the East Coast shellfish toxins to
weak cation exchange resins has led to variations of the fluorometric assay (Buckley et al., 1976, 1978).

The present study undertook the application of fluorometric
techniques to New England shellfish.   The assay developed has been
compared to both the mouse bioassay and another published fluor-
metric method.

MATERIALS AND METHODS

STX Reference Standard

The reference standard of purified shellfish poison (STX, 100
µg per ml) was provided by Dr. E. J. Schantz.   The STX was stored
in the freezer and diluted with distilled water prior to use.

Shellfish Extracts

Frozen toxic samples of the soft-shell clam, *Mya arenaria*,
and the blue mussel, *Mytilus edulis*, collected during the 1972
Central New England red tide were available whole and were thawed

out prior to use. Acid extracts of whole shellfish were prepared by the standard method by homogenizing 100 g shucked, drained shellfish meat with 100 ml 0.1 N HCl in a blender (Osterizer Classic VIII). The homogenate was boiled gently 5 minutes and allowed to cool to room temperature. Either 5.0 N HCl or 0.1 N NaOH was added to lower or raise, respectively, the pH to 4.0 to 4.5. The final volume was adjusted to 200 ml with distilled water. The supernate obtained after centrifugation (2920 x g, head 211, International Clinical Centrifuge) provided the final acid extract used in subsequent experimentation.

Mouse Bioassay

The mouse bioassay served as the standardized method to quantitate the toxin content of samples. The mouse bioassays, strain C57BL/6J (Jackson Laboratories), were performed in Dr. Sasner's laboratory.

Solution Fluorometry

Solution fluorometry was carried out using a Farrand Spectrofluorometer MK-1 (IP28 detector). All measurements used matched, square fluorescence cells (Fisher). The power source was turned on 15 minutes before ignition of the xenon lamp which was allowed to warm up an additional 15 minutes before any fluorescent readings were attempted. Excitation and emission wavelengths were set by manually adjusting the proper monochromator. Fluorescence was read directly by adjusting the range sensitivity and noting the needle deflection on the microammeter scale. Excitation and emission wavelength scans were carried out by setting one of the monochromators at a fixed wavelength, engaging the motor drive for the other monochromator, and setting the drive speed to 100 nm per minute. The attachment of a strip chart recorder (Sargent Welch XKR) operating at 100 mV and a chart speed of 1 cm per minute produced a synchronous scan of 100 nm per cm.

STX Fluorescence

Dilutions of a stock solution of STX were used to establish a STX fluorescence curve. Alkaline oxidation was carried out by mixing 2.0 ml STX solution with 2.0 ml 1.0 N NaOH and 0.6 ml 1% (v/v) H_2O_2 and incubating the mixture at room temperature for 40 minutes. The final fluorescent product was made by the addition of 0.2 ml glacial acetic acid. Fluorescence at 381 nm was measured for the samples excited at 332 nm. A blank, consisting of 0.6 ml distilled H_2O in place of H_2O_2, was subtracted from each sample value to obtain the standard curve.

Shellfish Assays

        Aliquots of the shellfish acid extracts were taken to the 2.0
ml assay volume with distilled water and hydrogen peroxide oxidation
was carried out as outlined above for the STX standard. If a linear
relationship between ml extract in assay and fluorescence intensity
at 381 nm due to 332 nm excitation was not observed, further 1:10
or 1:100 dilutions of the original extract in distilled H_2O were
made and aliquots of this, taken to 2.0 ml assay volume with distilled H_2O, were used in the oxidation reaction. For each sample
fluorescence reading, the fluorescence of a blank containing the
same amount of extract, but distilled H_2O in place of 1% H_2O_2, was
subtracted. When linearity was established, the slope of ml extract
vs. μamps of sample minus blank was related to the standard curve
to obtain μg STX per ml extract. By accounting for any extract
dilutions and the g shellfish meat per ml extract, the toxin content
was reported as μg STX per 100 g meat. In addition to determining
the toxicity graphically, the area under the curve of a fluorometric
scan was determined with a planimeter (Gelman). The area, with the
area of the blank subtracted, is compared to the area under a STX
scan, 0.0024 μg STX per vernier unit (1 vernier unit = 0.01 sq. in.)
with a range setting of 0.3. For comparisons with a published
fluorometric method, 2.0 g homogenized shellfish meat was assayed
by the modified procedure of Bates and Rapoport (Bates and Rapoport,
1975).

RESULTS AND DISCUSSION

        The development of a chemical assay for STX in West Coast
shellfish using alkaline hydrogen peroxide oxidation has suggested
an assay technique 100 times more sensitive than the existing mouse
bioassay (Bates and Rapoport, 1975). The method developed for East
Coast shellfish was based on the establishment of a STX standard
curve. The excitation and emission spectra of the fluorescent product of 2.0 μg STX is shown in Figure 1. The excitation maximum
of 332 nm and the emission maximum of 381 nm can be compared to the
literature values of 330 nm excitation and 380 nm emission (Bates
and Rapoport, 1975). In both spectra, the blank did not contribute
significantly to the fluorescence intensities.

        The stability and fluorescent yield of the STX fluorophor was
investigated. It was found that photodecomposition was minimized
and fluorescent yield increased if the oxidation reaction was carried out in the dark at room temperature. The fluorescent product
was not stable at basic pH, an oxidation reaction of pH 11.8 to
12.6 being·optimal, but the fluorophor was stabilized upon acidification with acetic acid, pH 4.5 being optimal, if the fluorescence
intensity was measured within 60 minutes.

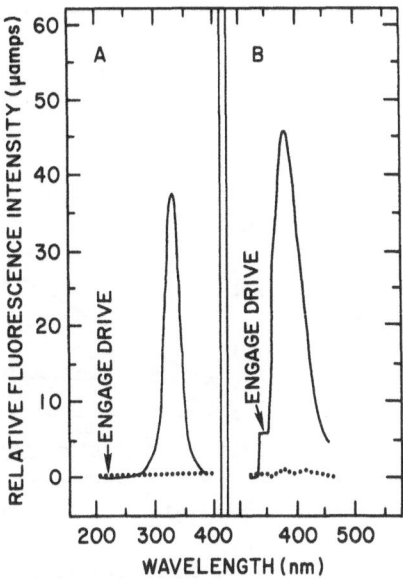

Fig. 1.   Spectral scans of the acid fluorescent H_2O_2 oxidation
          product of 2.0 µg STX. A. Excitation scan with
          emission = 400 nm. B. Emission scan with excitation =
          332 nm.   Slit widths = 5 to 10, Range = 1.0, Scan =
          100 nm/min. Sample (————) and Blank (·····).

     The STX fluorophor emission maximum and fluorescence intensity
varied with both the drive speed of the emission wavelength monochromator and the monochromator slit width combinations. An optimal intensity was given by the drive speed of 100 nm/minute. The
5 to 10 nm slit width combination (exciter-emission monochromator
slit widths, respectively) was chosen so that the irradiation
falling on the sample to excite it was limited to a 5 nm spectral
bandwidth and the emission to the photo-multiplier was maximized
to collect a 10 nm spectral bandwidth.

     The standard curve in Figure 2 shows 0.005 to 0.01 µg STX as
a lower limit of sensitivity. As expected, contribution of the
blank increased at the very dilute STX concentrations. The slope
of this curve, 0.1869 µamps per µg STX, was used to check instru-
mentation and assay procedures. The reciprocal slope, 5.3505 µg
STX per µamp, was used to calculate toxicity in samples of unknown
STX content. By this procedure, toxin could be assayed for as low
as 0.005 µg STX. This was considerably more sensitive than the
mouse bioassay which detected toxin to the level of 0.3 µg STX, a
lower limit which in practice cannot be attained since the presence
of sodium ions in shellfish extracts counteracts the effect of low
levels of STX (Schantz et al., 1958).

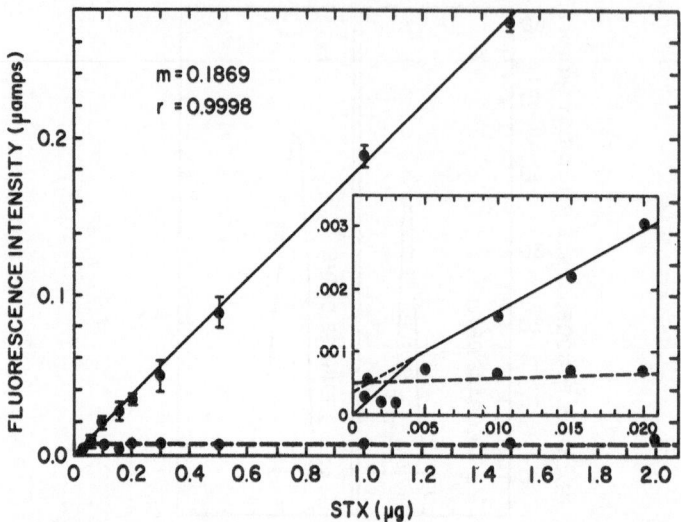

Fig. 2.   Calibration curve of MK-1 with STX using 5 to 10 nm slits.
Sample minus blank ± S.E.M. for n = 5 (●──●), and blank
(●---●).   Inset shows expanded scale for 0 to 0.02 µg
STX.

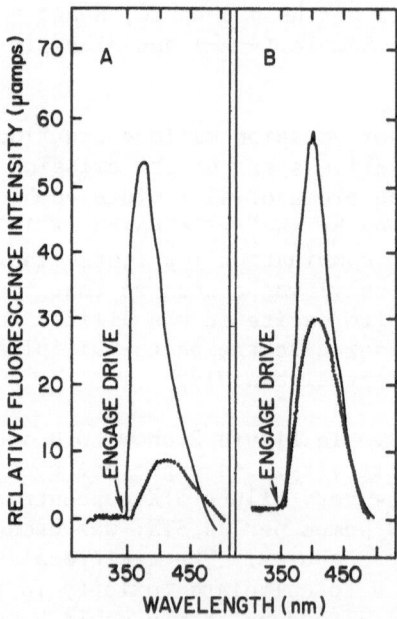

Fig. 3.   Emissions scans of toxic shellfish acid extracts.
A. *Mya arenaria*, Range 0.3, 0.5 ml HCl extract.
B. *Mytilus edulis*, Range 0.3, 0.3 ml HCl extract.
Excitation = 332 nm, Scan = 100 nm/min, Slit widths =
5 to 10 nm.   Sample (───) and Blank (·····).

The emission scans of the fluorophors of toxic M. *arenaria* and M. *edulis* shown in Figure 3, suggested that fluorescence measurements of the products from dilutions of acid extracts can be graphed and linearity between concentration and fluorescence, as sample minus blank, established. Although the emission maximum varied with the extract, 385 nm and 390 nm for M. *arenaria* and M. *edulis*, respectively, measurements of fluorescence intensity at 381 nm with 332 nm excitation would enable standardization and referral to STX fluorescence intensity. The difference in maxima was likely due to the pigmentation present in the extracts.

Table 1 compares the tube assay presented here with the mouse bioassay and the modified Bates and Rapoport determinations. The S.E.M. values are from 3 to 11% of the mean with the tube assay and 4 to 45% of the mean with the Bates and Rapoport determinations. With highly toxic M. *edulis* samples, the Bates and Rapoport modified method shows greater variance and generally tends to underestimate the toxin content. Discrepancies of the Bates and Rapoport method may arise because the method was developed for West Coast shellfish. There are more toxins involved in East Coast shellfish. The Bates and Rapoport method reported, to a large extent, quantitation of non-toxic samples (Bates and Rapoport, 1975).

Table 1. Comparison of Solution Fluorometric Assay with Mouse Bioassay[a]

Sample	μg STX/100 g meat			
	Mouse Bioassay	HCl Extract		Bates-Rapoport
		Graph	Scan	
M. arenaria	<40	0	0.86	38.19 (5.09)
	<40	11.98 (0.53)	0	0
	170.70	273.24 (20.05)	195.52	163.69 (32.72)
	232.32	397.95 (17.26)	242.24	385.29 (27.18)
	2161.28	2123.51	—	—
M. edulis	<40	43.32 (4.79)	0	4.91 (1.44)
	<40	21.23 (1.56)	0	9.33 (4.18)
	259.00	380.01 (15.30)	270.40	93.12 (11.63)
	561.40	344.89 (11.69)	516.26	298.43 (12.53)

[a] Values in table are determined graphically by calculating slope of linear part of curve and is given as Mean μg STX/100 g meat (± S.E.M. for n = 4). Values determined by scan are calculated from area under fluorescence curve using Gelman planimeter. A (—) indicates no result calculated.

For shellfish having low toxin content, underestimation of the toxicity by as much as 60% by the mouse bioassay has been reported (McFarren, 1975). The fluorometric method developed here can elim- inate this large error as well as report toxicity values below 40 μg STX per 100 g meat. The best correlation between mouse bioassay and solution fluorometry was achieved with fluorescence scans and calculation of the toxicity from the area under the emission peak. The calculation of toxicity by areas is applicable at low as well as at high toxicity levels. Determination of toxicity by area under the fluorescence peak necessitated that the concentration of extract selected for the fluorescence reaction and emission scan be one that was on the linear portion of the concentration vs. fluo- rescence curve. In all cases 0.5 or 0.3 ml of the acid extract was adequate to produce the values in Table 1.

The representation of toxin content as STX is justified by the similar fluorescence yields of the major toxin of New England shellfish and STX. This justification is borne out by the similar fluorescence scans of STX and extracts of *M. arenaria* or *M. edulis*. The correlation between the mouse bioassay and the fluorometric tube assay has suggested that the solution fluorometric assay could be a sensitive method for determining the toxin content in shell- fish extracts.

ACKNOWLEDGMENTS

This work was supported by grant number 04-7-158-44034 from the Office of Sea Grant Programs to the University of New Hampshire and constituted part of the support received by Nancy H. Shoptaugh for dissertation work from which this paper was obtained.

REFERENCES

Bates, H. A., R. Kostriken, and H. Rapoport. 1978. A chemical assay for saxitoxin. Improvements and modifications. J. Agr. Food Chem. *26*:252-254.
Bates, H. A., and H. Rapoport. 1975. A chemical assay for saxi- toxin, the paralytic shellfish poison. J. Agr. Food Chem. *23*:237-239.
Buckley, L. J., M. Ikawa, and J. J. Sasner, Jr. 1976. Isolation of *Gonyaulax tamarensis* toxins from soft-shell clams (*Mya arenaria*) and a thin-layer chromatographic-fluorometric method for their detection. J. Agr. Food Chem. *24*:107-111.
Buckley, L. J., Y. Oshima, and Y. Shimizu. 1978. Construction of a paralytic shellfish toxin analyzer and its application. Anal. Biochem. *85*:157-164.

McFarren, E. F. 1959. Report on collaborative studies of the bio-assay for paralytic shellfish poison. J. Assoc. Off. Agr. Chemists *42*:263-271.

Mosher, H. S., F. A. Fuhrman, H. D. Buchwald, and H. G. Fischer. 1964. Tarichatoxin-tetrodotoxin: a potent neurotoxin. Science (N.Y.) *144*:1100-1110.

Schantz, E. J. 1961. II. Some chemical and physical properties of paralytic shellfish poisons related to toxicity. J. Med. Pharm. Chem. *4*:459-468.

Schantz, E. J., E. F. McFarren, M. L. Schafer, and K. H. Lewis. 1958. Purified shellfish poison for bioassay standardiza-tion. J. Assoc. Off. Agr. Chemists *41*:160-168.

Shoptaugh, N. H., L. J. Buckley, M. Ikawa, and J. J. Sasner, Jr. 1978. Detection of *Gonyaulax* toxins and other guanidine compounds on thin-layer silica gel chromatograms. Toxicon *16*:509-513.

Sommer, H., and K. F. Meyer. 1937. Paralytic shellfish poisoning. Arch. Pathol. *24*:560-598.

SIMPLIFIED MONITORING OF ANATOXIN-A BY REVERSE-PHASE HIGH

PERFORMANCE LIQUID CHROMATOGRAPHY AND THE SUB-ACUTE

EFFECTS OF ANATOXIN-A IN RATS

N. B. Astrachan[1] and B. G. Archer[2]

University of Idaho
Moscow, Idaho

ABSTRACT

    Anatoxin-a is a potent neuromuscular blocking agent produced
by certain strains of the blue-green algae *Anabaena flos-aquae*.
Its sporadic and unpredictable production during algal blooms,
occasionally causing animal deaths, led to a concern that animals
and humans may ingest sub-chronic doses of anatoxin-a, resulting
in delayed toxicity. A method to detect and monitor anatoxin-a in
natural water bodies was therefore a primary concern. Toxic *A.
flos-aquae* was cultivated and the toxin extracted by centrifugation
and solvent extraction procedures. Anatoxin-a was then analyzed by
reverse-phase high performance liquid chromatography (HPLC) using
an isocratic gradient. A useful concentration range where detector
response remained linear to toxin concentration went from 1 to 500
ppm. Combined with the extraction procedure, which averaged 90%
recoveries, an overall assay sensitivity of 0.1 ppm was achieved.

    Anatoxin-a in drinking water was then administered to rats at
0 ppm, 0.5 ppm, and 5 ppm for 7 weeks. (5 ppm equals approximately
1/20 the acute oral LD_{50}). In a second study, anatoxin-a was
injected intraperitoneally (i.p.) to rats at 0 mg/kg and 0.062 mg/kg
daily for 21 days. (0.062 mg/kg equals approximately 1/4 the acute
i.p. LD_{50}). No changes were seen in body weight, food consumption,

---

[1]Present address: Stauffer Chemical Co., Farmington, Connecticut.

[2]Present address: Altex Scientific Inc., Berkeley, California.

or serum enzymes (SGPT, GGTP alkaline phosphatase, and cholinester-
ase) in either study. Gross and microscopic examination of the
tissues showed no differences between treated and control groups.
This indicates that anatoxin-a does not cause apparent toxic signs
in doses less than those causing acute effects.

INTRODUCTION

Anatoxin-a is one of the major toxins produced sporadically
and unpredictably by the blue-green alga *Anabaena flos-aquae*. This
usually occurs in the late summer months when warm waters promote
dense algal blooms in lakes, reservoirs, and other bodies of water
(Prescott, 1968). *Anabaena* can then accumulate along the shoreline
in concentrations great enough to kill animals drinking the contaminated water. Over the past 100 years, the number of domestic and
wild animal deaths from *A. flos-aquae* poisoning has sometimes num-
bered in the thousands (Gorham, 1964).

Anatoxin-a (Figure 1) is a basic, cyclic amine which is stable
in aqueous solution (Devlin et al., 1977) and acts as a potent
neuromuscular blocking agent, causing death by respiratory arrest
in a manner similar to decamethonium bromide (Figure 1) (Carmichael
et al., 1975). It possesses a structural similarity to cocaine
(Figure 1), but does not share the same pharmacological actions.
A similarity was also noticed between anatoxin-a and acetylcholine
(Figure 1) in the spacial arrangements and molecular distances of
the molecules. Anatoxin-a is a very potent toxin, having an acute
intraperitoneal LD_{50} of approximately 0.25 mg/kg in the laboratory
rat and mouse (Carmichael et al., 1975). Animals injected at that
does level can die within 3 minutes preceded by symptoms of ataxia
and convulsions.

No human deaths have been attributed to anatoxin-a poisoning;
however, cases of acute gastrointestinal disease, allergic dermatitis, and general malaise have been documented and attributed to
Anabaena flos-aquae (Schwimmer and Schwimmer, 1968). This led to
the concern that possible ingestion of sub-acute doses of anatoxin-
a might cause delayed toxicological symptoms in man and animals, a
likely possibility based on the common occurrence of *A. flos-aquae*
in reservoirs and other water supplies. A simple method was therefore developed to detect and monitor anatoxin-a in aqueous samples.
In addition, anatoxin-a was administered to rats in sub-acute doses
to assess the possible sub-acute toxicological effects.

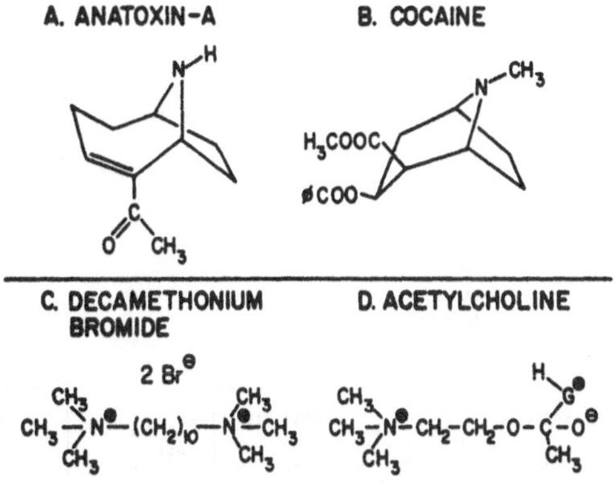

Fig. 1. Structure of: (A) Anatoxin-a; (B) Cocaine; (C) Decamethonium bromide; (D) Acetylcholine.

MATERIALS AND METHODS

Isolation and Purification of Anatoxin-a

NRC-44-1 strain of *A. flos-aquae* was grown in ASM-1-Tr media
(Carmichael and Gorham, 1974) at 22°C, illuminated continuously by
4 rows of 40 watt cool-white fluorescent tubes at a distance of 30
cm, and aerated at 1 to liters per minute with air filtered through
0.4 μ Nuclepore ^{R} filters. Ten liters of sterile media in 12 liter
pyrex carboys were inoculated at initial cell densities of 10^5 per
milliliter. Exponential cell growth, reaching plateau cell densities of 2 to 5 x 10^7 per milliliter at 12 to 14 days was observed
at which time the cultures contained 1 to 2 mg toxin per liter.
Because most of the toxin was released from the cells, only the
media was used as a source of toxin. At maturity, the cells were
separated from the media after adjusting to pH 4 by centrifugation
in a CEPA Schnell Model LE continuous flow centrifuge operated at
40,000 rpm and a flow rate of 300 ml/minute. Twenty liter volumes
of clarified media were adjusted to pH 10.5 and delivered at 20
ml/min through a 4 x 30 mm reverse phase column (XAD-2, Applied
Science Laboratories) which absorbed the toxin. The XAD-2 column
had been prepared by washing with acetone and 1 mM Na_2CO_3. Following absorption of the toxin the column was washed with 2 liters of
1 mM Na_2CO_3 and the toxin eluted with 1.5 liters of 0.01 M HCl in
ethanol. The ethanolic eluate was evaporated under reduced pressure to a small volume and the concentrate was in turn extracted
with water, clarified by centrifugation, extracted with hexane;
adjusted to pH 11 and partitioned 3 times against equal volumes of
chloroform. The chloroform extracts were pooled and shaken with
equal volumes of 1 mM HCl three times which finally were evaporated
under reduced pressure to remove residual chloroform and concentrate

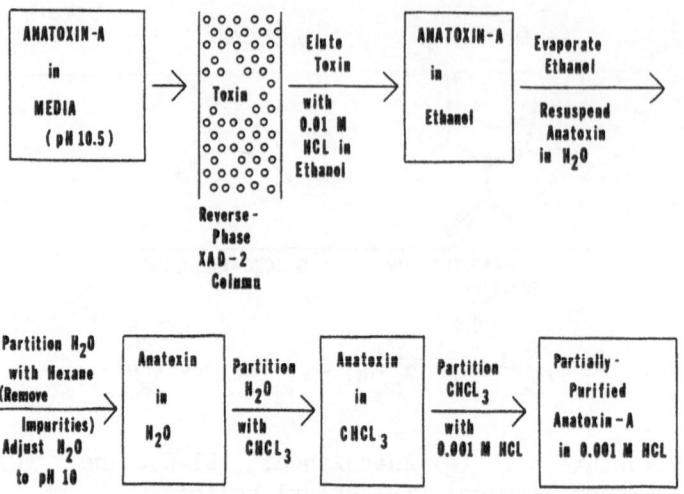

Fig. 2. Method to extract and purify anatoxin-a.

the toxin (Figure 2). Recoveries of 75%, verified by mouse bioassay, were achieved. This partially pure toxin, which was used
in toxicity experiments without further purification, had a UV
absorbance spectrum which qualitatively indicated that anatoxin-a
was the principle UV-absorbing component. Contaminants were, however, detected by thin layer chromatography (see below). No degradation of toxin occurred during storage at 4°C in dilute aqueous
acid for one year.

To obtain an analytical standard the toxin was further purified by chromatography on a 2.5 x 20 cm silica gel 60 (E.M. Merck)
column. Samples were applied in either tetrahydrofuran or iso-
propanol and eluted with a methanol gradient. The purity of
recovered toxin was assessed by thin layer chromatography on
silica gel with a mobile phase of chloroform-methanol-water (60:
40:4), UV absorbance spectroscopy, and high pressure liquid chromatography (see below). Concentrations of pure anatoxin-a were
determined by absorbance at 227 nm using a molar absorptivity of
$10^4/M/m$ (Devlin et al., 1977).

Liquid Chromatographic Assay for Anatoxin-a

One hundred ml of algal suspension was made alkaline by adding
2 ml of 0.5 M Na_2CO_3, shaken with 100 ml chloroform, clarified by
centrifugation, and extracted with 10 ml 0.01 N HCl (Figure 3).
Ten μl of acidic extract was injected onto a 3 x 250 mm Spherisorb
ODS 10 μ column with a fixed mobile phase of 30% 0.01 M $HClO_4$ (pH
3) 70% methanol at a flow rate of 1.2 ml/min using a Spectra Physics

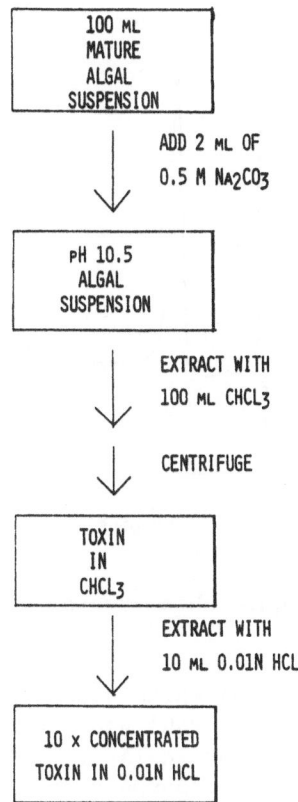

Fig. 3.   Extraction-concentration procedure for HPLC analysis of
anatoxin.

model 3500B chromatograph. The column effluent was monitored at
227 nm using a Spectra Physics model 770 variable wavelength detector connected to a Hewlett Packard model 3380A integrator.

Subacute Oral Toxicity

Three groups of 20 180 g female Sprague-Dawley rats were
maintained individually in wire-bottom cages and fed ground Purina
rat chow ad libidum. Two test groups received 5.1 ppm or 0.51 ppm
anatoxin-a in their drinking water and a control group received no
toxin. Water bottles were autoclaved and refilled with fresh
toxin-containing water every other day. In a preliminary study of
water consumption, rats were found to drink approximately 0.1 ml/g
body weight per day. With this estimate, it was calculated that
rats consuming normal amounts of water containing 5.1 ppm would
receive 0.51 mg/kg body weight or approximately 0.08 oral LD_{50}.
Weekly body weights were recorded throughout the study. All rats
were heart-bled under methoxyflurane anesthesia on days 0, 7, 14,

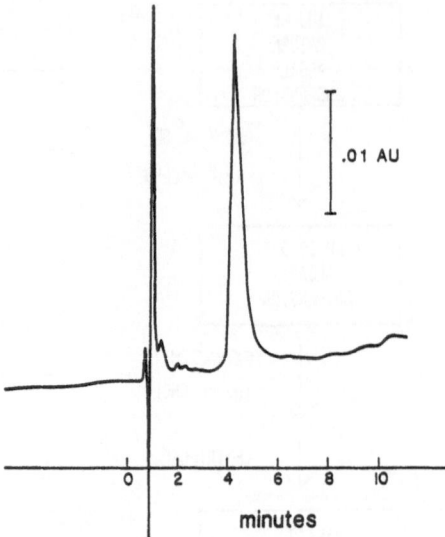

Fig. 4. High pressure liquid chromatogram of 0.69 µg anatoxin-a.

27, 41, and 54. Red and total white blood cell counts were determined using a Clay-Adams model 2401 blood cell counter. Serum
activities of alkaline phosphatase (AP), glutamid pyruvic transaminase (GPT), gamma glutamyl transpeptidase (GGTP), and choline
esterase (CE) were measured using reagent kits from Worthington
Biochemicals (AP, GPT, GGTP) and Sigma Chemical Company (CE) as
recommended. Spectrophotometric measurements were recorded on a
Cary 219 instrument fitted with jacketed cuvette to maintain constant temperature. At seven weeks, all rats were killed by cervical dislocation and examined for gross lesions. Liver, kidneys,
and spleens were weighed. Tissues from these organs and adrenals,
heart, lungs, and brain were frozen and also fixed in 5% formalin
for histologic examination.

Subacute Parenteral Toxicity

     Test and control groups of 18 180 g female Sprague-Dawley rats
were maintained as described. A 0.016 mg sample of toxin (about
1/4 i.p. LD_{50}) in 0.2 ml of 1 mM HCl was administered by intraperitoneal injection daily for 21 days to the test rats. Control rats
were similarly injected with 1 mM HCl for the same period. AP and
GPT were measured as described on days 0, 7, 14, and 21. Whole
blood and brain acetylcholinesterase activities were measured using
a kinetic assay according to Ellman et al. (1961). Food consumption, growth rate, and organ weights were measured as before and
post-mortem examinations of organs for gross and microscopic
lesions were also done.

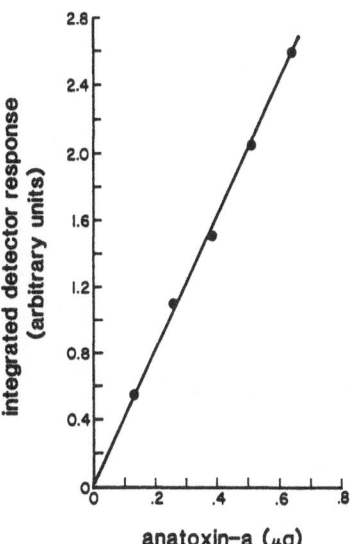

Fig. 5. Proportionality of absorbance peak area to amount of anatoxin-a.

RESULTS

Assay for Anatoxin-a

The typical chromatogram shown in Figure 4 illustrates the clean separation of anatoxin-a achieved by reverse-phase chromatography. A slight decrease in retention time with increasing amounts of toxin was noticed, but did not affect the linear dependence of detector response on concentration as shown in Figure 5. The least amount of toxin quantifiable was 0.01 µg and the detector response remained proportional to amounts of toxin greater than 5 µg. Thus, a useful concentration range using a 10 µl injection loop is 1 to 500 ppm. Combined with the extraction procedure which affected a 10-fold concentration and had an average recovery of 90%, an overall assay sensitivity of 0.1 ppm was achieved.

Subacute Toxicities of Anatoxin-a

No apparent differences between treated and control groups were detected in either the oral or parenteral study. Figures 6 and 7 illustrate the close parallels in hematological and enzymatic parameters measured. Serum activities of GGTP (not shown in Figure 6) remained at or below the detection limit of the assay (approximately 1 IU/liter in all groups throughout the treatment

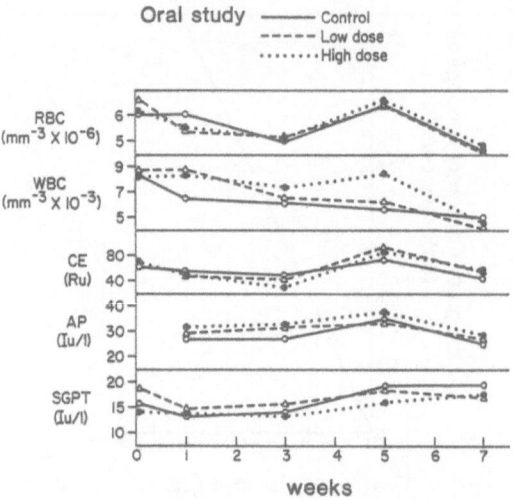

Fig. 6.   Mean values for red and total white cell counts, serum
          choline esterase (CE), alkaline phosphatase (AP), and
          glutamic pyruvate transaminase (SGPT) of rats provided
          water containing anatoxin-a. Data from rats reading
          5.1, 0.51 and no anatoxin-a are represented by the
          dotted, dashed, and solid lines, respectively.

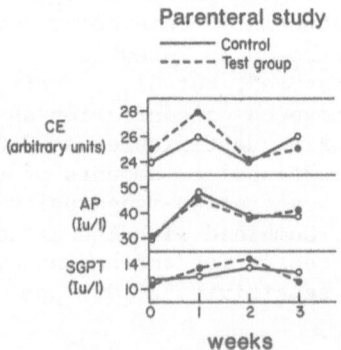

Fig. 7.   Mean values for serums choline esterase, alkaline
          phosphatase, and glutamic pyruvate transaminase of
          rats administered anatoxin-a by intraperitoneal injec-
          tion. The dashed line represents data from rats given
          0.25 LD_{50} per day; the solid line represents data from
          control rats.

period. Behavior, food consumption, body weight, and organ weights
were also similar. No significant lesions or abnormalities in
organs or tissues were revealed by gross and microscopic examination.

DISCUSSION

The assay for detecting and monitoring anatoxin-a in aqueous
samples proved to be a simple, accurate, and effective method to
detect the toxin at concentrations ranging from 600 ppm to 0.1 ppm
using the extraction procedure described. The significance of this
methodology lies in the ability to determine potentially hazardous
levels of anatoxin-a in municipal water supplies. The availability
of a very sensitive assay for anatoxin-a may also stimulate and
facilitate further research in this area.

Because consumption of approximately 0.08 oral LD_{50} anatoxin-a
per day by rats for seven weeks caused no apparent effects, the
effective dose was increased to 0.25 LD_{50} per day and the route of
administration changed to intraperitoneal injection. Even when
this amount of toxin was injected daily for 21 days, however, no
changes in the monitored parameters were noticed. The only published data to which our results may be compared are those of Carmichael et al. (1977) in which 4 oral doses, each approximately
1/8 to 1/4 of an oral LD_{50}, of lyophilized A. flos-aquae were administered to a calf over a 90 minute period. While mild signs appeared
after the second dose, death apparently required the cumulative
effect of all 4 doses. In contrast, our data indicate that 0.25
LD_{50} given repeatedly at 24 hour intervals is not cumulative, suggesting that anatoxin-a does not cause significant effects on the
nervous system in amounts less than those causing acute signs, and
also causes no toxic effects besides its specific neuromuscular
depolarizing activity. The parameters monitored during the periods
of toxin administration were non-specific, however, and could remain
unaffected while more subtle effects not observable in this study
were developing. Toxic responses developing too slowly to have been
detected in this study are also possible, but in view of the sporadic
and transient nature of toxin production in nature, the possibility
that a delayed toxic response due to anatoxin-a in the environment
is a significant hazard seems remote.

ACKNOWLEDGMENTS

We thank Dr. Wayne Carmichael for generously providing the
strain of toxic alga, Dr. R. Heimsch and Dr. W. Funk for support
and encouragement, Dr. L. Koller for examining tissue sections,
and the Stauffer Chemical Company toxicology staff for their advice

and assistance. This project was supported by award SER 77-06943
from the National Science Foundation.

REFERENCES

Carmichael, W. W., D. F. Biggs, and P. R. Gorham. 1975. Toxi-
    cology and pharmacological action of *Anabaena flos-aquae*
    toxin. Science *187*:542-544.
Carmichael, W. W., and P. R. Gorham. 1974. An improved method
    for obtaining axenic clones of planktonic blue-green algae.
    J. Phycol. *10*:238-240.
Carmichael, W. W., and P. R. Gorham. 1978. Anatoxins from clones
    of *Anabaena flos-aquae* isolated from lakes of western
    Canada. Mitt. Int. Ver. Limnol. *21*:285-295.
Carmichael, W. W., P. R. Gorham, and D. F. Biggs. 1977. Two
    laboratory case studies on the oral toxicity to calves of
    the freshwater cyanophyte (blue-green alga) *Anabaena flos-
    aquae* NRC-44-1. Can. Vet. J. *18*:71-75.
Devlin, J. P., O. E. Edwards, P. R. Gorham, N. R. Hunter, R. K.
    Pike, and B. Stavric. 1977. Anatoxin-a, a toxic alkaloid
    from *Anabaena flos-aquae* NRC-44h. Can. J. Chem. *55*:1367-
    1371.
Ellman, G. L., K. D. Courtney, V. Andres, Jr., and R. M. Feather-
    stone. 1961. A new and rapid colorimetric determination
    of acetylcholinesterase activity. Biochem. Pharmacol.
    7:88-95.
Gorham, P. R. 1964. Toxic algae. Pages 307-336 *in* D. F. Jackson,
    ed., Algae and Man. Plenum Press, New York.
Prescott, G. W. 1968. The Algae, a Review. Houghton Mifflin
    Company, Boston. 436 pp.
Schwimmer, M., and D. Schwimmer. 1968. Medical aspects of phy-
    cology. Pages 279-358 *in* D. F. Jackson, ed., Algae, Man,
    and the Environment. Syracuse University Press, New York.
    369 pp.

ISOLATION AND CHARACTERIZATION OF ENDOTOXIN FROM

CYANOBACTERIA (BLUE-GREEN ALGAE)

Georg Keleti, Jan L. Sykora, Laurie A. Maiolie,
Dennis L. Doerfler, and Iain M. Campbell

University of Pittsburgh
Graduate School of Public Health
Pittsburgh, Pennsylvania

ABSTRACT

A Sewickley, Pennsylvania, epidemic of water-borne gastroenteritis of unknown etiology which occurred in 1975 was characterized
by high concentration of *Schizothrix calcicola* (Cyanobacteria) in
the corresponding drinking water system. As a consequence, the
most common cyanobacterial contaminants of drinking water were
tested for endotoxin (lipopolysaccharide - LPS) content. LPS was
isolated and characterized from *Schizothrix calcicola*, *Anabaena
flos-aquae* (UTEX 1444), *Oscillatoria tenuis*, and *Oscillatoria
brevis*. The isolation was performed by using Westphal's modified
phenol-water extraction method. Subsequently, the glucan contaminating LPS was eliminated by enzymatic hydrolysis by cellulase.
The resulting macromolecular substances were composed of a polysaccharide moiety and a lipid part. Contrary to Wang and Hill's
report that *Anabaena flos-aquae* A37 contains a polysaccharide
without covalently bound lipid, *Anabaena flos-aquae* (UTEX 1444)
contained a true LPS. No endotoxin could be isolated from the
toxic strain *Anabaena flos-aquae* NRC-44-1 and preliminarily from
Anabaena cylindrica.

All examined cyanobacterial LPS contained glucose, xylose,
mannose, and rhamnose. The sugar and fatty acid composition varied
among the cyanobacterial species and differed substantially from
LPS of Enterobacteriaceae. Two identical unidentified fast running
sugars detected by paper chromatography were found in *Oscillatoria
brevis* and *O. tenuis* which suggests a close relationship between
the two species. The isolated cyanobacterial LPS's always induced

Limulus amoebocyte lysate gelation, and were not toxic to mice
except for the LPS from *O. brevis*. *S. calcicola* and *A. flos-aquae*
LPS had a positive Schwartzman reaction.

The detection of distinct O-antigen chemotypes in the tested
species of Cyanobacteria indicates the chemical composition of LPS
could be used for taxonomic classification of this group of microorganisms.

INTRODUCTION

A water-borne outbreak of gastroenteritis affected approximately 5,000 persons in Sewickley, Pennsylvania during August 1975.
Extensive microbiological and chemical analyses of specimens
obtained from patients and of water samples failed to identify a
causative agent. Investigation of the Sewickley water system, however, revealed an accumulation of *Schizothrix calcicola* in the open
finished water reservoirs (Lippy and Erb, 1976). Even one month
after the outbreak, the water in the reservoir with the longest
detention time was still contaminated by this species (400,000 cells
per ml). The presence of high concentrations of endotoxins (2,500
ng/ml) has led to a speculation on a possible relationship between
the consumption of LPS derived from *S. calcicola* and the reported
illness. Under normal operating conditions, concentrations of *S.
calcicola* in this reservoir were much lower, ranging from 0 to
1,200 cells/ml, and the corresponding LPS concentrations were
between 25 to 250 ng/ml.

Stanier and Cohen-Bazire (1977) in their excellent review of
Cyanobacteria reported that the LPS components of the outer membrane
layer have been isolated from a considerable number of Cyanobacteria.
The blue-green algae have shown a general chemical resemblance to
those of other Gram-negative bacteria; they contained a variety of
sugars and amide-linked β-hydroxy-fatty acids. These authors
(Stanier and Cohen-Bazire, 1977) affirmed further that, though
limited, the qualitative data on carbohydrate composition of cyanobacterial LPS reveal a considerable diversity and may well provide
useful chemotaxonomic markers.

Weckesser et al. (1979) reported that LPS was isolated and
chemically characterized from the chroococcacean, oscillatorian,
and heterocystous subgroups of Cyanobacteria.

Endotoxins are generally found in the outer membrane of the
cell wall of Gram-negative bacteria where they form complexes with
proteins and phospholipids. They represent the O-antigens (somatic)
of bacteria and are thus the chemical basis for serologic classification of Gram-negative bacteria. Furthermore, they are receptors
for many bacteriophages and therefore play an important role in

bacteriophage typing. Both mentioned properties of LPS are due to the polysaccharide moiety. Additionally, and very importantly, the lipid moiety of LPS is responsible for the endotoxic activities.

It has been found that LPS, especially from enterobacteria, are composed of three structural regions, I, II, and III. Region I is represented by the O-specific polysaccharide. This consists of oligosaccharide repeating units. This structural region of the LPS is distinguished by its great variability: many different sugar residues may be present in many combinations and with many kinds of glycosidic linkages.

Region II is an oligosaccharide called the "common core". It has been shown that a number of different core oligosaccharides exist which are structurally closely related.

Region III is the lipid moiety of the LPS which is termed lipid A. The LPS are anchored in the outer membrane via their lipid component. Lipid A is a glycophospholipid containing glucosamine, fatty acids, and phosphate (Jann and Westphal, 1975).

Many Gram-negative bacterial mutants possess only glycolipids, and the O-specific polysaccharide is absent. Sometimes even a part of the core oligosaccharide is lacking.

LPS exhibits a wide spectrum of biological activities (Table 1). Many of these, such as local and generalized Schwartzman phenomenon and depression of blood pressure leading to shock and death and partially pyrogenicity, are of a physiopathological nature and are harmful acitivites. LPS also exhibit other activities, such as adjuvantivity, induction of nonspecific resistance to infection, and tumor necrotic activity, which may be looked upon as beneficial. Most, if not all of these activities are dependent upon structures of the lipid A component. Therefore, if free lipid A is introduced into suitable biological systems, mainly experimental animals and preferably into species of high sensitivity, all of the biological effects expected from whole LPS will occur. In all cases, the acitivity of lipid A in the form of lipid A/BSA (bovine serum albumin) complexes is qualitatively indistinguishable from that of intact lipopolysaccharides.

The presence of endotoxin in the human environment can cause medical problems; for example, LPS can evoke endotoxemia, mostly in immunosuppressed and debilitated patients (e.g., those suffering from cancer, organ transplants, diabetes) and in infants. The presence of endotoxins in drinking water and the effect of LPS introduced into the blood stream have been observed by Hindman et al. (1975) during an epidemic of pyrogenic reactions among kidney dialysis patients.

Table 1. Biological Activities of Free Lipid A[a]

---

1. Pyrogenicity

2. Lethal toxicity in mice

3. Leukopenia

4. Leukocytosis

5. Local Schwartzman reaction

6. Bone marrow necrosis

7. Embryonic bone resorption

8. Complement activation

9. Depression of blood pressure

10. Platelet aggregation

11. Hageman factor activation

12. Induction of plasminogen activator

13. *Limulus* lysate gelation

14. Toxicity enhanced by BCG

15. Toxicity enhanced by adrenalectomy

16. Enhanced dermal reactivity to epinephrine

17. Induction of nonspecific resistance to infection

18. Induction of tolerance to endotoxin

19. Induction of early refractory state to temperature change

20. Adjuvance activity

21. Mitogenic activity for cells

22. Tumor necrotic activity

23. Macrophage activation

24. Induction of colonic stimulating factor

25. Induction of IgG synthesis in newborn mice

26. Induction of prostaglandin synthesis

27. Induction of interferon production

28. Induction of tumor-necrotizing factor

29. Induction of mouse liver pyruvate kinase

30. Type C RNA virus release from mouse spleen cells

31. Helper activity for friend spleen focus-forming virus in mice

32. Inhibition of phosphoenolpyruvate carboxykinase

33. Hypothermia in mice

---

[a]Galanos et al., 1977.

Two outbreaks of endotoxic reactions (fever, chills, and occasionally hypotension) have been reported in Michigan hospitals and the third in Massachusetts recently. The outbreaks were associated with the reuse of hospital "sterilized" cardiac catheters (Morbidity and Mortality Weekly-Report, 1979a, 1979b). The presence of endotoxin in humidifiers and in cotton plants can be related to the development of "humidifier disease" and clinical symptoms among office employees and cotton mill workers respectively

(Nowotny, 1969; Schiller et al., 1976). It is, therefore, important to study exhaustively the source of endotoxins (not only hetero-trophic Gram-negative bacteria), their chemical and biological characterization, and the mode of dispersal in the human environ-ment.

In this work, the most common cyanobacterial contaminants of drinking water were mass cultured. These strains were then examined for the presence of lipopolysaccharides and the isolated endotoxin was chemically and biologically characterized.

MATERIALS AND METHODS

Cyanobacterial Cultures

A brown strain of *Schizothrix calcicola* was isolated from Mosquito Creek Reservoir, near Youngstown, Ohio. *Anabaena flos-aquae* toxic strain NRC-44-1 was obtained from Dr. Wayne W. Carmichael, Wright State University, Dayton, Ohio. *Anabaena flos-aquae* UTEX 1444, *Oscillatoria brevis* UTEX B1567, *Oscillatoria tenuis* UTEX 1506, and *Anabaena cylindrica* UTEX B1611 were obtained from the culture collection of the University of Texas at Austin.

Cultivation

Schizothrix calcicola and *Oscillatoria tenuis* were cultured in Allen's medium; all the other investigated Cyanobacteria were grown in ASM-1-Tricine medium (Carmichael and Gorham, 1974). All the algal species were mass cultured in 9 liter sterilized bottles at 24°C, aerated with compressed air and illuminated with fluores-cent light at a daily 16 hour cycle as described by Keleti et al. (1979).

The heterotrophic bacterial contamination of Cyanobacteria was determined by plating serial dilutions (0.85% saline) of the cultures on Trypticase soy agar (BBL Microbiological Systems) in duplicate. The plates were incubated for 72 hours at 20° and 35°C. Because the majority of the bacteria contaminating *S. calcicola* were from the family Enterobacteriaceae, a laboratory strain of *Salmonella typhimurium* was selected as the biomass indicator. The bacterial biomass was calculated utilizing a 24-hour culture of this strain at 35°C in Soybean casein digest broth as a standard. A serial of dilutions in physiological saline containing this organism was placed on Trypticase soy agar medium, and the number of colonies of *S. typhimurium* present in the culture was deter-mined. One liter of the same culture was centrifuged, washed three times, and lyophilized, and the dry weight of bacteria was established. Based on this standard, the average count of hetero-trophic bacteria contaminating the cyanobacterial cultures consti-tuted only 0.03% of the total biomass.

Isolation and Purification of LPS

As seen from Table 2, the harvested cells were washed three times in sterile, distilled water and the resultant sediment was uniformly suspended in distilled water and freeze-dried. The lyophilized cyanobacterial cells were stirred 2 hours with dis-tilled water (350 ml H_2O per 20 g dried Cyanobacteria) and extracted twice for 20 minutes with 45% phenol solution at 68°C according to Westphal and Jann (1965). The extracts were cooled to 10°C and centrifuged. The aqueous layers and possibly the phenol layers from the two extractions were combined, and dialyzed against distilled water until all traces of phenol were removed, and then the remain-ing solution was centrifuged. The supernatant was freeze-dried and the contaminating glucan was eliminated by enzymatic hydrolysis with cellulase (two treatments with .25 mg enzyme per mg crude LPS) as described by Volk (1968). After two days dialysis against tap water, the nucleic acids were removed by ultracentrifuging three times at 105,000 x g for 4 hours. The sediment was resuspended in distilled water and centrifuged. The supernatant, representing the purified LPS, was lyophilized and the resulting LPS contained about 3% nucleic acids. The whole extraction procedure, except glucan the elimination step, is described in detail by Keleti and Lederer (1974).

Chemical Analysis

Sugar, fatty acid, hydroxyfatty acid separation and analysis, total carbohydrates, phosphorus, total aminosugars, heptoses, and 2-keto-3-deoxyoctonate (KDO) were determined as described in Keleti and Lederer's handbook (1974). Lipid A was separated in chloroform: methanol:H_2O (65:25) on a thin layer of silica gel according to Schiller et al. (1976).

Biological Characterization

The lethal effect of LPS was tested in duplicate by intraper-itoneal injections into five 20 g Swiss-Webster female mice for each concentration (from 500 µg to 5 mg). Schwartzman phenomemon was performed as described by Nowotny (1969), *Limulus* amoebocyte lysate test according to Sykora et al. (1980), and the rabbit illeal fluid loop accumulation test by the procedure of De and Chatterje (1953).

Table 2. Methods of Isolation of LPS from Cyanobacteria

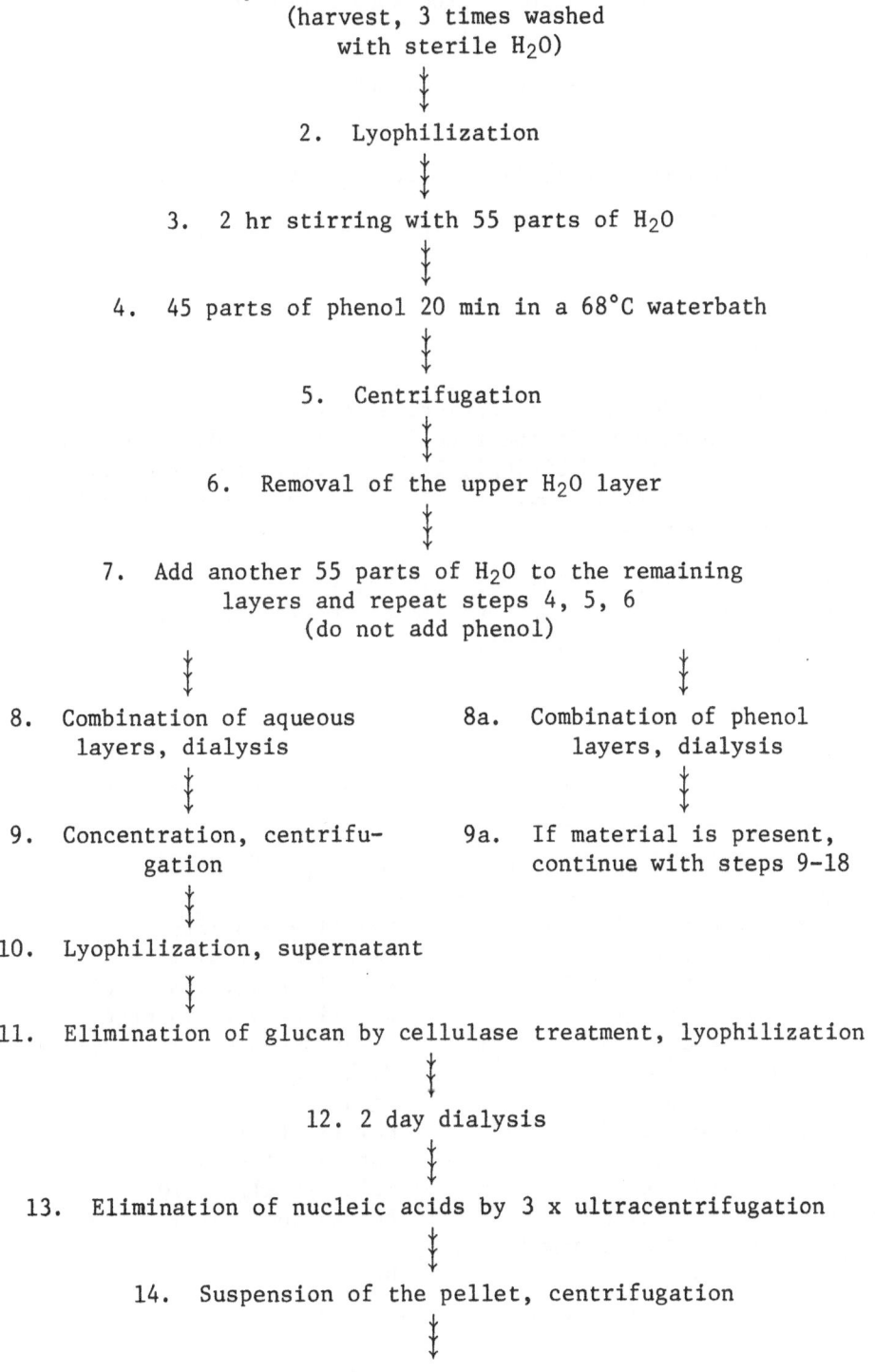

(Table 2 continued)

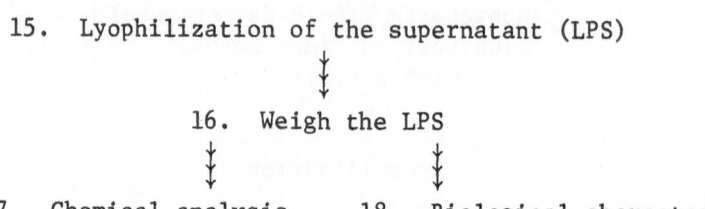

RESULTS AND DISCUSSION

We have now definite biochemical evidence that several Cyanobacteria have endotoxin in their cell wall (Table 3). Endotoxins
have been chemically and biologically characterized from not less
than six cyanobacterial genera and from twelve species (four isolated in our laboratory).

Weckesser et al. (1979) reported the presence and composition
of lipopolysaccharides in 10 *Synechoccus* PCC strains and in 4
Synechocystis strains (Table 4).

Table 3. List of Cyanobacteria from which Endotoxin was Isolated

Species	Author(s)
Anabaena flos-aquae UTEX 1444	Keleti and Sykora (in prep)
Anabaena PCC 7118 (*A. variabilis*)	Weckesser et al., 1974
Oscillatoria brevis UTEX B1567	Keleti et al. (in prep)
Oscillatoria tenuis UTEX 1506	Keleti et al. (in prep)
Phormidium africanum	Mikheyskaya et al., 1977
Phormidium laminosum	Mikheyskaya et al., 1977
Phormidium uncinatum	Mikheyskaya et al., 1977
Schizothrix calcicola UPB 1	Keleti et al., 1979
Synechococcus PCC 6301 KM	Katz et al., 1977
(*Anacystis nidulans*)	Weise et al., 1970
Synechococcus PCC 73109	Buttke and Ingram, 1975
(*Agmenellum quadruplicatum*)	
Synechocystis sp.	Drews et al., 1976

Table 4. Composition of Lipopolysaccharides of Cyanobacteria

Columns are grouped as follows — *Synechococcus* PCC[a]: 6907, 6307, 6911, 6603, 6908, 6311, 6301, 6312, 6910, 73109; *Synechococcus* PCC[b]: 6803, 6714, 6807, 6308; *Phormidium* sp.[c]

Component	6907	6307	6911	6603	6908	6311	6301	6312	6910	73109	6803	6714	6807	6308	Phorm.
2-0-methyl-xylose												+			
2,3,4-Tri-0-methyl-arabinose													+		
2-0-methyl-mannose													+	+	
3-0-methyl-D-mannose							+	+	+	+					
4-0-methyl-D-mannose							+	+	+	+					
2-0-methyl-fucose												+			
2,3,-Di-0-methyl-fucose												+			
6-0-methyl-heptose													+		
Xylose															+
Arabinose									+						+
Glucose	+	+	+	+	+	+	+	+	+	+	+	+	+	+	+
Galactose	+	+	+	+	+	+	+	+	+		+	+	+	+	+
Mannose	+	+	+	+	+	+	+	+	+	+	+	+	+	+	+
Heptose													+	+	
Fucose	+	+	+	+	+	+	+	+			+	+	+	+	
Rhamnose	+	+	+						+				+	+	+
Tyvelose	+	+	+												
Unidentified															+
2-Keto-3-deoxyoctonate						+	+	+	+	+					+
Uronic acids													+	+	
Galactosamine															+
Glucosamine	+	+	+	+	+	+	+	+	+	+	+	+	+		+
Mannosamine	+	+	+	+	+	+			+						
Unknown amino sugar	+	+	+												
Myristic acid	+	+	+		+								+	+	
Palmitic acid	+	+	+	+	+	+	+	+	+	+	+	+	+	+	+
Stearic acid	+	+	+	+	+	+	+	+			+	+	+	+	+
Oleic acid															+
Behenic acid									+						
Beta-hydroxymyristic acid	+	+	+		+	+	+	+	+	+	+	+	+	+	
Beta-hydroxypalmitic acid	+	+	+	+	+	+	+	+	+	+	+	+	+	+	
Beta-hydroxystearic acid									+						
branched β-OH fatty acids	+	+	+	+	+	+	+	+			+	+	+	+	

[a] Weckesser et al., 1979.
[b] Weckesser et al., 1979.
[c] Mikheyskaya et al., 1977.

The initial investigations of the chemical and bacterial characteristics of cyanobacterial LPS belong to the German workers Weise (1970), Drews (1970, 1974, 1976, 1979), B. Jann (1970), K. Jann (1970), Katz (1970, 1976, 1978), Weckesser (1974, 1976, 1977, 1979), Mayer (1974, 1979), Fromme (1974), and Schmidt (1976). *Synechococcus* PCC6301 is the synonym for *Anacystis nidulans* KM, *Synechococcus* PCC73109 for *Agmenellum quadruplicatum* and *Anabaena* PCC7118 is the synonym for *Anabaena variabilis*. The two genera *Synechococcus* and *Synechocystis*, represented by numerous strains were incorrectly designated *Aphanocapsa* in the initial study (Stanier and Cohen-Bazire, 1977). The lipopolysaccharides from *Agmenellum quadruplicatum* were isolated and chemically and biologically characterized by the American investigators Buttke and Ingram (1975), whereas endotoxins from the non-heterocystous filamentous Cyanobacteria, namely *Phormidium africanum*, *P. laminosum*, and *P. uncinatum*, were described by a Russian group, Mikheyskaya et al. (1977).

Sugar, Phosphorus and Protein Analysis of Cyanobacterial LPS

In 1979, Keleti et al. reported the composition and biological properties of LPS isolated from *Schizothrix calcicola*, the most common cyanobacterium contaminating drinking water systems in southwestern Pennsylvania. A further gas liquid chromatographic analysis has verified the presence of the previously isolated sugars, with a vast majority of rhamnose as well as two unidentified sugars not observed before.

There is some evidence that not all the species of Cyanobacteria contain LPS. Wang and Hill (1977) believe that the occurrence of LPS among Cyanobacteria cannot be generalized. They isolated only a polysaccharide from *Anabaena flos-aquae* A37; the lipid A moiety was completely absent.

Hot phenol-water extraction of *Anabaena flos-aquae* NRC-44-1 (toxic strain) did not detect any LPS, either in the aqueous layer or in the phenol layer. Only a polysaccharide was isolated from this species which was composed of glucosamine, galactose, mannose, xylose, rhamnose, and an unknown sugar which moved faster chromatographically. The lipid A part was completely absent.

An additional strain of this species (*A. flos-aquae* UTEX 1444), however, was cultured, harvested and treated and LPS was isolated by the technique described above. Lyophilized Cyanobacteria (18.9 g) yielded 0.16 g (0.86%) purified LPS. Analysis by paper chromatography and gas-liquid chromatography revealed the following sugars (Table 5): glucosamine, rhamnose, fucose, xylose, glucose, mannose, and galactose.

Table 5. Sugar, Phosphorus and Protein Analysis of *Schizothrix calcicola*, *Anabaena flos-aquae*, *Oscillatoria brevis*, and *Oscillatoria tenuis*

Cyanobacterial Strain	Sugars													Total amino-sugars %	Total carbohydrates %	Phosphorus %	Protein %
	Xylose	Glucose	Galactose	Mannose	Heptose %	KDO %	Fucose	Rhamnose	Fast running sugar I	Fast running sugar II	Unidentified I	Unidentified II	Glucosamine				
Schizothrix calcicola	+	+	+	+	0	0		+			+	+	+	2	63	<0.1	7.8
Anabaena flos-aquae UTEX 1444	+	+	+	+	0	0.16	+	+				+	+	2.1	63	0	12.5
Oscillatoria brevis	+	+	+	+	N.D.	N.D.		+	+	+				0	N.D.	N.D.	N.D.
Oscillatoria tenuis	+	+	+	+	1.12	0		+	+	+			+	.2	58	0	6.1

N.D. = Not Done

Interestingly, LPS from *Anabaena flos-aquae* UTEX 1444 did not
contain any heptose and possessed only a very low amount of KDO
(0.16%) as did other cyanobacterial endotoxins (from 0.5 to 1.8%).
In addition phosphorus appears to be absent in the lipid A moiety
such as from *A. nidulans* (Katz et al., 1977). These findings suggest that the "common core" region of cyanobacterial LPS and the
structure of lipid A region are different from those in Gram-negative bacteria. The contamination of LPS by proteins, even after
prolonged phenol extraction, remained relatively high (12.5%).
Mikheyskaya et al.'s (1977) LPS preparations from *Phormidium* sp.
contained not less than 7% to 20% proteins.

Healey affirmed (1973) that *Anabaena flos-aquae* UTEX 1444
originating from the Indiana culture collection displayed under
culture conditions characteristics similar to those of *Anabaena
variabilis*; the author therefore used the latter name. Tison and
Lingg (1979) showed that cultures of *Anabaena flos-aquae* UTEX 1444
were identified by Healey (1973) as *Anabaena variabilis* and the
authors therefore used the latter name. Weckesser et al. (1974)
analyzed the isolated LPS from *Anabaena* PCC7118, which is a synonym for *Anabaena variabilis*, and found L-acofriose, L-rhamnose,
D-mannose, D-glucose, D-galactose, and D-glucosamine in the sugar
part of LPS. From the lipid A part, palmitic acid, an unidentified fatty acid, β-hydroxymyristic acid, β-hydroxypalmitic acid,
and β-hydroxystearic acid were isolated.

As shown in Tables 5 and 6, LPS from *Anabaena flos-aquae* also
contained xylose, fucose, and a low amount of KDO. The fatty acid
complement contained not only palmitic acid and the three β-hydroxy-
fatty acids mentioned above, but also 10 long chain saturated and
unsaturated fatty acids which were not present in *Anabaena variabilis* LPS. A repeated precise chemical analysis of the LPS of
the two strains and a determination of the percentage of guanine
and cytosine will clarify whether *Anabaena flos-aquae* UTEX 1444 is
indeed *Anabaena variabilis*. Weckesser et al (1979) stated that
the occurrence of LPS among the Cyanobacteria cannot be generalized
mainly for the pleurocapsulean, the oscillatorian, and the heterocystous groups. LPS, however, was isolated in our laboratory from
Oscillatoria brevis and *Oscillatoria tenuis*. The yield was low,
0.3% for *Oscillatoria brevis* and 0.1% for *Oscillatoria tenuis*.

The sugars found in the *Oscillatoria* indicate that the two
species are closely related but still unique. The two unidentified, but chromatographically identical sugars (same Rf value),
one of which is most probably a pentasugar, were found in the two
Oscillatoria species. LPS from *O. tenuis* contains glucosamine
whereas this compound is absent in the endotoxin derived from *O.
brevis*. Aminosugars are only seldom absent in bacterial LPS.

Table 6. Fatty Acid Analysis of *Schizothrix calcicola*, *Anabaena flos-aquae*, *Oscillatoria brevis*, and *Oscillatoria tenuis*

Cyanobacterial Strain	Fatty Acids																		
	Lauric	Myristic	Pentadecanoic	Palmitic	Palmitoleic	Heptadecanoic	Stearic	Oleic	Linoleic	Linolenic	Nonadecanoic	Arachidic	Behenic	Lignoceric	Cerotic	β-hydroxylauric	β-hydroxymyristic	β-hydroxypalmitic	β-hydoxystearic
Schizothrix calcicola		+	+	+	+		+	+	+							+		+	
Anabaena flos-aquae UTEX 1444		+	+	+	+	+	+	+	+	+	+		+				+	+	+
Oscillatoria brevis	+	+	+	+	+	+	+	+	+		+	+	+	+	+				
Oscillatoria tenuis				+	+		+	+	+	+							+	+	+

Oscillatoria tenuis LPS had 1.12% L-glycero-D-mannoheptose, but KDO and phosphorus, which are usually present in LPS from Gram-negative bacteria, were absent. The contamination of protein was much lower than the content of protein in *Anabaena flos-aquae* UTEX 1444 LPS. A true lipopolysaccharide contains a polysaccharide with covalently bound lipid.

As already shown (Keleti et al., 1979), *Schizothrix calcicola* LPS contained eight long chain fatty acids. Surprisingly, the dominant component of lipid A was β-hydroxypalmitic acid and not β-hydroxymyristic acid, as commonly found in LPS from Gram-negative bacteria.

The fatty acids present in *Anabaena flos-aquae* LPS are described in Table 6. The majority of the fatty acids were palmitic, stearic and oleic acid. *Oscillatoria brevis* did not contain any β-hydroxyfatty acids and these were also absent from *Phormidium* sp. (Mikheyskaya, 1977). The related *Oscillatoria tenuis* LPS, however, contained β-hydroxymyristic acid, β-hydroxypalmitic acid, and β-hydroxystearic acid.

The lipid moiety of the LPS molecule derived from *S. calcicola* is more complex than previously thought. Silica gel thin layer chromatography performed on prepared LPS from *S. calcicola* revealed a very low concentration of lipid A containing only four separable components. Schiller et al. (1976) detected in *Salmonella minnesota* R595 lipid A at least 20 separable components.

Isolation of LPS from *Anabaena cylindrica* has to date been unsuccessful. This species contains large quantities of glucan and LPS seems to be absent. Additional isolation attempts will be performed with this species, since Dunn and Volk (1970) found that the walls of the vegetative cells and heterocysts of *Anabaena cylindrica* contained not less than 62% carbohydrates (glucose, mannose, and xylose).

Biological Characterization

The lethal effect of the purified endotoxin was tested in duplicate by intraperitoneal injections into five 20 g Swiss-Webster female mice for each concentration (500 μg, 1 mg, 5 mg). The endotoxin was suspended in PBS with 0.1% trithylamine.

The isolated and purified LPS from *Schizothrix calcicola* and *Anabaena flos-aquae* was non-toxic to mice when injected intraperitoneally even at concentrations as high as 5000 μg per mouse. Similarly, Weise et al. (1970) showed that LPS isolated from the cyanobacterium *Synechococcus nidulans* was not toxic to mice.

Table 7.   Biological Characterization of *Schizothrix calcicola*,
           Anabaena flos-aquae UTEX 1444, *Oscillatoria tenuis*,
           and *O. brevis*

Cyanobacterial Strain	Lethality in Mice	Schwartzman Reaction	LAL Gelation (Lowest Conc. Pos. LPS)	Rabbit Loop Assay 500 µg
S. calcicola	No death	40 µg and 80 µg Pos.	500 ng	Neg.[a]
Anabaena flos-aquae UTEX 1444	No death	40 µg and 80 µg Weakly Pos.	12.5 ng	Neg.[a]
O. brevis	LD_{50} 3.83 mg/mouse	N.D.[b]	10,000 ng	N.D.
O. tenuis	N.D.	20 µg, 40 µg, and 80 µg Negative	100 ng	N.D.

[a]Preliminary.

[b]N.D. = Not Done.

Mikheyskaya et al. (1977) have demonstrated that glucan-free LPS
from *Phormidium* species did not show any toxicity to mice either.

    Katz et al. (1977) used adrenalectomized mice for the lethal
toxicity test of *Anacystis nidulans*, strain KM. The LD_{50} of
electrodialyzed lipopolysaccharide in its triethylamine form was
found to be about 50 µg/adrenalectomized mouse. The respective
value for *Salmonella minnesota* endotoxin amounts to 0.06 µg per
adrenalectomized mouse. *Oscillatoria brevis* LPS had an LD_{50} of
3.830 µg per mouse, whereas LPS from Gram-negative bacteria from
various smooth strains of *Salmonella* has an LD_{50} of 300 to 500 µg.
The *O. brevis* LPS is the first cyanobacterial endotoxin which has
caused death in mice.

    Positive Schwartzman reaction was observed with LPS from
Schizothrix calcicola and *Anabaena flos-aquae*, but not with LPS
from *Oscillatoria brevis*. The lowest concentration of cyanobacterial LPS, causing LAL gelation, was 12.5 ng for *Anabaena flos-*

aquae, 100 ng for *Oscillatoria tenuis*, 500 ng for *Schizothrix calcicola*, and 10,000 ng for *Oscillatoria brevis*. *Klebsiella* or *E. coli* LPS evoke LAL gelation in a concentration from 0.125 to 2.5 ng.

The preliminary rabbit loop assay was negative with 500 µg *Szhizothrix calcicola* or *Anabaena flos-aquae* LPS, 7 hours after the injection of LPS into the small intestine. Generally, all examined biological reactions of cyanobacterial LPS were much weaker than the same tests with endotoxin from Gram-negative bacteria (Table 7).

Considering the broad spectrum in morphological and physiological differences of the many taxonomic groups of Cyanobacteria, it is a wide open field for studies on the composition of their cell walls. The isolation and the subsequent precise chemical analysis of LPS from all main groups of Cyanobacteria will shed more light on the taxonomy of this group.

ACKNOWLEDGMENTS

The work was supported by an Environmental Protection Agency research grant No. R805754010.

REFERENCES

Buttke, T. M., and L. O. Ingram. 1975. Comparison of lipopolysaccharides from *Agmenellum quadruplicatum* to *Escherichia coli* and *Salmonella typhimurium* by using thin-layer chromatography. J. Bacteriol. *124*:1566-1573.

Carmichael, W. W., and P. R. Gorham. 1974. An improved method for obtaining axenic clones of planktonic blue-green algae. J. Phycol. *10*:238-240.

De, S. W., and D. W. Chatterje. 1953. An experimental study of the mechanism of action of Cholerae on the intestinal mucous membrane. J. Pathol. Bacteriol. *66*:559-562.

Drews, G., W. Schmidt, A. Katz, and W. Weckesser. 1976. Proc. Sect. Int. Symp. Photosynthetic Prokaryotes, Dundee.

Dunn, J. H., and D. P. Volk. 1970. Composition of the cellular envelopes of *Anabaena cylindrica*. J. Bacteriol. *103*:153-158.

Galanos, G., O. Lüderitz, E. T. Rietschel, and O. Westphal. 1977. Newer aspects of the chemistry and biology of bacterial lipopolysaccharides, with special reference to their lipid A component. Pages 239-335 *in* J. W. Goodwin, ed., Int. Rev. Biochem. The Biochemistry of Lipids II, Vol. 14.

Healey, F. T. 1973. Characteristics of phosphorus deficiency in *Anabaena*. J. Phycol. *9*:1383-1394.

Hindman, S. H., M. S. Favero, and A. Peterson. 1975. Pyrogenic reactions during haemodialysis caused by extra-mural endotoxins. Lancet *18*:732-737.

Jann, K., and O. Westphal. 1975. Biological polysaccharides. Pages 1-125 *in* The Antigen, Vol. 3.

Katz, A., J. Weckesser, and J. Drews. 1977. Chemical and biological studies on the lipopolysaccharide (O-antigen) of *Anacystis nidulans*. Archiv. Microbiol. *113*:247-256.

Keleti, G., and W. H. Lederer. 1974. Handbook of Micromethods for the Biological Sciences. Van Nostrand Reinhold Company, New York.

Keleti, G., J. L. Sykora, E. C. Lippy, and M. A. Shapiro. 1979. Composition and biological properties of lipopolysaccharides isolated from *Schizothrix calcicola* (Ag.) Gomont (Cyanobacteria). Appl. Environ. Microbiol. *38*(3):471-477.

Lippy, E. C., and J. Erb. 1976. Gastrointestinal illness at Sewickley, Pa. J. Am. Water Works Assoc. *68*:606-610.

Lundholm, W., and R. Rylander. 1979. Particle size and bacteria content in airborne cotton dust. Cotton Dust Proceedings.

Mikheyskaya, L. V., R. G. Ovodova, and S. Y. Ovodov. 1977. Isolation and characteristics of lipopolysaccharides from cell walls of blue-green algae of the genus *Phormidium*. J. Bacteriol. *130*:1-3.

MMWR. Morbidity and Mortality Weekly Reports. 1979a. Endotoxic reactions associated with the reuse of cardiac catheters - Michigan. *28*:189.

MMWR. Morbidity and Mortality Weekly Reports. 1979b. Endotoxic reactions associated with the reuse of cardiac catheters - Massachusetts. *23*:25-27.

Nowotny, A. 1969. Basic Exercises in Immunochemistry. Springer Verlag, New York. 197 pp.

Rylander, R., P. Haglind, M. Lundholm, J. Mattsby, and K. Stenquist. 1978. Humidifier fever and endotoxin exposure. Clinical Allergy *8*(5):511-516.

Schiller, J. G., R. Ribovich, D. S. Feingold, and J. S. Youngner. 1976. Interferon production in mice by components of *Salmonella minnesota* R 595 lipid A. Infect. Immun. *14*:586-589.

Stanier, R. Y., and G. Cohen-Bazire. 1977. Phototrophic prokaryotes: the Cyanobacteria. Ann. Rev. Microbiol. *31*:225-274.

Sykora, J. L., G. Keleti, R. Roche, D. R. Volk, G. P. Kay, R. A. Burgess, M. A. Shapiro, and E. C. Lippy. 1980. Endotoxins, algae and *Limulus* amoebocyte lysate test in drinking water. Water Research *14*:829-839.

Tison, D. L., and A. J. Lingg. 1979. Dissolved organic matter utilization and oxygen uptake in algal-bacterial microcosms. Can. J. Microbiol. *25*:1315-1320.

Volk, W. V.   1968.   Isolation of D-galacturonic acid 1-phosphate
        from hydrolysates of cell wall lipopolysaccharide extracted
        from *Xanthomonas comprestris*.   J. Bacteriol. *95*:782-786.

Wang, A. W., and A. Hill.   1977.   Chemical analysis of the phenol
        water-extractable material from *Anabaena flos-aquae*.   J.
        Bacteriol. *130*:558-570.

Weckesser, J., G. Drews, and H. Mayer.   1979.   Lipopolysaccharides
        of phytosynthetic prokaryotes.   Ann. Rev. Microbiol. *32*:215-
        239.

Weckesser, J., A. Katz, G. Drews, H. Mayer, and I. Fromme.   1974.
        Lipopolysaccharide containing L-acofriose in the filamentous
        blue-green alga *Anabaena variabilis*.   J. Bacteriol. *120*:672-
        678.

Weise, G., G. Drews, B. Jann, and K. Jann.   1970.   Identification
        and analysis of a lipopolysaccharide in cell walls of the
        blue-green alga *Anacystis nidulans*.   Arch. Mikrobiol. *71*:89-
        98.

Westphal, O., and K. Jann.   1965.   Bacterial lipopolysaccharides,
        extraction with phenol water and further applications of
        the procedure.   Methods Carbohydr. Chem. *5*:83-91.

THE DIRECTORY TO
TOXIC BLUE-GREEN ALGAE (CYANOBACTERIA)
LITERATURE

Compiled by
Wayne W. Carmichael
and
Peter E. Bent

1980

Department of Biological Sciences
Wright State University
Dayton, Ohio 45435

ABSTRACT

This literature search was conducted in order to provide a
comprehensive list of the publications concerning toxic blue-green
algae. Although originally intended for distribution to participants at the 1980 conference on The Water Environment: Algal
Toxins and Health, it is hoped that this directory will be of use
to all researchers interested in the subject. Our sources for
this directory were several review papers, computer searches, and
personal communications from conference participants and others
around the world. Many important articles are published in
journals which may not be widely distributed. We hope this
directory will help lead researchers to literature they might
otherwise have missed. Interesting developments are taking place
in toxic blue-green algae research. Good communication between
researchers all over the world is needed now more than ever.

It is expected that the directory will be updated as research in
this area continues. Additions and/or corrections will be greatly
appreciated and incorporated into future directories.

Aitkin, R., and B. Keed. 1974. Control for toxic algae in dam
    water? Research News. Agric. Gaz. N. S. W. *85*(2):45.
Alam, M., M. Ikawa, J. J. Sasner, Jr., and P. J. Sawyer. 1973.
    Purification of *Aphanizomenon flos-aquae* toxin and its
    chemical and physiological properties. Toxicon *11*:65-72.
Alam, M., Y. Shimizu, M. Ikawa, and J. J. Sasner, Jr. 1978. Rein-
    vestigation of the toxins from the blue-green alga, *Aphani-
    zomenon flos-aquae*, by a high performance chromatographic
    method. J. Environ. Sci. Health *A13*(7):493-499.
Amann, M. 1977. Untersuchungen über ein pteridiz als bestandteil
    des toxischen prinzips aus *Synechococcus*. D. Naturioiss.
    Dissertation. Eberhard-Karls Universitat, Tübingen. 64 pp.
Aplin, T. E. H. 1967. Water blooms. J. Agric. West. Aust. *8*(11):
    472-474.
Aplin, T. E. H., and D. C. Main. 1976. Toxic water blooms. West.
    Aust. Dep. Agric. Bull. No. 3940:3-4.
Arthur, J. C. 1885. Second report on some algae of Minnesota
    supposed to be poisonous. Page 109 *in* University of Minne-
    sota, Department of Agriculture Report 1881/86, Biennial
    Report of the Board of Regents No. 4, Suppl. No. 1. Idem.
    Bull. Minn. Acad. Nat. Sci. *3*:97.
Ashworth, C. T., and M. F. Mason. 1946. Observations on the
    pathological changes produced by a toxic substance present
    in blue-green algae (*Microcystis aeruginosa*). Am. J. Pathol.
    22:369-383.
Ayres, S. 1949. Discussion of W. M. Sams' Seabather's eruption.
    Arch. Dermatol. Syphilol. *60*:236.
Aziz, K. M. S. 1974. Diarrhea toxin obtained from a waterbloom-
    producing species, *Microcystis aeruginosa* Kütz. Science
    183:1206-1207.
Bailey, J. W. 1959. Algae can poison cattle. Jersey J. *6*:5.
Banner, A. H. 1959. A dermititis-producing alga in Hawaii.
    Hawaii Med. J. *19*:35-36.
Banner, A. H. 1967. Marine toxins from the Pacific. I. Advances
    in the investigation of fish toxins. Pages 157-165 *in* F. E.
    Russell and P. R. Saunders, eds., Animal Toxins. Pergamon
    Press, New York.
Banner, A. H., P. J. Scheuer, S. Sasaki, P. Helfrich, and C. B.
    Alender. 1960. Observations on ciguatera-type toxin in
    fish. Ann. N. Y. Acad. Sci. *90*:770-787.
Barker, F. J. 1977. Are algae toxic to honey bees. Ariz. Acad.
    Sci. *12*(2):84-85.
Barnum, D. A., J. A. Henderson, and A. G. Steward. 1950. Algae
    poisoning in Ontario. Milk Producer *25*:312.
Bent, P. E., and W. W. Carmichael. 1980. Isolation and charac-
    terization of peptide toxins from fresh-water cyanophytes
    Anabaena flos-aquae and *Microcystis aeruginosa*. Ohio J.
    Sci. *80*:66.
Berlin, R. 1948. Haff disease in Sweden. Acta Med. Scand. *129*:
    560.

Biggs, D. F., and W. F. Dryden. 1977. Action of anatoxin-I at
    the neuromuscular junction. Proc. West. Pharmacol. Soc.
    20:461-466.

Bishop, C. T., E. F. L. J. Anet, and P. R. Gorham. 1959. Isola-
    tion and identification of the fast-death factor in *Micro-
    cystis aeruginosa* NRC-1. Can. J. Biochem. Physiol. *37*:453-
    471.

Bossenmaier, E. F., T. A. Olson, E. Rueger, and W. H. Marsh. 1954.
    Some field and laboratory aspects of duck sickness at White-
    water Lake, Manitoba. Pages 163-175 *in* Trans. 19th North
    American Wildlife Conference, March 8-10. Wildlife Manage-
    ment Inst., Washington, D. C.

Braginskii, L. P. 1955. Otoksichnosti sine-zdenykh vodorosley
    (On the toxicity of blue-green algae). Priroda (Mosc.)
    44(1):117.

Branco, S. M. 1959. Algas toxicas controle das toxinas em aguas
    de abasticimento (Toxic algae control of toxins in waste-
    water). Rev. Dept. Aguas Esgatos Sao Paulo (Brasil) *20*(33):
    21-30; *20*(34):29-42.

Brandenburg, T. O., and F. M. Shigley. 1947. "Water bloom" as a
    cause of poisoning in livestock in North Dakota. J. Am.
    Vet. Med. Assoc. *110*:384.

Buttke, T. M., and L. O. Ingram. 1975. Comparison of lipopoly-
    saccharides from *Agmenellum quadruplicatum* or *Escherichia
    coli* and *Salmonella typhimurium* by using thin-layer chroma-
    tography. J. Bacteriol. *124*:1566-1573.

Campbell, H. F., O. E. Edwards, J. W. Elder, and R. J. Kolt. 1979.
    Total synthesis of DL-anatoxin-a and DL-isoanatoxin-a. Pol.
    J. Chem. *53*:27-36.

Campbell, H. F., O. E. Edwards, and R. Kolt. 1977. Synthesis of
    noranatoxin-A and anatoxin-A. Can. J. Chem. *55*:1372-1379.

Cannon, D., J. C. Dean, and I. C. Smalls. 1972. Growth and con-
    trol of two species of *Anabaena* (Cyanophyta) in a water
    supply/reservoir. Bull. Aust. Soc. Limnol. *3*:15.

Cardellina, J. H., II, F. J. Marner, and R. E. Moore, 1979. Sea-
    weed dermatitis: structure of lyngbyatoxin A. Science
    204:193-195.

Carmichael, W. W. 1974. *Anabaena flos-aquae* Toxin: Its Toxicol-
    ogy and Mechanism of Action. Ph.D. Thesis. University of
    Alberta, Edmonton, Alberta. 134 pp.

Carmichael, W. W., and D. F. Biggs. 1978. Muscle sensitivity
    differences in two avian species to anatoxin-a produced by
    the freshwater cyanophyte *Anabaena flos-aquae* NRC-44-1.
    Can. J. Zool. *56*(3):510-512.

Carmichael, W. W., D. F. Biggs, and P. R. Gorham. 1975. Toxicol-
    ogy and pharmacological action of *Anabaena flos-aquae* toxin.
    Science *187*:542-544.

Carmichael, W. W., D. F. Biggs, and M. A. Peterson. 1979. Pharma-
    cology of anatoxin-a, produced by the freshwater cyanophyte
    Anabaena flos-aquae NRC-44-1. Toxicon *17*:229-236.

Carmichael, W. W., and P. R. Gorham. 1977. Factors influencing the toxicity and animal susceptibility of *Anabaena flos-aquae* (Cyanophyta) blooms. J. Phycol. *13*(2):97-101.

Carmichael, W. W., and P. R. Gorham. 1978. Anatoxins from clones of *Anabaena flos-aquae* isolated from lakes of western Canada. Mitt. Int. Ver. Limnol. *21*:285-295.

Carmichael, W. W., and P. R. Gorham. 1980. Freshwater cyanophyte toxins: types and their effects on the use of micro-algae biomass. (In press) *in* G. Shelef and C. J. Soeder, eds., The Production and Use of Micro-Algae Biomass. Elsevier/North Holland, New York.

Carmichael, W. W., P. R. Gorham, and D. F. Biggs. 1977. Two laboratory case studies on the oral toxicity to calves of the freshwater cyanophyte (blue-green alga) *Anabaena flos-aquae* NRC-44-1. Can. Vet. J. *18*:71-75.

Chamberlain, W. J. 1948. Effects of algae on water supply. University of Queensland, Brisbane, Department of Chemistry. Papers, Vol. 1, No. 29. 80 pp., 60 plates.

Chaput, M., and G. A. Grant. 1958. Toxic algae. III. Screening of a number of species. Can. Defence Research Board, Defence Research Laboratories. Report No. 279, Project No. D52-20-20-18. Ottawa. 6 pp.

Cohen, S. G., and C. B. Rief. 1953. Cutaneous sensitization to blue-green algae. J. Allergy *24*:452-457.

Collins, M. 1978. Algal toxins. Microbiol. Rev. *42*(4):725-746.

Cotton, H. L. 1914. Algae poisoning. Am. J. Vet. Med. *9*:903.

Dahl, J. 1968. Giftige blågrønnalger i ferskvand. Skr. Danm. Fiskeri- og Havundersøgelser *28*:47-52.

Davidson, F. F. 1959. Poisoning of wild and domestic animals by a toxic waterbloom of *Nostoc rivulare* Kütz. J. Am. Water Works Assoc. *51*:1277.

Deem, A. W., and F. Thorp. 1939. Toxic algae in Colorado. J. Am. Vet. Med. Assoc. *95*:542.

Devlin, J. P., O. E. Edwards, P. R. Gorham, N. R. Hunter, R. K. Pike, and B. Stavric. 1977. Anatoxin-a, a toxic alkaloid from *Anabaena flos-aquae* NRC-44h. Can. J. Chem. *55*:1367-1371.

Dillenberg, H. O. 1959. Toxic waterbloom in Saskatchewan. Presented before the 14th Annual Meeting INCDNCM, August 26-29. Washington State College, Pullman, Washington.

Dillenberg, H. O., and M. K. Dehnel. 1960. Toxic waterbloom in Saskatchewan, 1959. Can. Med. Assoc. J. *83*:1151.

Dillenberg, H. O., and M. K. Dehnel. 1961. "Waterbloom poisoning." Fast and "slow death" factors isolated from blue-green algae at Canadian NRC Laboratories. World-Wide Abstr. Gen. Med. *4*(4):20.

Durrell, L. W., and A. W. Deem. 1940. Toxic algae in Colorado. J. Colo. Wyo. Acad. Sci. *2*(6):18.

Elleman, T. C., I. R. Falconer, A. R. B. Jackson, and M. T. Runnegar. 1978. Isolation, characterization, and pathology of the toxin from a *Microcystis aeruginosa* (= *Anacystis cyanea*) bloom. Aust. J. Biol. Sci. *31*:209-218.

Eloff, K. 1980. A preliminary study of the toxins of different *Microcystis* strains. S. Afr. J. Sci. (Submitted for publication).

Elster, H. J. 1955. Toxische wirkungen des phytoplanktons (Toxic effects of phytoplankton). Naturwiss Rundsch. *8*:318.

Everist, S. 1974. Poisonous plants of Australia. Angus and Robertson, Sydney. Pages 566-569.

Firkins, G. S. 1953. Toxic algae poisoning. Iowa State Coll. Vet. *15*:151.

Fitch, C. P., L. M. Bishop, and W. L. Boyd. 1934. "Water bloom" as a cause of poisoning in domestic animals. Cornell Vet. *24*(1):30-39.

Francis, G. 1878. Poisonous Australian lake. Nature (London) *18*:11-12.

Gentile, J. H. 1971. Blue-green and green algal toxins. Pages 27-66 *in* S. Kadis, A. Ciegler and S. J. Ajl, eds., Microbial Toxins, Vol. VII, Algal and Fungal Toxins. Academic Press, London. 401 pp.

Gentile, J. H., and T. E. Maloney. 1969. Toxicity and environmental requirements of a strain of *Aphanizomenon flos-aquae* (L.) Ralfs. Can. J. Microbiol. *15*:165-173.

Gilliam, W. G. 1925. The effect on livestock of water contaminated with fresh water algae. J. Am. Vet. Med. Assoc. *67*(n.s.20): 780.

Golecki, J. R., and G. Drews. 1974. The structure of the blue-green algal cell wall. Cell walls of *Anabaena variabilis,* revealed by freeze etchings before and after extraction. Cytobiologie *8*:213-227.

Gorham, P. R. 1960. Toxic waterblooms of blue-green algae. Can. Vet. J. *1*:235-245.

Gorham, P. R. 1962. Laboratory studies on the toxins produced by waterblooms of blue-green algae. Am. J. Public Health *52*(12):2100-2105.

Gorham, P. R. 1964a. Toxic algae. Pages 307-336 *in* D. F. Jackson, ed., Algae and Man. Plenum Press, New York. 434 pp.

Gorham, P. R. 1964b. Toxic algae as a public health hazard. J. Am. Water Works Assoc. *56*:1481-1488.

Gorham, P. R. 1965. Toxic waterblooms of blue-green algae. Biol. Prob. Water Pollution, Third Sem., U. S. Public Health Serv. Bull. 999-WP-25:37-44.

Gorham, P. R., and W. W. Carmichael. 1979. Phycotoxins from blue-green algae. Pure Appl. Chem. *52*(1):165-174.

Gorham, P. R., and W. W. Carmichael. 1980. Toxic substances from freshwater algae. Prog. Water Technol. *12*:189-198.

Gorham, P. R., J. McLachlan, U. T. Hammer, and W. K. Kim. 1964.
Isolation and culture of toxic strains of *Anabaena flos-aquae* (Lyngb.) de Bréb. Int. Assoc. Theor. Appl. Limnol.
15:796-804.

Gorman, R. C. 1963. Control of algae in water supplies. J.
Agric. West. Aust. *4*(3):149-150.

Gortner, R. A. 1935. On toxic waterbloom. Page 473 *in* J. E.
Tilden, ed., The Algae and Their Life Relations. Univ. of
Minnesota Press, Minneapolis, Minnesota.

Grant, G. A. 1953. Toxic algae. I. Development of toxicity in
blue-green algae. Can. Defence Research Board, Defence
Research Chemical Laboratories. Report No. 124, Project
No. D52-20-20-18. Ottawa. 11 pp.

Grant, G. A., and E. O. Hughes. 1953. Development of toxicity in
blue-green algae. Can. J. Public Health *44*:334-339.

Grauer, F. H. 1959. Dermatitis escharotica caused by a marine
alga. Hawaii Med. J. *19*:32-36.

Grauer, F. H., and H. L. Arnold, Jr. 1961. Seaweed dermatitis:
first report of a dermatitis-producing marine alga. Arch.
Dermatol. *84*:720-732.

Grigor, L. V., Y. A. Kirpenko, V. M. Orlovskii, and V. V. Stankevich.
1977. Antimicrobial effect of toxic metabolites of some
blue-green algae. Gidrobiol. Zh. *13*:57-62.

Hammer, U. T. 1968. Toxic blue-green algae in Saskatchewan. Can.
Vet. J. *9*:221-229.

Hardin, F. F. 1961. Seabather's eruption. J. Med. Assoc. (Georgia)
50:450.

Hashimoto, Y., N. Fusetani, and K. Nozawa. 1972. Screening of the
toxic algae on coral reefs. Pages 569-572 *in* Proc. 7th Int.
Seaweed Symp., Sapporo, 1971.

Hashimoto, Y., H. Kamiya, K. Nozawa, and K. Yamazato. 1975. Occurrence of a toxic blue-green alga inducing skin dermatitis in
Okinawa. Toxicon *13*:95-96.

Hashimoto, Y., H. Kamiya, K. Yamazato, and K. Nozawa. 1976. Occur-
rence of a toxic blue-green alga inducing skin dermatitis in
Okinawa. Pages 333-338 *in* A. Ohsaka, K. Hayashi and Y.
Sawai, eds., Animal, Plant, and Microbial Toxins, Vol. I.
Plenum Press, New York.

Healey, J. S. 1968. Water containing algae may be dangerous to
stock. Agric. Gaz. N. S. W. *79*(3):157.

Heaney, S. I. 1971. The toxicity of *Microcystis aeruginosa* Kütz.
from some English reservoirs. Water Treat. Exam. *20*:235-244.

Hindersson, R. 1933. Förgiftning av nøtkreatur genom
sötvattensplankton (Poisoning of cattle by fresh-water
plankton). Finsk. Veterinartidskrift *39*:179.

Hoffmann, J. R. H. 1976. Removal of *Microcystis* toxins in water
purification processes. Water S. A. (Pretoria) *2*:58-60.

Howard, N. J., and A. E. Berry. 1933. Algal nuisances in surface
waters. Can. Public Health J. *24*:377.

Huber, C. S. 1972. The crystal structure and absolute configura-
     tion of 2,9-diacetyl-9-azabicyclo (4,2,1) non-2,3-ene. Acta
     Crystallogr. Sect. B 28:2577-2582.

Hughes, E. O., P. R. Gorham, and A. Zehnder. 1958. Toxicity of a
     unialgal culture of *Microcystis aeruginosa*. Can. J. Micro-
     biol. 4:225-236.

Hurst, E. 1942. The poison plants of New South Wales. Pages
     423-424 *in* The Poison Plants Committee. Snelling Printing
     Works, Syndey.

Ilyaletoinova, S. G., and A. M. Dubitskii. 1972. Method of deter-
     mining the toxicity of blue-green algae in relation to
     Culicidae larvae. Izv. Akad. Nauk. Kaz. SSR. Ser. Biol. Nauk.
     2:39-42.

Ingram, W. M., and G. W. Prescott. 1954. Toxic fresh-water algae.
     Am. Mid. Nat. 52:75-87.

Jackim, E., and J. Gentile. 1968. Toxins of a blue-green alga:
     similarity to saxitoxin. Science 162:915-916.

Kato, Y., and P. J. Scheuer. 1975. The aplysiatoxins. Pure Appl.
     Chem. 41:1-14.

Kato, Y., and P. J. Scheuer. 1976. The aplysiatoxins; reactions
     with acid and oxidants. Pure Appl. Chem. 48:29-33.

Keleti, G., J. L. Sykora, E. C. Lippy, and M. A. Shapiro. 1979.
     Composition and biological properties of lipopolysaccharides
     isolated from *Schizothrix calcicola* (Ag.) Gomont (Cyanobac-
     teria). Appl. Environ. Microbiol. 38(3):471-477.

Kirpenko, Y. A., I. I. Perevozchenko, L. A. Sirenko, and L. F.
     Lukina. 1975. Isolation of toxin from blue-green algae
     biomass and some of its physico-chemical properties. Dopov.
     Akad. Nauk Ukr. Rsr. Ser. B. Pages 359-361.

Kirpenko, Y. A., L. A. Sirenko, V. M. Orlovskii, and L. F. Lukina.
     1977. Toxic blue-green algae and culture of the organism.
     In A. V. Topachevskiy and E. E. Kvetnetskia, eds., Naukogo
     Dumka. Ukranian Scientific Publishing House, Kiev. 252 pp.
     (Russian).

Konst, H., P. D. McKercher, P. R. Gorham, A. Robertson, and J.
     Howell. 1965. Symptoms and pathology produced by toxic
     Microcystis aeruginosa NRC-1 in laboratory and domestic
     animals. Can. J. Comp. Med. Vet. Sci. 29:221-228.

Kristiansen, J. 1969. Rød vandblomst forarsaget af blågrønalge.
     Ferskvandsfiskeribladet 67:88-92.

Lippy, E. C., and J. Erb. 1976. Gastrointestinal illness at
     Sewickley, Pa. J. Am. Water Works Assoc. 68:606-610.

Louw, P. G. J. 1950. The active constituent of the poisonous
     algae, *Microcystis toxica* Stephens. S. Afr. Ind. Chemist
     4:62-66.

MacDonald, D. W. 1960. Algal poisoning in beef cattle. Can. Vet.
     J. 1:108.

Mackenthun, K. M., E. F. Herman, and A. F. Bartsch. 1948. A heavy
     mortality of fishes resulting from the decomposition of algae
     in the Yahara River, Wisconsin. Trans. Am. Fish. Soc. 75:175.

MacKinnon, A. F. 1950. Report on algae poisoning. Can. J. Comp.
    Med. *14*:208.
Mahprevskaya, A. Y. 1979. On causes of fish kills under the
    influence of blue-green algae. Abstracts, Soc. Int. Limnol.
    Workshop on Hypertrophic Ecosystems, September. Vaxjo,
    Sweden.
Main, D. C., P. H. Berry, R. L. Peet, and J. P. Robertson. 1977.
    Sheep mortalities associated with the blue-green alga
    Nodularia spumigena. Aust. Vet. J. *53*:578-581.
Maloney, T. E., and R. A. Carnes. 1966. Toxicity of a *Microcystis*
    waterbloom from an Ohio pond. Ohio J. Sci. *66*(5):514-517.
Mason, M. F., and R. E. Wheeler. 1942. Observations upon the
    toxicity of blue-green algae. Am. Soc. Biol. Chem. 36th
    Ann. Meeting, Boston. Fed. Proc. *1*:124.
Mathew, J. M. 1930. Hume Reservoir and algal infestation.
    Commonw. Eng. *17*:401-405.
May, V. 1970. A toxic alga in New South Wales and its distribu-
    tion. Contrib. N. S. W. Natl. Herb. *4*(3):84-86.
May, V. 1971. Forecasts of poisonous alga outbreaks? Research
    Pars. Agric. Gaz. N. S. W. Page 116.
May, V. 1972. Blue-green algal blooms at Braidwood, New South Wales
    (Australia). N. S. W. Dep. Agric. Sci. Bull. *82*:1-45.
May, V. 1974. Suppression of blue-green algal blooms in Braidwood
    Lagoon with alum. J. Aust. Inst. Agric. Sci. *40*:54-57.
May, V. 1976. Control of algae blooms by removal of nutrients.
    Proc. Aust. Water Resources Council. Aust. Gov. Public
    Serv., Canberra. Pages 79-82.
May, V. 1978a. Areas of recurrence of toxic algae within Burrin-
    juck Dam, New South Wales, Australia. Telopea *1*(5):295-313.
May, V. 1978b. Ferric alum can reduce toxic algae in farm dams.
    Agric. Gaz. N. S. W. *89*(4):6-7.
May, V., and H. Baker. 1978. Reduction of toxic algae in farm
    dams by ferric alum. N. S. W. Dep. Agric. Tech. Bull. *19*:
    1-16.
May, V., and E. J. McBarron. 1973. Occurrence of the blue-green
    alga, *Anabaena circinalis* Rabenh. in New South Wales and
    toxicity to mice and bees. J. Aust. Inst. Agric. Sci. *39*:
    264-266.
McBarron, E. 1976. Medical and veterinary aspects of plant
    poisons in N. S. W. N. S. W. Dep. Agric. 243 pp.
McBarron, E. J., and V. May. 1966. Poisoning of sheep in New
    South Wales by the blue-green alga, *Anacystis cyanea* (Kützetz.)
    Dr. and Dail. Aust. Vet. J. *42*:449-453.
McBarron, E. J., R. I. Walker, I. Gardner, and K. H. Walker. 1975.
    Toxicity to livestock of the blue-green alga *Anabaena cir-
    cinalis*. Aust. Vet. J. *51*:587.
McDowell, M. E. et al. 1960. Algae feeding in humans: accepta-
    bility, digestibility and toxicity. Abstr. 44th Ann. Meet-
    ing Fed. Am. Soc. Exp. Biol., Chicago. Fed. Proc. *19*(11):
    319.

McIvor, R. A., and G. A. Grant. 1955. Toxic algae. II. Prelim-
    inary attempts to prepare toxic extracts. Can. Defence
    Research Board, Defence Research Chem. Laboratories. Report
    No. 190, Project No. D52-20-20-18. Ottawa. 7 pp.
McLeod, J. A., and G. F. Bondar. 1952. A case of suspected algal
    poisoning in Manitoba. Can. J. Public Health 43:347.
Mikhaylev, V. V., and D. L. Tephji. 1961. Otoksichnosti sine-
    nelenykh vodorosley reki Volgi (On the toxicity of blue-
    green algae of the Volga River). Zcologicheskii Zh. (Mosc.)
    40:1619.
Mills, J. G., and J. T. Wyatt. 1974. Ostracod reactions to non-
    toxic and toxic algae. Oecologia (Berl.) 17:171-177.
Moikeha, S. N., and G. W. Chu. 1971. Dermatitis-producing alga
    Lyngbya majuscula Gomont in Hawaii. II. Biological prop-
    erties of the toxic factor. J. Phycol. 7:8-13.
Moikeha, S. N., G. W. Chu, and L. R. Berger. 1971. Dermatitis-
    producing alga Lyngbya majuscula Gomont in Hawaii. I. Iso-
    lation and chemical characterization of the toxic factor.
    J. Phycol. 7:4-8.
Moore, G. T. 1900. Algae as a cause of the contamination of
    drinking water. Am. J. Pharm. 72:25.
Moore, R. E. 1977. Toxins from blue-green algae. Bioscience
    27(12):797-802.
Mulhean, C. J. 1959. Beware algae! They can poison livestock.
    J. Agric. S. Aust. 62(9):406-408.
Mullor, J. B. 1945. Algas toxicas (Toxic algae). Rev. Sanid.,
    Asistencia Social Trabajo (Santa Fe, Argent.) 1:95.
Mullor, J. B., and A. M. Wachs. 1948. Algunas caracteristicas
    del alga toxica Anabaena venenosa (Some characteristics of
    toxic alga Anabaena venenosa). Congreso Sudamericano
    Quimica. 4th Trabajos Presentados 1:326-327.
Murthy, J. R., and J. B. Capindale. 1970. A new isolation and
    structure for the endotoxin from Microcystis aeruginosa
    NRC-1. Can. J. Biochem. 48:508-510.
Mynderse, J. S., and R. E. Moore. 1978. Toxins from blue-green
    algae: structures of oscillatoxin A and three related
    bromine-containing toxins. J. Org. Chem. 43:2301-2303.
Mynderse, J. S., R. E. Moore, M. Kashiwage, and T. R. Norton.
    1977. Antileukemia activity in the Oscillatoriaceae: iso-
    lation of debromoaplysiatoxin from Lyngbya. Science 196:
    538-540.
Nygård, J. J. 1977. Toksiske blågrønnalger i ferskvann. Norsk
    Institutt for Vannforskning. Årbok 1976, 17-24, Oslo 1977.
O'Donoghue, J. G., and G. S. Wilton. 1951. Algal poisoning in
    Alberta. Can. J. Comp. Med. 15:193.
Olrik, K. 1976. Giftige alger. Vand 4:1-7.
Olson, T. A. 1949. History of toxic plankton and associated
    phenomena. Algae-laden water causes death of domestic
    animals; nature of poison. Sewage Works Eng. 20(2):71.

Olson, T. A. 1951. Toxic plankton. Pages 86-95 *in* Proceedings of Inservice Training Course in Water Works Problems, February 15-16. University of Michigan, School of Public Health, Ann Arbor, Michigan.

Olson, T. A. 1952. Toxic plankton. Water Sewage Works *99*:75.

Olson, T. A. 1955. Studies of algae poisoning. With special reference to the relationship of this phenomenon to losses of wildfowl and other birds. Flicker (Minneapolis) *27*:105.

Olson, T. A. 1960. Water poisoning - a study of poisonous algae blooms in Minnesota. Am. J. Public Health *50*:883.

Olson, T. A. 1964. Blue-greens. Pages 349-356 *in* J. P. Lindurska, ed., Waterfowl Tomorrow. U. S. Dept. Int. Fish. and Wildlife Serv. U. S. Gov. Printing Office, Washington.

Orlovskii, V. M., and Y. A. Kirpenko. 1976. Biologically active metabolites of blue-green algae and their effect on experimental animals. Gig. Sanit. *3*:13-17.

Østrøm, B. 1976. Fertilization of the Baltic by nitrogen fixation in the blue-green alga *Nodularia spumigena*. Remote Sens. Environ. *4*:305-310.

Payne, B. W., and E. D. F. Williams. 1977. An investigation of virus-like inclusions in *Microcystis toxica* by means of digital image analysis. Electron Microsc. Soc. South. Afr. *7*:107-108.

Porter, E. D. Investigations of supposed poisonous vegetation in the waters of some of the lakes of Minnesota. Page 95 *in* Dep. of Agric. Report 1881/86, Biennial Report of the Board of Regents, No. 4., Suppl. No. 1.

Porter, K. G., and J. D. Orcutt. 1978. Nutritional adequacy, manageability and toxicity as factors that determine the food quality of green and blue-green algae for *Daphnia*. (In press) *in* Evolution and Ecology of Zooplankton Communities. Univ. Press of New England.

Powell, R. C., E. M. Nevels, and M. E. McDowell. 1961. Algae feeding in humans. J. Nutr. *75*(1):7.

Prescott, G. W. 1933. Some effects of the blue-green algae, *Aphanizomenon flos-aquae*, on lake fish. Collect. Net *8*:77-80.

Prescott, G. W. 1938. Objectionable algae and their control in lakes and reservoirs. Reprint: Louisiana Municipal Rev. Vol. 1, Nos. 2 and 3 (July/August and September/October). Shreveport.

Prescott, G. W. 1948. Objectionable algae with reference to the killing of fish and other animals. Hydrobiologia *1*:1-13.

Rabin, R., and A. Darbre. 1975. An improved extraction procedure for the endotoxin from *Microcystis aeruginosa* NRC-1. Biochem. Soc. Trans. *3*:428-430.

Ransom, R. E., T. A. Nerad, and P. G. Meier. 1978. Acute toxicity of some blue-green algae to the protozoan *Paramecium caudatum*. J. Phycol. *14*:114-116.

Remer, F. 1943. Cited in A. H. Quin, Sheep poisoned by algae.
    J. Am. Vet. Med. Assoc. *102*:299.
Rose, E. F. 1953. Toxic algae in Iowa lakes. Proc. Iowa Acad.
    Sci. *60*:738.
Runnegar, M. T. C., and I. R. Falconer. 1975. Isolation of toxin
    from a naturally occurring algal bloom of *Microcystis
    aeruginosa* (= *Anacystis cyanea*). Proc. Aust. Biochem. Soc.
    8:5.
Runnegar, M. T. C., and I. R. Falconer. 1980. Variation with
    temperature of toxin production in cultures of the blue-
    green alga, *Microcystis aeruginosa*. Aust. Soc. Aquat. Bot.
    Phycol. Inaugural Meeting. (In press).
Saint George-Grambauer, T. D. 1957. Plants poisonous to live-
    stock. J. Agric. S. Aust. *60*(7):298.
Sams, W. M. 1949. Seabather's eruption. Arch. Dermatol. Syphilol.
    60:227.
Sawyer, P. J., J. H. Gentile, and J. J. Sasner, Jr. 1968. Demon-
    stration of a toxin from *Aphanizomenon flos-aquae* (L.) Ralfs.
    Can. J. Microbiol. *14*:1199-1204.
Schantz, E. J., V. E. Ghazarossian, H. K. Schnoes, F. M. Strong,
    J. P. Springer, J. O. Pezzanite, and J. Clardy. 1975. The
    structure of saxitoxin. J. Am. Chem. Soc. *97*:1238-1239.
Schmidt, W., G. Drews, J. Weckesser, I. Fromme, and D. Borowiak.
    1980. Characterization of the lipopolysaccharides from
    eight strains of the cyanobacterium *Synechococcus*. Arch.
    Microbiol. *127*:209-215.
Schmidt, W., G. Drews, J. Weckesser, and H. Mayer. 1980. Lipopoly-
    saccharides in four strains of the unicellular cyanobacterium
    Synechocystis. Arch. Microbiol. *127*:222.
Schwimmer, D., and M. Schwimmer. 1964. Algae and medicine. Pages
    368-412 *in* D. F. Jackson, ed., Algae and Man. Plenum Press,
    New York. 434 pp.
Schwimmer, M. 1964. Human gastro-intestinal, respiratory, and
    skin disorders associated with algae. Tabular data and
    special references. Pages 94-96 *in* K. M. Mackenthun, W. M.
    Ingram and R. Porges, eds., Limnological Aspects of Recre-
    ational Lakes. Division of Water Supply and Pollution
    Control, U. S. Public Health Service Publication No. 1167.
    Washington, D. C.
Schwimmer, M., and D. Schwimmer. 1955. The Role of Algae and
    Plankton in Medicine. Grune and Stratten, New York. 85 pp.
Schwimmer, M., and D. Schwimmer. 1968. Medical aspects of phycol-
    ogy. Pages 279-358 *in* D. F. Jackson, ed., Algae, Man and
    the Environment. Syracuse University Press, New York.
    369 pp.
Scott, R. M. 1952. Algal toxins. Public Works *85*:54.
Scott, W. E. 1980. Large scale cultivation of toxic *Microcystis*.
    Water Report C. S. I. R. No. 11. Pretoria.
Senior, V. E. 1960. Algal poisoning in Saskatchewan. Can. J.
    Comp. Med. *24*:26.

Seydel, E. 1913. Fischsterben durch wasserblüte (Fish death due to waterbloom). Mitt. Fischerei-Ver. Brandenburg n.s. 5(9): 87.

Shelubsky (Shilo), M. 1950. Observations on the properties of a toxin produced by *Microcystis*. Int. Assoc. Theor. Appl. Limnol. *11*:362.

Shilo, M. 1972. Toxigenic algae. Pages 233-265 *in* O. J. D. Hockenhull, ed., Progress in Industrial Microbiology, Vol. 11. Churchill Livingstone Press, Edinburgh.

Simpson, B., and P. R. Gorham. 1958. Source of the fast-death factor produced by unialgal *Microcystis aeruginosa* NRC-1. Abstr. Phycol. Soc. Am. News Bull. *11*:59-60.

Sirenko, L. A. 1979. Toxicity fluctuations and factors determining them. Proc. Soc. Int. Limnol. Workshop of Hypertrophic Ecosystems, September. Vaxjo, Sweden.

Sirenko, L. A., Y. A. Kirpenko, L. F. Lukina, O. V. Kovalenko, and L. M. Zimovets. 1976a. Toxicity of blue-green algae causative agents of water bloom. Gidrobiol. Zh. *12*:22-28.

Sirenko, L. A., Y. A. Kirpenko, L. F. Lukina, O. V. Kovalenko, and L. M. Zimovets. 1976b. Toxicity of blue-green algae the causative agents of the blooming of water. Gidrobiol. Zh. *12*:13-18.

Skulberg, O. M. 1956. Vannblomst av blågrønne alger som årsak til forgiftning av dyr og mennesker. Fauna, No. 1:19-26.

Skulberg, O. M. 1972. Blågrønnalger i norske vannforekomster, mulige konsekvenser av sunnhetsmessig betydning for mennesker og dyr. Tidsskr. Norske Laegeforen. *92*(12):851-854. (English summary).

Skulberg, O. M. 1979. Toxic effects of blue-green algae, first case of *Microcystis* poisoning reported from Norway. Norwegian Institute for Water Research (NIVA), Temarapport No. 4. Olso, Norway.

Snell, T. W. 1980. Blue-green algae and selection in rotifer populations. Oecologia (Berl.) *46*:343-346.

Solly, W. W. 1966. A bloom of the alga *Nodularia*. Newsl. Aust. Soc. Limnol. *5*(2):28-29.

Spivak, C. E., B. Witkop, and E. X. Albuquerque. 1980. Anatoxin-a: a novel, potent agonist at the nicotinic receptor. Mol. Pharmacol. *18*:384-394.

Stalker, M. On the Waterville cattle's disease. Page 105 *in* University of Minnesota, Department of Agriculture Report 1881/86, Biennial Report of the Board of Regents No. 4, Suppl. No. 1.

Stangenberg, M. 1968. Toxic effects of *Microcystis aeruginosa* Kg. extracts on *Daphnia longispina* O. F. Müller and *Eucypris virens* Jurine. Hydrobiologia *32*:81-88.

Stavric, B., and P. R. Gorham. 1966. Toxic factors from *Anabaena flos-aquae* (Lyngb.) de Bréb. clone NRC-44h. Page 21 *in* Abstracts, Proc. Can. Soc. Plant Physiol. Ann. Meeting, University of British Columbia, Vancouver.

Stephens, E. L. 1948. *Microcystis toxica* sp. nov.: a poisonous
    alga from the Transvaal and Orange Free State. Hydrobiologia
    1:14.

Stephens, E. L. 1949. *Microcystis toxica* sp. nov.: a poisonous
    alga from the Transvaal and Orange Free State. Trans. R.
    Soc. S. Afr. *32*(1):105-112.

Stewart, A. G., D. A. Barnum, and J. A. Henderson. 1950. Algal
    poisoning in Ontario. Can. J. Comp. Med. *14*:197.

Steyn, D. G. 1943. Poisoning of animals by algae on dams and
    pans. Farming S. Afr. *18*:489.

Steyn, D. G. 1944a. Poisonous and non-poisonous algae (waterbloom
    scum) in dams and pans. Farming S. Afr. *19*:465.

Steyn, D. G. 1944b. Vergiftiging deur slyk (algae) op damme en
    panne (Poisoning by algae on dams and pans). S. Afr. Med.
    J. *18*:378.

Steyn, D. G. 1945a. Poisoning of animals and human beings by
    algae. S. Afr. J. Sci. *41*:243.

Steyn, D. G. 1945b. Poisoning of animals by algae (scum and water-
    bloom) in dams and pans. Union of South Africa, Department
    of Agriculture and Forestry, Government Printer, Pretoria.
    9 pp.

Sykora, J. L., R. Roche, F. L. Kriess, M. A. Barath, D. Volk, J. T.
    Coyne, R. R. Jackson, and G. Keleti. 1978. Algae in open
    drinking water reservoirs and toxicity of *Shizothrix calci-
    cola*. Health Effects Research Laboratory, Environmental
    Protection Agency, Interim Report for 1977-78. Grant No.
    80536810. Cincinnati, Ohio. 94 pp.

Telitchenko, M. M., and M. Gusev. 1965. O toksichnoste sinezelenykh
    vodorosley (Toxicity of blue-green algae). Dokl. Akad. Nauk
    SSSR *160*:1424.

Tetz, V. I. 1964. O toksichnosti sinezelenykh vodorosley (Toxicity
    of blue-green algae). Gig. Sanit. *29*:106.

Thomson, W. K. 1958. Toxic algae. V. Study of toxic bacterial
    contaminants. Can. Defence Research Board, Defence Research
    Kingston Laboratory. Report No. 63, Project No. D52-20-20-18.
    Ottawa.

Thomson, W. K., A. C. Laing, and G. A. Grant. 1957. Toxic algae.
    IV. Isolation of toxic bacterial contaminants. Can. Defence
    Research Board, Defence Research Kingston Laboratory. Report
    No. 51, Project No. D52-20-20-18. Ottawa. 9 pp.

Tisdale, E. S. 1931a. Epidemic of intestinal disorders in
    Charleston, W. Va., occurring simultaneously with unprec-
    edented water supply conditions. Am. J. Public Health *21*:
    198.

Tisdale, E. S. 1931b. The 1930-31 drought and its effect upon
    water supply. Am. J. Public Health *21*:198-200.

Toerien, D. F., W. E. Scott, and M. J. Pitout. 1976. *Microcystis*
    toxins: isolation, identification, implications. Water S.
    A. (Pretoria) *2*:160-162.

Tustin, R. C., S. J. van Rensburg, and J. N. Eloff. 1973. Hepatic damage in the primate following ingestion of toxic algae. Pages 383-385 *in* S. J. Saunders and J. Terblanche, eds., Liver: Proceedings International Liver Congress. Pitman Medical, London.

Veldee, M. V. 1931. Epidemiological study of suspected water-borne gastroenteritis. Am. J. Public Health *21*:1227-1235.

Vinberg, G. G. 1955. Toxic phytoplankton. Translated from: Uspekhi Sovr. Biologii 38, *2*(5):216-226. National Research Council of Canada. Technical Translation TT-549. Ottawa. 25 pp.

Weckesser, J., A. Katz, G. Drews, H. Mayer, and I. Fromme. 1974. Lipopolysaccharide containing L-acofriose in the filamentous blue-green alga *Anabaena variabilis*. J. Bacteriol. *120*:672-678.

Wheeler, R. E., J. B. Lackey, and S. A. Schott. 1943. A contribution on the toxicity of algae. J. Am. Vet. Med. Assoc. *102*: 230.

Willén, E. 1976. Phytoplankton and environmental factors in Lake Hjälmaren. SNV PM 718/NLU Rapport *87*:69-70.

Wilmot Enterprise. 1925. October 1. Farmer tells some news (on stock poisoning in Big Stone Lake). Wilmot, South Dakota.

Wilmot Enterprise. 1925. October 1. One hundred twenty seven hogs, 4 cows die after drinking water from (Big Stone) lake, stock stricken, last Saturday, all die in short time, lake water sent in for analysis. Wilmot, South Dakota.

Woodcock, E. F. 1927. Plants of Michigan poisonous to livestock. J. Am. Vet. Med. Assoc. *25*:475.

Zehnder, A., E. O. Hughes, and P. R. Gorham. 1956. Giftige blau-algen (Toxic blue algae). Abstr. Verh. Schweiz. Naturforsch. Ges. *136*:126.

Zillberg, B. 1966. Gastroenteritis in Salisbury European children - a five year study. Centr. Afr. J. Med. *12*:164-168.

LIST OF CONTRIBUTORS

ALAM, Maktoob
Department of Medicinal
   Chemistry and Pharmacognosy
College of Pharmacy
University of Houston
Houston, Texas 77004

ALLEN, E. A. Dale
Department of Botany
University of Alberta
Edmonton, Alberta T6G 2E9
Canada

ARCHER, B. G.
Altex Scientific Inc.
Berkeley, California

ASTRACHAN (JONES), Nori B.
Toxicology Department
Stauffer Chemical Company
400 Farmington Avenue
Farmington, Connecticut 06032

AZIZ, K. M. S.
International Centre for
   Diarrhoeal Disease Research
G. P. O. Box 128
Dacca 2
Bangladesh

BAHLS, Loren L.
Montana Water Quality Bureau
Helena, Montana

BARLOW, Deryl J.
Department of Botany
University of Natal
P. O. Box 375
Pietermaritzburg, 3200
Republic of South Africa

BENT, Peter E.
Department of Biological Sciences
Wright State University
Dayton, Ohio 45435

BIGGS, David F.
Faculty of Pharmacy and
   Pharmaceutical Sciences
University of Alberta
Edmonton, Alberta T6G 2N8
Canada

BILLINGS, Wayne H.
Pennsylvania Department of
   Environmental Resources
Bureau of Community Environmental
   Control
480 Clearview Lane
Stroudsburg, Pennsylvania 18360

CAMPBELL, Iain M.
Graduate School of Public Health
University of Pittsburgh
Pittsburgh, Pennsylvania 15261

CARMICHAEL, Wayne W.
Department of Biological Sciences
Wright State University
Dayton, Ohio 45435

CARTER, Philip W.
Department of Biochemistry
University of New Hampshire
Durham, New Hampshire 03824

COHEN, W. S.
School of Biological Sciences
University of Kentucky
Lexington, Kentucky 40506

COLLINS, Michael D.
Environmental Health
  Surveillance Center
University of Missouri
Columbia, Missouri 65201

COURT, Gary J.
Biology Department
Brookhaven National Laboratory
Upton, New York 11973

DOERFLER, Dennis L.
Graduate School of Public Health
University of Pittsburgh
Pittsburgh, Pennsylvania 15261

ECKER, Marilyn M.
Department of Zoology
University of New Hampshire
Durham, New Hampshire 03824

ELOFF, J. N.
Department of Botany
University of the Orange Free
  State
P. O. Box 339
Bloemfontein, 9300
Republic of South Africa

ESTERVIG, David
Dalton Research Center
University of Missouri
Columbia, Missouri

EULER, Kenneth L.
Department of Medicinal
  Chemistry and Pharmacognosy
College of Pharmacy
University of Houston
Houston, Texas 77004

FALCONER, I. R.
Department of Biochemistry and
  Nutrition
University of New England
Armidale, New South Wales, 2351
Australia

FOXALL, Thomas L.
Department of Zoology
University of New Hampshire
Durham, New Hampshire 03824

GARRO, Frank M.
Dalton Research Center
University of Missouri
Columbia, Missouri

GERBA, Charles P.
Department of Virology and
  Epidemiology
Baylor College of Medicine
National Center for Groundwater
  Research
Houston, Texas 77030

GLASSER, Stephen
U. S. Forest Service
Gallatin National Forest
Bozeman, Montana

GORHAM, Paul R.
Department of Botany
University of Alberta
Edmonton, Alberta T6G 2E9
Canada

GOWANS, C. S.
Biology Department
University of Missouri
Columbia, Missouri

GOYAL, Sager M.
Department of Virology and
  Epidemiology
Baylor College of Medicine
National Center for Groundwater
  Research
Houston, Texas 77030

HARRIS, D. O.
School of Biological Sciences
University of Kentucky
Lexington, Kentucky 40506

HAUMAN, John H.
National Institute for Water
  Research
Council for Scientific and
  Industrial Research
P. O. Box 395
Pretoria, 0001
Republic of South Africa

HORPESTAD, Abe
Montana Water Quality Bureau
Helena, Montana

IKAWA, Miyoshi
Department of Biochemistry
University of New Hampshire
Durham, New Hampshire 03824

JUDAY, Richard E.
Department of Chemistry
University of Montana
Missoula, Montana 59812

KELETI, Georg
Graduate School of Public Health
University of Pittsburgh
Pittsburgh, Pennsylvania 15261

KELLER, Edward J.
Department of Chemistry
University of Montana
Missoula, Montana 59812

KIRPENKO, N. I.
Institute of Hydrobiology
Ukrainian Academy of Sciences
Str. Vladimirskaya, 44
P. O. Box 252003
Kiev, U. S. S. R.

KIRPENKO, Yu. A.
Institute of Hydrobiology
Ukrainian Academy of Sciences
Str. Vladimirskaya, 44
P. O. Box 252003
Kiev, U. S. S. R.

KRÜGER, G. H. J.
Department of Botany
University of the Orange Free
  State
P. O. Box 339
Bloemfontein, 9300
Republic of South Africa

KYCIA, J. Helen
Biology Department
Brookhaven National Laboratory
Upton, New York 11973

LINCOLN, Edward P.
Agricultural Engineering Dept.
University of Florida
Gainesville, Florida 32611

MAIBACH, Howard I.
Department of Dermatology
University of California
San Francisco, California

MAIOLIE, Laurie A.
Graduate School of Public Health
University of Pittsburgh
Pittsburgh, Pennsylvania 15261

MAY, Valerie
National Herbarium of New South
  Wales
Royal Botanic Gardens
Sydney, New South Wales, 2000
Australia

MOORE, Richard E.
Department of Chemistry
University of Hawaii
Honolulu, Hawaii 96822

NICHOLS, Donald G.
Department of Biology
Eastern Washington University
Cheney, Washington 99004

ØSTENSVIK, Øyvin
Dept. of Food Hygiene
The Veterinary College of Norway
Oslo, Norway

PATTERSON, Gregory M. L.
School of Biological Sciences
University of Kentucky
Lexington, Kentucky 40506

RUNNEGAR, Maria T.
Department of Biochemistry and
   Nutrition
University of New England
Armidale, New South Wales, 2351
Australia

SASNER, John J., Jr.
Department of Zoology
University of New Hampshire
Durham, New Hampshire 03824

SCHANTZ, Edward J.
Department of Food Microbiology
   and Toxicology
Food Research Institute
University of Wisconsin
Madison, Wisconsin 53706

SCOTT, William E.
National Institute for Water
   Research
Council for Scientific and
   Industrial Research
P. O. Box 395
Pretoria, 0001
Republic of South Africa

SHILO, Moshe
Division of Microbial and
   Molecular Ecology
Institute of Life Sciences
Hebrew University
Jerusalem, Israel

SHOPTAUGH, Nancy H.
Food Research Institute
University of Wisconsin
Madison, Wisconsin 53706

SIEGELMAN, Harold W.
Biology Department
Brookhaven National Laboratory
Upton, New York 11973

SIRENKO, L. A.
Institute of Hydrobiology
Ukrainian Academy of Sciences
Str. Vladimirskaya, 44
P. O. Box 252003
Kiev, U. S. S. R.

SKULBERG, Olav M.
Norwegian Institute for Water
   Research
P. O. Box 333
Oslo 3, Norway

SØLI, Nils E.
Department of Pharmacology and
   Toxicology
The Veterinary College of Norway
Oslo, Norway

SOLTERO, Raymond A.
Department of Biology
Eastern Washington University
Cheney, Washington 99004

SWANSON, Tracey
Dalton Research Center
University of Missouri
Columbia, Missouri

SYKORA, Jan L.
Graduate School of Public Health
University of Pittsburgh
Pittsburgh, Pennsylvania 15261

VAN DER WESTHUIZEN, A. J.
Department of Botany
University of the Orange
  Free State
P. O. Box 339
Bloemfontein, 9300
Republic of South Africa

WATSON, Winsor H.
Department of Zoology
University of New Hampshire
Durham, New Hampshire 03824

WEGENER, Karin
Department of Zoology
University of New Hampshire
Durham, New Hampshire 03824